Bioinorganic Chemistry

Bioinorganic Chemistry

Edited by James Denton

SYRAWOOD
PUBLISHING HOUSE

New York

Published by Syrawood Publishing House,
750 Third Avenue, 9th Floor,
New York, NY 10017, USA
www.syrawoodpublishinghouse.com

Bioinorganic Chemistry
Edited by James Denton

International Standard Book Number: 978-1-68286-848-5 (Hardback)

Cataloging-in-Publication Data

Bioinorganic chemistry / edited by James Denton.
    p. cm.
Includes bibliographical references and index.
ISBN 978-1-68286-848-5
1. Bioinorganic chemistry. 2. Chemistry, Inorganic. 3. Biochemistry. I. Denton, James.
QP531 .B56 2020
572.51--dc23

TABLE OF CONTENTS

PREFACE

The main aim of this book is to educate learners and enhance their research focus by presenting diverse topics covering this vast field. This is an advanced book which compiles significant studies by distinguished experts in the area of analysis. This book addresses successive solutions to the challenges arising in the area of application, along with it; the book provides scope for future developments.

Bioinorganic chemistry is the science that studies the role of metals in biology. This field involves the study of the behavior of metalloproteins and artificially introduced metals in medicine and toxicology. Some of the focus areas in bioinorganic chemistry are the study of transport and storage of metal ions, enzymology, bioorganometallic chemistry, toxicity, oxygen transport and activation proteins, metals in medicine, etc. Many reactions involve water and metal ions that are usually at the catalytic centers for these enzymes. Aerobic life has an extensive use of metals like copper, iron and manganese. Heme is used in the form of hemoglobin by red blood cells for the transport of oxygen. Some metalloproteins protect biological systems from the potentially harmful effects of reactive oxygen-containing molecules. This book aims to shed light on some of the unexplored aspects of bioinorganic chemistry and the recent researches in this field. It elucidates new techniques and their applications in a multidisciplinary manner. Researchers and students in this field will be assisted by this book.

It was a great honour to edit this book, though there were challenges, as it involved a lot of communication and networking between me and the editorial team. However, the end result was this all-inclusive book covering diverse themes in the field.

Finally, it is important to acknowledge the efforts of the contributors for their excellent chapters, through which a wide variety of issues have been addressed. I would also like to thank my colleagues for their valuable feedback during the making of this book.

Editor

Optimization Studies on Recovery of Metals from Printed Circuit Board Waste

P. Sivakumar ⓘD, D. Prabhakaran, and M. Thirumarimurugan

Department of Chemical Engineering, Coimbatore Institute of Technology, Coimbatore 641 014, Tamil Nadu, India

Correspondence should be addressed to P. Sivakumar; chemsiva13@gmail.com

Academic Editor: Albrecht Messerschmidt

The aim of the study was to recover copper and lead metal from waste printed circuit boards (PCBs). The electrowinning method is found to be an effective recycling process to recover copper and lead metal from printed circuit board wastes. In order to simplify the process with affordable equipment, a simple ammonical leaching operation method was adopted. The selected PCBs were incinerated into fine ash powder at 500°C for 1 hour in the pyrolysis reactor. Then, the fine ash powder was subjected to acid-leaching process to recover the metals with varying conditions like acid-base concentration, electrode combination, and leaching time. The relative electrolysis solution of 0.1 M lead nitrate for lead and 0.1 M copper sulphate for copper was used to extract metals from PCBs at room temperature. The amount of lead and copper extracted from the process was determined by an atomic absorption spectrophotometer, and results found were 73.29% and 82.17%, respectively. Further, the optimum conditions for the recovery of metals were determined by using RSM software. The results showed that the percentage of lead and copper recovery were 78.25% and 89.1% should be 4 hrs 10 A/dm^2.

1. Introduction

Recycling of e-waste is an important subject not only from the point of waste treatment but also from the recovery aspect of valuable materials [1–4]. Among the resources in e-waste, metals contribute more than 95% of the materials market value. Hence, the recovery of valuable metals is the inherent motive in e-waste disposal. In the past decades, many techniques for recovering valuable metals from e-waste have been developed such as gravity separation, magnetic separation, and electrostatic separation [5] synthesis of CuCl with e-waste, separation of PCBs with organic solvent method [6, 7], cyanide and noncyanide lixiviants leaching methods, ammonium persulfate leaching bioleaching methods [8–10], or a combination of these approaches. Among those methods, hydrometallurgical methods are more accurate, predictable, and controllable [11]. Therefore, hydrometallurgical techniques are most active in the research of valuable metal recovery from electronic scraps in the past two decades. However, traditional hydrometallurgical methods are acid dependent, time-consuming, and inefficient for simultaneous recovery of precious metals. Remarkably, a large amount of corrosive or toxic reagents, such as aqua regia, nitric acid, cyanide and halide, are consumed, producing large quantities of toxic and corrosive fumes or solution [12, 13]. Therefore, it is necessary to seek a more environmental friendly method for the recovery of valuable metals from e-wastes. Hydrometallurgical methods are used in the upgrading and refining stages of the recycling chain [14–16]. In this research article, the recovery of lead and copper metals from e-waste is widely investigated. The PCBs were converted into fine ash powder and subjected to electrowinning process for the recovery of metals. The experimental results were determined by EDS and AAS, respectively. Furthermore, the experimental results are validated through RSM software at different parameters like acid-base concentration, electrode combination, and leaching time [17–22].

2. Materials and Methods

2.1. Materials. The computer PCBs were collected from various sources for the recovery of metals. The collected PCBs were crushed using roll crusher and powdered by a hammer mill. The crushed PCBs were incarnated through pyrolysis to avoid side reaction in the leaching process with the electrolyte solution. The optimum condition of the pyrolysis reactor was 500°C in atmospheric pressure for 1 h where the epoxy resins and polymers were volatized at the temperature less than 500°C. The volatized contents were condensed and collected separately. The ferrous materials present in the obtained ash were separated by a magnetic separator.

2.2. Electrowinning Process.

The fine ash powder was treated with aqua regia solution (3 : 1 ratio of HCl and HNO_3) in the incineration chamber in order to avoid the liberation of toxic fumes. Then the precipitated salts obtained from the leaching was analyzed by EDS to determine the composition of metal present in the salts (Figure 1). The electrowinning setup consists of bath arrangement and amplifier. The bath having two slots for the anode and cathode fixing and the electrode is connected with amplifier, and the current density was varied through the amplifier (Figure 2).

2.3. Extraction Process of Lead.

About 25 g of incinerated fine ash was added into the acid bath followed by the addition of ammonical electrolyte solution. The current density was set to 1 to 10 (A/dm^2). The solution was agitated at regular interval to get an effective electrodeposition:

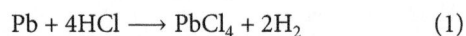

$$Pb + 4HCl \longrightarrow PbCl_4 + 2H_2 \qquad (1)$$

After the stipulated time of operation, pure lead was deposited on lead cathode. The deposited elements were scrapped and stored in an air tight container. The recovered lead quantitated from the EDS method. The spent acid left with mud filtered at pH 6–10 was stored in a glass container for further treatment.

2.4. Extraction Process of Copper.

About 25 g of incinerated fine ash was added into the acid bath followed by the addition of ammonical electrolyte solution. The current density was set to 1 to 10 (A/dm^2). The solution was agitated at regular intervals to get an effective electrodeposition. After the stipulated time of operation, pure copper (cupric) was deposited on the cathode and impure copper (cuprous ion) were deposited on the anode. The deposited elements were scrapped and stored in an air tight container. The recovered copper quantitated from the EDS method. The spent acid left with mud (nonleached elements) was filtered (pH–8.4) and were stored in a glass container for further treatment (Figure 3):

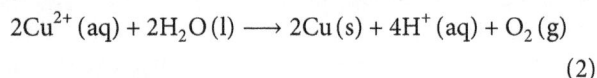

$$2Cu^{2+}(aq) + 2H_2O(l) \longrightarrow 2Cu(s) + 4H^+(aq) + O_2(g) \qquad (2)$$

The spent solution collected from the electrodeposition was neutralized to 6.9 for the safe disposal as per the standard. Moreover, the presence of any metal in the spent solution was analyzed by Fourier-transform infrared spectroscopy. The results (Figure 4) show that the metallic traces were found to be absent which confirms that all the metals recovered from the ashes deposited on the electrode.

3. Results and Discussion

3.1. RSM for Lead. The response surface methodology (RSM) is a statistical modeling technique employed for multiple regression analysis using quantitative data obtained from designed experiments to solve multivariable equations (Table 1). The response surfaces can be visualized as three-dimensional plots that exhibit the response as a function of two factors while keeping the other factors constant. In this above plot, the red zone corresponds to the extract percentage above 85%, yellow zone shows 60 to 70%, and the blue zone confirms below 40% extraction of lead (Figures 5 and 5(a)). The regression equation for the RSM data plots for the lead is

$$\begin{aligned} Extract = {}& 66.36 + 9.0175 * A + 7.37375 * B + 6.42375 * C \\ & + 1.235 * AB + 0.17 * AC + (-7.1525 * BC) \\ & + (-18.6187 * A^2) + (-1.67125 * B^2) \\ & + (-1.81625 * C^2). \end{aligned}$$

$$(3)$$

The model as a function of coded factor could be utilized to predict the response of each parameter within the given limit. Here, the maximum limit of process parameters (factors) is termed (coded) as +1 and minimum limit is terms (coded) as −1. The modifed equation or coded equation is very much useful in order to find the comparative effect of the process parameters by relating the coefficient of factors. The final equation in terms of actual factors is

$$\begin{aligned} Extract = {}& -70.2665 + 5.06454 * CD + 0.201188 * solvent \\ & + 25.6766 * time + 0.000914815 * CD * solvent \\ & + 0.0125926 * CD * time \\ & + (-0.0317889 * solvent * time) \\ & + (-0.229861 * CD^2) + (-7.42778e - 05 * solvent^2) \\ & + (-0.807222 * time^2). \end{aligned}$$

$$(4)$$

Equation (4) in terms of process parameters could be utilized to predict the response for the provided levels of each parameter (Table 2). In this equation, the original units of each parameters should be considered for each levels. In order to evaluate the comparative effect of each factor, the above equation should not be considered since the coefficients are balanced to embrace the units of each parameters. Also, the intercept does not fall at design space center.

3.2. Analysis of Variance (ANOVA). Analysis of variance is used to determine the significant effects of process variables on current efficiency (Table 3) along with the factor coding. The sum of squares is found to be Type III—partial derived from the ANOVA quadratic model. The model F value of 4.43 implies the model is significant. A minimum value of 3.12% is possible for the F value due to noise. p values less

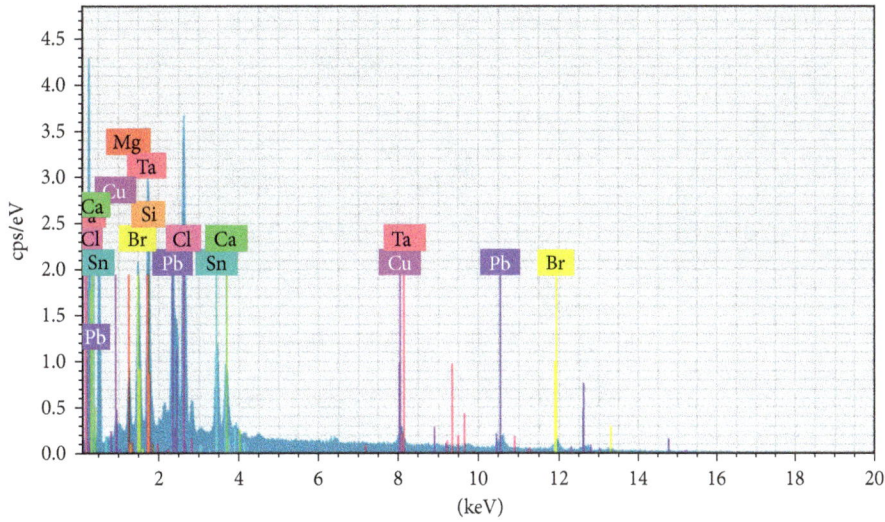

FIGURE 1: Initial analysis of raw materials.

FIGURE 2: Experimental setup of electrowinning process.

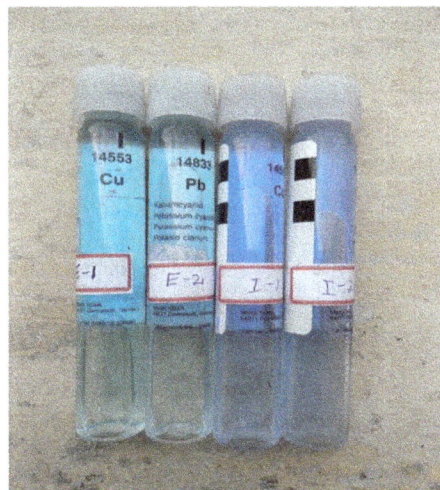

FIGURE 3: Bath solutions of copper and lead.

than 0.0500 indicate model terms are significant. In this case A, A^2 are significant model terms. Values greater than 0.1000 indicate the model terms are not significant. If there are many insignificant model terms (not counting those required to support hierarchy), model reduction may improve the model. The lack of fit F value of 63.27 implies the lack of

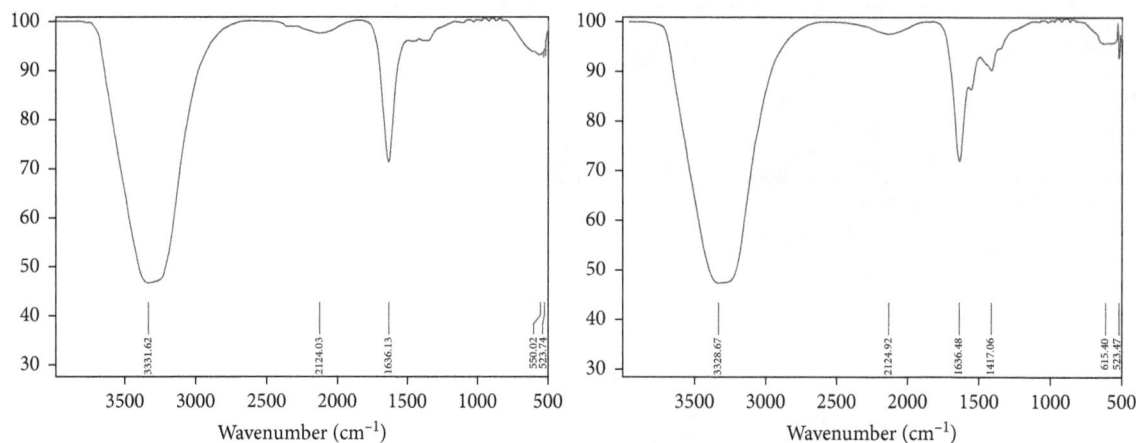

FIGURE 4: FTIR analysis of bath solution.

TABLE 1: RSM parameters for lead extraction.

Std	Run	Factor 1 A: CD A/dm^2	Factor 2 B: solvent ml	Factor 3 C: time Hrs
1	8	1	400	2.5
2	9	19	400	2.5
3	17	1	700	2.5
4	12	19	700	2.5
5	1	1	550	1
6	14	19	550	1
7	6	1	550	4
8	5	19	550	4
9	4	10	400	1
10	10	10	700	1
11	7	10	400	4
12	11	10	700	4
13	2	10	550	2.5
14	16	10	550	2.5
15	13	10	550	2.5
16	15	10	550	2.5
17	3	10	550	2.5

fit is significant. There is only a 0.08% chance that a lack of fit F value could be large that could occur due to noise. The coefficient represents the expected change in response per unit change in the factor value, when all remaining factors were constant. The intercept in an orthogonal design is the overall average response of all the runs. The coefficients are adjustments around the average factor settings. When the factors are orthogonal, the variance inflation factors (VIFs) are 1; VIFs greater than 1 indicate multicolinearity; the higher the VIF, the more severe the correlation of factors. As a rough rule, VIFs less than 10 are tolerable. Hence, from the data obtained (Table 4), the VIF values of lead are found to be tolerable.

3.3. Model Terms. For a standard deviation of 1, the power calculations are performed using response type "continuous," and parameters are $\Delta = 2$ and $\sigma = 1$. The power is evaluated over −1 to +1 coded factor space. From (Table 5), the standard errors should be similar to each other in

a balanced design. The ideal VIF value should be 1, VIFs above 10 are cause for concern, and VIFs above 100 are cause for alarm, indicating coefficients are poorly estimated due to multicolinearity, where ideal Ri^2 is 0.0. High Ri^2 means terms are correlated with each other, possibly leading to poor models. If the design has multilinear constraints, then multicolinearity will exist to a greater degree. This inflates the VIFs and the Ri^2, rendering these statistics would not perform well. Hence, FDS could be used. Power is an inappropriate tool to evaluate response surface designs. Use prediction-based metrics provided in this program via fraction of design space (FDS) statistics.

3.4. Fit Statistics. A negative predicted R^2 implies that the overall mean may be a better predictor of the response than the current model. In some cases, a higher order model may also predict better. Adeq. precision measures the signal to noise ratio. A ratio greater than 4 is desirable. The ratio of 5.915 indicates an adequate signal. This model can be used to navigate the design space. The optimization of current efficiency is shown in Figure 6. From the results, it is observed that 69% of lead extract is obtained at current density = 10 A dm^{-2}, solvent ratio = 5 : 2, and the electrolysis time = 4 hours (Figures 7 and 8). The significance of regression coefficients were analyzed using the p-test and t-test. The p values are used to check the effect of interaction among the variables. A larger magnitude of t-value and a smaller magnitude of p value are significant in the corresponding coefficient term. The coefficient of current efficiency and the corresponding t and p values are shown in Table 6. Finally, the coefficients in the interaction terms for current density-electrolysis time is significant compared to current density-solvent ratio, and current density-electrolysis time.

3.5. RSM for Copper. The regression equation for the RSM data plots for the copper is in terms of coded factors form as follows:

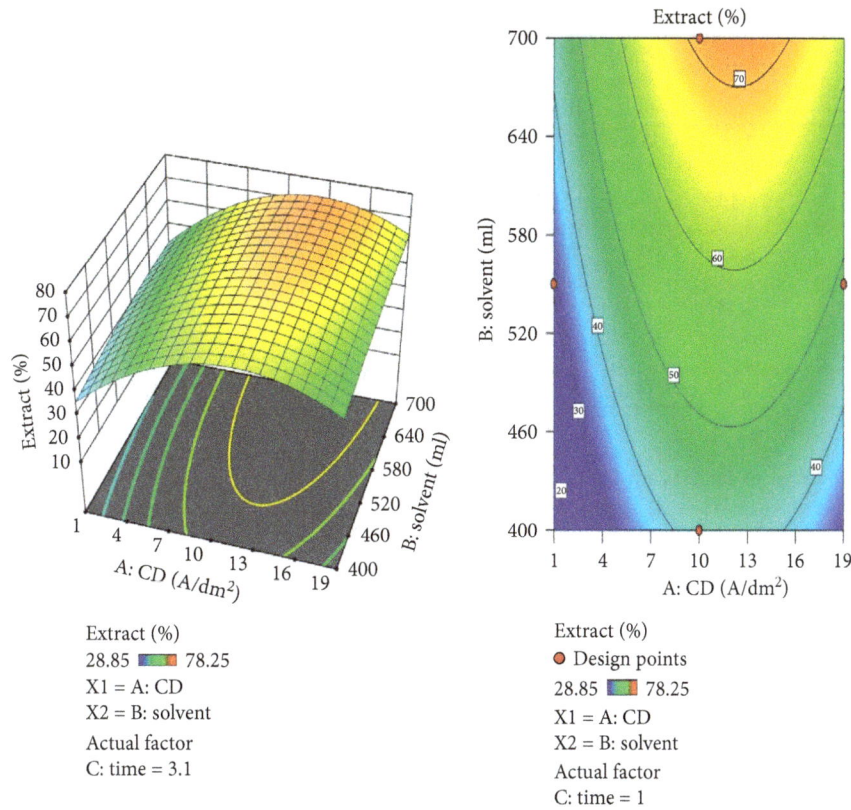

FIGURE 5: Contour plot for recovery of Lead.

Extract (%)
28.85 ▮▮▮ 78.25
X1 = A: CD
X2 = B: solvent
Actual factor
C: time = 3.1

Extract (%)
⬤ Design points
28.85 ▮▮▮ 78.25
X1 = A: CD
X2 = B: solvent
Actual factor
C: time = 1

TABLE 2: Box–Behnken experimental design table for recovery of lead.

Std	Run	Factor 1 A: CD A/dm²	Factor 2 B: solvent ml	Factor 3 C: time Hrs
1	8	1	400	2.5
2	9	19	400	2.5
3	17	1	700	2.5
4	12	19	700	2.5
5	1	1	550	1
6	14	19	550	1
7	6	1	550	4
8	5	19	550	4
9	4	10	400	1
10	10	10	700	1
11	7	10	400	4
12	11	10	700	4
13	2	10	550	2.5
14	16	10	550	2.5
15	13	10	550	2.5
16	15	10	550	2.5
17	3	10	550	2.5

$$\text{Extract}\,(E) = 48 + 24 * A + 8.25 * B + 15.5 * C + 0.5 * AB$$
$$+ 7.5 * AC + 2.5 * BC + \left(-6.5 * A^2\right) + 8 * B^2$$
$$+ 1.5 * C^2.$$

(5)

TABLE 3: ANOVA quadratic model for lead.

Source	Sum of squares	DOF	Mean square	F value	p value	
Model	3152.45	9	350.27	4.43	0.0312	Significant
A-CD	650.52	1	650.52	8.23	0.0240	
B-solvent	434.98	1	434.98	5.51	0.0514	
C-time	330.12	1	330.12	4.18	0.0802	
AB	6.10	1	6.10	0.0772	0.7891	
AC	0.1156	1	0.1156	0.0015	0.9706	
BC	204.63	1	204.63	2.59	0.1516	
A^2	1459.61	1	1459.61	18.47	0.0036	
B^2	11.76	1	11.76	0.1489	0.7111	
C^2	13.89	1	13.89	0.1758	0.6876	
Residual	553.05	7	79.01			
Lack of fit	541.64	3	180.55	63.27	0.0008	Significant
Pure error	11.41	4	2.85			
Total	3705.50	16				

The model (Equation 5) as a function of coded factor could be utilized to predict the response of each parameter within the given limit. Here, the maximum limit of process parameters (factors) is termed(coded) as +1 and minimum limit is termed (coded) as −1. The modified equation or coded equation is very much useful in order to find the comparative effect of the process parameters by relating the coefficient of factors (Table 7).

The final equation in terms of actual factors is

TABLE 4: Coefficients in terms of coded factors for lead.

Factor	Coefficient estimate	DOF	Standard error	95% CI low	95% CI high	VIF
Intercept	66.36	1	3.98	56.96	75.76	
A-CD	9.02	1	3.14	1.59	16.45	1.0000
B-solvent	7.37	1	3.14	−0.0573	14.80	1.0000
C-time	6.42	1	3.14	−1.01	13.85	1.0000
AB	1.24	1	4.44	−9.27	11.74	1.0000
AC	0.1700	1	4.44	−10.34	10.68	1.0000
BC	−7.15	1	4.44	−17.66	3.36	1.0000
A^2	−18.62	1	4.33	−28.86	−8.38	1.01
B^2	−1.67	1	4.33	−11.91	8.57	1.01
C^2	−1.82	1	4.33	−12.06	8.43	1.01

TABLE 5: Model terms in RSM for lead.

Term	Standard error	VIF	Ri^2	Power (%)
A	0.3536	1	0.0000	68.1
B	0.3536	1	0.0000	68.1
C	0.3536	1	0.0000	68.1
AB	0.5000	1	0.0000	40.8
AC	0.5000	1	0.0000	40.8
BC	0.5000	1	0.0000	40.8
A^2	0.4873	1.00588	0.0058	93.8
B^2	0.4873	1.00588	0.0058	93.8
C^2	0.4873	1.00588	0.0058	93.8

$$\begin{aligned}
\text{Extract} = {}& 191.503 + 2.32716 * \text{CD} + (−0.773056 * \text{solvent}) \\
& + (−11.3333 * \text{time}) + 0.000555556 * \text{CD} * \text{solvent} \\
& + 0.833333 * \text{CD} * \text{time} + 0.025 * \text{solvent} * \text{time} \\
& + \left(−0.0802469 * \text{CD}^2\right) + 0.0008 * \text{solvent}^2 \\
& + 1.5 * \text{time}^2.
\end{aligned}$$

$$(6)$$

Equation (5) in terms of process parameters could be utilized to predict the response for the provided levels of each parameter. In this equation, the original units of each parameters should be considered for each levels. In order to evaluate the comparative effect of each factor, the above equation should not be considered since the coefficients are balanced to embrace the units of each parameters. Also, the intercept does not falls at design space center (Table 8). In this contour plot, the red zone indicates extract percentages above 85%. And yellow and blue zones indicate 60 to 70% and below 40% extraction of copper (Figures 9 and 9(a)).

3.6. Analysis of Variance (ANOVA).
Analysis of variance is used to determine the significant effects of process variables on current efficiency along with the factor coding. The sum of squares is found to be Type III—partial derived from the ANOVA quadratic model. The model F value of 155.08 in the Table 9 implies the model is significant. A minimum value of 0.01% is possible for the F value due to noise. P values less than 0.0500 indicate model terms are significant.

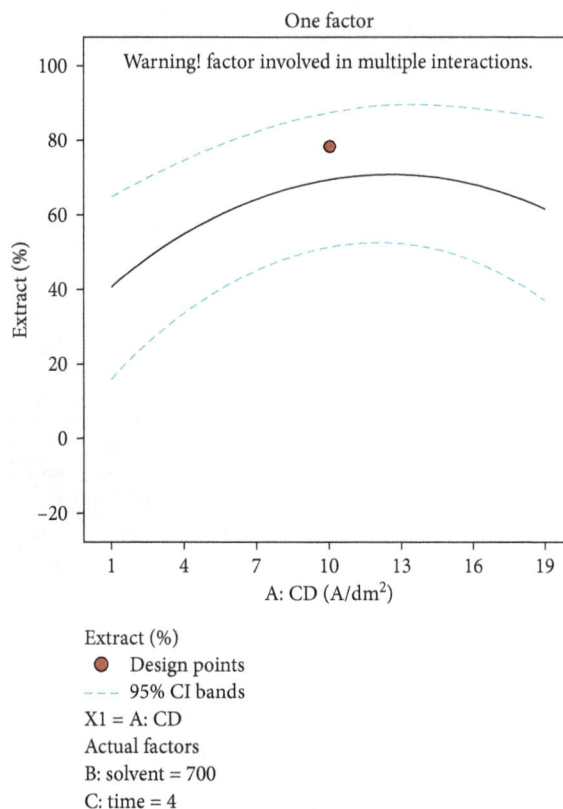

FIGURE 6: Current density vs extract % for lead.

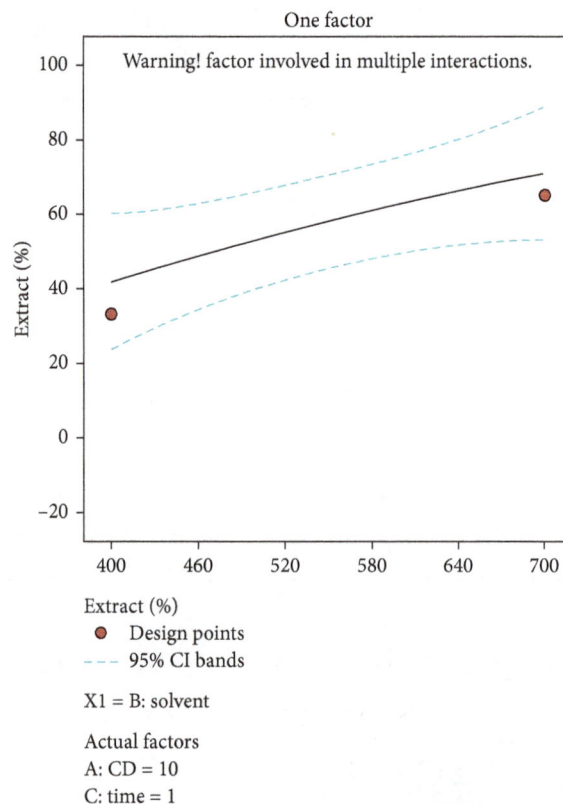

FIGURE 7: Solvent vs extract % for lead.

One factor

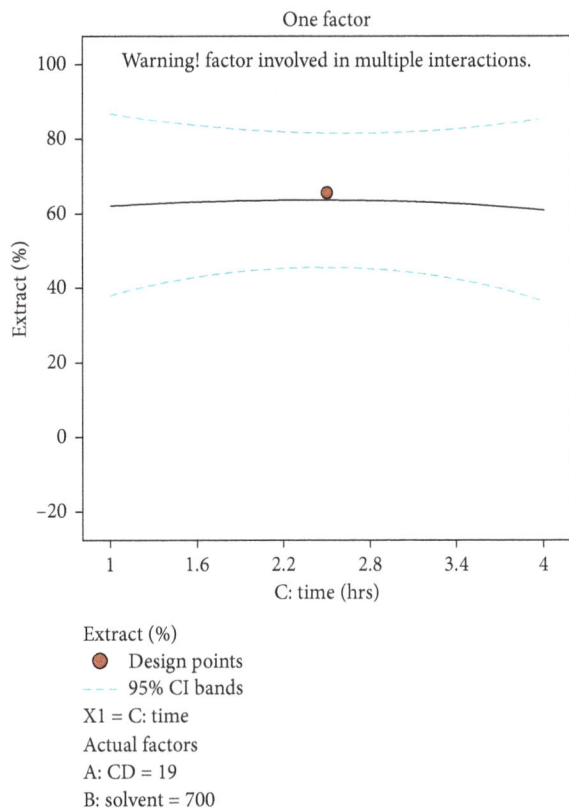

FIGURE 8: Time vs extract % for lead.

TABLE 6: Fit statistics.

Std. dev.	8.89
Mean	55.96
CV (%)	15.88
R^2	0.8507
Adjusted R^2	0.6589
Predicted R^2	−1.3436
Adeq. precision	5.9146

TABLE 7: RSM parameters for copper extraction.

Std	Run	Factor 1 A: CD A/dm^2	Factor 2 B: solvent ml	Factor 3 C: time Hrs
1	12	1	400	2
2	14	19	400	2
3	2	1	600	2
4	4	19	600	2
5	18	1	500	1
6	13	19	500	1
7	16	1	500	3
8	11	19	500	3
9	8	10	400	1
10	1	10	600	1
11	6	10	400	3
12	7	10	600	3
13	5	10	500	2
14	9	10	500	2
15	17	10	500	2
16	3	10	500	2
17	15	10	500	2

TABLE 8: Box–Behnken experimental design table for recovery of copper.

Std	Run	Factor 1 A: CD A/dm^2	Factor 2 B: solvent ml	Factor 3 C: time Hrs
1	12	1	400	2
2	14	19	400	2
3	2	1	600	2
4	4	19	600	2
5	18	1	500	1
6	13	19	500	1
7	16	1	500	3
8	11	19	500	3
9	8	10	400	1
10	1	10	600	1
11	6	10	400	3
12	7	10	600	3
13	5	10	500	2
14	9	10	500	2
15	17	10	500	2
16	3	10	500	2
17	15	10	500	2
18	10	10	500	2

In this case, A, B, C, AC, A^2, and B^2 are significant model terms. Values greater than 0.1000 indicate the model terms are not significant. If there are many insignificant model terms (not counting those required to support hierarchy), model reduction may improve the model. The lack of fit F value is nil that implies the lack of fit is significant. The coefficient represents the expected change in response per unit change in factor value, when all remaining factors were constant. The intercept in an orthogonal design is the overall average response of all the runs. The coefficients are adjustments around the average factor settings. When the factors are orthogonal, the VIFs are 1; VIFs greater than 1 indicate multicolinearity; the higher the VIF, the more severe the correlation of factors. As a rough rule, VIFs less than 10 are tolerable. Hence, from the data obtained (Table 10), the VIF Values of lead are found to be tolerable.

3.7. Model Terms. For a standard deviation of 1 the power calculations are performed using response type "continuous," and the parameters are $\Delta = 2$ and $\sigma = 1$. The power is evaluated over −1 to +1 coded factor space (Table 11). The standard errors should be similar to each other in a balanced design. The ideal VIF value should be 1, VIFs above 10 are cause for concern and VIFs above 100 are cause for alarm, indicating coefficients are poorly estimated due to multicolinearity, where ideal Ri^2 is 0.0. High Ri^2 means terms are correlated with each other, possibly leading to poor models. If the design has multilinear constraints, then multicolinearity will exist to a greater degree. This inflates the VIFs and the Ri^2, rendering these statistics would not perform well. Hence, FDS could be used. Power is an inappropriate tool to evaluate response surface designs. Use prediction-based metrics provided in this program via fraction of design space (FDS) statistics.

Extract (%)
● Design points above predicted value
○ Design points below predicted value

12 ▭ 89
X1 = A: CD
X2 = B: solvent
Actual factor
C: time = 3

Extract (%)
● Design points

12 ▭ 89
X1 = A: CD
X2 = B: solvent

Actual factor
C: time = 3

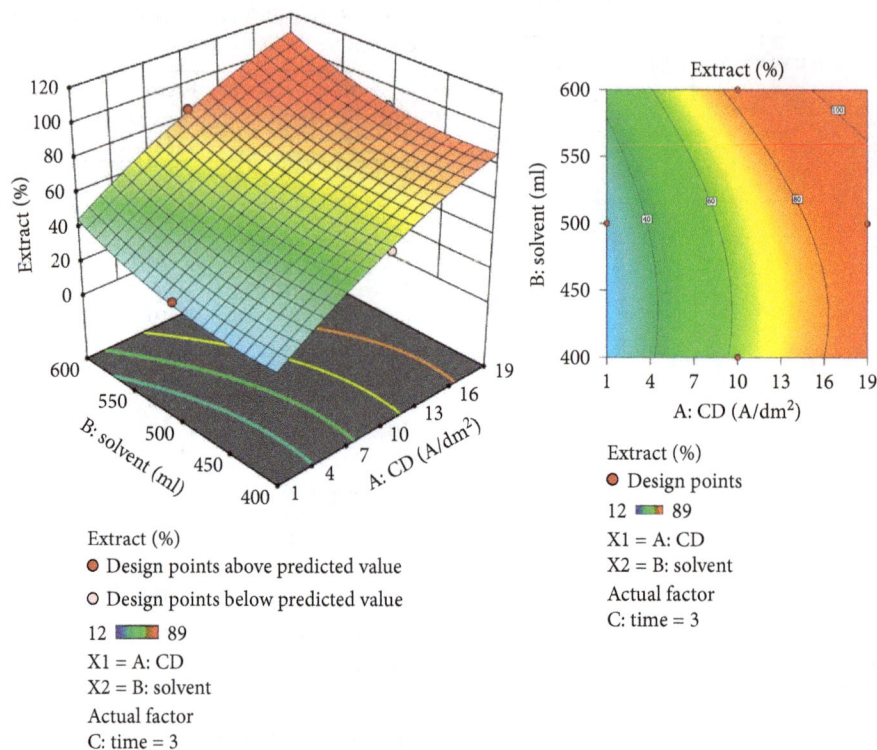

FIGURE 9: Contour plot for recovery of copper.

TABLE 9: ANOVA quadratic model for copper.

Source	Sum of squares	DOF	Mean square	F value	p value	
Model	7763.50	9	862.61	155.08	<0.0001	Significant
A-CD	4608.00	1	4608.00	828.40	<0.0001	
B-solvent	544.50	1	544.50	97.89	<0.0001	
C-time	1922.00	1	1922.00	345.53	<0.0001	
AB	1.0000	1	1.0000	0.1798	0.6827	
AC	225.00	1	225.00	40.45	0.0002	
BC	25.00	1	25.00	4.49	0.0668	
A^2	184.36	1	184.36	33.14	0.0004	
B^2	279.27	1	279.27	50.21	0.0001	
C^2	9.82	1	9.82	1.77	0.2206	
Residual	44.50	8	5.56			
Lack of fit	44.50	3	14.83			
Pure error	0.0000	5	0.0000			
Total	7808.00	17				

TABLE 10: Coefficients in terms of coded factors for copper.

Factor	Coefficient estimate	DOF	Standard error	95% CI low	95% CI high	VIF
Intercept	48.00	1	0.9629	45.78	50.22	
A-CD	24.00	1	0.8339	22.08	25.92	1.0000
B-solvent	8.25	1	0.8339	6.33	10.17	1.0000
C-time	15.50	1	0.8339	13.58	17.42	1.0000
AB	0.5000	1	1.18	−2.22	3.22	1.0000
AC	7.50	1	1.18	4.78	10.22	1.0000
BC	2.50	1	1.18	−0.2193	5.22	1.0000
A^2	−6.50	1	1.13	−9.10	−3.90	1.02
B^2	8.00	1	1.13	5.40	10.60	1.02
C^2	1.50	1	1.13	−1.10	4.10	1.02

TABLE 11: Model terms in RSM for copper.

Term	Standard error	VIF	Ri^2	Power (%)
A	0.3536	1	0.0000	69.8
B	0.3536	1	0.0000	69.8
C	0.3536	1	0.0000	69.8
AB	0.5000	1	0.0000	42.1
AC	0.5000	1	0.0000	42.1
BC	0.5000	1	0.0000	42.1
A^2	0.4787	1.01852	0.0182	95.4
B^2	0.4787	1.01852	0.0182	95.4
C^2	0.4787	1.01852	0.0182	95.4

3.8. Fit Statistics. A predicted R^2 implies that the overall mean may be a better predictor of the response than the current model. In some cases, a higher order model may also predict better. Adeq. precision measures the signal to noise ratio. A ratio greater than 4 is desirable. A ratio of 44.9 indicates an adequate signal. This model can be used to navigate the design space. The optimization of current efficiency is shown in Figure 10. The optimum extraction of 69% Cu is obtained at current density = 19 A dm^{-2}, solvent ratio = 5 : 2, and electrolysis time = 4 hour (Figures 11 and 12). The significance of regression coefficients was analyzed using the p-test and t-test. The p values are used to check the

effect of interaction among the variables. A larger magnitude of t-value and a smaller magnitude of p value are significant in the corresponding coefficient term. The coefficient of

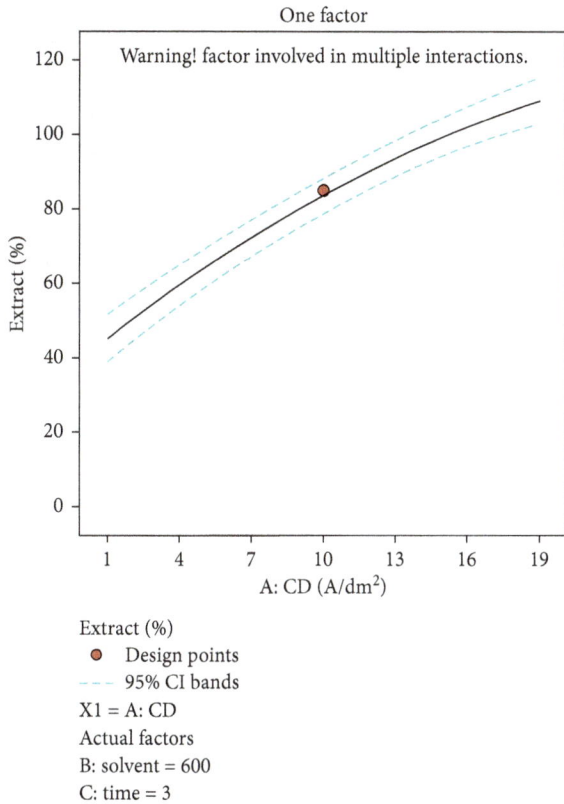

FIGURE 10: Current density vs extract % for copper.

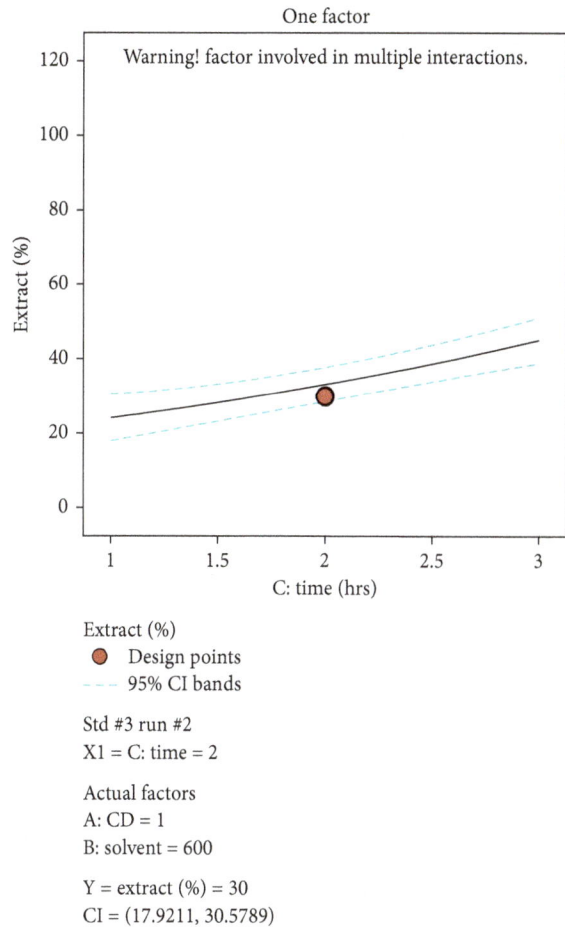

FIGURE 12: Time vs extract % for lead.

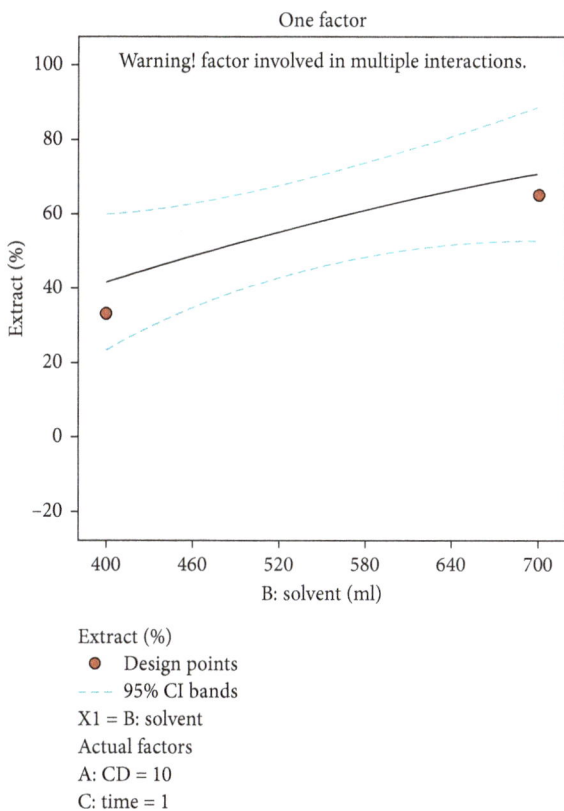

FIGURE 11: Solvent vs extract % for lead.

TABLE 12: Fit statistics.

Std. dev.	2.36
Mean	49.33
CV (%)	4.78
R^2	0.9943
Adjusted R^2	0.9879
Predicted R^2	0.9088
Adeq. Precision	44.9395

current efficiency and the corresponding t and p values are shown in (Table 12). Finally, the coefficients in the interaction terms for current density-electrolysis time is significant compared to current density-solvent ratio and current density-electrolysis time.

4. Conclusion

The ammonia-lead nitrate and ammonia-copper sulphate system have been employed as a leaching agent for recovery of lead and copper from scraped printed circuit board wastes. A two-stage leaching was employed, wherein the first stage consisted of leaching the scrap board with 0.1 M Pb $(NO_3)_2$ and 0.1 M $CuSO_4$ which results in the selective dissolution of lead and copper leaching rate, and other metals was found in lower amounts, respectively. The undissolved residue portion from the leaching stage containing nickel, tin, and silica were leached out in respective treatments. The current efficiency was found to increase with current density and concentration ratio with the contact time in acid bath. Hence, 73.29% lead and 82.17% copper have been successfully recovered from the electrolysis process. And, also by RSM Software prediction, the recovery of lead and copper are as 78.25% and 89.1%, respectively. In addition to the quadratic model equation, ANOVA, model terms, and fit statistics were also tested for the experimental conditions.

Acknowledgments

The authors convey sincere thanks to the Management and Principal of Coimbatore Institute of Technology, Coimbatore-641014, for sponsoring the project through the TEQIP-II fund.

References

[1] A. Mecucci and K. Scott, "Leaching and electrochemical recovery of copper, lead and tin from scrap printed circuit boards," *Journal of Chemical Technology and Biotechnology*, vol. 77, no. 4, pp. 449–457, 2002.

[2] B.-S. Kim, J.-C. Lee, S.-P. Seo, Y.-K. Park, and H. Y. Sohn, "A process for extracting precious metals from spent printed circuit boards and automobile catalysts," *Journal of Minerals, Metals and Materials*, vol. 56, no. 12, pp. 55–58, 2004.

[3] J. Cui and L. Zhang, "Metallurgical recovery of metals from electronic waste: a review," *Journal of Hazardous materials*, vol. 158, no. 2-3, pp. 228–256, 2008.

[4] F.-R. Xiu and F.-S. Zhang, "Recovery of copper and lead from waste printed circuit boards by supercritical water oxidation combined with electrokinetic process," *Journal of Hazardous materials*, vol. 165, no. 1-3, pp. 1002–1007, 2009.

[5] J. Guo and Z. Xu, "Recycling of non-metallic fractions from waste printed circuit boards: a review," *Journal of Hazardous materials*, vol. 168, no. 2-3, pp. 567–590, 2009.

[6] Y. J. Park and D. J. Fray, "Recovery of high purity precious metals from printed circuit boards," *Journal of Hazardous materials*, vol. 164, no. 2-3, pp. 1152–1158, 2009.

[7] Z.-F. Cao, H. Zhong, G.-Y. Liu, and S.-J. Zhao, "Techniques of Copper recovery from Mexican copper oxide ore," *Mining Science and Technology*, vol. 19, no. 1, pp. 45–48, 2009.

[8] B. Bayat and B. Sari, "Comparative evaluation of microbial and chemical leaching processes for heavy metal removal from dewatered metal plating sludge," *Journal of Hazardous materials*, vol. 174, no. 1-3, pp. 763–769, 2010.

[9] H. Z. Mousavi, A. Hosseynifar, V. Jahed, and S. A. M. Dehghani, "Removal of lead from aqueous solution using waste tire rubber ash as an adsorbent," *Brazilian Journal of Chemical Engineering*, vol. 27, no. 1, pp. 79–87, 2010.

[10] M. Yilmaz, T. Tay, M. Kivanc, and H. Turk, "Removal of corper(II) ions from aqueous solution by a lactic acid bacterium," *Brazilian Journal of Chemical Engineering*, vol. 27, no. 2, pp. 309–314, 2010.

[11] J. L. P. Sheeja and P. Selvapathy, "Comparative study on the removal efficiency of cadmium and lead using hydroxide and sulfide precipitation with the complexing agents," *International Journal of Current Research in Chemistry and Pharmaceutical Sciences*, vol. 1, no. 6, pp. 38–42, 2014.

[12] L. H. Yamane, V. T. de Moraes, D. C. R. Espinosa, and J. A. S. Tenório, "Recycling of WEEE: characterization of spent printed circuit boards from mobile phones and computers," *Waste Management*, vol. 31, no. 12, pp. 2553–2558, 2011.

[13] M. Delfini, M. Ferrini, A. Manni, P. Massacci, L. Piga, and A. Scoppettuolo, "Optimization of precious metal recovery from waste electrical and electronic equipment boards," *Journal of Environmental Protection*, vol. 2, no. 6, pp. 675–682, 2011.

[14] D. Pant, D. Joshi, M. K. Upreti, and R. K. Kotnala, "Chemical and biological extraction of metals present in E-waste: a hybrid technology," *Waste Management*, vol. 32, no. 5, pp. 979–990, 2012.

[15] E. Nishikawa, A. F. A. Neto, and M. G. A. Vieira, "Equilibrium and thermodynamic studies of zinc adsorption on expanded vermiculite," *Adsorption Science and Technology*, vol. 30, no. 8-9, pp. 759–772, 2012.

[16] J. Sohaili, S. K. Muniyandi, and S. S. Mohamad, "A review on printed circuit boards waste recycling technologies and reuse of recovered nonmetallic materials," *International Journal of Scientific and Engineering Research*, vol. 3, no. 2, pp. 138–144, 2012.

[17] B. Ali, M. M. Salarirad, and F. Veglio, "Process development for recovery of copper and precious metals from waste printed circuits board with emphasize on palladium and gold leaching and precipitation," *Waste Management*, vol. 33, no. 11, pp. 2354–2363, 2013.

[18] B. Oleksiak, G. Siwiec, and A. B. Grzechnik, "Recovery of precious metals from waste materials by the method of flotation process," *Metalurgija*, vol. 52, no. 1, pp. 107–110, 2013.

[19] A. Vidyadhar and A. Das, "Enrichment implication of froth flotation kinetics in the separation and recovery of metal values from printed circuit boards," *Separation and Purification Technology*, vol. 118, pp. 305–312, 2013.

[20] S. Fogarasi, F. Imre-Lucaci, A. Imre-Lucaci, and I. Petru, "Copper recovery and gold enrichment from waste printed circuit boards by mediated electrochemical oxidation," *Waste Management*, vol. 273, pp. 215–221, 2014.

[21] U. Jadhav and H. Hocheng, "Hydrometallurgical recovery of metals from large printed circuit board pieces," *Scientific Reports*, vol. 5, no. 1, article 14574, 2015.

[22] J. Sheeja, "Studies on the removal of chromium with the complexing agents," *Oriental Journal of Chemistry*, vol. 32, no. 4, pp. 2209–2213, 2016.

Adsorptive Removal of Hexavalent Chromium by Diphenylcarbazide-Grafted *Macadamia* Nutshell Powder

Londolani C. Maremeni,[1] **Sekomeng J. Modise** ⓘ,[2]
Fanyana M. Mtunzi ⓘ,[1] **Michael J. Klink** ⓘ,[1] **and Vusumzi E. Pakade** ⓘ[1]

[1]*Department of Chemistry, Faculty of Applied and Computer Sciences, Vaal University of Technology, Private Bag X021, Vanderbijlpark 1911, South Africa*
[2]*Institute of Chemical and Biotechnology, Faculty of Applied and Computer Sciences, Vaal University of Technology, Vanderbijlpark 1911, South Africa*

Correspondence should be addressed to Vusumzi E. Pakade; vusumzip@vut.ac.za

Academic Editor: Konstantinos Tsipis

Macadamia nutshell powder oxidized by hydrogen peroxide solutions (MHP) was functionalized by immobilizing 1,5′-diphenylcarbazide (DPC) on its surface. The effectiveness of grafting was confirmed by the Fourier transform infrared spectrum due to the presence of NH and C=C stretches at 3361, 1591, and 1486 cm^{-1}, respectively, on the grafted material which were absent in the nongrafted material. Thermogravimetric analysis revealed that the presence of DPC on the surface of *Macadamia* shells lowered the thermal stability from 300°C to about 180°C owing to the volatile nature of DPC. Surface roughness as a result of grafting was appreciated on the scanning electron microscopy images. Parameters influencing the adsorptive removal of Cr(VI) were examined and found to be optimal at pH 2, 120 min, 150 mg/L, and 2.5 g/L. Grafting MHP with DPC leads to an increase in the Langmuir monolayer capacity from 37.74 to 72.12 mg/g. Grafting MHP with DPC produced adsorbent with improved removal efficiency for Cr(VI).

1. Introduction

Natural biosorbents like mango kernels, *Macadamia* nutshells, coconut shells, pine cone, almond shells, sawdust, palm branches, and hazelnut consist in their plant cell walls chiefly the lignin, cellulose, and hemicellulose as the main structural components [1–5]. Consequently, these materials contain a cornucopia of surface functional groups including ketones, aldehydes, esters, ethers, and alcohols. The escalating use of natural biosorbents in adsorption for the remediation of metal pollutants has been largely due to their abundant availability, inexpensiveness, biodegradability, easy desorption, good reusability, and the diverse functional groups they possess which are needed for metal abstraction [6–8]. Further, these materials are regarded as cheap because they require minimum processing. However, when used as adsorbents, some shortfalls like poor surface areas and low

adsorption capacities are experienced probably because of the highly cross-linked polymeric nature of the said materials [1, 4, 9]. In addition, upon being used as adsorbents, there is a tendency of small organic molecules trapped within the polymeric cross-linked chains being released into the environments causing high biological oxygen demand and chemical oxygen demand [9]. These drawbacks have restricted the utilization of biosorbents in adsorption in their pristine form.

To address the downfalls, researchers have explored various pretreatment techniques such as (i) biological treatment with fungi or bacteria, (ii) chemical methods using acid, alkaline, oxidation, or solvent extraction, (iii) physical methods through sonication, pyrolysis, or mechanical agitation, and (iv) physicochemical methods, for example, steam explosion [5, 10]. During pretreatment, surface chemical properties of an adsorbent could be altered through masking or elimination of certain groups or exposure to more adsorption sites [7, 11]. Greater separation of components is achieved much faster and conveniently with chemical pretreatment methods than biological methods [5]. Depending on the intended use of the biomass and the properties desired, various chemical agents, H_2O_2 [7], ozone [12], H_3PO_4 [13], HNO_3 [14], NaOH [4], and so on, have been reconnoitred for either hydrolysis, oxidation, or delignification [5]. Oxidative chemical treatment could lead to increased oxygenated functionalities like COOH, C–O–C, CO, and OH which may then increase metal ion complexation, ion exchange, and chelation [5, 15].

To supplement pretreatment methods for improved performance, further modifications (grafting, cross-linking, or polymerization) could be carried out on the surface of adsorbents. Acrylic acid monomer units were grafted on the surface of coir pith, and an adsorption capacity increase from 165 to 196 mg/g was reported [16]. The improved performance was attributed to the increased density of –COOH adsorption sites on the material surface. Behbahani et al. [17] grafted 1,5′-diphenylcarbazide on the surface of multiwalled carbon nanotubes for the extraction of Cd(II) ions from water samples and food products. Extraction efficiency greater than 97% with a limit of detection (LOD) of 0.05 ng/mL was achieved. Grafting or polymerization introduces specific functional groups, NH, SH, OH, or COOH, on the surface of adsorbents. In a recent study [18], it was shown that modification of natural *Populus tremula* fibers with amino ligands led to increased adsorption capacity of acid blue 25 from 22.33 mg/g for unmodified to 48 mg/g for hydrazine fiber and 67 mg/g for ethylenediamine fiber. Masau stones were chemically modified with diethylenetriamine through a cross-linking protocol, and 87.32 mg/g adsorption capacity was reported [19]. The presence of protonated amino groups at acidic conditions was credited to the biosorption-coupled reduction removal mechanism of Cr(VI). Similarly, Pakade et al. [20] achieved high adsorption capacity of 145.5 mg/g following modification of *Macadamia* nutshell-based carbon with diethylenetriamine and triethylamine ligands. From the mentioned studies, it is obvious that Cr(VI) has a high affinity for amine functional groups, and it is this narrative that led to the present study where *Macadamia* nutshells were functionalized, for the first time, with 1,5′-diphenylcarbazide for the eradication of hexavalent chromium from aqueous solution.

Hexavalent chromium [Cr(VI)] and trivalent chromium [Cr(III)] are the most stable forms of Cr in the environment. Cr(VI) is a toxic strong oxidant capable of penetrating the biological cell membranes due to its similarity to sulphates in structure [21]. On the other hand, Cr(III) is less toxic and is regarded as a micronutrient at minute concentrations but could be toxic if concentrations are high [22]. Exposure to Cr(VI) compounds may lead to health detrimental effects like lung cancer, kidney damage, epigastric pain, nausea, vomiting, and even death [23, 24]. Careless disposal and improper treatment of effluents from leather tanning, electroplating, textile dyeing, mining, and wood preservation industries lead to contamination of the environment by Cr(VI) compounds. Owing to their advantages of simplicity,

low cost, and fine-tuning of functional groups, adsorption methods have found more applications for the elimination of toxic metals. Therefore, the current study aimed to employ *Macadamia* nutshells oxidized with H_2O_2 in removing Cr(VI) from aqueous solution and also investigate whether any amelioration in adsorption performance would be achieved following functionalization of H_2O_2-treated materials with 1,5′-diphenylcarbazide. Confirmation of the grafting was sought after using the Fourier transform infrared spectroscopy, thermal analysis, and scanning electron microscopy. Adsorption was evaluated by varying pH, time, concentration, and dosage, and various models were used to substantiate the adsorption mechanism involved.

2. Materials and Methods

2.1. Chemicals and Instrumentation. Reagent grade potassium dichromate, stannous chloride, ethyl chloride, methanol, toluene, hydrogen peroxide (50% wt.), sodium hydroxide, hydrochloric acid, and 1,5′-diphenylcarbazide were purchased from LabChem and Merck Chemical Co. (Johannesburg, South Africa) and were used without further purification. pH adjustments were monitored by HI 2210 from Hanna Instruments (Johannesburg, South Africa), while ultrapure water used for all reactions was obtained from Siemens LaboStar equipment (Warrendale, Pennsylvania, USA). Batch adsorption experiments were conducted in duplicate on an end-over-end Labcon 3100U electrical shaker (Maraisburg, South Africa). Sample preparation and analysis of chromium in solution before and after adsorption was conducted as detailed in our previous studies [4]. *Macadamia* nutshells were kindly donated by Eastern Produce Estates SA (Pty) Ltd from Tzaneen and Danroc (Pty) Ltd from Barbaton, South Africa.

2.2. Preparation of Adsorbents

2.2.1. Chemical Pretreatment. Following removal of debris and dirt, *Macadamia* nutshells were washed with running tap water and then dried for 24 h in an oven at 105°C. The dried nutshells were then ground, milled, and sieved collecting between 90 μm and 150 μm screens. The collected sample was designated as raw *Macadamia* nutshells (RMNs). Then, the RMN was oxidized and/or bleached with 20%, 35%, and 50% (v/v) H_2O_2 solution, resulting in the formation of more oxygen-bearing functional groups on its surface. The resultant materials were labeled 20 MHP, 35 MHP, and 50 MHP to correspond to the different H_2O_2 concentrations used for oxidation of *Macadamia* nutshell powder. The milling process not only aids with particle size reduction but also could reduce the crystallinity and the degree of polymerization of the polysaccharides in biomass [5] leading to improved adsorption performance.

2.2.2. Functionalization of MHP with Diphenylcarbazide. A method adapted from Behbahani et al. [17] was used with modifications. Briefly, one gram of 20 MHP was suspended in 100 mL of dry ethyl chloride ($CH_3CH_2Cl_2$) under nitrogen (N_2) atmosphere in a 250 mL round-neck flask equipped

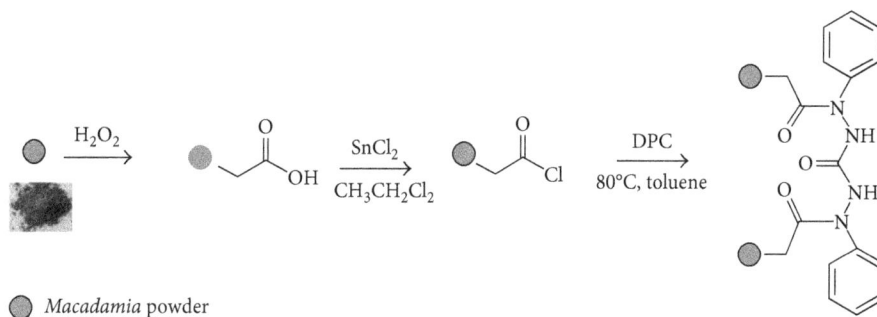

FIGURE 1: Preparation of diphenylcarbazide-grafted adsorbents.

with a magnetic stirrer and a condenser. About 2 g of stannous chloride was added to the mixture. Following 24 h of reflux, the solvent ($CH_3CH_2Cl_2$) was removed under reduced pressure and the resultant solid was suspended in toluene. About 5 g of excess diphenylcarbazide (DPC) was added to the reaction vessel, and the contents were refluxed for further 24 h. On completion, the solid was filtered, washed with toluene and methanol, and then dried at room temperature. The product was labeled 20 MHPD to signify DPC grafting on 20 MHP. The same procedure was followed but 20 MHP was replaced with 35 MHP and 50 MHP with the resolution products labeled 35 MHPD and 50 MHPD, respectively. Figure 1 shows the preparation procedure.

2.3. Characterization of Adsorbents.

Scanning electron microscopic images were obtained from an FEI Quanta 200 SEM (FEI, Hillsboro, OR, USA) and Jeol IT-300 tungsten scanning electron microscopy equipped with secondary backscattering electron detectors and an Oxford energy dispersive X-ray (EDX) analysis external probe. Samples were coated with gold using a Quorum Q150R sputter coater (Quorum Technologies Ltd, East Sussex, UK). The coating thickness was precisely controlled at 5 nm using the film thickness monitor option of the Quorum Q150R sputter. Functional group analysis on the surface of adsorbents was obtained from a PerkinElmer Spectrum 400 FT-IR/FT-NIR spectrometer (Waltham, MA, USA) recording from 4000 to 500 cm^{-1}. Thermal analysis of adsorbents was elucidated using a PerkinElmer TGA 4000 thermogravimetric analyzer (Waltham, USA). Centrifugation was accomplished employing a CL10 ThermoScientific centrifuge (Johannesburg, South Africa).

2.4. Adsorption Studies.

The *Macadamia* adsorbents (MHP and MHPD) were evaluated for their Cr(VI) removal efficacy through batch experiments conducted in duplicate. Various parameters that influence adsorption including the effect of pH (pH 1 to pH 12), contact time (20–180 min), initial adsorbate concentration (25 to 150 mg/L), and adsorbent dosage (0.63 to 10.63 g/L) were studied. In a typical experiment setup, 0.05 g of adsorbent was charged to a solution of Cr(VI) (20 mL of 100 mg/L) contained in a 100 mL glass bottle, and the pH was adjusted with HCl or NaOH diluted solutions. The contents were allowed to react for 120 min on the electrical shaker followed by solid/liquid separation through centrifugation. The concentration of Cr(VI) and total Cr was measured with UV-Vis spectrophotometer T80$^+$ (PG Instruments) and atomic absorption spectroscopy (AA-7000 from Shimadzu, Kyoto, Japan), respectively, as detailed in our previous studies [4, 25]. The performance of adsorption was evaluated by calculating the removal percentage (%R) and maximum adsorption capacity (q_e (mg/g)) using (1) and (2), respectively:

$$\%R = \frac{(C_0 - C_e)}{C_0} \times 100, \tag{1}$$

$$q_e = \frac{V(C_0 - C_e)}{m}, \tag{2}$$

where C_e is the adsorbate equilibrium concentration (mg/L), C_0 is the adsorbate initial concentration (mg/L), m is the mass of adsorbent (g), and V is the volume of solution used in adsorption (L).

3. Results and Discussion

3.1. Characterization of Materials

3.1.1. Fourier Transform Infrared Spectroscopy (FTIR) Analysis.

Figure 2(a) depicts the Fourier transform infrared spectroscopy (FTIR) spectra of RMN, 20 MHP, 35 MHP, and 50 MHP adsorbents. The hydrogen peroxide-oxidized *Macadamia* nutshells (MHP) have a broad vibrational stretch at 3330 cm^{-1} accredited to the presence of bonded OH groups of the cellulose, C=O of carboxyl functional groups at 1728 cm^{-1}, and the –C–O stretch of the primary alcohol at 1029 cm^{-1}. The H_2O_2 treatment was mild as the structural backbone of the RMN was not greatly altered, but peak shift characteristics of treatment were observed. In accordance with the results reported during chemical treatment of grape peelings with H_2O_2 [7] and lightweight expanded clay aggregate adsorbent with aqueous solution of magnesium chloride and hydrogen peroxide [26], no new peaks were observed suggesting a minimal influence of H_2O_2 on the chemical structure of biosorbents. The C=O band at 1712 cm^{-1} in the RMN shifted to 1728 cm^{-1} in MHP adsorbents accompanied by a change in peak shape implying that the H_2O_2 yielded C=O of a different chemical environment to the RMN. In addition, the C–O band of primary alcohols also shifted from 1031 to 1029 cm^{-1} after peroxide

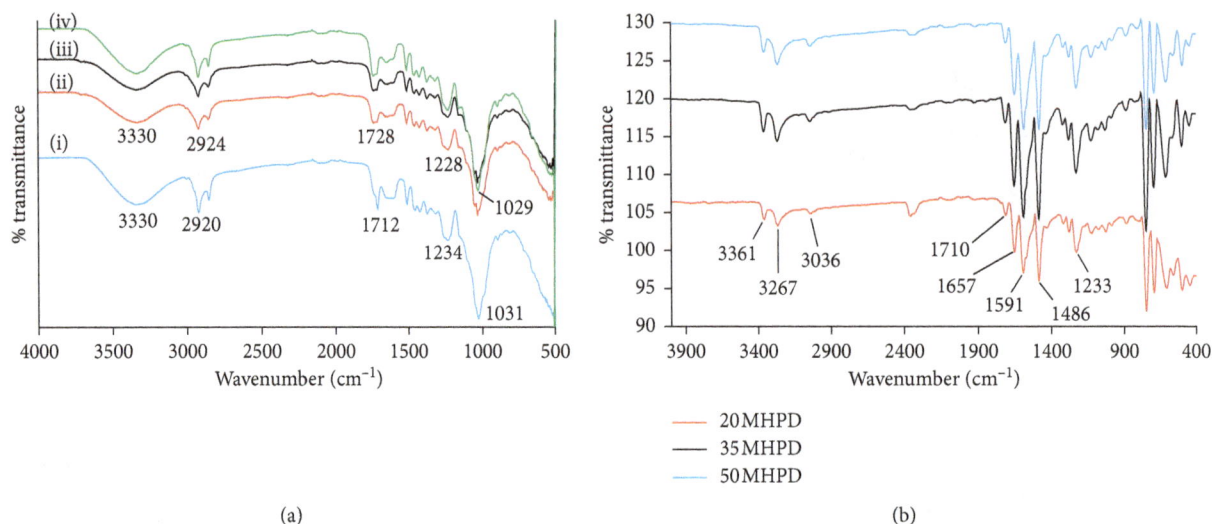

FIGURE 2: (a) FTIR spectra of RMN (i), 20 MHP (ii), 35 MHP (iii), and 50 MHP (iv) adsorbents. (b) FTIR spectra of 20 MHPD, 35 MHPD, and 50 MHPD adsorbents.

treatment. The –C–O–C vibrations shifted from 1234 to 1228 cm^{-1} after chemical treatment of the RMN with H_2O_2 [27]. Immobilization of DPC on MHP resulted in adsorbents with different functional groups as depicted in Figure 2(b). New absorption peaks at 3361 cm^{-1} attributed to the –NH stretch of DPC, –C=O of the amide at 1657 cm^{-1}, –CH stretch of the aromatic ring at 3036 cm^{-1}, –CN stretch at 1233 cm^{-1}, –CH bending at 743 cm^{-1}, and –C=C vibrations of the benzene ring at 1591 and 1486 cm^{-1} were observed. In addition, absorption peaks due to the parent material were still present: carboxyl –C=O at 1710 cm^{-1} and primary alcohol –C–O stretch at 1100 cm^{-1} [18]. All these observations pointed to the effectiveness of modification and immobilization of the DPC ligand.

3.1.2. Thermogravimetric Analysis (TGA) and Differential Thermal Analysis (DTA).

Figure 3 shows the TGA and DTA thermograms for the RMN and MHP. According to Paduraru et al. [28], thermal degradation of biomass follows four distinct stages, namely, the moisture evolution and decomposition of hemicellulose, cellulose, and lignin. These different decomposition stages were also observed with our materials. In all thermograms, the first decomposition observed from 50°C to 90°C was attributed to the loss of moisture and sorbed water. The second decomposition with the maximum mass loss rate at 318°C, 303°C, 326°C, and 317°C for RMN, 20 MHP, 35 MHP, and 50 MHP, respectively, was due to degradation of hemicellulose structures. The third decomposition observed at 385°C (RMN), 392°C (20 MHP), 374°C (35 MHP), and 382°C (50 MHP) was accredited to cellulose structure disintegration. The final degradation which took place beyond hemicellulose degradation was attributed to the slow decomposition of lignin. In addition, the RMN, 20 MHP, and 35 MHP displayed loss of volatile components, probably CH_4, H_2, CO_2, or CO, at about 250°C, and this was absent in 50 MHP adsorbent probably because the 50% (v/v) H_2O_2 treatment was too

harsh and eliminated all the volatile components Yang et al. [29] also observed evolution of volatile compounds (CH_4, H_2, CO_2, and CO) at temperatures below 300°C during pyrolysis of hemicellulose. The exact degradation temperatures of cellulose and hemicellulose were different in all materials indicating that the treatment influenced the structure and surface properties of these materials.

The DPC-grafted materials exhibited distinct thermograms compared to the H_2O_2-treated materials. The thermograms are shown in Figure 4. In all thermograms, loss of moisture and adsorbed water was observed at about 80°C. The more pronounced degradation at about 205°C was attributed to the decomposition of DPC from MHPD. Another notable peak was at 270°C attributable to the decomposition of cellulosic structures. Slow degradation of lignin can be appreciated from 335°C onwards. Clearly, the 50 MHPD exhibited slightly different decomposition peak shapes. Dhakal et al. [30] alluded that the difference in peak shapes and decomposition temperatures showed that the structural backbone of the adsorbents was affected by the chemical treatment to which they were exposed to.

3.1.3. Scanning Electron Microscopy (SEM).

The morphological properties of the adsorbent surface were elucidated using scanning electron microscopy (SEM). The SEM images for MHP and MHPD adsorbents are shown in Figure 5. RMN SEM was typical of plant material SEM with rough surface texture and scaling but with no observable porosity [4]. Upon treatment with H_2O_2, some surface porosity was observed as round spots in MHP samples. Grafting MHP with DPC yielded materials with much rougher texture, but pores disappeared. The differences in surface topographies between MHP and MHPD adsorbents proved that functionalization did occur. The disappearance of open pores due to grafting was observed elsewhere [31], while Albadarin et al. [19] also noted differences in surface topographies of

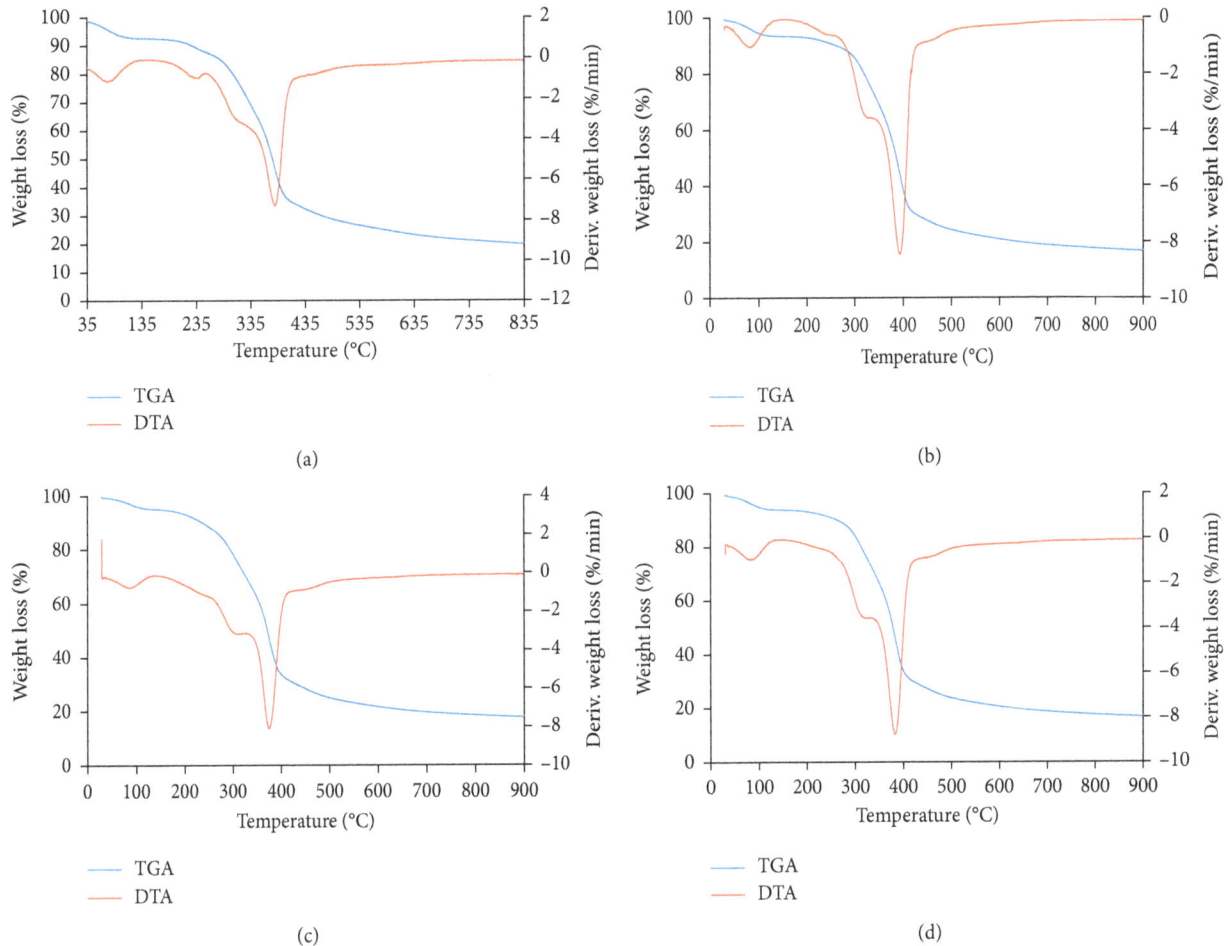

FIGURE 3: (a)TGA and DTA thermograms for RMN (a), 20 MHP (b), 35 MHP (c), and 50 MHP (d) adsorbents.

amine-modified masau stones incurred by chemical treatment.

3.1.4. Brunauer–Emmett–Teller (BET) Analysis.

The Brunauer–Emmett–Teller (BET) surface analysis was carried out to assess the pore volume, surface area, and pore size of the adsorbents. Table 1 lists the BET, surface area, pore volume and pore diameter results for MHP and MHPD adsorbents. Typical of biomass, the surface area values were relatively small ranging from 0.0063 to 0.5093 m^2/g. It seems that grafting of DPC on MHP decreased the surface areas as in the case of 35 MHP (1.0019 m^2/g) and 35 MHPD (0.2034 m^2/g) as a result of surface coverage. On the contrary, the surface area for 20 MHP increased to 0.5093 m^2/g after DPC grafting. Surface areas in the range of 2.8 to 6.3 m^2/g for palm branches [32] and from 3.1477 to 3.6672 m^2/g for peat and coconut fibers [33] have been reported.

3.2. Adsorption Experiments

3.2.1. Effect of pH.

The effect of pH on the removal of hexavalent chromium by 20 MHP, 35 MHP, 50 MHP, 20 MHPD, 35 MHPD, and 50 MHPD was carried out in batch adsorption experiments by varying the pH from 1 to 12 while

the initial concentration of Cr(VI) was 150 mg/L and the adsorbent dose was 2.5 g/L. The results are displayed as %R versus initial pH in Figure 6. Both the MHP and the MHPD adsorbents exhibited a similar trend where %R decreased as the initial pH of the solution was increased from 1 to 12. Highest removal of Cr(VI) took place at pH 1, but there was a clear separation in terms of performance between the MHP and MHPD adsorbents, with the latter exhibiting superior removal starting at 90% and decreasing to 79%, while the former ranged from 50% to 4% removal. That is, all these adsorbents showed a strong dependence in solution pH where maximum adsorption occurred at acidic conditions [4, 34]. The high removal at low pH is the result of electrostatic attraction between positively charged adsorbent sites (protonation) ($COOH_2^+$ and OH_2^+ for MHP and NH^+ for MHPD) and the hydrogen chromate ions [35]. Hydrogen chromate ($HCrO_4^-$) is usually the most dominant Cr(VI) species at acidic conditions and under oxidizing conditions. On the other hand, dichromate ($Cr_2O_7^{2-}$) is stable at weakly acidic and low oxidizing conditions, but as the pH is increased towards the basic region, the equilibrium tends to shift to chromate (CrO_4^{2-}) [25, 32]. The decline in %R as the pH was increased could be due to the depletion of protons resulting in less protonated sites and more OH^- groups that can lead to competition with chromate ions for adsorption

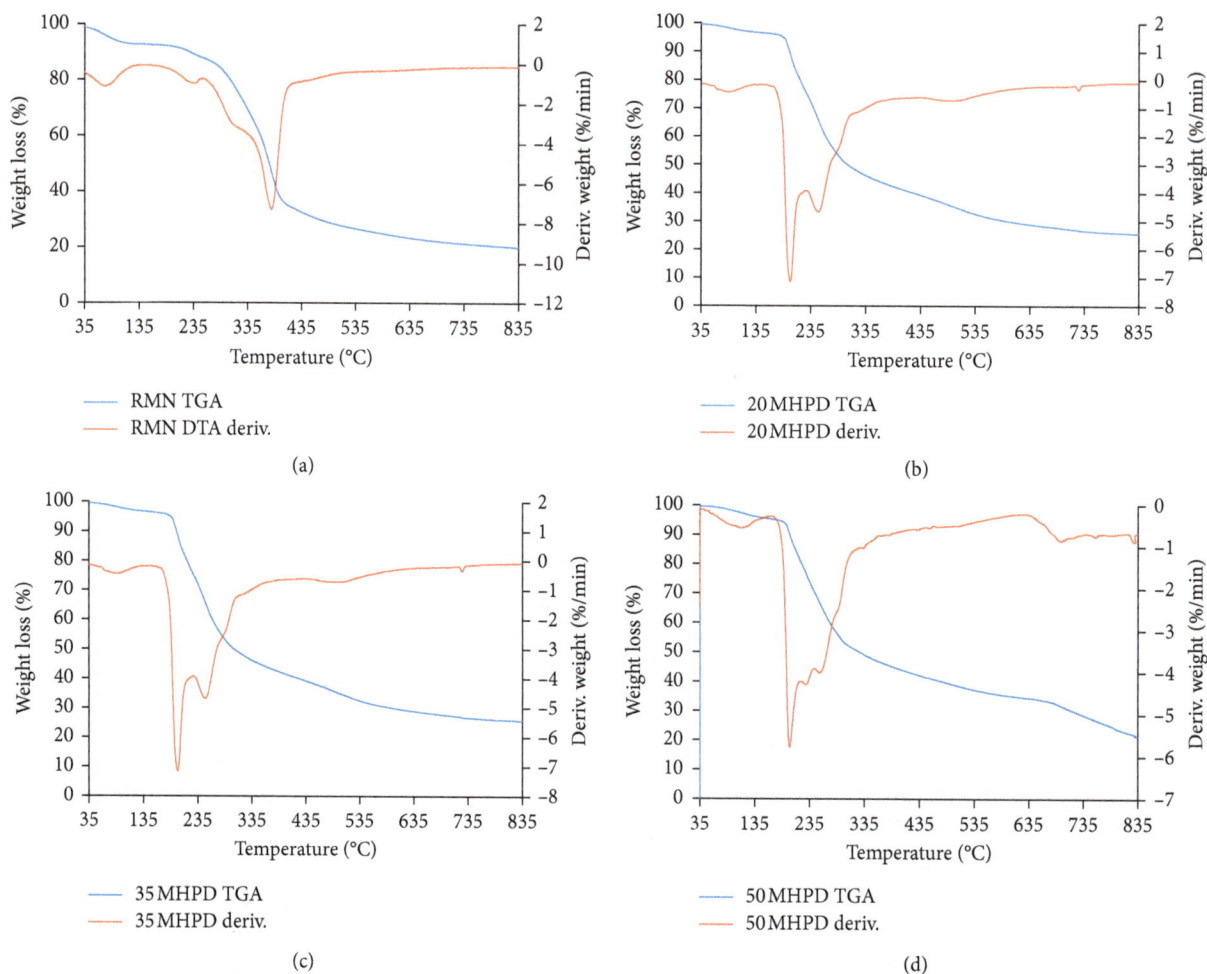

FIGURE 4: TGA and DTA thermograms for RMN (a), 20 MHPD (b), 35 MHPD (c), and 50 MHPD (d) adsorbents.

sites. Similar pH efficiency trends have been reported by several researchers [20, 31, 35–37]. Besides the predicted electrostatic attraction of Cr(VI) to the cationic group mechanism, the reduction of Cr(VI) to Cr(III) cannot be ruled out because it has been demonstrated in the past that *Macadamia* nutshell-based adsorbents favored adsorption-coupled reduction uptake of Cr(VI) [4, 38].

3.2.2. Effect of Contact Time and Kinetics. To investigate the feasibility and efficiency of Cr(VI) adsorption onto MHP and MHPD adsorbents, contact time was varied from 20 to 180 min, while all other parameters were kept constant. The % removal of Cr(VI) by MHP and MHPD as a function of time is shown in Figure 7. Two distinct sets of results were observed where the MHP adsorbents performed poorly compared to MHPD. The %R for MHP adsorbents increased from 12 to 70% as time was varied from 20 to 180 min and seemed to not have reached equilibrium even after 180 min of contact time. This could mean that the process of adsorption by MHP adsorbents was quite slow as there were adsorption sites not accessible after 180 min of reaction time. The order of adsorption efficiency was 50 MHP > 35 MHP > 20 MHP. Grafting MHP with DPC yielded adsorbents with better performance in terms of %R

and reaction kinetics. The %R of MHPD was much higher than that of MHP ranging from 65 to 98%, while the saturation was achieved only after 40 min. In some of these curves, the three stages of adsorption processes were notable, that is, the boundary layer saturation (20 to 30 min), diffusion into internal pores (30 to 40 min), and equilibration (beyond 40 min). The high affinity of the adsorbate to adsorption sites (physical adsorption or ion exchange) was associated with the higher sorption rate at initial times [34]. MHPD adsorbents possessed more functional groups with better accessibility due to the high affinity of Cr(VI) for protonated amino groups. The order of efficiency was 20 MHPD > 50 MHPD > 35 MHPD.

3.2.3. Kinetic Parameters. Pseudo-first-order [39] and pseudo-second-order [40] kinetic models were employed to study the kinetic parameters governing the adsorption of Cr(VI) to MHP and MHPD adsorbents. The time dependency data were fitted into nonlinear pseudo-first-order (PFO) and pseudo-second-order (PSO) equations (3) and (4). The results as illustrated in Table 2 revealed that all the adsorbents, except 20 MHP and 35 MHPD, followed the PSO rate model as judged by the higher coefficient of correlation (R^2), lower X^2,

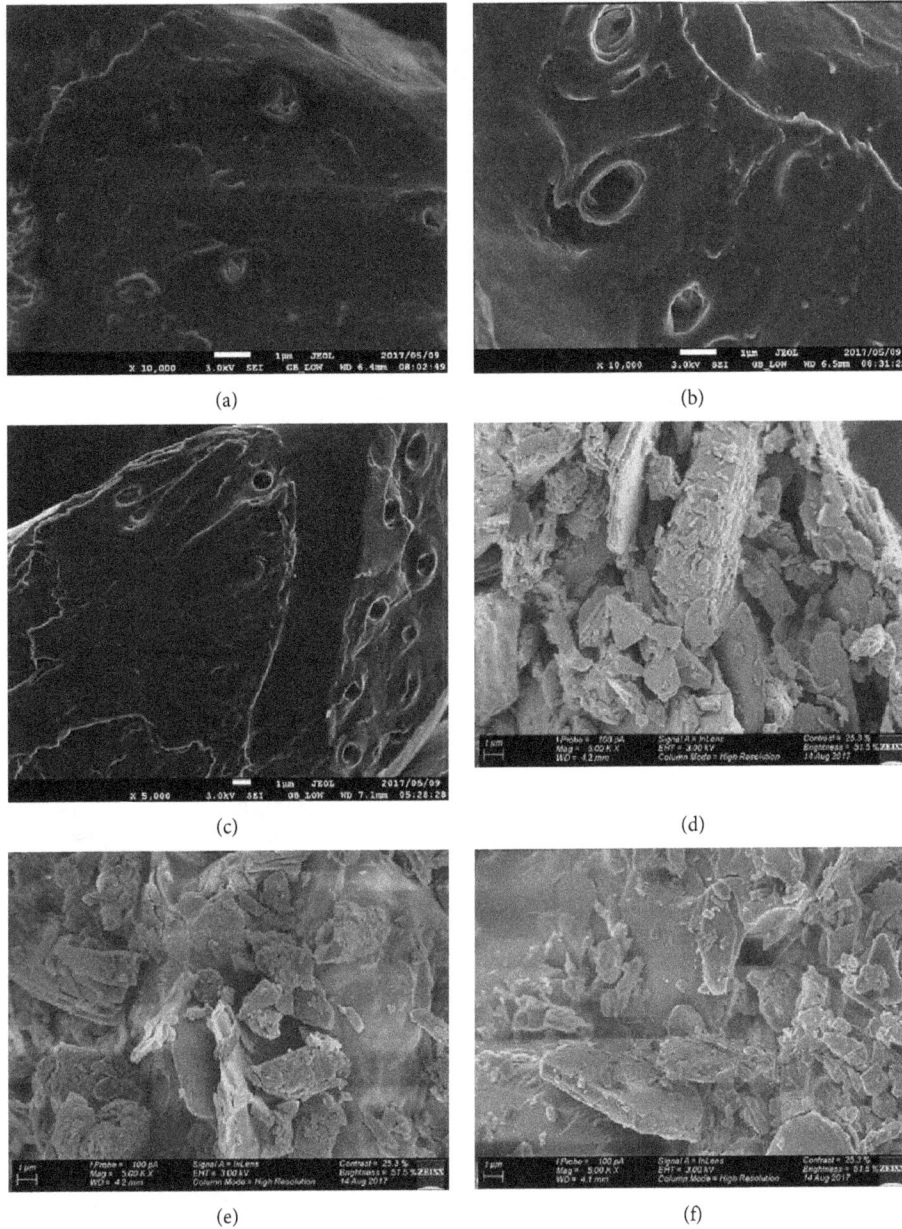

(a)

(b)

(c)

(d)

(e)

(f)

FIGURE 5: Scanning electron microscopic images for 20 MHP (a), 35 MHP (b), 50 MHP (c), 20 MHPD (d), 35 MHPD (e), and 50 MHPD (f).

and closeness of q_t to q_e values in PSO in comparison to the PFO model. It could then be inferred that PSO represented the adsorption of Cr(VI) by all MHP and MHPD adsorbents barring the 20 MHP and 35 MHPD which obeyed PFO. The adsorbents obeying PSO were predicted to favor chemisorption [31, 33], while those described by PFO favored physisorption [18]. It has been argued by Albadarin et al. [19] that the PFO and PSO kinetic models are empirical equations and therefore not fully account for the chemical and physical interactions of adsorbates. The complex nature of Cr(VI) interaction with adsorbents (i.e., adsorption/reduction) makes it difficult to explain the mechanism with only one model as is the case with 20 MHP. The k values were relatively small, implying that the sorption process required more time to approach saturation [19].

$$q_t = q_e \left(1 - \exp^{-k}1^t\right), \tag{3}$$

$$q_t = \frac{tk_2 \cdot q_e^2}{\left(1 + k_2 tq_e\right)}, \tag{4}$$

where k_1 (1/min) is the PFO rate constant, q_t (mg/g) is the amount of Cr(VI) (mg) adsorbed by the adsorbent (g) at time t, and k_2 (g/mg min) is the PSO rate constant.

3.2.4. Intraparticle and Liquid Film Diffusion. Intraparticle and/or film diffusion (external mass transfer) represent possible mechanisms by which adsorption of adsorbates onto porous materials could take place [41]. The intraparticle diffusion model proposed by Weber et al. [42] represented

TABLE 1: BET surface characterization.

Adsorbents	Surface characterization		Pore size (nm)
	Pore volume (cm^3/g)	BET surface area (m^2/g)	
20 MHP	—	0.0063	—
35 MHP	0.002717	1.0019	10.85013
50 MHP	—	0.5093	—
20 MHPD	—	0.1047	—
35 MHPD	—	0.2034	—
50 MHPD	—	0.0647	—

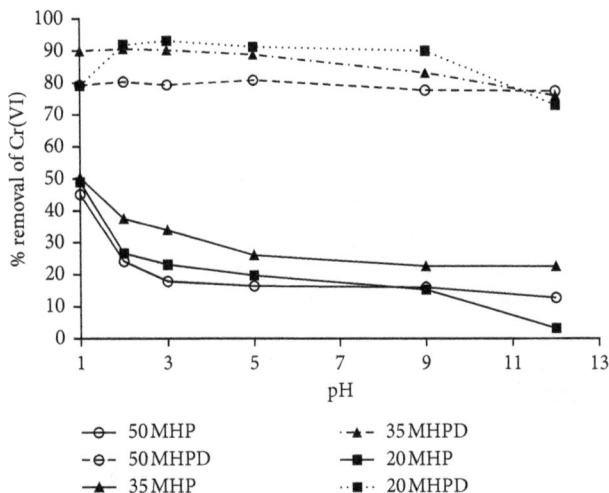

FIGURE 6: Effect of initial solution pH on the removal of Cr(VI) by MHP and MHPD adsorbents (conditions: initial Cr(VI) concentration 150 mg/L; dosage 2.5 g/L; and time 120 min).

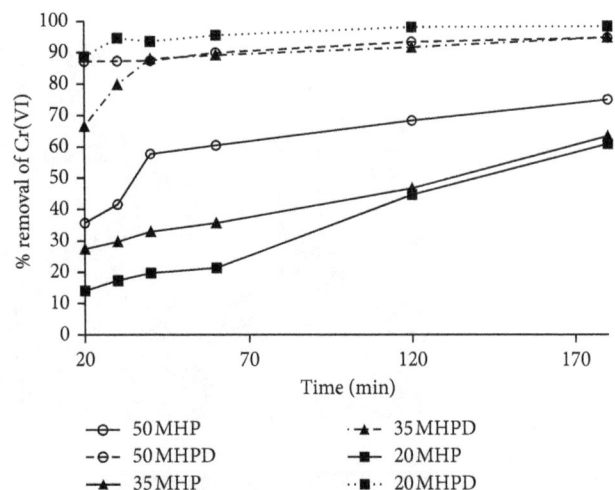

FIGURE 7: Effect of contact time for the removal of Cr(VI) by MHP and MHPD adsorbents (conditions: initial Cr(VI) concentration 150 mg/L; dosage 2.5 g/L; and pH 1).

here by (5) was used to determine the rate-controlling steps and mechanism of adsorption of Cr(VI). In addition, liquid film diffusion was evaluated by fitting the data into (6).

$$q_\mathrm{t} = k_\mathrm{d} t^{0.5} + C, \qquad (5)$$

$$\mathrm{Ln}\left(1 - \frac{q_\mathrm{t}}{q_\mathrm{e}}\right) = -k_\mathrm{fd} t, \qquad (6)$$

where k_d (mg/(g·min$^{0.5}$)) is the rate constant for intraparticle diffusion, k_fd is the film diffusion rate constant, and the boundary layer thickness is represented by the intercept C (mg/g). According to literature [43–45], intraparticle diffusion is deemed the only rate-controlling step when the plots of q_t versus $t^{0.5}$ yield a straight line passing through the origin. It can be seen in Figures 8(a) and 8(c) that the adsorbents exhibited different mechanisms, that is, linear line over the entire time interval (20 MHP, 35 MHP, 20 MHPD, and 50 MHPD) and multilinearity (50 MHP and 35 MHPD). Even those that were straight lines, none passed through the origin but implied the dominance of the intraparticle diffusion mechanism and insignificant mass transfer resistance. The curves with multilinearity showed the existence of film diffusion (the first steep curve) and intraparticle diffusion (the second flat curve). Assessment of the liquid thin-film diffusion model (Figures 8(b) and 8(d)) shows great linearity

with $R^2 > 0.98$, but lines did not pass through the origin point suggesting that film diffusion was also not the only rate-limiting step.

3.2.5. Effect of Initial Concentration and Adsorption Isotherms. Figure 9 shows the adsorption capacity (q_e) as a function of initial adsorbate concentration. It can be appreciated that the adsorption capacities increased as the initial concentration of Cr(VI) was increased from 20 to 180 mg/L. The magnitudes of q_e varied as follows: 4.9 to 15 mg/g for 20 MHP, 5.1 to 20 mg/g for 35 MHP, 5.1 to 40 mg/g for 50 MHP, 6 to 70 mg/g for 20 MHPD, 7 to 50 mg/g for 35 MHPD, and 7 to 74 mg/g for 50 MHPD. These results clearly showed that the adsorption of Cr(VI) was concentration dependent, that no saturation was realized except 20 MHP and 35 MHP, and that MHPD adsorbents were superior to MHP adsorbents. The increase in sorption capacity as concentration was increased could be explained on the basis that, at low adsorbate concentrations, the number of available adsorption sites compared to Cr(VI) ions was higher leading to higher Cr(VI) removal, but as the initial concentration of Cr(VI) was increased, this ratio (adsorption site/Cr(VI) ion) decreased resulting in saturation, lesser %R, and higher adsorption capacities [25, 36, 46].

TABLE 2: PFO and PSO kinetic parameters for adsorption of Cr(VI) by MHP and MHPD.

Parameters	20 MHP	35 MHP	50 MHP	20 MHPD	35 MHPD	50 MHPD
q_t (mg/g)	18.11	18.96	22.5	58.77	56.89	56.81
q_e (mg/g)	21.26	17.05	21.77	57.96	55.80	53.58
k_1 (1/min)	0.0107	0.0224	0.0325	0.1286	0.0644	0.151
R^2	0.939	0.615	0.940	0.799	0.967	0.637
X^2	41.71	14.04	11.24	0.40	5.93	0.780
q_e (mg/g)	31.09	20.82	25.73	59.52	60.25	56.78
k_2 (g/mg min)	0.0025	0.00117	0.00147	0.0085	0.00198	0.00863
R^2	0.942	0.725	0.936	0.896	0.881	0.887
X^2	89.12	26.83	22.1	0.95	11.72	2.19

To account for the interaction between Cr(VI) and MHPD adsorbents, nonlinear Langmuir and Freundlich adsorption isotherms were used to model the adsorption equilibrium data from the effect of concentration. The Langmuir pertains to monolayer adsorption of adsorbates onto homogeneous adsorption sites, while the Freundlich advocates the multilayer adsorption mechanism on heterogeneous sites. Table 3 illustrates the adsorption constants obtained from the isotherms. Comparing the MHP adsorbents using the coefficient of determination (R^2) and the variance, it can be seen that the Freundlich model exhibited higher R^2 values and lower variance. In addition, the q_e and Langmuir q_m values were not closer together. Therefore, it was concluded that the data for MHP were best described by the Freundlich model, implicating that the mechanism of removal was a multilayer process on heterogeneous adsorption sites. The Freundlich $1/n_F$ values were less than unity indicating a feasible and favorable adsorption of Cr(VI) on the MHP surface.

3.2.6. Effect of Dosage and Adsorption Capacity Comparison.

Figures 10(a) and 10(b) portray the %R of Cr(VI) as a function of adsorbent dosage for MHP and MHPD adsorbents. In both cases, the %R of Cr(VI) increased with increasing dose levels from 0.63 to 10.63 g/L. The %R increased sharply from 50 to 88% as the dose was increased from 0.63 to 4.63 g/L and slowed down afterward almost reaching equilibrium for MHP adsorbents at 92% removal. For MHPD adsorbents, the %R increased steeply from 67 to 93% when the dosage was increased from 0.63 to 2.63 g/L and only increased to 95% afterward as equilibrium was attained. It is clear that MHPD adsorbents were superior to MHP adsorbents as only half the dosage (2.63 g/L) was needed to achieve the same maximum %R by MHP at 4.63 g/L. The increased removal at higher doses could be due to the presence of more adsorption sites leading to high ion exchange capacity and surface area [5, 34, 36]. In both cases, the adsorption capacity decreased as the dose was increased probably because there was overlapping resulting from overcrowding of adsorption sites [47].

The performance of the prepared adsorbents was demonstrated by comparing their adsorption capacities to those of other biosorbents found in the literature. The results illustrated in Table 4 revealed that adsorption capacities of adsorbents may vary widely making it difficult to compare because of different experimental conditions (dosage, pH, adsorbate concentration, stirring rate, shaking versus stirring, and column versus batch) used. Nonetheless, the present result was comparable to other adsorbents but lower than others.

3.3. Selectivity of Cr(VI) Ion.

Investigation of the effect of competing ions is paramount for any developed adsorbent because pollutants often exist together with a host of other chemicals including organic, inorganic, and biological materials. It has been shown that the presence of sulphates and nitrates impacted the removal of Cr(VI) negatively [15, 48, 49]. In this study, a concoction of a solution containing 150 mg/L of each of Cr(VI), SO_4^{2-}, Cl^-, and NO_3^- was prepared from their respective salts and used to investigate the effect of co-ions on adsorption of Cr(VI). Another solution containing only Cr(VI) was used as a control. The results depicting the %R of Cr(VI) in the presence and absence of co-ions are presented in Figure 11. The percent removal of Cr(VI) decreased from 80 to 79% for 20 MHP, 87 to 80% for 50 MHP, 95 to 93% for 20 MHPD, and 85 to 82% for 50 MHPD, while the %R was exactly the same for 35 MHPD but the removal was higher in the presence of co-ions when 35 MHP was used. The 50 MHPD adsorbents showed superior selectivity than 20 MHPD and 35 MHPD. The latter was negatively affected by the presence of competing ions, but still, the %R was still greater than 80%. The prepared adsorbents demonstrated good recognition of Cr(VI) in the presence of competing ions and can thus be recommended as alternative low-cost adsorbents for the mitigation of Cr(VI).

3.4. Application to Real-World Sample.

A wastewater sample with the chemical qualities Cr^{3+} (4 mg/L), Zn^{2+} (1 mg/L), Fe^{2+} (6 mg/L), Ni^{2+} (2 mg/L), and Cu^{2+} (31 mg/L) was used to investigate the performance of the developed adsorbents in a real-world sample. All were prepared from their chloride salts. The sample was spiked with 30 mg/L Cr(VI). Figure 12 shows the results of %R of Cr(VI) from the spiked solution by MHPD adsorbents. The efficiency in the removal of Cr(VI) by MHP adsorbents decreased in the order 20 MHP > 35 MHP > 50 MHP from 82%, 75%, and 53% removal, respectively. In contrast, the removal increased in the order 80%, 85%, and 88% for 20 MHPD, 35 MHPD, and 50 MHPD, respectively. The decrease in %R as the percent of

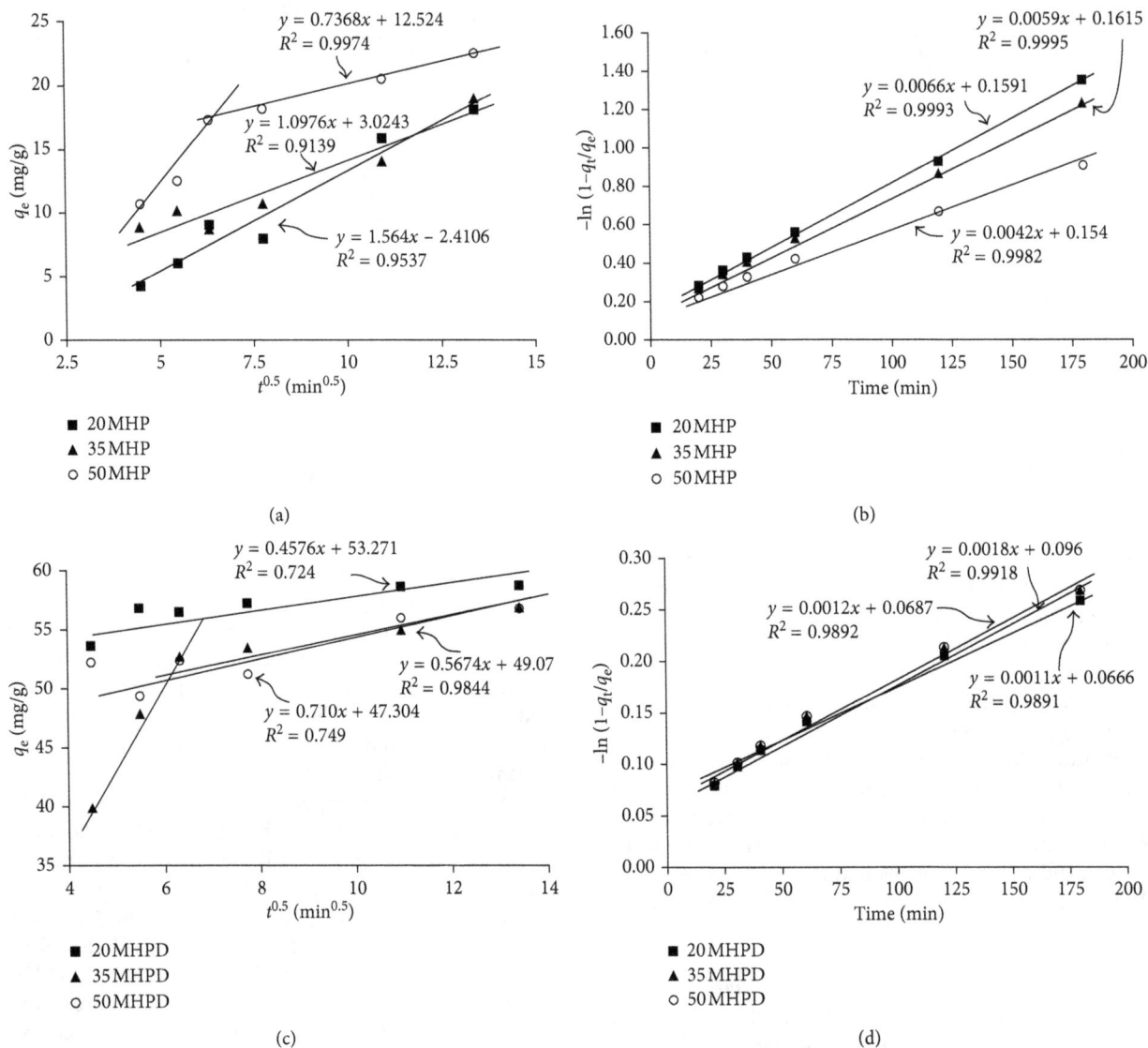

FIGURE 8: Intraparticle and liquid film diffusion of MHP (a, b) and MHPD (c, d) adsorbents.

H_2O_2 increased (20 MHP, 35 MHP, and 50 MHP) was probably due to the high density of oxygenated groups on the surface imparted by a high concentration of H_2O_2 leading to the high affinity for cations. On the other hand, the removal increased with increasing H_2O_2 concentration for MHPD because of higher concentration of DPC which favored removal of Cr(VI) at low pH due to protonation. The results revealed that the prepared adsorbents are suitable to be used for the remediation of water contaminated with metal ions as the %R was greater than 70% except for the 50 MHP adsorbents.

4. Conclusion

Three different concentrations of hydrogen peroxide (20, 35, and 50% vol.) were utilized to pretreat the *Macadamia* nutshell powder. The pretreatment imparts some oxygen-containing groups (C=O, OH, and C–O–C) on the surface of

the nutshells. It was demonstrated by FTIR that the H_2O_2 treatment resulted in shifting of C=O peak from 1712 to 1728 cm^{-1} and the C–O peak from 1031 to 1029 cm^{-1} in comparison to the RMN. Grafting of 1,5′-diphenylcarbazide on the RMN was confirmed with FTIR, SEM, and TGA. The FTIR showed new peaks at 3333 and 1657 cm^{-1} attributed to NH and amide C=O, respectively. Successful grafting was further corroborated by the TGA curves of MHPD showing high volatility due to the presence of DPC compared to MHP. SEM micrographs of MHPD exhibited much rougher surface than the MHP, and this was associated with grafting. The BET surface area ranged from 0.0063 to 0.5093 m^2/g, and these were typical of biomass materials. The adsorption efficiency of MHP treated with 50% H_2O_2 improved from 37.74 to 72.12 mg/g for the grafted materials. The improvement in adsorption capacity validated the efficiency of grafting. The adsorption process was best described by Langmuir and PSO. In addition, it was shown that

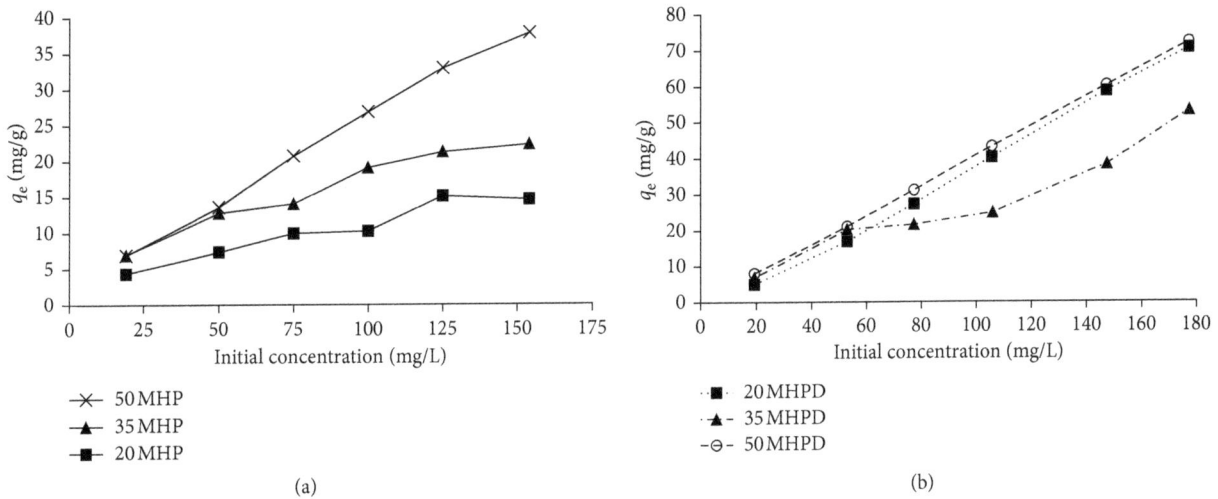

(a)

(b)

FIGURE 9: Effect of initial solution concentration for the removal of Cr(VI) by MHP and MHPD adsorbents (conditions: contact time 180 min; dosage 2.5 g/L; and pH 2).

TABLE 3: Freundlich and Langmuir adsorption constants for MHP.

Models	Parameters	20 MHP	35 MHP	50 MHP
Langmuir $q_e = q_{max} \cdot b \cdot C_e / (1 + bC_e)$	q_e (mg/g)	14.60	22.24	37.74
	q_m (mg/g)	21.77	27.76	60.59
	b (L/mg)	0.0173	0.0391	0.0189
	R^2	0.873	0.935	0.961
	Var*	2.72	2.81	6.67
Freundlich $q_e = K_F \cdot C_e^{1/nf}$	K_F (mg^{1-1n}L$^{1/n}$/g)	1.33	3.27	3.39
	n_F	1.96	2.33	1.84
	R^2	0.906	0.947	0.974
	Var*	2.02	2.29	4.39

*Sum of square errors divided by degrees of freedom.

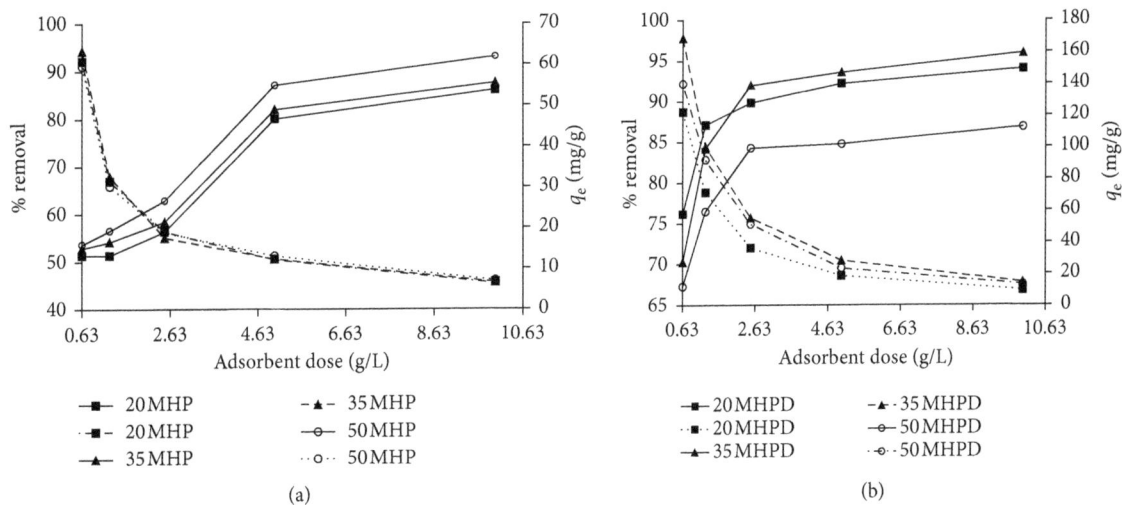

(a)

(b)

FIGURE 10: Effect of adsorbent dosage for the removal of Cr(VI) by MHP and MHPD adsorbents (conditions: initial Cr(VI) concentration 150 mg/L; pH 2; and time 120 min).

TABLE 4: Comparison of adsorption capacities.

Adsorbents	Pretreatment	Functionalization	pH	q_m (mg/g)	References
Potato peels	HCl	—	2.5	3.28	[45]
Raw rutin	Extraction/isolation	—	3	26.3	[34]
Rutin resin	Formaldehyde/HNO$_3$	—	3	41.6	[34]
Banana peels	HCl/NaOH/H$_2$O$_2$	Acrylonitrile grafted	3	6.17	[31]
Palm branches	H$_2$SO$_4$	—	2	25	[32]
Palm branches	Acetic acid	Chitosan	6	55	[32]
Palm branches	H$_2$SO$_4$	Cationic surfactant	6	41.7	[32]
Grape peelings	H$_2$O$_2$	—	5.5	39.06	[7]
Banana peels	—	—	5	3	[15]
Orange peels	—	—	3	9	[15]
Coir pith	—	Acrylic acid	2	165	[16]
Masau stones	NaOH	Epichlorohydrin/diethylenetriamine	3.5	87.33	[19]
Coir pith	—	Acrylic acid	2	196	[16]
Macadamia nutshell	H$_2$O$_2$	—	1	37.74	This study
Macadamia nutshell	H$_2$O$_2$	Diphenylcarbazide	1	72.12	This study

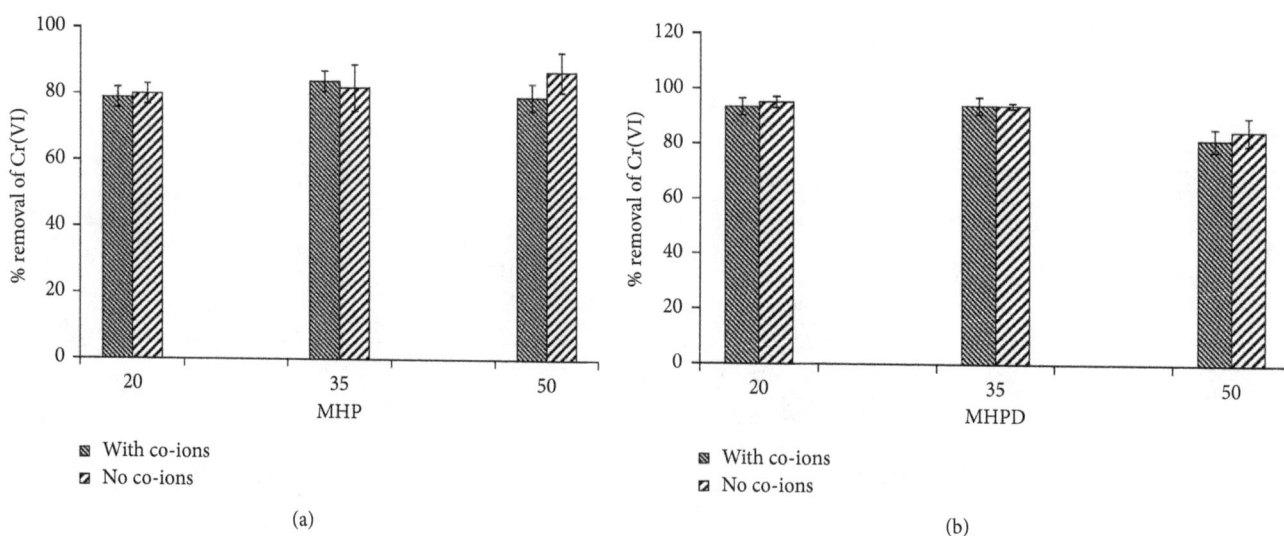

(a)

(b)

FIGURE 11: The effect of co-ions on the removal of Cr(VI) by MHP (a) and MHPD (b) adsorbents.

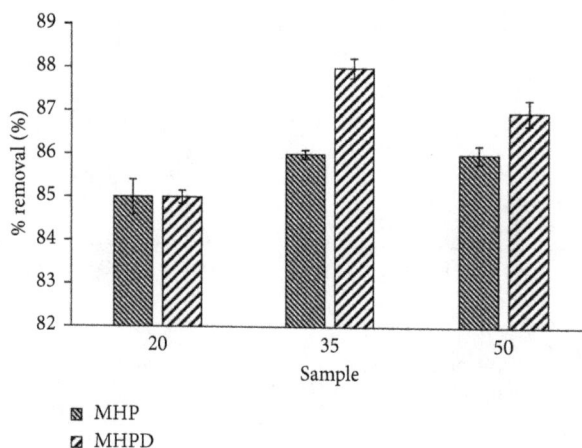

FIGURE 12: Real sample characteristics and removal of Cr(VI) from the real sample by MHP and MHPD adsorbents.

intraparticle diffusion was not the only rate-controlling step. The adsorption of Cr(VI) by MHPD was less affected by the presence of competing ions as it was shown in selectivity studies and application to real-world sample.

Abbreviations

DPC:	1,5′-diphenylcarbazide
MHP:	*Macadamia* nutshell powder oxidized by hydrogen peroxide solutions
20 MHP:	*Macadamia* nutshell powder oxidized by 20% (v/v) hydrogen peroxide solutions
35 MHP:	*Macadamia* nutshell powder oxidized by 35% (v/v) hydrogen peroxide solutions
50 MHP:	*Macadamia* nutshell powder oxidized by 50% (v/v) hydrogen peroxide solutions
20 MHPD:	20 MHP adsorbent grafted with DPC
35 MHPD:	35 MHP adsorbent grafted with DPC
50 MHPD:	50 MHP adsorbent grafted with DPC
BET:	Brunauer–Emmett–Teller
DTA:	Differential thermal analysis
EDX:	Energy dispersive X-ray
FTIR:	Fourier transform infrared spectroscopy
LOD:	Limit of detection
PFO:	Pseudo-first order
PSO:	Pseudo-second order
RMN:	Raw *Macadamia* nutshell
SEM:	Scanning electron microscopy
TGA:	Thermogravimetric analysis.

Additional Points

Highlights. *Macadamia* nutshell powder was oxidized with hydrogen peroxide. The peroxide-modified materials were grafted with diphenylcarbazide. Grafting with diphenylcarbazide improved the adsorption capacity. Presence of co-ions did not influence the removal efficiency. Adsorbents demonstrated good recoveries when subjected to real-world sample.

Acknowledgments

The National Research Foundation (TTK13061018779 and TTK160510164648) and the Vaal University of Technology financial support are gratefully appreciated.

References

[1] W. S. W. Ngah and M. A. K. M. Hanafiah, "Removal of heavy metal ions from wastewater by chemically modified plant wastes as adsorbents: a review," *Bioresource Technology*, vol. 99, no. 10, pp. 3935–3948, 2008.

[2] J. L. Gardea-Torresdey, M. K. Becker-Hapak, J. M. Hosea, and D. W. Darnall, "Effect of chemical modification of algal carboxyl groups on metal ion binding," *Environmental Science & Technology*, vol. 24, no. 9, pp. 1372–1378, 1990.

[3] H. Kołoczek, J. Chwastowski, and W. Żukowski, "Peat and coconut fiber as biofilters for chromium adsorption from contaminated wastewaters," *Environmental Science and Pollution Research*, vol. 23, no. 1, pp. 527–534, 2016.

[4] V. E. Pakade, T. D. Ntuli, and A. E. Ofomaja, "Biosorption of hexavalent chromium from aqueous solutions by *Macadamia* nutshell powder," *Applied Water Science*, vol. 7, no. 6, pp. 3015–3030, 2017.

[5] R. Kumar, R. K. Sharma, and A. P. Singh, "Cellulose-based grafted biosorbents-journey from lignocellulose biomass to toxic metal ions sorption applications-a review," *Journal of Molecular Liquids*, vol. 232, pp. 62–93, 2017.

[6] S. J. Köhler, P. Cubillas, J. D. Rodriguez-Blanco, C. Bauer, and M. Prieto, "Removal of cadmium from wastewaters by aragonite shells and the influence of other divalent cations," *Environmental Science & Technology*, vol. 41, no. 1, pp. 112–118, 2007.

[7] E. Rosales, J. Meijide, T. Tavares, M. Pazos, and M. A. Sanromán, "Grapefruit peelings as a promising biosorbent for the removal of leather dyes and hexavalent chromium," *Process Safety and Environmental Protection*, vol. 101, pp. 61–71, 2016.

[8] X. Liu, Z. Q. Chen, B. Han, C. L. Su, Q. Han, and W. Z. Chen, "Biosorption of copper ions from aqueous solution using rape straw powders: optimization, equilibrium and kinetic studies," *Ecotoxicology and Environmental Safety*, vol. 150, pp. 251–259, 2018.

[9] A. Abdolali, W. S. Guo, H. H. Ngo, S. S. Chen, N. C. Nguyen, and K. L. Tung, "Typical lignocellulosic wastes and by-products for biosorption process in water and wastewater treatment: a critical review," *Bioresource Technology*, vol. 160, pp. 57–66, 2014.

[10] J. Putro, F. Soetaredjo, S. Lin, Y. Ju, and S. Ismadji, "Pretreatment and conversion of lignocellulose biomass into valuable chemicals," *RSC Advances*, vol. 6, no. 52, pp. 46834–46852, 2016.

[11] R. H. Vieira and B. Volesky, "Biosorption: a solution to pollution?," *International Microbiology*, vol. 3, no. 1, pp. 17–24, 2000.

[12] Z. Kádár, N. Schultz-Jensen, J. S. Jensen, M. A. T. Hansen, F. Leipold, and A.-B. Bjerre, "Enhanced ethanol production by removal of cutin and epicuticular waxes of wheat straw by plasma assisted pretreatment," *Biomass and Bioenergy*, vol. 81, pp. 26–30, 2015.

[13] Y. H. P. Zhang, S. Y. Ding, J. R. Mielenz et al., "Fractionating recalcitrant lignocellulose at modest reaction conditions," *Biotechnology and Bioengineering*, vol. 97, no. 2, pp. 214–223, 2007.

[14] M. E. Himmel, W. S. Adney, J. O. Baker et al., "Advanced bioethanol production technologies: a perspective," *ACS Symposium Series*, vol. 666, pp. 2–45, 1997.

[15] K. Pakshirajan, A. N. Worku, M. A. Acheampong, H. J. Lubberding, and P. N. L. Lens, "Cr(III) and Cr(VI) removal from aqueous solutions by cheaply available fruit waste and algal biomass," *Applied Biochemistry and Biotechnology*, vol. 170, no. 3, pp. 498–513, 2013.

[16] P. Suksabye and P. Thiravetyan, "Cr(VI) adsorption from electroplating plating wastewater by chemically modified coir pith," *Journal of Environmental Management*, vol. 102, pp. 1–8, 2012.

[17] M. Behbahani, A. Bagheri, M. M. Amini et al., "Application of multiwalled carbon nanotubes modified by diphenylcarbazide for selective solid phase extraction of ultra-traces Cd(II) in water samples and food products," *Food Chemistry*, vol. 141, no. 1, pp. 48–53, 2013.

[18] N. Tka, M. Jabli, T. A. Saleh, and G. A. Salman, "Amines modified fibers obtained from natural *Populus tremula* and

their rapid biosorption of Acid Blue 25," *Journal of Molecular Liquids*, vol. 250, pp. 423–432, 2018.

[19] A. B. Albadarin, S. Solomon, T. A. Kurniawan, C. Mangwandi, and G. Walker, "Single, simultaneous and consecutive biosorption of Cr(VI) and Orange II onto chemically modified masau stones," *Journal of Environmental Management*, vol. 204, pp. 365–374, 2017.

[20] V. E. Pakade, L. C. Maremeni, T. D. Ntuli, and N. T. Tavengwa, "Application of quaternized activated carbon derived from *Macadamia* nutshells for the removal of hexavalent chromium from aqueous solutions," *South African Journal of Chemistry*, vol. 69, pp. 180–188, 2016.

[21] I. Polowczyk, B. F. Urbano, B. L. Rivas, M. Bryjak, and N. Kabay, "Equilibrium and kinetic study of chromium sorption on resins with quaternary ammonium and N-methyl-D-glucamine groups," *Chemical Engineering Journal*, vol. 284, pp. 395–404, 2016.

[22] K. Selvarag, S. Manonmani, and S. Pattabhi, "Removal of hexavalent chromium using distillery sludge," *Bioresource Technology*, vol. 89, no. 2, pp. 207–211, 2003.

[23] E. Browning, *Chromium in Toxicity of Industrial Metals*, Butterworths and Co., London, UK, 2nd edition, 1969.

[24] M. Cieslak-Golonka, "Toxic and mutagenic effects of chromium(VI)," *Polyhedron*, vol. 15, no. 21, pp. 3667–3689, 1995.

[25] V. E. Pakade, O. B. Nchoe, L. Hlungwane, and N. T. Tavengwa, "Sequestration of hexavalent chromium from aqueous solutions by activated carbon derived from *Macadamia* nutshells," *Water Science and Technology*, vol. 75, no. 1, pp. 196–206, 2017.

[26] E. M. Kalhori, K. Yetilmezsoy, N. Uygur, M. Zarrabi, and R. M. A. Shmeis, "Modeling of adsorption of toxic chromium on natural and surface modified lightweight expanded clay aggregate (LECA)," *Applied Surface Science*, vol. 287, pp. 428–442, 2013.

[27] A. Demirbas, "Mechanisms of liquefaction and pyrolysis reactions of biomass," *Energy Conversion and Management*, vol. 41, no. 6, pp. 633–646, 2000.

[28] C. Paduraru, L. Tofan, C. Teodosiu, I. Bunia, N. Tudorachi, and O. Toma, "Biosorption of zinc(II) on rapeseed waste: equilibrium studies and thermogravimetric investigations," *Process Safety and Environmental Protection*, vol. 94, pp. 18–28, 2015.

[29] H. Yang, R. Yan, H. Chen, D. H. Lee, and C. Zheng, "Characteristics of hemicellulose, cellulose and lignin pyrolysis," *Fuel*, vol. 86, no. 12-13, pp. 1781–1788, 2007.

[30] R. P. Dhakal, K. N. Ghimire, and K. Inoue, "Adsorptive separation of heavy metals from an aquatic environment using orange waste," *Hydrometallurgy*, vol. 79, no. 3-4, pp. 182–190, 2005.

[31] A. Ali, K. Saeed, and F. Mabood, "Removal of chromium (VI) from aqueous medium using chemically modified banana peels as efficient low-cost adsorbent," *Alexandria Engineering Journal*, vol. 55, no. 3, pp. 2933–2942, 2016.

[32] M. M. Shouman, N. A. Fathy, S. A. Khedr, and A. A. Attia, "Comparative biosorption studies of hexavalent chromium ion onto raw and modified branches," *Advances in Physical Chemistry*, vol. 2013, Article ID 159712, 9 pages, 2013.

[33] J. Chwastowski, P. Staroń, H. Kołoczek, and M. Banach, "Adsorption of hexavalent chromium from aqueous solutions using Canadian peat and coconut fiber," *Journal of Molecular Liquids*, vol. 248, pp. 981–989, 2017.

[34] N. A. Fathy, S. T. El-Wakeel, and R. R. Abd El-Latif, "Biosorption and desorption studies on chromium(VI) by novel biosorbents of raw rutin and rutin resin," *Journal of Environmental Chemical Engineering*, vol. 3, no. 2, pp. 1137–1145, 2015.

[35] S. O. Owalude and A. C. Tella, "Removal of hexavalent chromium from aqueous solutions by adsorption on modified groundnut hull," *Beni-Suef University Journal of Basic and Applied Sciences*, vol. 5, no. 4, pp. 377–388, 2017.

[36] M. Jain, V. K. Garg, and K. Kadirvelu, "Adsorption of hexavalent chromium from aqueous medium onto carbonaceous adsorbents prepared from waste biomass," *Journal of Environmental Management*, vol. 91, no. 4, pp. 949–957, 2010.

[37] Y. Sun, Q. Yue, B. Gao, Y. Gao, Q. Li, and Y. Wang, "Adsorption of hexavalent chromium on *Arundo donax* Linn activated carbon amine-crosslinked copolymer," *Chemical Engineering Journal*, vol. 217, pp. 240–247, 2013.

[38] L. Hlungwane, E. L. Viljoen, and V. E. Pakade, "*Macadamia* nutshells derived activated carbon and attapulgite clay combination for synergistic removal of Cr(VI) and Cr(III)," *Adsorption Science & Technology*, vol. 36, no. 1-2, pp. 713–731, 2017.

[39] S. Lagergren, "Zur theorie der sogenannten adsorption geloster stoffe," *Kungliga Sevenska Vetenskapas Akademiens*, vol. 24, pp. 1–39, 1898.

[40] Y. S. Ho and G. McKay, "Pseudo-second order model for sorption processes," *Process Biochemistry*, vol. 34, no. 5, pp. 451–465, 1999.

[41] R. L. Tseng and S. K. Tseng, "Pore structure and adsorption performance of the KOH activated carbons prepared from corncob," *Journal of Colloid and Interface Science*, vol. 87, no. 2, pp. 428–437, 2005.

[42] W. J. Weber, J. C. Morris, and J. Sanit, "Kinetics of adsorption on carbon from solution," *Journal of the Sanitary Engineering Division-Civil Engineering*, vol. 89, no. 2, pp. 31–60, 1963.

[43] V. J. P. Poots, G. McKay, and J. J. Healy, "The removal of acid dye from effluent, using natural adsorbents: I peat," *Water Research*, vol. 10, no. 12, pp. 1061–1066, 1976.

[44] P. Saha, "Assessment of the removal of methylene blue dye using tamarind fruit shell as biosorbent," *Water, Air, & Soil Pollution*, vol. 213, no. 1–4, pp. 287–299, 2010.

[45] K. M. Doke and E. M. Khan, "Equilibrium, kinetic and diffusion mechanism of Cr(VI) adsorption onto activated carbon derived from wood apple shell," *Arabian Journal of Chemistry*, vol. 10, pp. S252–S260, 2017.

[46] F. Mutongo, O. Kuipa, and P. K. Kuipa, "Removal of Cr(VI) from aqueous solutions using powder of potato peelings as a low cost sorbent," *Bioinorganic Chemistry and Applications*, vol. 2014, Article ID 973153, 7 pages, 2014.

[47] C. Namasivayam, K. Kadirvelu, and M. Kumuthu, "Removal of direct red and acid brilliant blue by adsorption on to banana pith," *Bioresource Technology*, vol. 64, no. 1, pp. 77–79, 1998.

[48] V. Neagu and S. Mikhalovsky, "Removal of hexavalent chromium by new quaternized crosslinked poly(4-vinylpyridines)," *Journal of Hazardous Materials*, vol. 183, no. 1-3, pp. 533–540, 2010.

[49] V. Pakade, E. Cukrowska, J. Darkwa, N. Torto, and L. Chimuka, "Selective removal of chromium (VI) from sulphates and other metal anions using an ion-imprinted polymer," *Water SA*, vol. 37, no. 4, pp. 529–537, 2011.

Synthesis, Characterization, and Antimicrobial Activity of Novel Sulfonated Copper-Triazine Complexes

Supun Katugampala,[1] **Inoka C. Perera,**[2] **Chandrika Nanayakkara,**[3] **and Theshini Perera** ⓘ[1]

[1]*Department of Chemistry, University of Sri Jayewardenepura, Nugegoda, Sri Lanka*
[2]*Department of Zoology and Environment Science, University of Colombo, Colombo, Sri Lanka*
[3]*Department of Plant Science, University of Colombo, Colombo, Sri Lanka*

Correspondence should be addressed to Theshini Perera; theshi@sjp.ac.lk

Academic Editor: Aurel Tabacaru

Metallotriazine complexes possess interesting biological and medicinal properties, and the present study focuses on the synthesis, characterization, and antimicrobial activity of four novel copper-triazine derivatives in search of potent antibacterial and antifungal drug leads. In this study, 3-(2-pyridyl)-5,6-diphenyl-1,2,4-triazine-4,4'-disulfonic acid monosodium salt (L1, ferrozine) and 3-(2-pyridyl)-5,6-di(2-furyl)-1,2,4-triazine-5,5'-disulfonic acid disodium salt (L2, ferene) have been used as ligands to study the complexation towards copper(II). The synthesized complexes, $[CuCl_2(ferrozine)] \cdot 7H_2O \cdot MeOH$ (**1**), $[CuCl_2(ferrozine)_2] \cdot 5H_2O \cdot MeOH$ (**2**), $[CuCl_2(ferene)] \cdot H_2O \cdot MeOH$ (**3**), and $[CuCl_2(ferene)_2] \cdot H_2O \cdot MeOH$ (**4**), have been characterized spectroscopically, and preliminary bioassays have been carried out. FTIR spectroscopic data have shown that N=N and C=N stretching frequencies of complexes have been shifted towards lower frequencies in comparison with that of the ligands, confirming new bond formation between Cu and N, which in turn lowers the strength of N=N and C=N bonds. In addition, a bathochromic shift has been observed for UV-visible spectra of complexes (**1**), (**2**), (**3**), and (**4**). Furthermore, elemental analysis data have been useful to obtain empirical formulas of these complexes and to establish the purity of each complex. Complexes (**1**) and (**2**) have shown antibacterial activity for both *S. aureus* (ATCC® 25923) and *E. coli* (ATCC® 25922) at 1 mg/disc concentration, and ferrozine has shown a larger inhibition zone against the clinical sample of *C. albicans* at 1 mg/disc concentration in comparison with the positive control, fluconazole.

1. Introduction

Transition metals have numerous and unique biological, chemical, and physical properties due to the availability of d electrons in valance shells. Much attention has been focused on copper complexes due to their various potential biological activities [1–4] out of which antimicrobial [5] and antiviral activities is paramount [6–15].

Since triazine is a well-known natural material which possesses many biological properties [16–21], it is not surprising that organometallic complexes of triazine with first row transition metals (Mn [22, 23], Co [24, 25], Ni [24, 25], Cu [22, 24–28], and Zn [25]), with second row transition metals (Ru [29], Pd [30], Ag [31], and Cd [32]), and with third row transition metals (Re [33] and Pt [34–36]) have

been synthesized, and their activities explored as catalysts [37] and biological agents such as antibacterial [25], anticancer [29, 36], antifouling [24], antifungal [33], anti-HIV [35], antimicrobial [25], antiproliferative [26, 34], antiviral [28, 35], and DNA binding [26, 29, 30] agents.

Even though many reports exist of metal complexes of triazine derivatives as detailed above, metal complexes containing the pyridyl-1,2,4-triazine core are relatively unexplored. Platinum(II) complexes of sulphonated 2-pyridyl-1,2,4-triazine have been reported to possess anti-HIV activity [35]. A copper(II) complex bearing 2,4,6-tris(2-pyridyl)-1,3,5-triazine ligand has been reported to bind DNA in a moderately strong way exhibiting significantly better anticancer activity against breast cancer in comparison with cisplatin [26]. An octahedral complex of rhenium(V), ML1L2L3L4

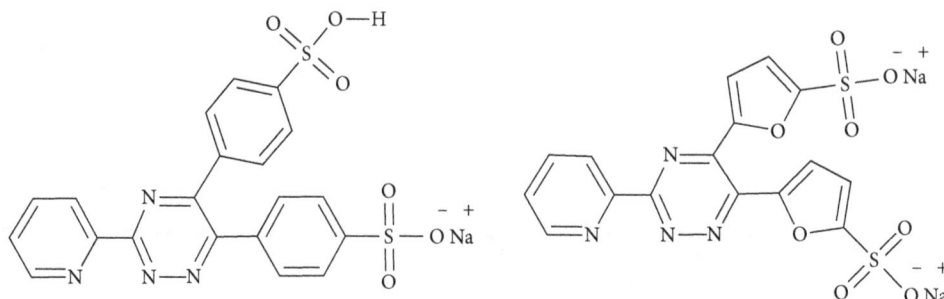

FIGURE 1: Structure of 3-(2-pyridyl)-5,6-diphenyl-1,2,4-triazine-p,p'-disulfonic acid monosodium salt (L1) (a) and 3-(2-pyridyl)-5,6-di(2-furyl)-1,2,4-triazine-5,5'-disulfonic acid disodium salt (L2) (b).

(where L1 = oxo, L2 = chloride, L3 = triphenylphosphine, and L4 = 3-hydrazino-5,6-diphenyl-1,2,4-triazine), has shown comparable antifungal activity against *Alternaria alternata* and *Aspergillus niger* [33]. We ourselves have explored the potential of using rhenium complexes of ferene and ferrozine (Figure 1) as biological imaging agents [38]. In our most recent work, we have commented on the possible use of the scaffold of sulfonated pyridyl triazine complexes being utilized as serum albumin transporters [39]. As such, it seems prudent to now explore its binding towards copper.

Thus, the current study explores the synthesis of four novel water-soluble complexes of the type, ML_nCl_2 (Figure 2) (where M = Cu^{2+}, L = 3-(2-pyridyl)-5,6-diphenyl-1,2,4-triazine-4',4''-disulfonic acid sodium salt/3-(2-pyridyl)-5,6-di(2-furyl)-1,2,4-triazine-5',5''-disulfonic acid disodium salt, and n = 1/2), their chemical characterization, and preliminary tests to assess antimicrobial activity of above synthesized complexes as well as of the ligands.

2. Experimental

2.1. Materials Used.

All chemicals and reagents used for the synthesis were commercially available and used without further purification. 3-(2-Pyridyl)-5,6-diphenyl-1,2,4-triazine-4,4'-disulfonic acid monosodium salt (ferrozine), 3-(2-pyridyl)-5,6-di(2-furyl)-1,2,4-triazine-5,5'-disulfonic acid disodium salt (ferene), and methanol ACS reagent (assay ≥99.8%) were purchased from Sigma-Aldrich, and copper (II) chloride dihydrate was purchased from Research-Lab Fine Chem Industries. Mueller-Hinton agar was purchased from Hardy Diagnostics, USA. Sodium chloride, sodium hydroxide, and dextrose were purchased from HiMedia Laboratories. The bacteria were obtained by the Industrial Technology Institute, Colombo.

2.2. Instrumentation.

Elemental analysis was carried out on PerkinElmer 2400 Series II CHNS/O Elemental Analyzer at Atlantic Microlabs, USA. IR spectra were recorded using Thermo Scientific NICOLET iS10 spectrophotometer in the spectral range 4000–650 cm^{-1} for both ligands and complexes. Thermo Spectronic Helios alpha UV-Vis double-beam spectrophotometer was used to measure the absorbance in the range of 190–1100 nm, and baseline correction was performed using matched quartz cuvettes. High-resolution mass spectra were recorded on an Agilent 6210 ESI TOF LCMS mass spectrometer.

2.3. Synthesis

2.3.1. Preparation of [CuCl₂(ferrozine)]·7H₂O·MeOH (1).

A solution of ferrozine (0.25 mmol, 0.1269 g) in methanol (8.0 cm³) was added to copper chloride dihydrate (0.25 mmol, 0.0435 g) in methanol (2.0 cm³). Then the resulting mixture was stirred for 2 hours at room temperature and progression of reaction checked using TLC. A light green colour crystalline precipitate was obtained after 2 days and collected by filtration (yield: 0.1264 g, 64%). IR (ATR; v/cm^{-1}): 1596.84(m) and 1498.22(s), $v_{C=N}$ and $v_{N=N}$. UV-Vis (MeOH; λ_{max} [nm]): 205, 242, 298, and 327. Anal. Calc. for $C_{20}H_{13}Cl_2CuN_4NaO_6S_2$·7H₂O·CH₃OH: C, 32.12; H, 3.98; N, 7.14. Found: C: 31.68%, H: 3.80%, and N: 7.42%. ESI-MS (m/z): $[M - H]^-$ calcd for $C_{20}H_{13}ClCuN_4O_6S_2$, 565.9179; found, 565.9188.

2.3.2. Preparation of [CuCl₂(ferrozine)2]·5H₂O·MeOH (2).

A procedure similar to that given above was followed using copper chloride dihydrate (0.25 mmol, 0.0435 g) and ferrozine (0.50 mmol, 0.2538 g). The resulting mixture was stirred for 5 hours. A dark green colour crystalline precipitate was obtained after 2 days and collected by filtration (yield: 0.1937 g, 62%). IR (ATR; v/cm^{-1}): 1595.69(m) and 1498.50(s), $v_{C=N}$ and $v_{N=N}$. UV-Vis (MeOH; λ_{max} [nm]): 213, 240, 301, and 334. Anal. Calc. for $C_{40}H_{26}Cl_2CuN_8Na_2O_{12}S_4$·5H₂O·CH₃OH: C, 39.66; H, 3.25; N, 9.03. Found: C: 39.29%, H: 3.76%, N: 9.23%. ESI-MS (m/z): $[M - H]^-$ calcd for $C_{40}H_{26}CuN_8O_{12}S_4$, 999.9833; found, 999.9776.

2.3.3. Preparation of [CuCl₂(ferene)]·H₂O·MeOH (3).

A solution of ferene (0.25 mmol, 0.1236 g) in methanol (8.0 cm³) was added to copper chloride dihydrate (0.25 mmol, 0.0435 g) in methanol (2.0 cm³). Then the resulting mixture was stirred for 6 hours at room temperature and progression of reaction checked using TLC technique initially and at the end. A yellow colour crystalline precipitate was obtained after 1 day and collected by filtration (yield: 0.1183 g, 75%). IR (ATR; v/cm^{-1}): 1567.49(m) and 1499.15(s), $v_{C=N}$ and $v_{N=N}$. UV-Vis (MeOH; λ_{max} [nm]): 202, 239, 338, and 371. Anal. Calc. for $C_{16}H_8Cl_2CuN_4O_8S_2$·H₂O·CH₃OH: C, 32.16; H, 2.54; N, 8.82.

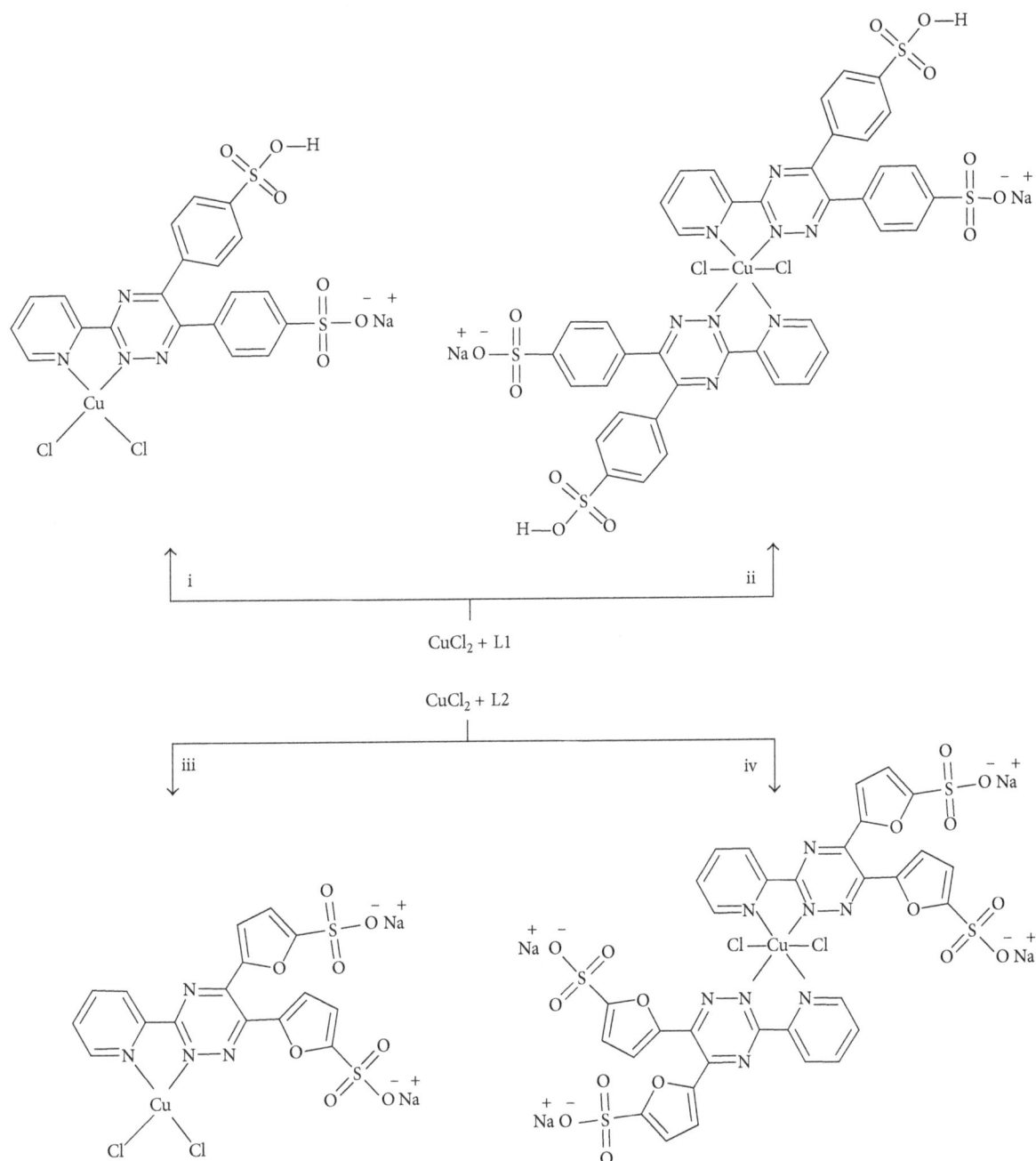

FIGURE 2: Synthetic routes for ML1Cl$_2$ (complex (**1**)) (i), M(L1)$_2$Cl$_2$ (complex (**2**)) (ii), ML2Cl$_2$ (complex (**3**)) (iii), and M(L2)$_2$Cl$_2$ (complex (**4**)) (iv) complexes. NB: L1 = 3-(2-pyridyl)-5,6-diphenyl-1,2,4-triazine-p,p'-disulfonic acid monosodium salt; L2 = 3-(2-pyridyl)-5,6-di(2-furyl)-1,2,4-triazine-5,5′-disulfonic acid disodium salt. Solvent molecules in complexes (**1**)–(**4**) have been omitted for clarity. Molar ratios of reactants: (i) CuCl$_2$: L1 = 1 : 1, (ii) CuCl$_2$: L1 = 1 : 2, (iii) CuCl$_2$: L2 = 1 : 1, and (iv) CuCl$_2$: L2 = 1 : 2.

Found: C: 32.12%, H: 2.76%, N: 9.29%. ESI-MS (m/z): [M]$^-$ calcd for C$_{16}$H$_8$CuN$_4$O$_8$S$_2$, 510.9085; found, 510.9084.

2.3.4. Preparation of [CuCl$_2$(ferene)$_2$]·H$_2$O·MeOH (4).

A procedure similar to above was followed using copper chloride dihydrate (0.25 mmol, 0.0435 g) and ferene (0.50 mmol, 0.2472 g). The resulting mixture was stirred for 5 hours. A brown-yellow colour crystalline precipitate was obtained after 1 day and collected by filtration (yield: 0.1912 g, 65%). IR (ATR; ν/cm^{-1}): 1569.82(m) and 1494.40(s), $\nu_{C=N}$

and $\nu_{N=N}$. UV-Vis (MeOH; λ_{max} [nm]): 208, 246, 338 and 371. Anal. Calc. for C$_{32}$H$_{16}$Cl$_2$CuN$_8$Na$_4$O$_{16}$S$_4$·H$_2$O·CH$_3$OH: C, 33.78; H, 1.89; N, 9.56. Found: C: 33.76%, H: 2.42%, N: 9.58%.

2.4. Antimicrobial Assay.

Compounds were tested against Gram-positive *Staphylococcus aureus* ATCC® 25923 and Gram-negative *Escherichia coli* ATCC® 25922 bacterial species and a clinical isolate of *Candida albicans* as a fungal species. Antimicrobial assay was performed by a standard disk diffusion assay [40] where the inhibition zones were measured and expressed as a mean of three replicates. Gentamycin and flucanazole were used as positive controls, and methanol was used as the negative control.

3. Results and Discussion

3.1. Synthesis. Copper chloride and the relevant ligands were used in 1 : 1 and 1 : 2 ratios to synthesize the desired metal complexes (Figure 2). Thin-layer chromatography (TLC) was initially used to monitor the progress of reaction, and visualization of spots was done using an iodine bath.

3.2. FTIR Analysis. FTIR data were recorded for dried crystals of ligands and complexes (1)–(4), and literature values were utilized where relevant [41]. The stretching frequency of the pyridine ring ($\nu_{C=N}$) and stretching frequency of the triazine ring ($\nu_{N=N}$) are considered mostly, because their values change upon formation of new bonds serving as good indicators of complex formation.

Stretching frequencies of N=N and C=N in complexes (1) and (2) have shifted to lower frequencies as expected, compared to those values of the free ferrozine ligand, due to σ donation of N lone pair which lowers strength of N=N and C=N bonds (Table 1). Furthermore, a broad band around 3400–3300 cm^{-1} was observed due to OH groups from methanol or water.

Similarly, stretching frequencies of N=N and C=N in complexes (3) and (4) were observed at lower frequencies in comparison with those of the free ferrozine ligand (Table 1), and a broad band was observed around 3400–3300 cm^{-1} due to OH groups of solvent.

3.3. UV-Visible Spectroscopy. UV-Vis spectra of reactants and complexes (1, 2, 3, and 4) were recorded in methanol at room temperature (Figure 3, Table S1, Supplementary Materials). The absorption wavelengths of complexes (1)–(4) have shifted towards longer wavelengths (bathochromic shift) compared to the wavelengths of the reactants (copper, ferrozine, and ferene). Both ferrozine and ferene have aromatic ring systems, and π–π^* transitions are thus possible [42]. These results are in agreement with those previously reported for zinc complexes of ferene and ferrozine [39] where a bathochromic shift was observed for both mono and bis complexes in comparison with that of the free ligand.

3.4. Elemental Analysis. Empirical formulas related to experimental values aided in obtaining the exact molecular formulas of all four complexes (Table 2). It can be seen that experimental values are within ±0.4% of expected values indicating purity of the synthesized complexes.

3.5. Antimicrobial Activity. All four complexes and ligands were studied *in vitro* for their antimicrobial activity against Gram-positive *Staphylococcus aureus* ATCC® 25923 and negative bacteria *Escherichia coli* ATCC® 25922 as well as the unicellular fungal species, *Candida albicans*. Inhibition zones were obtained by adding a concentration of 1 mg/disc, and the diameters of the zones are given in Table 3 for bacteria and Table 4 for fungi.

TABLE 1: FTIR data comparison chart of complexes (1)–(4) in comparison with those of free ligands.

	$\nu_{C=N}$ (cm^{-1})	$\nu_{N=N}$ (cm^{-1})
Ferrozine	1608	1503
Complex (1)	1596	1498
Complex (2)	1595	1498
Ferene	1589	1507
Complex (3)	1567	1499
Complex (4)	1570	1494

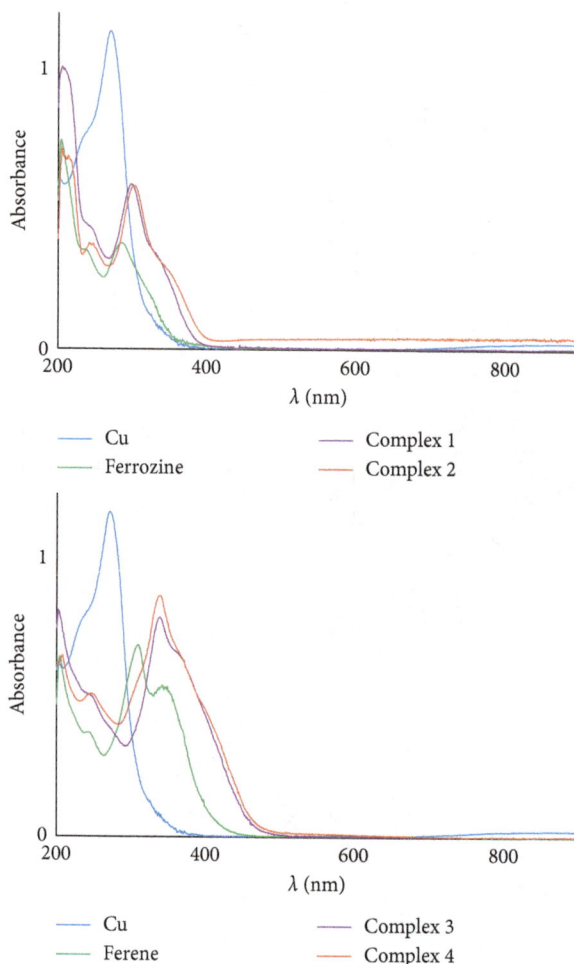

FIGURE 3: UV-visible spectra recorded in methanol of ferrozine, complexes (1) and (2) (a) and ferene, complexes (3) and (4) (b).

TABLE 2: Elemental analysis data of complexes.

Complex	Value	C (%)	H (%)	N (%)
(1)	Calculated	32.12	3.98	7.14
	Experimental	31.68	3.80	7.42
(2)	Calculated	39.66	3.25	9.03
	Experimental	39.29	3.76	9.23
(3)	Calculated	32.16	2.54	8.82
	Experimental	32.12	2.76	9.29
(4)	Calculated	33.78	1.89	9.55
	Experimental	33.76	2.42	9.58

TABLE 3: Mean inhibition zone diameter at 1 mg/disc of complexes (1) and (2) and at 20 μg/disc of gentamicin.

	Mean inhibition zone diameter ± SEM (mm)	
	S. aureus ATCC® 25923	E. coli ATCC® 25922
Complex (1)	8.75 ± 0.75	7.50 ± 1.00
Complex (2)	7.00 ± 0.00	7.75 ± 0.25
Positive control (gentamicin)	26.00 ± 1.50	30.75 ± 0.75
Negative control	ND	ND

ND, not detected.

TABLE 4: Mean inhibition zone diameter for *Candida albicans* at 1 mg/disc of ferrozine and at 1 mg/disc of fluconazole.

	Mean inhibition zone diameter ± SEM (mm)
Ferrozine	13.00 ± 2.00
Fluconazole	29.75 ± 0.25

Analysis of the inhibition zone diameter revealed that only complex (1) and complex (2) show moderate antibacterial activity when compared to the positive control. It is interesting to see that ferrozine ligand demonstrates antifungal activity.

Antimicrobial activity reported here is of moderate value. Further studies are warranted to optimize this system for greater activity.

4. Conclusions

We have described the synthesis of four novel water-soluble copper complexes bearing sulfonated pyridyl triazine ligands. FTIR spectroscopic data have confirmed the existence of Cu-N bonds in all four complexes because stretching frequencies of N=N and C=N complexes have been shifted towards lower frequencies in comparison with that of the ligands. In UV-Vis spectra, a bathochromic shift has been observed for complexes (1)–(4). Furthermore, elemental analysis data have been useful to obtain empirical formulas of these complexes and to establish the purity of each complex.

Preliminary bioassays in antimicrobial activity showed moderate antibacterial activity with complexes (1) and (2) whereas ferrozine showed antifungal activity against *Candida albicans*. To the best of our knowledge, we are the first to report on the antifungal activity of ferrozine. These findings provide a potential lead for antimicrobial drug development.

Acknowledgments

Financial assistance from the University of Sri Jayewardenepura (Grant no. ASP/01/RE/SCI/2015/19) is gratefully acknowledged. Clinical sample of *Candida albicans* was provided by the Department of Microbiology, Faculty of Medicine, University of Colombo. An earlier version of this work with two of the complexes detailed here was presented as an abstract at Chemistry in Sri Lanka 2015 as found in the link http://www.ichemc.edu.lk/wp-content/uploads/2015/10/vol-31-no.-2.pdf. The authors also acknowledge the Centre for Advanced Material Research of the University of Sri Jayewardenepura.

References

[1] A. Valent, M. Melník, D. Hudecová, B. Dudová, R. Kivekäs, and M. R. Sundberg, "Copper(II) salicylideneglycinate complexes as potential antimicrobial agents," *Inorganica Chimica Acta*, vol. 340, pp. 15–20, 2002.

[2] M. Ibrahim, F. Wang, M. M. Lou et al., "Copper as an antibacterial agent for human pathogenic multidrug resistant *Burkholderia cepacia* complex bacteria," *Journal of Bioscience and Bioengineering*, vol. 112, no. 6, pp. 570–576, 2011.

[3] A. Latif Abuhijleh and C. Woods, "Synthesis, characterization, and oxidase activities of copper(II) complexes of the anticonvulsant drug valproate," *Journal of Inorganic Biochemistry*, vol. 64, no. 1, pp. 55–67, 1996.

[4] B. S. Creaven, B. Duff, D. A. Egan et al., "Anticancer and antifungal activity of copper(II) complexes of quinolin-2(1H)-one-derived Schiff bases," *Inorganica Chimica Acta*, vol. 363, no. 14, pp. 4048–4058, 2010.

[5] H. I. Cervantes, J. A. Alvarez, J. M. Munoz, V. Arreguin, J. L. Mosqueda, and A. E. Macias, "Antimicrobial activity of copper against organisms in aqueous solution: a case for copper-based water pipelines in hospitals?," *American Journal of Infection Control*, vol. 41, no. 12, pp. e115–e118, 2013.

[6] N. Shionoiri, T. Sato, Y. Fujimori et al., "Investigation of the antiviral properties of copper iodide nanoparticles against feline calicivirus," *Journal of Bioscience and Bioengineering*, vol. 113, no. 5, pp. 580–586, 2012.

[7] G. Betanzos-Cabrera, F. J. Ramirez, J. L. Munoz, B. L. Barron, and R. Maldonado, "Inactivation of HSV-2 by ascorbate-Cu (II) and its protecting evaluation in CF-1 mice against encephalitis," *Journal of Virological Methods*, vol. 120, no. 2, pp. 161–165, 2004.

[8] J. L. Sagripanti, L. B. Routson, A. C. Bonifacino, and C. D. Lytle, "Mechanism of copper-mediated inactivation of herpes simplex virus," *Antimicrobial Agents and Chemotherapy*, vol. 41, no. 4, pp. 812–817, 1997.

[9] L. A. White, C. Y. Freeman, B. D. Forrester, and W. A. Chappell, "*In vitro* effect of ascorbic acid on infectivity of herpesviruses and paramyxoviruses," *Journal of Clinical Microbiology*, vol. 24, no. 4, pp. 527–531, 1986.

[10] G. Borkow, S. S. Zhou, T. Page, and J. Gabbay, "A novel anti-influenza copper oxide containing respiratory face mask," *PLoS One*, vol. 5, no. 6, p. e11295, 2010.

[11] M. Horie, H. Ogawa, Y. Yoshida et al., "Inactivation and morphological changes of avian influenza virus by copper ions," *Archives of Virology*, vol. 153, no. 8, pp. 1467–1472, 2008.

[12] J. O. Noyce, H. Michels, and C. W. Keevil, "Inactivation of influenza A virus on copper versus stainless steel surfaces," *Applied and Environmental Microbiology*, vol. 73, no. 8, pp. 2748–2750, 2007.

[13] J. I. Nieto-Juarez, K. Pierzchla, A. Sienkiewicz, and T. Kohn, "Inactivation of MS2 coliphage in Fenton and Fenton-like systems: role of transition metals, hydrogen peroxide and sunlight," *Environmental Science & Technology*, vol. 44, no. 9, pp. 3351–3356, 2010.

[14] M. T. Yahya, T. M. Straub, and C. P. Gerba, "Inactivation of coliphage MS-2 and poliovirus by copper, silver, and chlorine," *Canadian Journal of Microbiology*, vol. 38, no. 5, pp. 430–435, 1992.

[15] F. X. Abad, R. M. Pinto, J. M. Diez, and A. Bosch, "Disinfection of human enteric viruses in water by copper and silver in combination with low levels of chlorine," *Applied and Environmental Microbiology*, vol. 60, no. 7, pp. 2377–2383, 1994.

[16] R. Menicagli, S. Samaritani, G. Signore, F. Vaglini, and L. Dalla Via, "In vitro cytotoxic activities of 2-alkyl-4,6-diheteroalkyl-1,3,5-triazines: new molecules in anticancer research," *Journal of Medicinal Chemistry*, vol. 47, no. 19, pp. 4649–4652, 2004.

[17] S. Melato, D. Prosperi, P. Coghi, N. Basilico, and D. Monti, "A combinatorial approach to 2,4,6-trisubstituted triazines with potent antimalarial activity: combining conventional synthesis and microwave-assistance," *ChemMedChem*, vol. 3, no. 6, pp. 873–876, 2008.

[18] C. Zhou, J. Min, Z. Liu et al., "Synthesis and biological evaluation of novel 1,3,5-triazine derivatives as antimicrobial agents," *Bioorganic & Medicinal Chemistry Letters*, vol. 18, no. 4, pp. 1308–1311, 2008.

[19] K. Srinivas, U. Srinivas, K. Bhanuprakash, K. Harakishore, U. S. Murthy, and V. J. Rao, "Synthesis and antibacterial activity of various substituted s-triazines," *European Journal of Medicinal Chemistry*, vol. 41, no. 11, pp. 1240–1246, 2006.

[20] A. Baliani, G. J. Bueno, M. L. Stewart et al., "Design and synthesis of a series of melamine-based nitroheterocycles with activity against trypanosomatid parasites," *Journal of Medicinal Chemistry*, vol. 48, no. 17, pp. 5570–5579, 2005.

[21] Y. Z. Xiong, F. E. Chen, J. Balzarini, E. De Clercq, and C. Pannecouque, "Non-nucleoside HIV-1 reverse transcriptase inhibitors. Part 11: structural modulations of diaryltriazines with potent anti-HIV activity," *European Journal of Medicinal Chemistry*, vol. 43, no. 6, pp. 1230–1236, 2008.

[22] J. G. Małecki, B. Machura, and A. Świtlicka, "X-ray studies, spectroscopic characterisation and DFT calculations for Mn (II), Ni(II) and Cu(II) complexes with 5,6-diphenyl-3-(2-pyridyl)-1,2,4-triazine," *Structural Chemistry*, vol. 22, no. 1, pp. 77–87, 2010.

[23] M. M. Najafpour, D. M. Boghaei, and V. McKee, "Synthesis, characterization, crystal structure and oxygen-evolution activity of a manganese(II) complex with 2,4,6-tris (2-pyridyl)-1,3,5-triazine," *Polyhedron*, vol. 29, no. 17, pp. 3246–3250, 2010.

[24] H. A. E. Hemaida, A. A. E. Dissouky, and S. M. M. Sadek, "Potential antifouling agents: copper, cobalt, and nickel complexes of 3-(2-acetyl pyridylidene) hydrazino-5,6-diphenyl-1,2,4-triazine," *Egyptian Journal of Aquatic Research*, vol. 31, pp. 45–56, 2005.

[25] K. Singh, Y. Kumar, P. Puri, C. Sharma, and K. R. Aneja, "Antimicrobial, spectral and thermal studies of divalent cobalt, nickel, copper and zinc complexes with triazole Schiff bases," *Arabian Journal of Chemistry*, vol. 10, pp. S978–S987, 2017.

[26] K. Abdi, H. Hadadzadeh, M. Salimi, J. Simpson, and A. D. Khalaji, "A mononuclear copper(II) complex based on the polypyridyl ligand 2,4,6-tris(2-pyridyl)-1,3,5-triazine (tptz), $[Cu(tptz)_2]^{2+}$: X-ray crystal structure, DNA binding and in vitro cell cytotoxicity," *Polyhedron*, vol. 44, no. 1, pp. 101–112, 2012.

[27] B. Machura, A. Świtlicka, R. Kruszynski, J. Mroziński, J. Kłak, and J. Kusz, "Coordination studies of 5,6-diphenyl-3-(2-pyridyl)-1,2,4-triazine towards Cu^{2+} cation. X-ray studies, spectroscopic characterization and DFT calculations," *Polyhedron*, vol. 27, no. 13, pp. 2959–2967, 2008.

[28] C. G. Palivan, H. M. N. Palivan, B. A. Goodman, and C. Cristescu, "ESR study of some asymmetric-triazine copper (II) complexes having high antiviral activity," *Applied Magnetic Resonance*, vol. 15, no. 3-4, pp. 477–488, 1998.

[29] N. Busto, J. Valladolid, M. Martinez-Alonso et al., "Anticancer activity and DNA binding of a bifunctional Ru(II) arene aquacomplex with the 2,4-diamino-6-(2-pyridyl)-1,3,5-triazine ligand," *Inorganic Chemistry*, vol. 52, no. 17, pp. 9962–9974, 2013.

[30] W. H. Al-Assy and M. M. Mostafa, "Comparative studies and modeling structures of two new isomers containing binuclear PdII complexes derived from 2,4,6-tri-(2-pyridyl)-1,3,5-triazine (TPTZ)," *Spectrochimica Acta Part A: Molecular and Biomolecular Spectroscopy*, vol. 120, pp. 568–573, 2014.

[31] M. M. Najafpour, M. Hołyńska, M. Amini, S. H. Kazemi, T. Lis, and M. Bagherzadeh, "Two new silver(I) complexes with 2,4,6-tris(2-pyridyl)-1,3,5-triazine (tptz): preparation, characterization, crystal structure and alcohol oxidation activity in the presence of oxone," *Polyhedron*, vol. 29, no. 14, pp. 2837–2843, 2010.

[32] F. Marandi, M. Jangholi, M. Hakimi, H. A. Rudbari, and G. Bruno, "Synthesis and crystal structures of the first cadmium complexes of 3,5,6-tris(2-pyridyl)-1,2,4-triazine ligand," *Journal of Molecular Structure*, vol. 1036, pp. 71–77, 2013.

[33] M. M. Mashaly, H. F. El-Shafiy, S. B. El-Maraghy, and H. A. Habib, "Synthesis, properties and thermal studies of oxorhenium(V) complexes with 3-hydrazino-5,6-diphenyl-1,2,4-triazine, benzimidazolethione and 2-hydrazinobenzimidazole. Mixed ligand complexes, pyrolytical products and biological activity," *Spectrochimica Acta Part A: Molecular and Biomolecular Spectroscopy*, vol. 61, no. 8, pp. 1853–1869, 2005.

[34] I. Łakomska, B. Golankiewicz, J. Wietrzyk et al., "Synthesis, spectroscopical characterization and the biological activity in vitro of new platinum(II) complexes with imidazo[1,5-a]-1,3,5-triazine derivatives and dimethylsulfoxide," *Inorganica Chimica Acta*, vol. 358, no. 6, pp. 1911–1917, 2005.

[35] A. N. Vzorov, D. Bhattacharyya, L. G. Marzilli, and R. W. Compans, "Prevention of HIV-1 infection by platinum triazines," *Antiviral Research*, vol. 65, no. 2, pp. 57–67, 2005.

[36] R. W. Sun, D. L. Ma, E. L. Wong, and C. M. Che, "Some uses of transition metal complexes as anti-cancer and anti-HIV agents," *Dalton Transactions*, no. 43, pp. 4884–4892, 2007.

[37] C. J. Cramer and D. G. Truhlar, "Density functional theory for transition metals and transition metal chemistry," *Physical Chemistry Chemical Physics*, vol. 11, no. 46, pp. 10757–10816, 2009.

[38] K. Ranasinghe, S. Handunnetti, I. C. Perera, and T. Perera, "Synthesis and characterization of novel rhenium(I) complexes towards potential biological imaging applications," *Chemistry Central Journal*, vol. 10, no. 1, p. 71, 2016.

[39] N. Abeydeera, I. C. Perera, and T. Perera, "Synthesis, characterization, and BSA-binding studies of novel sulfonated zinc-triazine complexes," *Bioinorganic Chemistry and Applications*, vol. 2018, Article ID 7563820, 7 pages, 2018.

[40] Institute CaLS, "Methods for dilution antimicrobial susceptibility tests for bacteria that grow aerobically; approved standard—seventh edition," in *M7–A7*, Clinical and Laboratory Standards Institute, Wayne, PA, USA, 2006.

[41] V. Béreau and J. Marrot, "Coordination studies of 5,6-diphenyl-3-(2-pyridyl)-1,2,4-triazine towards Zn^{2+} cation. Synthesis and characterization by X-ray diffraction and spectroscopic methods," *Comptes Rendus Chimie*, vol. 8, no. 6-7, pp. 1087–1092, 2005.

[42] D. Eastwood, R. L. Lidberg, and M. S. Dresselhaus, "Ultraviolet-visible fluorescence spectroscopy of selected polyaromatic hydrocarbons and organometallics on hexagonal graphite and boron nitride," *Chemistry of Materials*, vol. 6, no. 2, pp. 211–215, 1994.

4

A Review on Green Synthesis, Biomedical Applications, and Toxicity Studies of ZnO NPs

V. N. Kalpana and V. Devi Rajeswari ⓘD

Department of Biomedical Sciences, School of Biosciences and Technology, VIT, Vellore, Tamil Nadu, India

Correspondence should be addressed to V. Devi Rajeswari; vdevirajeswari@vit.ac.in

Academic Editor: Francesco Paolo Fanizzi

The advance of reliable and eco-friendly strategies for the development of nanoparticles is a fundamental key to the discipline of nanotechnology. Nanoparticles have been continuously evaluated and have been used in many industrial applications for a decade. In particular, the role of zinc oxide nanoparticles (ZnO NPs) has received a great interest because of various properties such as UV filter properties and photochemical, antifungal, high catalyst, and antimicrobial activities. Because of the high rate of poisonous chemicals and the extreme surroundings used within the chemical and physical methods, the green techniques have been adopted using plants, fungi, bacteria, and algae for the synthesis of nanoparticles. Therefore, this paper considers various green synthesis methods to provide the evidence of ZnO NP role to several applications, and in addition, biomedical applications and toxic effect were reviewed. Therefore, the paper used various secondary sources to collect the relevant review articles. From the findings, the green route of synthesis is rather safe and eco-friendly when compared to physical and chemical means of synthesis. On the other hand, its biomedical applications in this sector are increased day by day in various processes including bioimaging, drug delivery, biosensors, and gene delivery. With respect to its toxicity properties, ZnO NPs can act as smart weapons against multiple drug-resistant microorganisms and as a talented substitute for antibiotics.

1. Introduction

In science and technology, one among the rapidly developing concepts in the latest years is nanotechnology, which has brought tremendous development. The nanomaterial which comprises distinctive physicochemical properties has the potential to develop new systems, structures, devices, and nanoplatforms with impending bids in extensive variety of disciplines [1, 2]. Nanomaterials are particles that are in nanoscale size, and they are very small particles with improved thermal conductivity, catalytic reactivity, nonlinear optical performance, and chemical stability due to their large surface area-to-volume ratio [3]. This quality has attracted many researchers to locate novel techniques for their synthesis. Though conventional techniques (physical and chemical methods) use less time to synthesize bulk amount of nanoparticles, they require toxic chemicals like protective agents to maintain stability, which leads to toxicity in the

environment. Keeping this in mind, green technology by using plants is rising as an eco-friendly, nontoxic, and safe option, since plant extract-mediated biosynthesis of nanoparticles is economically advantageous and offers natural capping agents in the form of proteins [4]. To regulate chemical toxicity in the environment, biological synthesis of various metal oxide and metal nanoparticles through plant extraction is used, which is a marginal technique for regulating chemical synthesis, and it permits a distinct shape and size of nanoparticles with a meticulous synthesis [5].

For biomedical applications, improvement in biodegradable, functionalized, and biocompatible nanomaterials is being remained a tremendous vivacious area for research. Until now, among numerous other biomedical applications [6–9] well examined are paramagnetic nanoparticles [10, 11], quantum dots (QDs) [6, 12], nanoshells [13], and carbon nanotubes (CNTs) [14, 15]. An extensive variety of nanostructures could exhibit zinc oxide (ZnO), which has exclusive properties such as semiconducting, piezoelectric, and optical [16, 17]. Hence, nanomaterials based on ZnO are deliberated for extensive range of applications such as energy storage, nanosensors, cosmetic products, nano-optical devices, nanoelectronic devices, and so on [18–23]. Biodegradability and low toxicity are one among the most significant characteristics of ZnO nanomaterials. For adults, an indispensable trace element is Zn^{2+} and it is being elaborated into numerous features of metabolism. In acidic and strong basic conditions, ZnO NPs could be dissolved slowly. Solubilized ZnO nanoparticles have shown that the release of Zn^{2+} ions can exert stress on cells and have adverse impacts on different organisms [24]. The required properties of ZnO nanomaterials have attained increased concern towards biomedical applications [25]. The ZnO nanoparticle's toxic effect is due to their solubility. In the extracellular areas, ZnO nanoparticles are dissolved, which sequentially raise the level of intracellular $[Zn^{2+}]$. The dissolution of ZnO nanoparticles in the medium and the mechanism of increased intracellular $[Zn^{2+}]$ level are still speculative [26].

Holistically, this review will summarize the current status of the use of ZnO nanomaterials for biomedical applications, their green synthesis nature, and its toxic effect.

Organic nanoparticles and inorganic nanoparticles are the two categories of nanoparticles based on their components. Organic nanoparticles include carbon nanoparticles (fullerenes), whereas inorganic nanoparticles include magnetic nanoparticles, noble metal nanoparticles (gold and silver), and semiconductor nanoparticles (such as zinc oxide and titanium dioxide) [27]. Nanoparticles can also be categorized based on their origin, dimensions, and structural content.

Based on the origin of nanomaterial, it is categorized into natural nanomaterial and artificial nanomaterial [28]. On considering the dimensions of nanomaterial, it is categorized as zero-dimensional (0D), one-dimensional (1D), two-dimensional (2D), and three-dimensional (3-D) nanomaterials. The zero-dimensional nanomaterials have nanodimensions at all three directions; one-dimensional nanomaterials have only one nanodimension which is exterior to the nanometer range; and two-dimensional nanomaterials have two nanodimensions exterior to the nanometer range, whereas three-dimensional nanomaterials have all the nanodimensions exterior to the nanometer range. These comprise bulk materials developed with individual blocks that are in the nanometer scale (1–100 nm) [29].

According to the structural configuration and morphology, nanomaterials have been divided into amalgamated materials and nanodispersions. Extremely branched macromolecules are dendrimers with the dimensions in the nanometer scale. In the metal-based materials, the chief component for these particles is metal, where the nanomaterials comprised nanosilver, nanogold, metal oxides like titanium dioxide, and finally closely packed semiconductors such as quantum dots. The morphology of carbon-based nanomaterials is tubes, hollow spheres, or ellipsoids. The carbon nanomaterials that are spherical and ellipsoidal are referred as fullerenes and cylindrical ones are called as nanotubes [28].

2. Nanoparticle Synthesis Methods

Bottom-up and top-down are the two approaches recommended for the biosynthesis of nanoparticles [30]. In case of a bottom-up approach, the most important reaction occurred is oxidation/reduction. The synthesis of nanoparticles is currently an important area of research, which seeks an eco-friendly approach and green materials for current scenario [31]. The major steps involved in the preparation of nanoparticles that have to be evaluated from the point of green chemistry are (i) the solvent medium used for the synthesis, (ii) environmentally benign reducing agent, and (iii) the nontoxic material for the stabilization of the nanoparticles. The majority of the chemical and physical methods mentioned so far largely depend on organic solvents. This is principally due to the hydrophobicity of the capping agents used [32]. Synthesis with bio-organisms is compatible with the principles of green chemistry: (i) eco-friendly approach, (ii) the reducing agent used, and (iii) the capping agent in the reaction. The synthesis of inorganic metal oxide nanoparticles using biological elements has received immense attention due to their unusual properties (optical, electronic, chemical, etc.) [33].

3. Zinc Oxide Nanoparticles

Metal oxides play a very significant role in the science of materials, such as the production of microelectronic circuits, sensors, piezoelectric devices, fuel cells, surface passivation coatings, and corrosion catalysts. Metal oxides have also been used as absorbers of the environmental pollutant. In nanotechnology, oxide nanoparticles can show signs of unique chemical properties due to their limited size and high density of edges. An n-type semiconducting metal oxide is ZnO. Over the past few years, more interest is drawn towards zinc oxide NP since it has wider varieties of applications particularly in the fields of biomedical systems, optics, and electronics [34–40]. Among all these types of metal oxides, ZnO NPs attract much attention because of their stimulating properties [41] (such as the high direct bandwidth of 3.3 eV at room temperature and high excitation energies of 60 meV) [42], optical property, high catalytic activity, anti-inflammatory, wound healing, and UV filtering properties [1, 43–48]. Several authors have reported ZnO biosensors for cholesterol, enzyme biochemistry, and other biosensing applications [49, 50].

Zinc oxide as a nonhygroscopic and nontoxic, inorganic, polar, crystalline material is a very cheap, safe, and readily available, which has aroused great interest in various organic transformations, sensors, transparent conductors, and surface acoustic wave devices [51–53]. The ZnO NP is an exclusive material that has semiconducting, piezoelectric, and pyroelectric properties and has versatile applications in transparent electronics, UV light emitters, chemical sensors, spin electronics, personal care products, catalyst, coating, and paints [54, 55]. Due to these unique properties, ZnO NPs find applications in antireflection coatings, transparent electrodes in solar cells, UV light emitters, diode lasers, varistors, piezoelectric devices, spin electronics, surface acoustic wave propagators [52], as an antibacterial agent [56], as photonic material [57], and for gas sensing [58]. The biomolecules in the plant extract act as efficient capping agents, thus playing a major role in the NP synthesis. The capping agents seem to stabilize NPs through different mechanisms that include electrostatic stabilization, steric stabilization, stabilization by hydration forces, and stabilization using van der Waals forces. The stabilization of NPs is significant for its functions and its applications [59]. The utility of ZnO NPs in the field of food preservation and packaging industry when applied to biodegradable polymeric metrics and ZnO NPs has improved the quality of food and packaging mainly through three mechanisms, namely, the release of antimicrobial ions, destructing the integrity of the cells of bacteria, and formation of ROS due to light radiation [60]. Elmer and White reported the pesticidal properties of the ZnO by spraying synthesized ZnO on tomato and eggplant and it was noted that ZnO reduced disease estimate of 28% when compared to the control [61]. In vitro approaches use plant extracts to reduce zinc salt (zinc nitrate, sulfate, chloride, and many others) and endow with control over the size and shape of the nanoparticles. Fundamentally, the primary and secondary metabolites are present in plants, for example, saponins, tannins, starches, polypeptides, terpenoids, flavonoids, and phenolics, which act as reducing and capping agents. Mild solvents such as water, ethanol, and methanol are used for the extraction of the plant metabolites, which are allowed to react with the zinc salt solution in different conditions to achieve greatest yield [62–65].

4. Green Synthesis of ZnO NPs

Owing to the growing popularity of green methods, several methods have been implemented to produce ZnO NPs using different sources such as bacteria, fungus, algae, plants, and others. A list of tables was prepared to summarize the research carried out in this field (Table 1).

4.1. Plant-Mediated Biosynthesis of ZnO NPs. The synthesis of biological nanoparticles represents an alternative for the physical and chemical methods of nanoparticle formation. The majority of researchers focused on the green synthesis of nanoparticles for the formation of metal and oxide nanoparticles (Figure 1). The use of plants for the synthesis of nanoparticles is a rapid, low-cost, eco-friendly option and is safe for human use [31].

Vitex negundo plant extract was used to produce ZnO NPs with zinc nitrate hexahydrate as a precursor. The biosynthesized ZnO NPs showed antimicrobial activities against *E. coli* and *S. aureus* bacteria [77]. Dobrucka and Dugaszewska [78] used *Trifolium pratense* to synthesize ZnO NPs. The synthesized ZnO NPs were found to be of hexagonal shape and the sizes were found to be of 60–70 nm. Kalpana et al. [79] synthesized ZnO NPs using *Lagenaria siceraria* pulp extract. In addition, the author evaluated the biosynthesized ZnO nanoparticles for antidandruff, antimicrobial, and antiarthritic efficacy.

Dhanemozhi et al. [80] successfully synthesized the ZnO NPs from green tea leaf extract to evaluate their capacitance behavior for supercapacitor applications. ZnO NPs are known to be multifunctional inorganic nanoparticles with their major application in the treatment of urinary tract infection. Santhoshkumar et al. [81] synthesized the ZnO NPs using *Passiflora caerulea* leaf extract and tested against the pathogenic culture isolated from the urine of the patient suffering from urinary tract infection. The results showed that the synthesized ZnO NPs act as an antibacterial agent against urinary tract infection. Nava et al. [82] dealt with low-cost, nontoxic green synthesis of ZnO NPs prepared using *Camellia sinensis* extract. The efficiency of ZnO NPs as a photocatalyst for the degradation of various organic dyes such as methylene blue and methyl orange and their antioxidant activity by the DPPH assay has been studied by Siripireddy and Mandal [83] using *Eucalyptus globulus*. The synthesis of monophase crystalline ZnO nanoparticles with a size range of about 15.8 nm by the green novel and environmentally friendly pathway using the extract of *A. betulina* as an effective oxidizing/reducing agent has been demonstrated for the first time by Thema et al. [84]. Stable and spherical ZnO NPs were produced by using zinc nitrate and *Aloe vera* leaf extract. The various properties of ZnO NPs were characterized by the use of the UV-Vis spectrophotometer, FTIR, photoluminescence, XRD, SEM, and TEM analysis [43].

4.2. Microbe-Mediated Biosynthesis of ZnO NPs. Synthetic pathways of nanoparticles by microbes may involve combinations of basic cell biochemistry, the transport of ionic metals both in and out of cells, mechanism of resistance of microbes to toxic metals and activated metalbinding sites, ion accumulation metallic intracellular, and nucleation of metal oxides [85]. ZnO NPs were rapidly synthesized from a *Rhodococcus pyridinivorans* NT2 which were found to be moderately stable and roughly spherical with an average diameter of the particle of 100–120 nm [86]. *Serratia ureilytica* (HM475278)-mediated ZnO NPs have been reported by Dhandapani et al. [87]. ZnO NPs have aroused interest because of their many applications in the food industry. Selvarajan and Mohanasrinivasan [88] described an innovative method for the biosynthesis of ZnO NPs using a probiotic bacterium *Lactobacillus plantarum* VITES07. Kundu et al. [86] synthesized ZnO NPs from the zinc sulfate solution using an actinobacteria *Rhodococcus pyridinivorans* NT2. The synthesized ZnO NPs were explored for multifunctional textile finishing and for in vitro anticancer drug delivery in HT-29 colon carcinoma cell line. The studies of Shamsuzzaman et al. [89] have described the simple and green route for the biosynthesis of ZnO nanoparticles using *Candida albicans* as a capping and reducing agent. The author also used the synthesized ZnO NPs as a catalyst for the rapid and efficient synthesis of steroidal pyrazoline. Hussein et al. [85] reported *Bacillus cereus* as a biotemplate agent for the formation of ZnO NPs with raspberry and plate-like structures through a simple

TABLE 1: Green synthesis of ZnO NPs using various sources.

Type of green method	Applied material/organism	Particle size (nm)	Morphology of NPs	Activity carried out	References
Plant-mediated synthesis	*Limonia acidissima* (leaf)	12–53	Spherical	Antibacterial activity against *Mycobacterium tuberculosis*	[39]
	Euphorbia Jatropha (stem)	15	Hexagonal	Used as semiconductors	[66]
	Ceropegia candelabrum (leaf)	12–35	Hexagonal	Antibacterial potential against *Staphylococcus aureus, Bacillus subtilis, Escherichia coli, Salmonella typhi*	[67]
	Celosia argentea (leaves)	25	Spherical	Antibacterial potential against *Escherichia coli, Salmonella, Acetobacter*; drug delivery	[68]
	Couroupita guianensis (leaves)	57	Hexagonal unit cell	*Bacillus cereus, Klebsiella pneumonia, Escherichia coli, Mycobacterium luteus, V. cholerae*	[69]
	Allium cepa (bulb), *Allium sativum* (bulb), *Petroselinum crispum* (leaves)	70	Hexagonal wurtzite	Photodegradation of methylene blue	[48]
	Phyllanthus niruri (leaves)	25.61	Quasispherical	Catalytic activity	[70]
	Parthenium hysterophorus (leaves)	27–84	Spherical and hexagonal	*Aspergillus flavus, Aspergillus niger, Aspergillus fumigatus, Fusarium culmorum, Fusarium oxysporum*	[71]
	Solanum nigrum (leaves)	29	Quasispherical	*Staphylococcus aureus Salmonella paratyphi, Vibrio cholerae*	[72]
	Anisochilus carnosus (leaves)	20–40	Hexagonal wurtzite	*S. paratyphi, V. cholerae, S. aureus, and E. coli*	[37]
	Jacaranda mimosifolia (flower)	2–4	Hexagonal wurtzite	*Escherichia coli, Enterococcus faecium*	[73]
Seaweed-mediated synthesis	*Caulerpa peltata, Hypnea valencia, Sargassum myriocystum*	36	Rectangle, triangle, radial, and spherical	Antibacterial activity against *Staphylococcus aureus, Streptococcus mutans, Vibrio cholerae, Neisseria gonorrhoeae, Klebsiella pneumonia,* and antifungal activity against *Aspergillus niger* and *Candia* sp.	[74]
	Ulva lactuca	10–50	Hexagonal, rods, and rectangles	Photocatalytic, antibacterial, antibiofilm, and larvicidal activity	[75]
Microbe-mediated synthesis	*Aspergillus fumigatus*	60–80	Spherical	Antibacterial activity	[76]
	Aeromonas hydrophila	57	Spherical and oval	Antimicrobial activity	[41]

thermal decomposition of zinc acetate maintaining the original pH of the reaction mixtures. Baskar et al. [90] produced ZnO NPs using *Aspergillus terreus* filtrate synthesized extracellularly which were spherical with a range of 54.8 to 82.6 nm.

The nanoparticle's extracellular synthesis from the fungus is extremely beneficial due to economic viability, convenient downstream processing, and large-scale production [91]. Fungal strains are preferred over bacterial due to their enhanced metal bioaccumulation property and tolerance property [92]. From mycelia of *Aspergillus fumigatus*, the ZnO nanoparticles are synthesized [93]. By utilizing *Candida albicans*, the nanoparticles were synthesized with similar size ranging from 15 to 25 nm established through XRD, TEM, and SEM analysis [89].

5. Characterization of ZnO NPs

The synthesized nanoparticles are characterized by utilizing numerous techniques: FTIR (Fourier transform infrared spectroscopy), EDAX (energy dispersion analysis of X-ray), AFM (atomic force microscopy), XPS (X-ray photoelectron microscopy), ATR (attenuated total reflection), UV-DRS (UV-visible diffuse reflectance spectroscopy), XRD (X-ray diffractometer), TEM (transmission electron microscopy), TG-DTA (thermogravimetric-differential thermal analysis), DLS (dynamic light scattering), FE-SEM (field emission scanning electron microscopy), PL (photoluminescence analysis), Raman spectroscopy, and SEM (scanning electron microscopy) [94–96]. Plants are being examined extensively particularly that belong to Lamiaceae family such as *Vitex*

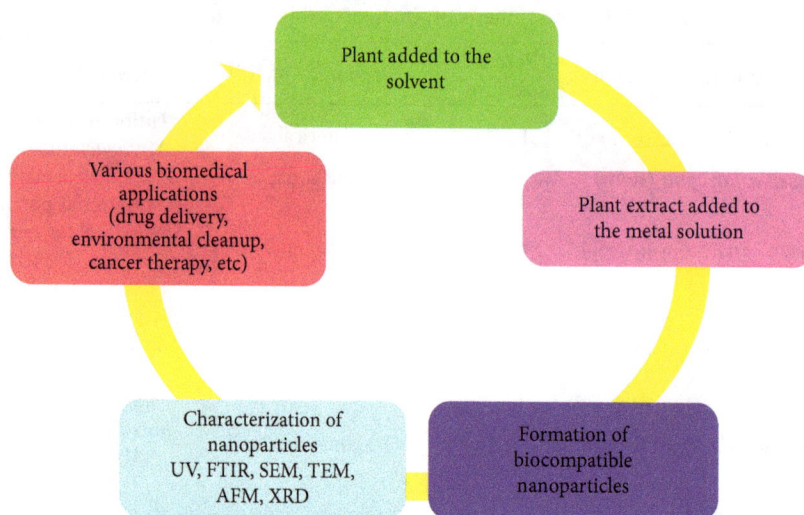

FIGURE 1: Plant-mediated biosynthesis of ZnO NPs.

negundo [97], *Plectranthus amboinicus* [98], and *Anisochilus carnosus* [37] which had the formation of NP with different shapes such as hexagonal, rod-shaped with agglomerates, quasispherical, and spherical, and further various sizes are also seen. From the outcome, it is clearly identified that the size of synthesized NPs is decreased on increasing the concentration of a plant extract [70, 77]. From the results, the size range is being observed and compared using various techniques such as TEM, XRD, and FE-SEM which had a closer range of values [77, 98], whereas SEM and EDAX had a similar result diverse from XRD results. Through Debye–Scherrer equation, synthesis of NPs from both *Vitex negundo* flower and leaf had a similar size of 38.17 nm, which was confirmed through XRD analysis [77]. For the synthesizing ZnO NPs, leaves of *Azadirachta indica* from Meliaceae family are generally being used [99, 100]. A similar size range of NPs was identified in every experiment, which was confirmed through the analysis of TEM and XRD with nanobuds, hexagonal disc shape, and spherical shape. From the studies, it is revealed that the formation of NPs is through the involvement of amine, carboxylic acid, carbonate moieties, alcohol, alkane, and amide, which was further confirmed by FTIR studies. *Aloe vera's* leaf peel and fresh leaf extract belong to Liliaceae family [101, 102]. Agglomerate formation was seen in the NP synthesis, which was extracted from *Moringa oleifera*, *Calotropis gigantea*, *Plectranthus amboinicus*, *Agathosma betulina*, *Nephelium lappaceum*, and *Pongamia pinnata*. To affirm the synthesis of NPs, UV-Vis spectrophotometry is employed, and the crystal NPs are obtained through centrifugation of mixture and drying the pellet in a hot air oven [95].

6. Biomedical Applications of ZnO NPs

6.1. Drug Delivery. The benefits of using ZnO NPs for the drug delivery were derived from their two main basic properties. First of all, due to their smaller size, nanoparticles can penetrate through smaller capillaries and are absorbed by the cells, allowing an efficient accumulation of drugs at

the target sites. Second, the use of biodegradable materials for the preparation of nanoparticles allows the prolonged discharge of drugs within the site targeted over a period of days or even weeks [103]. The role of synthesized ZnO NPs in drug release by using the drug metronidazole benzoate was studied [68] by observing its diffusion through egg membrane. Results revealed that the presence of ZnO NPs with the drug has much effect on the biological membrane.

6.2. Bioimaging of ZnO Particles. In preclinical research, fluorescence imaging is extensively utilized as it is convenient and inexpensive [104–107]. As ZnO nanomaterials have essential excitonic blue and near-UV emission, which has green luminescence associated with O_2 vacancies [108, 109], and for cellular imaging, there are numerous reports existing in previous studies on the utilization of ZnO nanomaterials. For cancer cell imaging, transferrin-conjugated green fluorescent ZnO NPs were utilized with least cytotoxicity [110]. ZnO nanomaterial's optical properties could be altered by adulterating with suitable elements [111]. According to a research, ZnO NPs were adulterated with various cations such as Co, Cu, or Ni, and in aqueous colloidal solutions, it was stabilized, which was employed in different cells for cellular imaging studies [112]. These tiny ZnO nanoparticles have a capability to penetrate it into the cell nucleus.

For biocompatibility and optical properties, heterostructural ZnO/Au nanocomposites, where Au NCs develop either along the nanorod surfaces or at the tip of ZnO nanorods, are synthesized and investigated [113]. For imaging of cancer cells in vitro, antiepidermal growth factor receptor antibody-conjugated ZnO nanorods were utilized in the latest research [114]. For optical imaging, QDs are extensively deliberated nanoparticles because of their numerous appealing optical properties [115–117]. It was identified that ZnO QDs are placed in the cytoplasm while applying for in vitro cell imaging, exhibiting stable luminescence under UV light in the absence of essential cytotoxicity. Same QDs were analyzed in a previous research which was trialed through mice after injecting intravenous and intradermal injections [118].

Every imaging technique has their own benefits and drawbacks [119]. Through multiple imaging modalities, nanomaterials could be functionalized to be detectable, which produce synergistic advantages. Nanomaterials are more appropriate for multimodality imaging while associated with small molecules since larger surface area provides higher sites for functionalization and also helps to engineer them for multimodal detection. In one particular research, Gd-doped ZnO QDs (with sizes of less than 6 nm) were emerged for both magnetic resonance imaging (MRI) and optical imaging [120]. Another study by Singh reported Fe_3O_4-ZnO core-shell magnetic QDs for potential cancer imaging and therapy.

There are better clinical relevance for single-photon emission computed tomography (SPECT) [122–126] and radionuclide-based imaging techniques, that is, PET [127–132], which are extensively utilized in the clinic than in optical imaging. PET and SPECT techniques are not only highly sensitive and quantitative but also have no tissue penetration limitation [133–136].

6.3. Drug Delivery with ZnO Nanomaterials.

ZnO nanomaterials are versatile nanoplatforms not only in bioimaging but also in a drug delivery application because of their versatile surface chemistry, large surface area, and phototoxic effect, along with others. Researches in vitro have identified that ZnO nanoparticles could be highly toxic either for cancer cells [126] or for bacteria and leukemic T cells [137].

Intrinsic blue fluorescence of ZnO QDs was smeared with folate-conjugated chitosan through electrostatic interaction, and by doxorubicin, it can be loaded at ~75% efficiency (extensively utilized chemotherapy drug is DOX) [138]. It was recommended that through hydrogen bonding and/or through collaborations with the ZnO QD surface, DOX was entrapped. But the aqueous stability of the ZnO QDs enriched the exterior chitosan layer because of the hydrophilicity and the charges. Conversely, at the normal physiological pH value of 7.4, DOX was released rapidly which requires to be improved for investigations in in vivo or in vitro researches.

In dendritic cell- (DC-) established cancer immunotherapy, one of the main complications is the improvement of delivery system which could provide the targeted antigens into DCs efficiently [139]. Due to extensive surface area, nanomaterials are challenging aspects for this application. To deliver carcinoembryonic antigen into DCs, Fe_3O_4-ZnO core-shell nanoparticles were produced with an average diameter of 16 nm which have the capability to help as imaging contrast agents [140].

6.4. Gene Delivery with ZnO Nanomaterials.

Over the last few years, gene therapy has involved substantial attention over cancer treatment [141]. Developing a safe gene vectors which could safeguard DNA from degradation as well as through high efficiency enabling cellular uptake of DNA is one of the foremost challenging aspects. For examining gene therapy application and gene delivery, extensive varieties of nanomaterials are utilized, which even comprise ZnO nanomaterial that had a positive outcome in numerous studies.

By a sequence of investigations, ZnO nanostructures, which are also like a three-dimensional tetrapod, were examined as gene vectors for delivering pEGFPN1 DNA (comprising the gene for green fluorescent protein) to A375 human melanoma cells [142, 143]. Through electrostatic interactions, the pDNA (plasmid DNA) was attached to ZnO nanostructures, and for gene delivery within the cells, the three needle-shaped legs preferred the internalization of the tips. It was observed that there was an absence for significant cytotoxicity which was reportedly attributed for the three-dimensional geometry.

For an efficient gene delivery, surface coating of nanomaterial acts as a significant role. According to an investigation, ZnO QDs were layered using positively charged poly(2-(dimethylamino)ethyl methacrylate) (PDMAEMA) polymers which are utilized for condensing pDNA for gene delivery [144]. The polymer-coated ZnO QDs presented fluorescence emission at 570 nm with a considerable amount of less than 20% which is capable of condensing large pDNA just like luciferase reporter gene. It was stated that COS-7 cells can be transfected proficiently with pDNA transmitting ZnO QDs with lower cytotoxicity. The ZnO QDs had a significantly decreased cytotoxicity in association with the application of PDMAEMA as the gene vector. The decrease in cytotoxicity was due to the existence of negatively charged polymethacrylate in the QDs which stabilized the positive charges.

6.5. Biosensors based on ZnO Nanomaterials.

Biosensors are extensively utilized in food industry, environmental monitoring, and healthcare and in chemical or biological analysis. Examples for biosensors are electrochemical, photometric, piezoelectric, and calorimetric among others when categorized based on the detection principles [145].

Nanomaterials, either as uncombined or as combination with biologically active substances, are attaining ever-increasing awareness because of their capability to deliver a suitable platform for developing high-performance biosensors, which is due to their distinctive features [23]. For instance, the higher surface area of nanomaterials could be utilized for immobilizing numerous biomolecules such as antibodies, enzymes, and other proteins. Moreover, they could be permitted for a direct electron transfer among the electrodes and the active sites of the biomolecules.

ZnO nanomaterials also provide numerous desirable traits apart from semiconducting properties such as biosensing, strong adsorption capability, high isoelectric point, and high catalytic efficiency (IEP; ~9.5) which are appropriate for adsorption of certain proteins such as antibodies and enzymes with less IEPs by electrostatic interaction [146]. Moreover, the favorable conditions of nanomaterials to be used in biosensors are lower toxicity, higher electron transfer capability, higher surface area, and better biocompatibility or stability [147]. The most widely stated ZnO-based biosensors are recognized for numerous small-molecule analytes such as cholesterol, phenol, urea, glucose, H_2O_2, and many more other things. Additionally, there are numerous biosensors for sensing other molecules and certain physical or chemical properties like pH [148, 149].

7. Toxicity Studies of ZnO NPs

Due to the increasing use of nanoparticles and their release in the environment, it is necessary to determine the toxicity of nanoparticles. Vicario-Pares et al. [150] conducted a toxicity study of three metal oxide nanoparticles, namely, CuO NPs (copper oxide nanoparticles), ZnO NPs, and TiO_2 NPs against zebra fish embryo. ZnO NPs were found to be less toxic than the ionic form of zinc, which exerts the highest toxicity. Studies of Zhu et al. [151] showed that ZnO NP toxicity is dose dependent. Similarly, Jeyabharathi et al. [152] evaluated the toxicity study of green synthesized ZnO NPs towards zebra fish embryos. The author synthesized ZnO NPs from *Amaranthus caudatus* leaf extract. Further, ZnO NPs were found to exhibit a higher antibacterial activity against *Staphylococcus epidermidis* and *Enterobacter aerogenes*. Results of the toxicity study show that ZnO NPs at a concentration of 10 mg/ml did not show any significant effect on survival and malformation in the zebra fish embryo. In a 90-day toxicity study, 100 nm ZnO NPs with different surface charges (negatively charged, ZnOAE100 [−] and positively charged ZnOAE100 [+]) were administered to Sprague Dawley rats to determine the toxic level and to identify target organs. Significant toxic effects were observed in both sexes at a concentration greater than 125 mg/kg. Also, there was an absence of adverse effect level at a concentration of about 31.25 mg/kg for both sexes [153].

8. Conclusions

Overall, from the study reviews, it is considered that green ZnO NP synthesis is much safer and environmentally friendly than the physical and chemical methods. ZnO NPs are one of the most important and versatile materials, due to their diverse properties, functionalities, various benefits, and applications to humans. The green sources act as a stabilizing and reducing agent for the synthesis of nanoparticles of controlled size and shape. Holistically, the ZnO NP application to crops increases the growth and yield in agriculture. As demand for food is increasing day by day, the yield of a staple crop is low. Thus, it is necessary to commercialize metal oxide nanoparticles for sustainable agriculture. On the other hand, its biomedical applications in this sector are increased day by day in various processes including bioimaging, drug delivery, biosensors, and gene delivery. With respect to its toxicity properties, ZnO NPs can act as smart weapons against multiple drug-resistant microorganisms and as a talented substitute for antibiotics. It is anticipated that this review could further streamline the research on innovative methodological and clinical correlations in this area. In the meantime, solutions to health problems will be

suggested by referring to this complex through scientific and research reports.

References

[1] H. Mirzaei and M. Darroudi, "Zinc oxide nanoparticles: biological synthesis and biomedical applications," *Ceramics International*, vol. 43, no. 1, pp. 907–914, 2017.

[2] S. C. C. Arruda, A. L. D. Silva, R. M. Galazzi, R. A. Azevedo, and M. A. Z. Arruda, "Nanoparticles applied to plant science: a review," *Talanta*, vol. 131, pp. 693–705, 2015.

[3] H. Agarwal, S. V. Kumar, and S. Rajeshkumar, "A review on green synthesis of zinc oxide nanoparticles–an eco-friendly approach," *Resource-Efficient Technologies*, vol. 3, no. 4, pp. 406–413, 2017.

[4] H. A. Salam, R. Sivaraj, and R. Venckatesh, "Green synthesis and characterization of zinc oxide nanoparticles from *Ocimum basilicum* L. var. purpurascens Benth.-*Lamiaceae* leaf extract," *Materials Letters*, vol. 131, pp. 16–18, 2014.

[5] P. T. Anastas and J. C. Warner, *Green Chemistry: Theory and Practice*, Oxford University Press, Oxford, UK, 2000.

[6] W. Cai and X. Chen, "Nanoplatforms for targeted molecular imaging in living subjects," *Small*, vol. 3, no. 11, pp. 1840–1854, 2007.

[7] H. Hong, Y. Zhang, J. Sun, and W. Cai, "Molecular imaging and therapy of cancer with radiolabeled nanoparticles," *Nano Today*, vol. 4, no. 5, pp. 399–413, 2009.

[8] Y. Zhang, H. Hong, D. V. Myklejord, and W. Cai, "Molecular imaging with SERS-active nanoparticles," *Small*, vol. 7, no. 23, pp. 3261–3269, 2011.

[9] M. V. Yigit and Z. Medarova, "In vivo and ex vivo applications of gold nanoparticles for biomedical SERS imagingi," *American Journal of Nuclear Medicine and Molecular Imaging*, vol. 2, no. 2, pp. 232–241, 2012.

[10] J. E. Hutchison, "Greener nanoscience: a proactive approach to advancing applications and reducing implications of nanotechnology," *ACS Nano*, vol. 2, no. 3, pp. 395–402, 2008.

[11] W. Cai, A. R. Hsu, Z.-B. Li, and X. Chen, "Are quantum dots ready for in vivo imaging in human subjects?," *Nanoscale Research Letters*, vol. 2, no. 6, pp. 265–281, 2007.

[12] W. Cai and H. Hong, "In a "nutshell": intrinsically radiolabeled quantum dots," *American Journal of Nuclear Medicine and Molecular Imaging*, vol. 2, no. 2, pp. 136–140, 2012.

[13] L. R. Hirsch, A. M. Gobin, A. R. Lowery et al., "Metal nanoshells," *Annals of Biomedical Engineering*, vol. 34, no. 1, pp. 15–22, 2006.

[14] L. Lacerda, A. Bianco, M. Prato, and K. Kostarelos, "Carbon nanotubes as nanomedicines: From toxicology to pharmacology☆," *Advanced Drug Delivery Reviews*, vol. 58, no. 14, pp. 1460–1470, 2006.

[15] H. Hong, T. Gao, and W. Cai, "Molecular imaging with single-walled carbon nanotubes," *Nano Today*, vol. 4, no. 3, pp. 252–261, 2009.

[16] Z. L. Wang, "Splendid one-dimensional nanostructures of zinc oxide: a new nanomaterial family for nanotechnology," *ACS Nano*, vol. 2, no. 10, pp. 1987–1992, 2008.

[17] P. Yang, R. Yan, and M. Fardy, "Semiconductor nanowire: what's next?," *Nano Letters*, vol. 10, no. 5, pp. 1529–1536, 2010.

[18] M. H. Huang, Y. Wu, H. Feick, N. Tran, E. Weber, and P. Yang, "Catalytic growth of zinc oxide nanowires by vapor transport," *Advanced Materials*, vol. 13, no. 2, pp. 113–116, 2001.

[19] Z. Fan and J. G. Lu, "Zinc oxide nanostructures: synthesis and properties," *Journal of Nanoscience and Nanotechnology*, vol. 5, no. 10, pp. 1561–1573, 2005.

[20] C. S. Lao, M.-C. Park, Q. Kuang et al., "Giant enhancement in UV response of ZnO nanobelts by polymer surface-functionalization," *Journal of the American Chemical Society*, vol. 129, no. 40, pp. 12096-12097, 2007.

[21] X. Wang, J. Liu, J. Song, and Z. L. Wang, "Integrated nanogenerators in biofluid," *Nano Letters*, vol. 7, no. 8, pp. 2475–2479, 2007.

[22] Y. Yang, W. Guo, Y. Zhang, Y. Ding, X. Wang, and Z. L. Wang, "Piezotronic effect on the output voltage of P3HT/ZnO micro/nanowire heterojunction solar cells," *Nano Letters*, vol. 11, no. 11, pp. 4812–4817, 2011.

[23] R. Yakimova, "ZnO materials and surface tailoring for biosensing," *Frontiers in Bioscience*, vol. E4, no. 1, pp. 254–278, 2012.

[24] J. Zhou, N. S. Xu, and Z. L. Wang, "Dissolving behavior and stability of ZnO wires in biofluids: a study on biodegradability and biocompatibility of ZnO nanostructures," *Advanced Materials*, vol. 18, no. 18, pp. 2432–2435, 2006.

[25] Y. Zhang, T. Nayak, H. Hong, and W. Cai, "Biomedical applications of zinc oxide nanomaterials," *Current Molecular Medicine*, vol. 13, no. 10, pp. 1633–1645, 2013.

[26] M. Pandurangan and D. H. Kim, "In vitro toxicity of zinc oxide nanoparticles: a review," *Journal of Nanoparticle Research*, vol. 17, no. 3, p. 158, 2015.

[27] V. N. Kalpana and V. D. Rajeswari, "Biosynthesis of metal and metal oxide nanoparticles for food packaging and preservation: a green expertise," *Food Biosynthesis*, pp. 293–316, 2017.

[28] A. Bose, *Emerging Trends of Nanotechnology in Pharmacy*, Pharmainfo, 2016, http://www.pharmainfo.net/book/emerging-trends-nanotechnology-pharmacy-1introduction-nanotechnology/classification.

[29] A. Alagarasi, *Introduction to Nanomaterials*, National Centre for Catalysis Research, Chennai, India, 2011, https://nccr.iitm.ac.in/2011.pdf.

[30] V. Thirumalai Arasu, D. Prabhu, and M. Soniya, "Stable silver nanoparticle synthesizing methods and its applications," *Journal of Biosciences Research*, vol. 1, pp. 259–270, 2010.

[31] G. Rajakumar, T. Gomathi, M. Thiruvengadam, V. D. Rajeswari, V. N. Kalpana, and I. M. Chung, "Evaluation of anti-cholinesterase, antibacterial and cytotoxic activities of green synthesized silver nanoparticles using from *Millettia pinnata* flower extract," *Microbial Pathogenesis*, vol. 103, pp. 123–128, 2017.

[32] P. Raveendran, J. Fu, and S. L. Wallen, "Completely "green" synthesis and stabilization of metal nanoparticles," *Journal of the American Chemical Society*, vol. 125, no. 46, pp. 13940-13941, 2003.

[33] P. Dhandapani, S. Maruthamuthu, and G. Rajagopal, "Biomediated synthesis of TiO$_2$ nanoparticles and its photocatalytic effect on aquatic biofilm," *Journal of Photochemistry and Photobiology B: Biology*, vol. 110, pp. 43–49, 2012.

[34] K. Prasad and A. K. Jha, "ZnO nanoparticles: synthesis and adsorption study," *Natural Science*, vol. 1, no. 2, pp. 129–135, 2009.

[35] S. Gunalan, R. Sivaraj, and V. Rajendran, "Green synthesized ZnO nanoparticles against bacterial and fungal pathogens," *Progress in Natural Science: Materials International*, vol. 22, no. 6, pp. 693–700, 2012.

[36] P. Vanathi, P. Rajiv, S. Narendhran, S. Rajeshwari, P. K. S. M. Rahman, and R. Venckatesh, "Biosynthesis and characterization of phyto mediated zinc oxide nanoparticles: a green chemistry approach," *Materials Letters*, vol. 134, pp. 13–15, 2014.

[37] M. Anbuvannan, M. Ramesh, G. Viruthagiri, N. Shanmugam, and N. Kannadasan, "*Anisochilus carnosus* leaf extract mediated synthesis of zinc oxide nanoparticles for antibacterial and photocatalytic activities," *Materials Science in Semiconductor Processing*, vol. 39, pp. 621–628, 2015a.

[38] M. Sundrarajan, S. Ambika, and K. Bharathi, "Plant-extract mediated synthesis of ZnO nanoparticles using *Pongamia pinnata* and their activity against pathogenic bacteria," *Advanced Powder Technology*, vol. 26, no. 5, pp. 1294–1299, 2015.

[39] B. N. Patil and T. C. Taranath, "*Limonia acidissima* L. leaf mediated synthesis of zinc oxide nanoparticles: a potent tool against *Mycobacterium tuberculosis*," *International Journal of Mycobacteriology*, vol. 5, no. 2, pp. 197–204, 2016.

[40] P. Jamdagni, P. Khatri, and J. S. Rana, "Green synthesis of zinc oxide nanoparticles using flower extract of *Nyctanthes arbor-tristis* and their antifungal activity," *Journal of King Saud University–Science*, vol. 30, no. 2, pp. 168–175, 2018.

[41] C. Jayaseelan, A. Abdul Rahuman, and A. Vishnu Kirthi, "Novel microbial route to synthesize ZnO nanoparticles using *Aeromonas hydrophila* and their activity against pathogenic bacteria and fungi," *Spectrochimica Acta Part A: Molecular and Biomolecular Spectroscopy*, vol. 90, pp. 78–84, 2012.

[42] M. R. Parra and F. Z. Haque, "Aqueous chemical route synthesis and the effect of calcination temperature on the structural and optical properties of ZnO nanoparticles," *Journal of Materials Research and Technology*, vol. 3, no. 4, pp. 363–369, 2014.

[43] G. Sangeetha, S. Rajeshwari, and R. Venckatesh, "Green synthesis of zinc oxide nanoparticles by aloe barbadensis miller leaf extract: Structure and optical properties," *Materials Research Bulletin*, vol. 46, no. 12, pp. 2560–2566, 2011.

[44] Z. Sheikhloo, M. Salouti, and F. Katiraee, "Biological synthesis of gold nanoparticles by fungus *Epicoccum nigrum*," *Journal of Cluster Science*, vol. 22, no. 4, pp. 661–665, 2011.

[45] E. D. Sherly, J. Judith Vijaya, N. Clament Sagaya Selvam, and L. John Kennedy, "Microwave assisted combustion synthesis of coupled ZnO–ZrO$_2$ nanoparticles and their role in the photocatalytic degradation of 2,4-dichlorophenol," *Ceramics International*, vol. 40, no. 4, pp. 5681–5691, 2014.

[46] K. Elumalai and S. Velmurugan, "Green synthesis, characterization and antimicrobial activities of zinc oxide nanoparticles from the leaf extract of *Azadirachta indica* (L.)," *Applied Surface Science*, vol. 345, pp. 329–336, 2015.

[47] V. Patel, D. Berthold, P. Puranik, and M. Gantar, "Screening of cyanobacteria and microalgae for their ability to synthesize silver nanoparticles with antibacterial activity," *Biotechnology Reports*, vol. 5, pp. 112–119, 2015.

[48] M. Stan, A. Popa, D. Toloman, A. Dehelean, I. Lung, and G. Katona, "Enhanced photocatalytic degradation properties of zinc oxide nanoparticles synthesized by using plant extracts," *Materials Science in Semiconductor Processing*, vol. 39, pp. 23–29, 2015.

[49] A. Umar, M. M. Rahman, M. Vaseem, and Y. B. Hahn, "Ultra-sensitive cholesterol biosensor based on low-temperature grown ZnO nanoparticles," *Electrochemistry Communications*, vol. 11, no. 1, pp. 118–121, 2009.

[50] J. Liu, C. Guo, C. M. Li et al., "Carbon-decorated ZnO nanowire array: a novel platform for direct electrochemistry of enzymes and biosensing applications," *Electrochemistry Communications*, vol. 11, no. 1, pp. 202–205, 2009.

[51] K. Bahrami, M. M. Khodaei, and A. Nejati, "One-pot synthesis of 1, 2, 4, 5-tetrasubstituted and 2, 4, 5-trisubstituted imidazoles by zinc oxide as efficient and reusable catalyst," *Monatshefte für Chemie–Chemical Monthly*, vol. 142, no. 2, pp. 159–162, 2011.

[52] R. Tayebee, F. Cheravi, M. Mirzaee, and M. M. Amini, "Commercial zinc oxide (Zn2+) as an efficient and environmentally benign catalyst for homogeneous benzoylation of hydroxyl functional groups," *Chinese Journal of Chemistry*, vol. 28, no. 7, pp. 1247–1252, 2010.

[53] C. R. Gorla, N. W. Emanetoglu, S. Liang et al., "Structural, optical, and surface acoustic wave properties of epitaxial ZnO films grown on (0112) sapphire by metalorganic chemical vapor deposition," *Journal of Applied Physics*, vol. 85, no. 5, pp. 2595–2602, 1999.

[54] M. S. Akhtar, S. Ameen, S. A. Ansari, and O. Yang, "Synthesis and characterization of ZnO nanorods and balls nanomaterials for dye sensitized solar cells," *Journal of Nanoengineering and Nanomanufacturing*, vol. 1, no. 1, pp. 71–76, 2011.

[55] N. P. Sasidharan, P. Chandran, and S. S. Khan, "Interaction of colloidal zinc oxide nanoparticles with bovine serum albumin and its adsorption isotherms and kinetics," *Colloids and Surfaces B: Biointerfaces*, vol. 102, pp. 195–201, 2013.

[56] L. Zhang, Y. Ding, M. Povey, and D. York, "ZnO nanofluids-a potential antibacterial agent," *Progress in Natural Science*, vol. 18, no. 8, pp. 939–944, 2008.

[57] J. Xie, H. Deng, Z. Q. Xu, Y. Li, and J. Huang, "Growth of ZnO photonic crystals by self-assembly," *Journal of Crystal Growth*, vol. 292, no. 2, pp. 227–229, 2006.

[58] C. Liewhiran and S. Phanichphant, "Improvement of flame-made ZnO nanoparticulate thick film morphology for ethanol sensing," *Sensors*, vol. 7, no. 5, pp. 650–675, 2007.

[59] B. Ajitha, Y. A. K. Reddy, P. S. Reddy, H. J. Jeon, and C. W. Ahn, "Role of capping agents in controlling silver nanoparticles size, antibacterial activity and potential application as optical hydrogen peroxide sensor," *RSC Advances*, vol. 6, no. 42, pp. 36171–36179, 2016.

[60] P. J. P. Espitia, N. D. F. F. Soares, J. S. Dos Reis Coimbra, N. J. De Andrade, R. S. Cruz, and E. A. A. Medeiros, "Zinc oxide nanoparticles: synthesis, antimicrobial activity and food packaging applications," *Food and Bioprocess Technology*, vol. 5, no. 5, pp. 1447–1464, 2012.

[61] W. H. Elmer and J. C. White, "The use of metallic oxide nanoparticles to enhance growth of tomatoes and eggplants in disease infested soil or soilless medium," *Environmental Science: Nano*, vol. 3, no. 5, pp. 1072–1079, 2016.

[62] N. A. N. Mohamad, N. A. Arham, J. Jai, and A. Hadi, "Plant extract as reducing agent in synthesis of metallic nanoparticles: a review," *Advanced Materials Research*, vol. 832, pp. 350–355, 2014.

[63] V. V. Makarov, A. J. Love, O. V. Sinitsyna et al., ""Green" nanotechnologies: synthesis of metal nanoparticles using plants," *Acta Naturae*, vol. 6, no. 1, p. 20, 2014.

[64] A. G. Ingale and A. N. Chaudhari, "Biogenic synthesis of nanoparticles and potential applications: an eco-friendly approach," *Journal of Nanomedicine and Nanotechnology*, vol. 4, no. 165, pp. 1–7, 2013.

[65] S. M. Roopan, V. D. Rajeswari, V. N. Kalpana, and G. Elango, "Biotechnology and pharmacological evaluation of Indian vegetable crop *Lagenaria siceraria*: an overview," *Applied Microbiology and Biotechnology*, vol. 100, no. 3, pp. 1153–1162, 2015.

[66] M. S. Geetha, H. Nagabhushana, and H. N. Shivananjaiah, "Green mediated synthesis and characterization of ZnO nanoparticles using *Euphorbia Jatropha* latex as reducing agent," *Journal of Science: Advanced Materials and Devices*, vol. 1, no. 3, pp. 301–310, 2016.

[67] M. Murali, C. Mahendra, Nagabhushan et al., "Antibacterial and antioxidant properties of biosynthesized zinc oxide nanoparticles from *Ceropegia candelabrum* L.–an endemic species," *Spectrochimica Acta Part A: Molecular and Biomolecular Spectroscopy*, vol. 179, no. 179, pp. 104–109, 2017.

[68] J. Vaishnav, V. Subha, S. Kirubanandan, M. Arulmozhi, and S. Renganathan, "Green synthesis of zinc oxide nanoparticles by *Celosia argentea* and its characterization," *Journal of Optoelectronic and Biomedical Materials*, vol. 9, pp. 59–71, 2017.

[69] G. Sathishkumar, C. Rajkuberan, K. Manikandan, S. Prabukumar, J. Daniel John, and S. Sivaramakrishnan, "Facile biosynthesis of antimicrobial zinc oxide (ZnO) nanoflakes using leaf extract of *Couroupita guianensis* Aubl," *Materials Letters*, vol. 188, pp. 383–386, 2017.

[70] M. Anbuvannan, M. Ramesh, G. Viruthagiri, N. Shanmugam, and N. Kannadasan, "Synthesis, characterization and photocatalytic activity of ZnO nanoparticles prepared by biological method," *Spectrochimica Acta Part A: Molecular and Biomolecular Spectroscopy*, vol. 143, pp. 304–308, 2015.

[71] P. Rajiv, S. Rajeshwari, and R. Venckatesh, "Bio-fabrication of zinc oxide nanoparticles using leaf extract of *Parthenium hysterophorus* L. and its size-dependent antifungal activity against plant fungal pathogens," *Spectrochimica Acta Part A: Molecular and Biomolecular Spectroscopy*, vol. 112, pp. 384–387, 2013.

[72] M. Ramesh, M. Anbuvannan, and G. Viruthagiri, "Green synthesis of ZnO nanoparticles using Solanum nigrum leaf extract and their antibacterial activity," *Spectrochimica Acta Part A: Molecular and Biomolecular Spectroscopy*, vol. 136, pp. 864–870, 2015.

[73] D. Sharma, M. I. Sabela, S. Kanchi et al., "Biosynthesis of ZnO nanoparticles using *Jacaranda mimosifolia* flowers extract: Synergistic antibacterial activity and molecular simulated facet specific adsorption studies," *Journal of Photochemistry and Photobiology B: Biology*, vol. 162, pp. 199–207, 2016.

[74] S. Nagarajan and K. A. Kuppusamy, "Extracellular synthesis of zinc oxide nanoparticle using seaweeds of gulf of Mannar, India," *Journal of Nanobiotechnology*, vol. 11, no. 1, p. 39, 2013.

[75] R. Ishwarya, B. Vaseeharan, S. Kalyani et al., "Facile green synthesis of zinc oxide nanoparticles using *Ulva lactuca* seaweed extract and evaluation of their photocatalytic, antibiofilm and insecticidal activity," *Journal of Photochemistry and Photobiology B: Biology*, vol. 178, pp. 249–258, 2018.

[76] A. Rajan, E. Cherian, and G. Baskar, "Biosynthesis of zinc oxide nanoparticles using *Aspergillus fumigatus* JCF and its antibacterial activity," *International Journal of Modern Science and Technology*, vol. 1, pp. 52–57, 2016.

[77] S. Ambika and M. Sundrarajan, "Antibacterial behaviour of *Vitex negundo* extract assisted ZnO nanoparticles against pathogenic bacteria," *Journal of Photochemistry and Photobiology B: Biology*, vol. 146, pp. 52–57, 2015.

[78] R. Dobrucka and J. Długaszewska, "Biosynthesis and antibacterial activity of ZnO nanoparticles using *Trifolium pratense* flower extract," *Saudi Journal of Biological Sciences*, vol. 23, no. 4, pp. 517–523, 2016.

[79] V. N. Kalpana, C. Payel, and V. Devi Rajeswari, "*Lagenaria siceraria* aided green synthesis of ZnO NPs: anti-dandruff, Anti-microbial and Anti-arthritic activity," *Research Journal of Chemistry and Environment*, vol. 21, no. 11, pp. 14–19, 2017.

[80] A. C. Dhanemozhi, V. Rajeswari, and S. Sathyajothi, "Green synthesis of zinc oxide nanoparticle using green tea leaf extract for supercapacitor application," *Mater Today: Proceedings*, vol. 4, no. 2, pp. 660–667, 2017.

[81] J. Santhoshkumar, S. V. Kumar, and S. Rajeshkumar, "Synthesis of zinc oxide nanoparticles using plant leaf extract against urinary tract infection pathogen," *Resource-Efficient Technologies*, vol. 3, no. 4, pp. 459–465, 2017.

[82] O. J. Nava, P. A. Luque, C. M. Gomez-Gutiérrez et al., "Influence of *Camellia sinensis* extract on zinc oxide nanoparticle green synthesis," *Journal of Molecular Structure*, vol. 1134, pp. 121–125, 2017.

[83] B. Siripireddy and B. K. Mandal, "Facile green synthesis of zinc oxide nanoparticles by *Eucalyptus globulus* and their photocatalytic and antioxidant activity," *Advanced Powder Technology*, vol. 28, no. 3, pp. 785–797, 2017.

[84] F. T. Thema, E. Manikandan, M. S. Dhlamini, and M. Maaza, "Green synthesis of ZnO nanoparticles via *Agathosma betulina* natural extract," *Materials Letters*, vol. 161, pp. 124–127, 2015.

[85] M. Z. Hussein, W. H. W. N. Azmin, M. Mustafa, and A. H. Yahaya, "*Bacillus cereus* as a biotemplating agent for the synthesis of zinc oxide with raspberry-and plate-like structures," *Journal of Inorganic Biochemistry*, vol. 103, no. 8, pp. 1145–1150, 2009.

[86] D. Kundu, C. Hazra, A. Chatterjee, A. Chaudhari, and S. Mishra, "Extracellular biosynthesis of zinc oxide nanoparticles using *Rhodococcus pyridinivorans* NT2: multifunctional textile finishing, biosafety evaluation and *in vitro* drug delivery in colon carcinoma," *Journal of Photochemistry and Photobiology B: Biology*, vol. 140, pp. 194–204, 2014.

[87] P. Dhandapani, A. S. Siddarth, S. Kamalasekaran, S. Maruthamuthu, and G. Rajagopal, "Bio-approach: ureolytic bacteria mediated synthesis of ZnO nanocrystals on cotton fabric and evaluation of their antibacterial properties," *Carbohydrate Polymers*, vol. 103, pp. 448–455, 2014.

[88] E. Selvarajan and V. Mohanasrinivasan, "Biosynthesis and characterization of ZnO nanoparticles using *Lactobacillus plantarum* VITES07," *Materials Letters*, vol. 112, pp. 180–182, 2013.

[89] A. Shamsuzzaman, A. Mashrai, H. Khanam, and R. N. Aljawfi, "Biological synthesis of ZnO nanoparticles using *C. albicans* and studying their catalytic performance in the synthesis of steroidal pyrazolines," *Arabian Journal of Chemistry*, vol. 10, pp. S1530–S1536, 2017.

[90] G. Baskar, J. Chandhuru, K. S. Fahad, and A. S. Praveen, "Mycological synthesis, characterization and antifungal activity of zinc oxide nanoparticles," *Asian Journal of Pharmacy and Technology*, vol. 3, no. 4, pp. 142–146, 2013.

[91] S. Azizi, M. B. Ahmad, F. Namvar, and R. Mohamad, "Green biosynthesis and characterization of zinc oxide nanoparticles using brown marine macroalga Sargassum muticum aqueous extract," *Materials Letters*, vol. 116, pp. 275–277, 2014.

[92] R. Pati, R. K. Mehta, and S. Mohanty, "Topical application of zinc oxide nanoparticles reduces bacterial skin infection in mice and exhibits antibacterial activity by inducing oxidative stress response and cell membrane disintegration in macrophages," *Nanomedicine: Nanotechnology, Biology and Medicine*, vol. 10, no. 6, pp. 1195–1208, 2014.

[93] K. Pavani, N. Kumar, and B. Sangameswaran, "Synthesis of lead nanoparticles by Aspergillus species," *Polish Journal of Microbiology*, vol. 61, no. 1, pp. 61–63, 2012.

[94] Y. A. Arfat, S. Benjakul, T. Prodpran, P. Sumpavapol, and P. Songtipya, "Properties and antimicrobial activity of fish protein isolate/fish skin gelatin film containing basil leaf essential oil and zinc oxide nanoparticles," *Food Hydrocolloids*, vol. 41, pp. 265–273, 2014.

[95] A. Yasmin, K. Ramesh, and S. Rajeshkumar, "Optimization and stabilization of gold nanoparticles by using herbal plant extract with microwave heating," *Nano Convergence*, vol. 1, no. 1, p. 12, 2014.

[96] S. Rajeshkumar, C. Malarkodi, M. Vanaja, and G. Annadurai, "Anticancer and enhanced antimicrobial activity of biosynthesizd silver nanoparticles against clinical pathogens," *Journal of Molecular Structure*, vol. 1116, pp. 165–173, 2016.

[97] L.-E. Shi, Z.-H. Li, W. Zheng, Y.-F. Zhao, Y.-F. Jin, and Z.-X. Tang, "Synthesis, antibacterial activity, antibacterial mechanism and food applications of ZnO nanoparticles: a review," *Food Additives and Contaminants: Part A*, vol. 31, no. 2, pp. 173–186, 2014.

[98] L. Fu and Z. Fu, "*Plectranthus amboinicus* leaf extract–assisted biosynthesis of ZnO nanoparticles and their photocatalytic activity," *Ceramics International*, vol. 41, no. 2, pp. 2492–2496, 2015.

[99] T. Bhuyan, K. Mishra, M. Khanuja, R. Prasad, and A. Varma, "Biosynthesis of zinc oxide nanoparticles from *Azadirachta indica* for antibacterial and photocatalytic applications," *Materials Science in Semiconductor Processing*, vol. 32, pp. 55–61, 2015.

[100] H. R. Madan, S. C. Sharma, Udayabhanu et al., "Facile green fabrication of nanostructure ZnO plates, bullets, flower, prismatic tip, closed pine cone: their antibacterial, antioxidant, photoluminescent and photocatalytic properties," *Spectrochimica Acta Part A: Molecular and Biomolecular Spectroscopy*, vol. 152, pp. 404–416, 2016.

[101] Y. Qian, J. Yao, M. Russel, K. Chen, and X. Wang, "Characterization of green synthesized nano-formulation (ZnO–*A. vera*) and their antibacterial activity against pathogens," *Environmental Toxicology and Pharmacology*, vol. 39, no. 2, pp. 736–746, 2015.

[102] K. Ali, S. Dwivedi, and A. Azam, "*Aloe vera* extract functionalized zinc oxide nanoparticles as nanoantibiotics against multi-drug resistant clinical bacterial isolates," *Journal of Colloid and Interface Science*, vol. 472, pp. 145–156, 2016.

[103] N. C. Shinde, N. J. Keskar, and P. D. Argade, "Nanoparticles: advances in drug delivery systems," *Research Journal of Pharmaceutical, Biological and Chemical Sciences*, vol. 3, no. 1, pp. 922–929, 2012.

[104] W. Cai and H. Hong, "Peptoid and positron emission tomography: an appealing combination," *American Journal of Nuclear Medicine and Molecular Imaging*, vol. 1, no. 1, pp. 76–79, 2011.

[105] X. Huang, S. Lee, and X. Chen, "Design of "smart" probes for optical imaging of apoptosis," *American Journal of Nuclear Medicine and Molecular Imaging*, vol. 1, no. 1, pp. 3–17, 2011.

[106] R. E. Wang, Y. Niu, H. Wu, M. N. Amin, and J. Cai, "Development of NGR peptide-based agents for tumor imaging," *American Journal of Nuclear Medicine and Molecular Imaging*, vol. 1, no. 1, pp. 36–46, 2011.

[107] D. L. Thorek, R. Robertson, W. A. Bacchus et al., "Cerenkov imaging–a new modality for molecular imaging," *American Journal of Nuclear Medicine and Molecular Imaging*, vol. 2, no. 2, pp. 163–173, 2012.

[108] K. Vanheusden, W. L. Warren, C. H. Seager, D. R. Tallant, J. A. Voigt, and B. E. Gnade, "Mechanisms behind green photoluminescence in ZnO phosphor powders," *Journal of Applied Physics*, vol. 79, no. 10, pp. 7983–7990, 1996.

[109] Y. W. Heo, D. P. Norton, and S. J. Pearton, "Origin of green luminescence in ZnO thin film grown by molecular-beam epitaxy," *Journal of Applied Physics*, vol. 98, no. 7, article 073502, 2005.

[110] S. Sudhagar, S. Sathya, K. Pandian, and B. S. Lakshmi, "Targeting and sensing cancer cells with ZnO nanoprobes in vitro," *Biotechnology Letters*, vol. 33, no. 9, pp. 1891–1896, 2011.

[111] F. Xue, J. Liang, and H. Han, "Synthesis and spectroscopic characterization of water-soluble Mn-doped ZnOxS1–x quantum dots," *Spectrochimica Acta Part A: Molecular and Biomolecular Spectroscopy*, vol. 83, no. 1, pp. 348–352, 2011.

[112] Y. L. Wu, S. Fu, A. I. Y. Tok et al., "A dual-colored biomarker made of doped ZnO nanocrystals," *Nanotechnology*, vol. 19, no. 34, article 345605, 2008.

[113] W. Q. Zhang, Y. Lu, T.-K. Zhang et al., "Controlled synthesis and biocompatibility of water-soluble ZnO nanorods/Au nanocomposites with tunable UV and visible emission intensity," *Journal of Physical Chemistry C*, vol. 112, no. 50, pp. 19872–19877, 2008.

[114] S. C. Yang, Y.-C. Shen, T.-C. Lu, T.-L. Yang, and J.-J. Huang, "Tumor detection strategy using ZnO light-emitting nanoprobes," *Nanotechnology*, vol. 23, no. 5, article 055202, 2012.

[115] X. Michalet, F. F. Pinaud1, L. A. Bentolila et al., "Quantum dots for live cells, in vivo imaging, and diagnostics," *Science*, vol. 307, no. 5709, pp. 538–544, 2005.

[116] I. L. Medintz, H. Tetsuo Uyeda, E. R. Goldman, and H. Mattoussi, "Quantum dot bioconjugates for imaging, labelling and sensing," *Nature Materials*, vol. 4, no. 6, pp. 435–446, 2006.

[117] W. Cai, Y. Zhang, and T. J. Kamp, "Imaging of induced pluripotent stem cells: from cellular reprogramming to transplantation," *American Journal of Nuclear Medicine and Molecular Imaging*, vol. 1, no. 1, pp. 18–28, 2011.

[118] Z. Y. Pan, J. Liang, Z.-Z. Zheng, H.-H. Wang, and H.-M. Xiong, "The application of ZnO luminescent nanoparticles in labeling mice," *Contrast Media and Molecular Imaging*, vol. 6, no. 4, pp. 328–330, 2011.

[119] T. F. Massoud and S. Gambhir, "Molecular imaging in living subjects: seeing fundamental biological processes in a new light," *Genes and Development*, vol. 17, no. 5, pp. 545–580, 2003.

[120] Y. Liu, K. Ai, Q. Yuan, and L. Lu, "Fluorescence-enhanced gadolinium-doped zinc oxide quantum dots for magnetic resonance and fluorescence imaging," *Biomaterials*, vol. 32, no. 4, pp. 1185–1192, 2011.

[121] S. Singh, "Multifunctional magnetic quantum dots for cancer theranostics," *Journal of Biomedical Nanotechnology*, vol. 7, no. 1, pp. 95–97, 2011.

[122] K. Murugan, M. Roni, C. Panneerselvam et al., "*Sargassum wightii*-synthesized ZnO nanoparticles reduce the fitness and reproduction of the malaria vector *Anopheles stephensi* and cotton bollworm *Helicoverpa armigera*," *Physiological and Molecular Plant Pathology*, vol. 101, 2017.

[123] P. R. Gandhi, C. Jayaseelan, R. R. Mary, D. Mathivanan, and S. R. Suseem, "Acaricidal, pediculicidal and larvicidal activity of synthesized ZnO nanoparticles using *Momordica charantia* leaf extract against blood feeding parasites," *Experimental Parasitology*, vol. 181, pp. 47–56, 2017.

[124] S. Rajeshwari, K. S. M. Rahman, P. Rajiv, and R. Venckatesh, "Biogenic zinc oxide nanoparticles synthesis using *Tabernaemontana divaricate* leaf extract and its anticancer activity against MCF-7 breast cancer cell lines," in *Proceedings of the International Conference on Agriculture, Biology and Environmental Science*, pp. 83–85, Rome Italy, September 2014.

[125] M. J. Akhtar, M. Ahamed, S. Kumar, M. M. Khan, J. Ahmad, and S. A. Alrokayan, "Zinc oxide nanoparticles selectively induce apoptosis in human cancer cells through reactive oxygen species," *International Journal of Nanomedicine*, vol. 7, p. 845, 2012.

[126] C. Hanley, J. Layne, A. Punnoose et al., "Preferential killing of cancer cells and activated human T cells using ZnO nanoparticles," *Nanotechnology*, vol. 19, no. 29, pp. 95–103, 2008.

[127] S. S. Gambhir, "Molecular imaging of cancer with positron emission tomography," *Nature Reviews Cancer*, vol. 2, no. 9, pp. 683–693, 2002.

[128] J. F. Eary, D. S. Hawkins, E. T. Rodler, and E. U. Conrad, "18F-FDG PET in sarcoma treatment response imaging," *American Journal of Nuclear Medicine and Molecular Imaging*, vol. 1, no. 1, pp. 47–53, 2011.

[129] A. Iagaru, "18F-FDG PET/CT: timing for evaluation of response to therapy remains a clinical challenge," *American Journal of Nuclear Medicine and Molecular Imaging*, vol. 1, no. 1, pp. 63-64, 2011.

[130] W. Vach, P. F. Høilund-Carlsen, B. M. Fischer, O. Gerke, and W. Weber, "How to study optimal timing of PET/CT for monitoring of cancer treatment," *American Journal of Nuclear Medicine and Molecular Imaging*, vol. 1, no. 1, pp. 54–62, 2011.

[131] M. M. Alauddin, "Positron emission tomography (PET) imaging with 18F-based radiotracers," *American Journal of Nuclear Medicine and Molecular Imaging*, vol. 2, no. 1, pp. 55–76, 2012.

[132] I. Grassi, C. Nanni, V. Allegri et al., "The clinical use of PET with 11C-acetate," *American Journal of Nuclear Medicine and Molecular Imaging*, vol. 2, no. 1, pp. 33–47, 2012.

[133] T. Buckle, N. S. van Berg, J. Kuil et al., "Non-invasive longitudinal imaging of tumor progression using an (111) indium labeled CXCR4 peptide antagonist," *American Journal of Nuclear Medicine and Molecular Imaging*, vol. 2, no. 1, pp. 99–109, 2012.

[134] M. D. Saint-Huberte, L. Brepoels, E. Devos et al., "Molecular imaging of therapy response with 18F-FLT and 18F-FDG following cyclophosphamide and mTOR inhibition," *American Journal of Nuclear Medicine and Molecular Imaging*, vol. 2, no. 1, pp. 110–121, 2012.

[135] M. Sun, D. Hoffman, G. Sundaresan, L. Yang, N. Lamichhane, and J. Zweit, "Synthesis and characterization of intrinsically radiolabeled quantum dots for bimodal detection," *American Journal of Nuclear Medicine and Molecular Imaging*, vol. 2, no. 2, pp. 122–135, 2012.

[136] Y. Zhang, T. R. Nayak, H. Hong, and W. Cai, "Graphene: a versatile nanoplatform for biomedical applications," *Nanoscale*, vol. 4, no. 13, p. 3833, 2012.

[137] H. Wang, D. Wingett, M. H. Engelhard et al., "Fluorescent dye encapsulated ZnO particles with cell-specific toxicity for

potential use in biomedical applications," *Journal of Materials Science: Materials in Medicine*, vol. 20, no. 1, pp. 11–22, 2009.

[138] Q. Yuan, S. Hein, and R. D. K. Misra, "New generation of chitosan-encapsulated ZnO quantum dots loaded with drug: synthesis, characterization and in vitro drug delivery response," *Acta Biomaterialia*, vol. 6, no. 7, pp. 2732–2739, 2010.

[139] C. G. Figdor, I. J. M. de Vries, W. Joost Lesterhuis, and C. J. M. Melief, "Dendritic cell immunotherapy: mapping the way," *Nature Medicine*, vol. 10, no. 5, pp. 475–480, 2004.

[140] N. H. Cho, T.-C. Cheong, J. H. Min et al., "A multifunctional core–shell nanoparticle for dendritic cell-based cancer immunotherapy," *Nature Nanotechnology*, vol. 6, no. 10, pp. 675–682, 2011.

[141] F. McCormick, "Cancer gene therapy: fringe or cutting edge?," *Nature Reviews Cancer*, vol. 1, no. 2, pp. 130–141, 2001.

[142] L. Nie, L. Gao, P. Feng et al., "Three-dimensional functionalized tetrapod-like ZnO nanostructures for plasmid DNA Delivery," *Small*, vol. 2, no. 5, pp. 621–625, 2006.

[143] L. Nie, L. Gao, X. Yan, and T. Wang, "Functionalized tetrapod-like ZnO nanostructures for plasmid DNA purification, polymerase chain reaction and delivery," *Nanotechnology*, vol. 18, no. 1, article 015101, 2007.

[144] P. Zhang and W. Liu, "ZnO QD@PMAA-co-PDMAEMA nonviral vector for plasmid DNA delivery and bioimaging," *Biomaterials*, vol. 31, no. 11, pp. 3087–3094, 2010.

[145] Z. Zhao, W. Lei, X. Zhang et al., "ZnO-based amperometric enzyme biosensors," *Sensors*, vol. 10, no. 2, pp. 1216–1231, 2010.

[146] J. X. Wang, X. W. Sun, A. Wei et al., "Zinc oxide nanocomb biosensor for glucose detection," *Applied Physics Letters*, vol. 88, no. 23, article 233106, 2006.

[147] S. A. Kumar and S. Chen, "Nanostructured zinc oxide particles in chemically modified electrodes for biosensor applications," *Analytical Letters*, vol. 41, no. 2, pp. 141–158, 2008.

[148] M. Willander and S. M. Al-Hilli, "ZnO nanorods as an intracellular sensor for pH measurements," *Micro and Nano Technologies in Bioanalysis*, vol. 102, pp. 187–200, 2009.

[149] V. Pachauri, A. Vlandas, K. Kern, and K. Balasubramanian, "Site-specific self-assembled liquid-gated ZnO nanowire transistors for sensing applications," *Small*, vol. 6, no. 4, pp. 589–594, 2010.

[150] U. Vicario-Pares, L. Castanaga, J. M. Lacave et al., "Comparative toxicity of metal oxide nanoparticles (CuO, ZnO and TiO2) to developing zebrafish embryos," *Journal of Nanoparticle Research*, vol. 16, no. 8, p. 2550, 2014.

[151] X. Zhu, L. Zhu, Z. Duan, R. Qi, Y. Li, and Y. Lang, "Comparative toxicity of several metal oxide nanoparticle aqueous suspensions to zebra fish (Danio rerio) early developmental stage," *Journal of Environmental Science and Health, Part A*, vol. 43, no. 3, pp. 278–284, 2008.

[152] S. Jeyabharathi, K. Kalishwaralal, K. Sundar, and A. Muthukumaran, "Synthesis of zinc oxide nanoparticles (ZnONPs) by aqueous extract of *Amaranthus caudatus* and evaluation of their toxicity and antimicrobial activity," *Materials Letters*, vol. 209, pp. 295–298, 2017.

[153] Y. R. Kim, J. I. Park, E. J. Lee et al., "Toxicity of 100 nm zinc oxide nanoparticles: a report of 90-day repeated oral administration in Sprague Dawley rats," *International Journal of Nanomedicine*, vol. 9, pp. 109–126, 2014.

5

Influence of the Number of Axial Bexarotene Ligands on the Cytotoxicity of Pt(IV) Analogs of Oxaliplatin

Yulia N. Nosova,[1] Ilia V. Zenin,[1] Varvara P. Maximova,[2]
Ekaterina M. Zhidkova,[2] Kirill I. Kirsanov,[2] Ekaterina A. Lesovaya,[2]
Anna A. Lobas,[3] Mikhail V. Gorshkov,[3] Olga N. Kovaleva,[4]
Elena R. Milaeva,[1] Markus Galanski,[5] Bernhard K. Keppler,[5] and Alexey A. Nazarov[1]

[1]Department of Medicinal Chemistry and Fine Organic Synthesis, Lomonosov Moscow State University, Leninskie Gory 1/3, Moscow 119991, Russia
[2]Blokhin Cancer Research Center, 24 Kashirskoye Shosse, Moscow 115478, Russia
[3]Institute for Energy Problems of Chemical Physics, Russian Academy of Sciences, Leninsky Pr. 38, Bld.2, Moscow 119334, Russia
[4]I.M. Sechenov First Moscow State Medical University, Moscow, Russia
[5]Faculty of Chemistry, Institute of Inorganic Chemistry, University of Vienna, Waehringer Str. 42, 1019 Vienna, Austria

Correspondence should be addressed to Alexey A. Nazarov; alexey.nazarov@me.com

Academic Editor: Giovanni Natile

We present the synthesis and cytotoxic potencies of new Pt(IV) complexes with bexarotene, an anticancer drug that induces cell differentiation and apoptosis via selective activation of retinoid X receptors. In these complexes bexarotene is positioned as an axial ligand. The complex of one bexarotene ligand attached to Pt(IV) oxaliplatin moiety was potent whereas its counterpart carrying two bexarotene ligands was inactive.

1. Introduction

The discovery of anticancer properties of platinum based complexes became a significant breakthrough in cancer treatment [1]. Currently cisplatin, carboplatin, and oxaliplatin remain major drugs for the first line treatment (alone and in combination) for a variety of malignancies including head and neck, testicular, breast, and ovarian tumors [2–5]. However, despite the success of platinum containing drugs, the intrinsic or acquired resistance, general toxicity, and other severe side effects are clinically unfavorable [4–6]. To overcome these problems novel strategies for the search of active antitumor compounds are being developed. Octahedral Pt(IV) complexes are of interest because of their kinetic inertness, low general toxicity, and possibility for oral administration [7, 8]. Satraplatin was the first Pt(IV) compound to enter phase III clinical trials as an oral drug for treatment of hormone-refractory prostate cancer. Unfortunately, this compound showed no convincing benefit for overall patient survival and was not approved by the FDA. Still, clinical trials of satraplatin in combination with different organic drugs continued [9, 10].

Combinations of two drugs in one molecule are extensively used in modern drug discovery and allow for control of activity, selectivity, and pharmacokinetics. The synthetic advantage of Pt(IV) complexes is the suitability for chemical modifications of axial positions which makes introduction of new active compounds relatively easy. Based on the proposed mechanism of action for Pt(IV) complexes, that is, activation by reduction, the release of the axial ligands can be useful for drug targeting and delivery to cancer cells [11]. This approach yielded a variety of promising complexes containing axial ligands such as folic acid [12], estradiol [13], short peptides [14], inhibitors of glutathione-S-transferase [15], pyruvate dehydrogenase kinase [16–18], histone deacetylase [19–21], cyclooxygenase [22–24], mitochondria associated hexokinase [25], or p53 activators [26].

Retinoids are biologically active analogs of vitamin A, which play an essential role in cell proliferation, differentiation, and apoptosis. Bexarotene, a selective agonist of retinoid X receptors, is used to treat cutaneous T-cell lymphoma by inducing cell differentiation and apoptosis and inhibiting metastasis [27–29]. Recently we have shown that introduction of bexarotene into Ru(II)-arene compounds resulted in highly cytotoxic agents [30]. Here, we report the synthesis, chemical characterization, and antiproliferative activity of Pt(IV) complexes with covalently attached axial ligand bexarotene.

2. Experimental Section

2.1. Materials. (OC-6-33)-(*trans*-1R,2R-Diaminocyclohexane)dihydroxido(oxalato)platinum(IV) [31], (OC-6-44)-acetato(*trans*-1R,2R-diaminocyclohexane)hydroxido(oxalato)platinum(IV) [32], and bexarotene [33] were synthesized in the Laboratory of Bioorganometallic Chemistry, Moscow State University. Oxalyl chloride was purchased from Fluka, St. Louis, USA.

2.2. Physical Measurements. NMR spectra were recorded on a Bruker FT-NMR Avance III 500 MHz instrument at 500.32 (^{1}H), 125.81 (^{13}C), 50.70 (^{15}N), and 107.57 (^{195}Pt) MHz. 2D NMR measurements were carried out using standard pulse programs. Chemical shifts were referenced relative to the solvent signal for ^{1}H and ^{13}C spectra. For ^{15}N and ^{195}Pt spectra, the external standards NH_4Cl and $K_2[PtCl_4]$ were used. ESI mass spectra were recorded on a LC/MSn ion trap mass spectrometer amaZon SL (Bruker, Bremen, Germany) with MeOH as a solvent. Elemental analysis was performed at Moscow State University with MicroCube Elementar analyzer.

2.3. Cell Lines and Culture Conditions. The MCF7, MCF7D (gift of N. I. Moiseeva), HaCat, A549, and SW480 cell lines were cultured in Dulbecco modified Eagle's medium (DMEM; PanEco, Russia) with 10% fetal bovine serum (HyClone, USA) and antibiotics (PanEco, Russia) in 5% CO_2, 37°C. The compounds were predissolved at 20 mM in dimethyl sulfoxide (DMSO) and added to the cell culture at the required concentration with maximum DMSO content of 0.5 v/v%. Cells in 96-well plates (7×10^3 cells/well) were treated with various concentrations of **3**, **4**, cisplatin, or bexarotene at 37°C for 72 h. Cell viability was determined using the MTT assay as follows: cells were incubated at 37°C for 4 h with 20 μl of 5 mg/ml solution of 3-(4,5-dimethylthiazol-2-yl)-2,5 diphenyltetrazolium bromide (Sigma-Aldrich, St. Louis, USA). The supernatant was discarded and formazan was dissolved in 150 μl of DMSO. The optical density of the solution was measured at 550 nm on a multiwell plate reader (Multiskan FC, Thermo Fisher Scientific, USA). The percentage of viable (i.e., MTT converting) cells was calculated from the absorbance of untreated cells (100%). Each experiment was repeated three times, and each concentration was tested in three replicates.

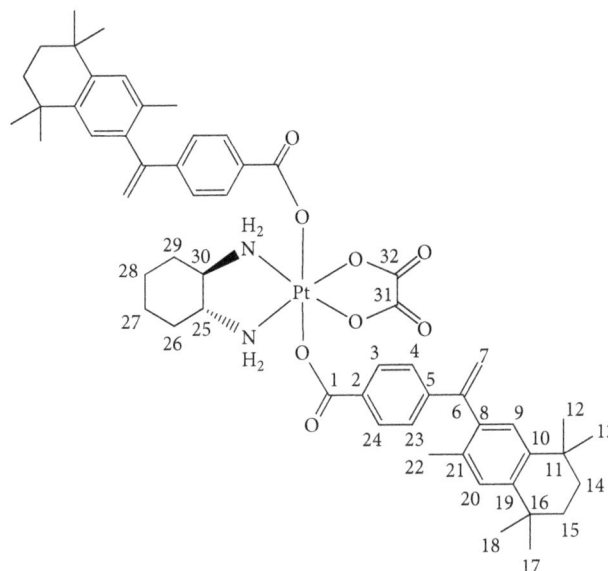

SCHEME 1

2.4. Synthesis

2.4.1. (OC-6-33)-(trans-1R,2R-Diaminocyclohexane)-bis(4-(1-(3,5,5,8,8-pentamethyl-5,6,7,8-tetrahydro-2-naphthyl)vinyl)benzoato)oxalatoplatinum(IV) 3. In Scheme 1, oxalyl chloride (2.56 mL, 30.0 mmol) followed by one-two drops of DMF was added to a stirred suspension of 4-(1-(3,5,5,8,8-pentamethyl-5,6,7,8-tetrahydro-2-naphthyl)vinyl)benzoic acid (413 mg, 1.19 mmol) in CH_2Cl_2 (60 mL). The reaction mixture was refluxed for 1 h until a clear solution was formed and then solvent and unreacted oxalyl chloride were removed under reduced pressure to yield the acid chloride as a pale yellow solid that was used without purification. A solution of 4-(1-(3,5,5,8,8-pentamethyl-5,6,7,8-tetrahydro-2-naphthyl)vinyl)benzoyl chloride in acetone (40 mL) was added to a stirred suspension of (OC-6-33)-(trans-1R,2R-diaminocyclohexane)dihydroxido(oxalato)platinum(IV) (100 mg, 0.23 mmol) and pyridine (193 μl, 2.4 mmol) in acetone (30 mL). The reaction mixture was stirred at room temperature for 12 h and concentrated to ~3 mL and the white precipitate formed was filtered off and washed with diethyl ether (2 × 5 mL). The compound was purified by column chromatography on Silicagel with acetone as eluent. The solvent was removed under reduced pressure; the compound was washed with diethyl ether (2 × 5 mL) and dried. Yield: 135 mg (53%), m.p. 213-214°C (decomp.). $C_{56}H_{68}N_2O_8Pt$ (1091.46): calcd. C 61.58, H 6.28, N 2.56; found C 61.25, H 5.89, N 2.42. ^{1}H NMR ([d6]-DMSO) δ: 8.51 (d, 2H, $J = 6.2$ Hz, NH₂), 8.23 (t, 2H, $J = 9.1$ Hz, NH₂), 7.84 (d, 4H, $J = 8.5$ Hz, H3, H24), 7.30 (d, 4H, $J = 8.5$ Hz, H4, H23), 7.15 (s, 2H, H20), 7.08 (s, 2H, H9), 5.90 (s, 2H, H7), 5.25 (s, 2H, H7), 2.82–2.74 (m, 2H, H25, H30), 2.17 (d, 2H, $J = 10.7$ Hz, H26, H29), 1.89 (s, 6H, H22), 1.66 (s, 8H, H14, H15), 1.58–1.48 (m, 4H, H26, H27, H28, H29), 1.27 (s, 12H, H17, H18), 1.25–1.18 (m, 14H, H12, H13, H27, H28) ppm. ^{13}C NMR ([d6]-DMSO) δ: 172.8 (C1), 164.5 (C31, C32), 148.8

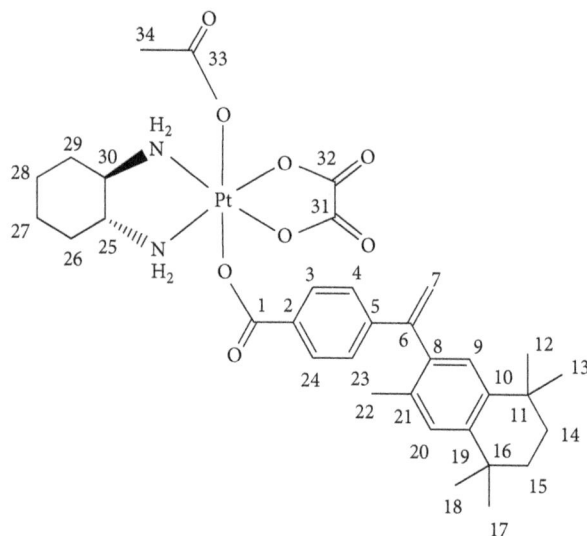

SCHEME 2

(C6), 144.2 (C19), 144.2 (C5), 142.3 (C10), 138.4 (C8), 132.5 (C21), 132.1 (C2), 130.2 (C3, C24), 128.3 (C20), 127.8 (C9), 126.4 (C4, C23), 117.4 (C7), 61.6 (C25, C30), 35.1 (C14/15), 35.1 (C14/15), 34.1 (C11/16), 34.0 (C11/16), 32.1 (C12, C13/C17, C18), 32.1 (C12, C13/C17, C18), 31.2 (C26, C29), 24.0 (C27, C28), 20.0 (C22) ppm. ^{15}N NMR ([d6]-DMSO) δ: −3.9 (NH$_2$) ppm. ^{195}Pt NMR ([d6]-DMSO) δ: 3228 ppm. ESI-MS: m/z = 1115 [M+Na$^+$]$^+$.

2.4.2. (OC-6-44)-acetato(trans-1R,2R-Diamineocyclohexane)-4-(1-(3,5,5,8,8-pentamethyl-5,6,7,8-tetrahydro-2-naphthyl)vinyl)benzoato)oxalatoplatinum(IV) 4. In Scheme 2, oxalyl chloride (1.32 mL, 15.4 mmol) and a catalytic amount of DMF were added to a suspension of 4-(1-(3,5,5,8,8-pentamethyl-5,6,7,8-tetrahydro-2-naphthyl)vinyl)benzoic acid (214 mg, 0.62 mmol) in CH$_2$Cl$_2$ (40 mL). The reaction mixture was refluxed for 1 h until a clear solution was formed. Solvent and unreacted oxalyl chloride were removed under reduced pressure to yield the acid chloride as a pale yellow solid that was used without purification. A solution of 4-(1-(3,5,5,8,8-pentamethyl-5,6,7,8-tetrahydro-2-naphthyl)vinyl)benzoyl chloride in acetone (20 mL) was added to a suspension of (OC-6-44)-acetato(trans-1R,2R-diaminocyclohexane)hydroxido(oxalato)platinum(IV) (115 mg, 0.24 mmol) and pyridine (100 μl, 1.24 mmol) in acetone (30 mL). The reaction mixture was stirred at room temperature for 12 h and a formed precipitate was filtered off, washed with diethyl ether (3 × 5 mL) and water (3 × 5 mL), and dried under reduced pressure. Yield: 70 mg (36%). m.p. 215-216°C (decomp.). C$_{34}$H$_{44}$N$_2$O$_8$Pt (803.81): calcd. C 50.80, H 5.52, N 3.49; found C 50.55, H 4.97, N 3.48. ^{1}H NMR ([d6]-DMSO) δ: 8.44 (brs, 2H, NH$_2$), 8.38 (t, 1H, J = 9.7 Hz, NH$_2$), 8.13 (t, 1H, J = 9.7 Hz, NH$_2$), 7.82 (d, 2H, J = 8.5 Hz, H3, H24), 7.28 (d, 2H, J = 8.5 Hz, H4, H23), 7.14 (s, 1H, H20), 7.07 (s, 1H, H9), 5.89 (s, 1H, H7), 5.24 (s, 1H, H7), 2.77–2.65 (m, 1H, H25/30), 2.66–2.57 (m, 1H, H25/30), 2.16–2.10 (m, 2H, H26, H29), 1.99 (s, 3H, H34) 1.88 (s, 3H, H22), 1.65 (s, 4H, H14, H15), 1.55–1.41 (m, 4H, H26, H27, H28, H29), 1.26 (s, 6H, H17, H18),

1.23 (s, 6H, H12, H13), 1.22–1.14 (m, 2H, H27, H28) ppm. ^{13}C NMR ([d6]-DMSO) δ: 179.0 (C33), 172.6 (C1), 164.2 (C31, C32), 148.8 (C6), 144.2 (C19), 144.1 (C5), 142.3 (C10), 138.4 (C8), 132.5 (C21), 132.2 (C2), 130.2 (C3, C24), 128.3 (C20), 127.8 (C9), 126.3 (C4, C23), 117.3 (C7), 61.8 (C25/30), 61.3 (C25/30), 35.1 (C14/15), 35.1 (C14/15), 34.1 (C16), 34.0 (C11), 32.1 (C12, C13/C17, C18), 32.1 (C12, C13/C17, C18), 31.3 (C26, C29), 24.0 (C27, 28), 23.5 (C34), 19.9 (C22) ppm. ^{15}N NMR ([d6]-DMSO) δ: −5.3 (NH$_2$), −5.1 (NH$_2$) ppm. ^{195}Pt NMR ([d6]-DMSO) δ: 3233 ppm. ESI-MS: m/z = 826 [M + Na$^+$]$^+$.

3. Results and Discussion

In order to obtain new Pt(IV) complexes with different number of bexarotene moieties as axial ligands, we used Pt(IV) analogs of oxaliplatin and the acid chloride of bexarotene (Scheme 3). Complexes **3** and **4** were prepared by reacting an excess of 4-(1-(3,5,5,8,8-pentamethyl-5,6,7,8-tetrahydro-2-naphthyl)vinyl)benzoyl chloride (prepared in situ from bexarotene and oxalyl chloride) with (OC-6-33)-(*trans*-1R,2R-diaminocyclohexane)dihydroxido(oxalato)platinum(IV) or (OC-6-44)-acetato(trans-1R,2R-diaminocyclohexane)hydroxido(oxalato) platinum(IV), respectively. Pyridine was used as an acceptor of HCl. Crude complexes were precipitated after concentration of the reaction mixture. Pyridine hydrochloride was removed by washing the precipitate with water to yield the pure complex.

Complexes **3** and **4** were characterized by ^{1}H, ^{13}C, ^{15}N, ^{195}Pt 1D, and 2D NMR spectroscopy, ESI mass spectrometry, and elemental analysis. In the ^{13}C{^{1}H} spectra we observed a shift of the carboxylic group that confirms the attachment of bexarotene to the platinum center. The nature of the axial ligand had only a minor influence on the ^{1}H and ^{13}C resonances in the oxaliplatin moiety [25, 36]. In ESI mass spectra the most abundant peaks were assigned to the [M+Na$^+$]$^+$ ion in the positive ion mode or to the [M−H$^+$]$^−$ ion in the negative ion mode for complex **4**, respectively; additionally minor peaks can be assigned to proton and potassium adducts. For all mass spectra the experimental isotopic patterns were in a good agreement with calculated isotopic distribution (Figure 1).

^{195}Pt NMR spectroscopy is a known method for monitoring the coordination sphere of the Pt(IV) center [34, 37]. For complexes **3** and **4**, the resonance at 3228 ppm and 3233 ppm, respectively, indicates a Pt(IV)N$_2$O$_4$ coordination sphere (Figure 2). As reported earlier, the nature of carboxylates in the axial position has no dramatic influence on the resonance in ^{195}Pt spectra [25, 36, 37].

The cytotoxicity of complexes **3** and **4**, bexarotene, and cisplatin was evaluated in the human MCF7 breast cancer cell line and its doxorubicin/cisplatin resistant subline MCF7D, as well as against colon carcinoma SW480 cells, non-small cell lung carcinoma A549, and immortalized human nonmalignant keratinocyte HaCat cell line using the colorimetric MTT-test after 72 h of incubation (Table 1). The cytotoxic potency of oxaliplatin, the Pt(II) precursor of the new complexes, was taken from literature.

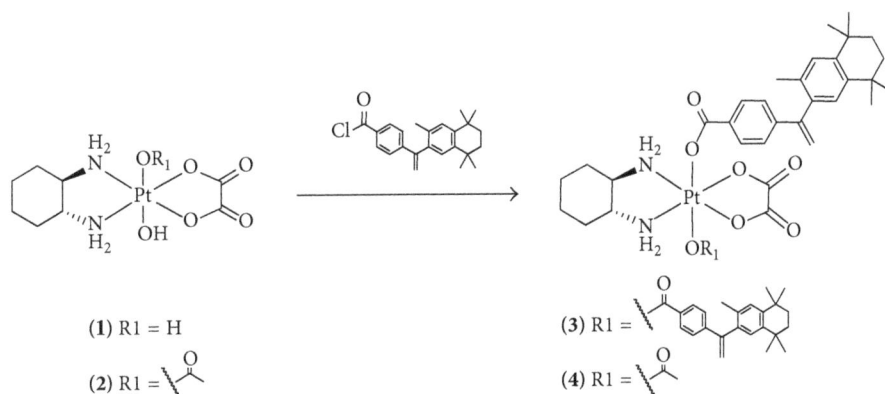

(1) R1 = H

(2) R1 = (acetyl group)

(3) R1 = (bexarotene acyl group)

(4) R1 = (acetyl group)

SCHEME 3: Synthesis of Pt(IV) complexes.

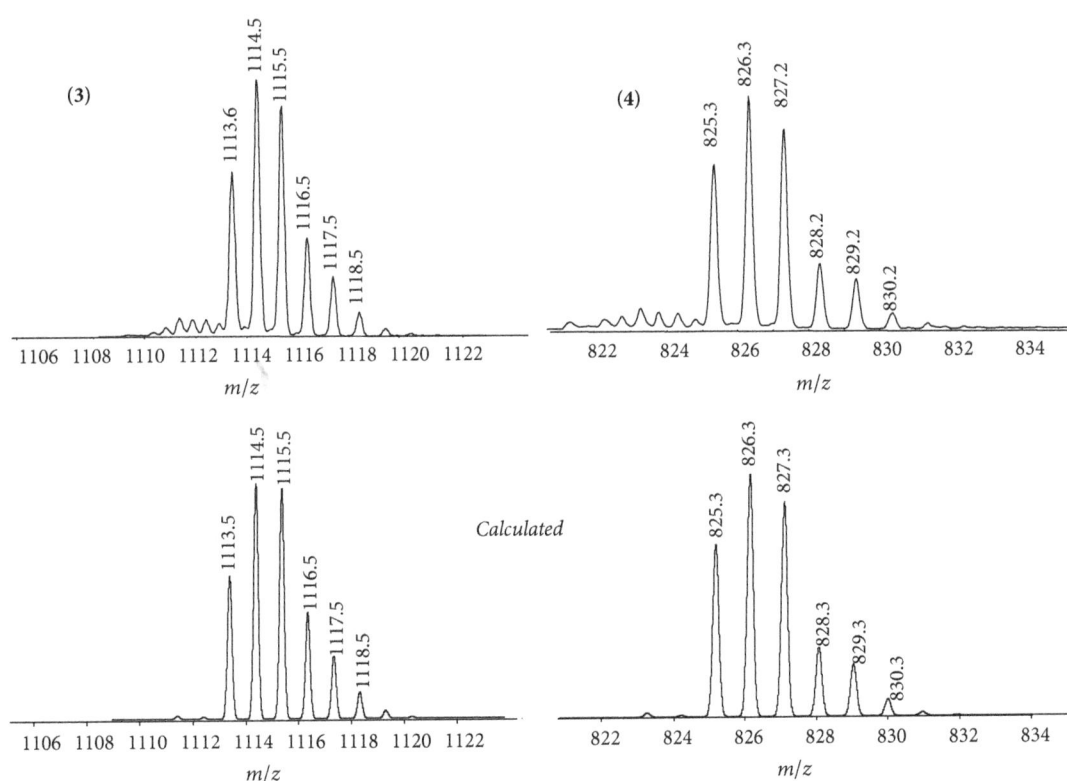

FIGURE 1: ESI-MS corresponding to $[M+Na^+]^+$ for complexes 3 and 4.

Complex 3 with two axial bexarotene ligands did not affect the viability of SW480, MCF7, MCF7D, and HaCat cells at concentrations $< 100\,\mu M$. Also, the cytotoxic effect against non-small cell lung carcinoma cell line A549 ($IC_{50} = 83 \pm 16\,\mu M$) was minor. In contrast, complex 4 with one bexarotene ligand showed a considerably higher cytotoxicity than cisplatin against SW480, HaCat, MCF7, MCF7D, and A549 cells. Complex 4 was notably more active than parent bexarotene and exhibited high sensitivity against breast cancer cells: the IC_{50} value in the MCF7 cell line was in the submicromolar range, providing a promising basis for further investigation (Table 1). Such a specificity (complex 4 is active, but 3 is not) is unexpected but rarely reported in the literature for Pt(IV) complexes with different ligands [37, 38], although

it is not a general rule. Recently we presented a similar design with lonidamine as biologically active component and no such specificity was observed [25].

4. Conclusions

A novel platinum(IV) complex featuring the oxaliplatin core and one axial bexarotene ligand exhibited high cytotoxicity against a panel of tumor cell lines. This complex is more active than cisplatin and preferential sensitivity of a breast cancer cell line to 4 compared to nonmalignant cells was found. Remarkably, complex 3 with two bexarotene ligands showed no activity in the tested cell lines. These results reveal a good

TABLE 1: Cytotoxicity of new complexes, bexarotene, cisplatin, and oxaliplatin.

Compounds	IC$_{50}$ (μM)[*]				
	SW480	A549	MCF7	MCF7D	HaCat
3	>100	83 ± 16	>100	>100	>100
4	11 ± 1.6	10 ± 1	0.47 ± 0.07	4.8 ± 0.5	8 ± 1.3
Bexarotene	80 ± 10	85 ± 9	67 ± 13	71 ± 21	>90
Cisplatin	14 ± 4.4	29.0 ± 10	14 ± 7	75 ± 5.8	30 ± 10
Oxaliplatin	0.9 ± 0.3 [34]	11.5 ± 3.9 [34]	43.8 [35]	—	—

[*]Data are mean ± standard deviation from 3 independent experiments; each drug concentration was tested in triplicate. —: no data.

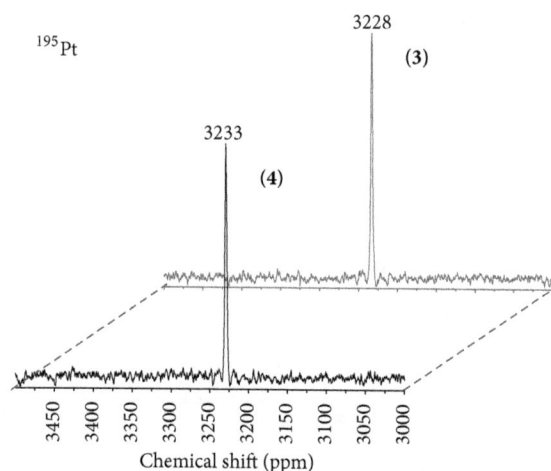

FIGURE 2: ^{195}Pt NMR complexes **3** and **4**.

potential for the use of bexarotene as ligand in the search for new metal-based anticancer compounds.

Acknowledgments

The authors thank the Russian Foundation for Basic Research (Grant no. 16-03-00743 (synthesis and characterization)) and Russian Science Foundation (Grant no. 14-13-00483 (biological study)) for financial support.

References

[1] B. Rosenberg, L. van Camp, and T. Krigas, "Inhibition of cell division in *Escherichia coli* by electrolysis products from a platinum electrode," *Nature*, vol. 205, no. 4972, pp. 698-699, 1965.

[2] M. A. Whiteside, C. J. Piyathilake, T. M. Bushell, and G. L. Johanning, "Intrinsic cisplatin resistance in lung and ovarian cancer cells propagating in medium acutely depleted of folate," *Nutrition and Cancer*, vol. 54, no. 2, pp. 274–284, 2006.

[3] Z. H. Siddik, "Cisplatin: mode of cytotoxic action and molecular basis of resistance," *Oncogene*, vol. 22, no. 47, pp. 7265–7279, 2003.

[4] A.-M. Florea and D. Büsselberg, "Cisplatin as an anti-tumor drug: cellular mechanisms of activity, drug resistance and induced side effects," *Cancers*, vol. 3, no. 1, pp. 1351–1371, 2011.

[5] A. S. Abu-Surrah and M. Kettunen, "Platinum group antitumor chemistry: design and development of new anticancer drugs complementary to cisplatin," *Current Medicinal Chemistry*, vol. 13, no. 11, pp. 1337–1357, 2006.

[6] M. J. Hannon, *Pure and Applied Chemistry*, vol. 79, no. 12, pp. 2243–2261, 2009.

[7] J. J. Wilson and S. J. Lippard, "Synthesis, characterization, and cytotoxicity of platinum(IV) carbamate complexes," *Inorganic Chemistry*, vol. 50, no. 7, pp. 3103–3115, 2011.

[8] J. J. Wilson and S. J. Lippard, "Synthetic methods for the preparation of platinum anticancer complexes," *Chemical Reviews*, vol. 114, no. 8, pp. 4470–4495, 2014.

[9] N. J. Wheate, S. Walker, G. E. Craig, and R. Oun, "The status of platinum anticancer drugs in the clinic and in clinical trials," *Dalton Transactions*, vol. 39, no. 35, pp. 8113–8127, 2010.

[10] G. Doshi, G. Sonpavde, and C. N. Sternberg, "Clinical and pharmacokinetic evaluation of satraplatin," *Expert Opinion on Drug Metabolism and Toxicology*, vol. 8, no. 1, pp. 103–111, 2012.

[11] D. Gibson, "Platinum(IV) anticancer prodrugs-hypotheses and facts," *Dalton Transactions*, vol. 45, no. 33, pp. 12983–12991, 2016.

[12] S. Dhar, Z. Liu, J. Thomale, H. Dai, and S. J. Lippard, "Targeted single-wall carbon nanotube-mediated Pt(IV) prodrug delivery using folate as a homing device," *Journal of the American Chemical Society*, vol. 130, no. 34, pp. 11467–11476, 2008.

[13] K. R. Barnes, A. Kutikov, and S. J. Lippard, "Synthesis, characterization, and cytotoxicity of a series of estrogen-tethered platinum(IV) complexes," *Chemistry and Biology*, vol. 11, no. 4, pp. 557–564, 2004.

[14] S. Mukhopadhyay, C. M. Barnés, A. Haskel, S. M. Short, K. R. Barnes, and S. J. Lippard, "Conjugated platinum(IV)-peptide complexes for targeting angiogenic tumor vasculature," *Bioconjugate Chemistry*, vol. 19, no. 1, pp. 39–49, 2008.

[15] H. A. Wee, I. Khalaila, C. S. Allardyce, L. Juillerat-Jeanneret, and P. J. Dyson, "Rational design of platinum(IV) compounds to overcome glutathione-S-transferase mediated drug resistance," *Journal of the American Chemical Society*, vol. 127, no. 5, pp. 1382-1383, 2005.

[16] J. Zajac, H. Kostrhunova, V. Novohradsky et al., "Potentiation of mitochondrial dysfunction in tumor cells by conjugates of metabolic modulator dichloroacetate with a Pt(IV) derivative of oxaliplatin," *Journal of Inorganic Biochemistry*, vol. 156, pp. 89–97, 2016.

[17] E. Wexselblatt, R. Raveendran, S. Salameh, A. Friedman-Ezra, E. Yavin, and D. Gibson, "On the stability of Pt(IV) prodrugs with haloacetato ligands in the axial positions," *Chemistry (Weinheim an der Bergstrasse, Germany)*, vol. 21, no. 7, pp. 3108–3114, 2015.

[18] T. C. Johnstone, N. Kulak, E. M. Pridgen, O. C. Farokhzad, R. Langer, and S. J. Lippard, "Nanoparticle encapsulation of mitaplatin and the effect thereof on in vivo properties," *ACS Nano*, vol. 7, no. 7, pp. 5675–5683, 2013.

[19] J. Yang, X. Sun, W. Mao, M. Sui, J. Tang, and Y. Shen, "Conjugate of Pt(IV)-histone deacetylase inhibitor as a prodrug for cancer chemotherapy," *Molecular Pharmaceutics*, vol. 9, no. 10, pp. 2793–2800, 2012.

[20] M. Alessio, I. Zanellato, I. Bonarrigo, E. Gabano, M. Ravera, and D. Osella, "Antiproliferative activity of Pt(IV)-bis(carboxylato) conjugates on malignant pleural mesothelioma cells," *Journal of Inorganic Biochemistry*, vol. 129, pp. 52–57, 2013.

[21] V. Novohradsky, L. Zerzankova, J. Stepankova et al., "Antitumor platinum(IV) derivatives of oxaliplatin with axial valproato ligands," *Journal of Inorganic Biochemistry*, vol. 140, pp. 72–79, 2014.

[22] Q. Cheng, H. Shi, H. Wang, Y. Min, J. Wang, and Y. Liu, "The ligation of aspirin to cisplatin demonstrates significant synergistic effects on tumor cells," *Chemical Communications*, vol. 50, no. 56, pp. 7427–7430, 2014.

[23] W. Neumann, B. C. Crews, L. J. Marnett, and E. Hey-Hawkins, "Conjugates of cisplatin and cyclooxygenase inhibitors as potent antitumor agents overcoming cisplatin resistance," *ChemMedChem*, vol. 9, no. 6, pp. 1150–1153, 2014.

[24] W. Neumann, B. C. Crews, M. B. Sárosi et al., "Conjugation of cisplatin analogues and cyclooxygenase inhibitors to overcome cisplatin resistance," *ChemMedChem*, vol. 10, no. 1, pp. 183–192, 2015.

[25] Y. N. Nosova, L. S. Foteeva, I. V. Zenin et al., "Enhancing the Cytotoxic Activity of Anticancer PtIV Complexes by Introduction of Lonidamine as an Axial Ligand," *European Journal of Inorganic Chemistry*, vol. 12, pp. 1785–1791, 2017.

[26] L. Ma, R. Ma, Y. Wang et al., "Chalcoplatin, a dual-targeting and p53 activator-containing anticancer platinum," *Chemical Communications*, vol. 51, no. 29, pp. 6301–6304, 2015.

[27] L. Qi, Y. Guo, P. Zhang, X. Cao, and Y. Luan, "Preventive and therapeutic effects of the retinoid x receptor agonist bexarotene on tumors," *Current Drug Metabolism*, vol. 17, no. 2, pp. 118–128, 2016.

[28] M. Duvic, A. G. Martin, Y. Kim et al., *Archives of Dermatology*, vol. 137, no. 5, pp. 581–593, 2001.

[29] W. Kempf, N. Kettelhack, M. Duvic, and G. Burg, "Topical and systemic retinoid therapy for cutaneous T-cell lymphoma," *Hematology/Oncology Clinics of North America*, vol. 17, no. 6, pp. 1405–1419, 2003.

[30] Y. N. Nosova, D. S. Karlov, S. A. Pisarev, I. A. Shutkov et al., "New highly cytotoxic organic and organometallic bexarotene derivatives," *Journal of Organometallic Chemistry*, vol. 839, pp. 91–97, 2017.

[31] G. M. Cruise, C. G. Harris, and M. J. Constant, "Polymers including active agents for occlusion of vascular sites and cavities within the body," US20150306227A1, 2015.

[32] B. Moreau, M. T. Bilodeau, K. Whalen et al., "Platinum compounds, compositions for Cancer therapy," WO2015200250A1, 2015.

[33] C. E. Wagner, P. W. Jurutka, P. A. Marshall et al., "Modeling, synthesis and biological evaluation of potential Retinoid X Receptor (RXR) selective agonists: Novel analogues of 4-[1-(3,5,5,8,8- pentamethyl-5,6,7,8-tetrahydro-2-naphthyl)ethynyl] benzoic acid (bexarotene)," *Journal of Medicinal Chemistry*, vol. 52, no. 19, pp. 5950–5966, 2009.

[34] U. Jungwirth, D. N. Xanthos, J. Gojo et al., "Anticancer activity of methyl-substituted oxaliplatin analogs," *Molecular Pharmacology*, vol. 81, no. 5, pp. 719–728, 2012.

[35] H. Xiao, D. Zhou, S. Liu et al., *Macromolecular Bioscience*, vol. 3, pp. 367–373, 2012.

[36] M. R. Reithofer, S. M. Valiahdi, M. A. Jakupec et al., "Novel di- and tetracarboxylatoplatinum(IV) complexes. Synthesis, Characterization, cytotoxic activity, and DNA platination," *Journal of Medicinal Chemistry*, vol. 50, no. 26, pp. 6692–6699, 2007.

[37] J. Banfić, A. A. Legin, M. A. Jakupec, M. Galanski, and B. K. Keppler, "Platinum(IV) complexes featuring one or two axial ferrocene bearing ligands - Synthesis, characterization, and cytotoxicity," *European Journal of Inorganic Chemistry*, vol. 3, pp. 484–492, 2014.

[38] S. G. Awuah, Y.-R. Zheng, P. M. Bruno, M. T. Hemann, and S. J. Lippard, "A Pt(IV) Pro-drug Preferentially Targets Indoleamine-2,3-dioxygenase, Providing Enhanced Ovarian Cancer Immuno-Chemotherapy," *Journal of the American Chemical Society*, vol. 137, no. 47, pp. 14854–14857, 2015.

6

Co(II) Coordination in Prokaryotic Zinc Finger Domains as Revealed by UV-Vis Spectroscopy

Valeria Sivo, Gianluca D'Abrosca, Luigi Russo, Rosa Iacovino, Paolo Vincenzo Pedone, Roberto Fattorusso, Carla Isernia, and Gaetano Malgieri

Department of Environmental, Biological and Pharmaceutical Science and Technology, University of Campania-Luigi Vanvitelli, Via Vivaldi 43, 81100 Caserta, Italy

Correspondence should be addressed to Carla Isernia; carla.isernia@unicampania.it
and Gaetano Malgieri; gaetano.malgieri@unicampania.it

Academic Editor: Spyros P. Perlepes

Co(II) electronic configuration allows its use as a spectroscopic probe in UV-Vis experiments to characterize the metal coordination sphere that is an essential component of the functional structure of zinc-binding proteins and to evaluate the metal ion affinities of these proteins. Here, exploiting the capability of the prokaryotic zinc finger to use different combinations of residues to properly coordinate the structural metal ion, we provide the UV-Vis characterization of Co(II) addition to Ros87 and its mutant Ros87_C27D which bears an unusual $CysAspHis_2$ coordination sphere. Zinc finger sites containing only one cysteine have been infrequently characterized. We show for the $CysAspHis_2$ coordination an intense d-d transition band, blue-shifted with respect to the Cys_2His_2 sphere. These data complemented by NMR and CD data demonstrate that the tetrahedral geometry of the metal site is retained also in the case of a single-cysteine coordination sphere.

1. Introduction

Metal ions in protein complexes exert many fundamental biological functions spanning from a simple structural role to direct participation in catalytic activities [1, 2]. Metalloproteins are, in fact, very abundant, and many of the biological metals have d-orbital electrons that consent them to experience different oxidation states. Moreover, transition metals allow d-orbital hybridization in complex with ligands and thus coordination of more ligands and a variety of coordination geometries [3]. In the different protein sites, metal ions can be found bound to endogenous (both backbone and side chain atoms of the polypeptide) or exogenous ligands (i.e., other molecules bound to the protein) [4, 5]. Many protein-bound metals are divalent ions, and the affinity evaluation of the protein for the metal has been the object of numerous studies [6–11]. The affinities measured in different buffers and at different pH values evidence their dependence upon the measurement conditions as well as the method used for the analysis. Affinity for a given metal ion, both native and exogenous, is certainly an essential information for

metalloproteins' complete characterization, and whatever be the used technique, it is well known that it crucially depends on the set of coordinating amino acids.

Among the metalloproteins and metal-binding domains, the zinc finger motif, characterized by the presence of a structural zinc ion, is surely the most emblematic [12–15] as it has been intensively studied for its known ubiquitous presence in the biological world (e.g., 3% of the genes of the human genome encode for zinc fingers containing proteins [16, 17]).

The zinc finger family is made up of several members that bind zinc with a different combination of cysteines and histidines. In the classical eukaryotic zinc fingers, also named "Kruppel ZF," two cysteines and two histidines bind zinc with high affinity. Four cysteine coordination sites and sites constituted by three Cys and one His can also be found to tightly bind the structural zinc ion, with this coordination being always essential for the domain folding [12–15].

The DNA-binding domain of the prokaryotic Cys_2His_2 zinc finger protein Ros (Ros87) folds in a domain that is structurally different and significantly larger than its eukaryotic counterpart. Ros87, held together by the structural zinc and by a 15-residue hydrophobic core, consists of 58 residues arranged in a $\beta\beta\beta\alpha\alpha$ topology [18]. Numerous Ros homologues (Ros/MucR family) have been identified [14,19–24], in which the coordination sphere appears to be composed of only one (the first) cysteine [19]. The second coordinating residue is usually an aspartate, indicating for this domain the possibility of a CysAspHis$_2$ coordination. The structural characterization of Ros87_C27D [25], an Ros87 mutant with an aspartate in the second coordinating position, has demonstrated that this residue surrogates the role of the second cysteine by monodentally coordinating the zinc ion; this mutation only slightly perturbs the functional structure of the domain.

The main issue when characterizing a zinc protein/peptide interaction is that Zn(II) is a $d10$ ion, spectroscopically silent. So, if a structural change accompanies the binding, folding or unfolding [8] can be followed with circular dichroism (CD) or nuclear magnetic resonance (NMR), but in general, the most diffuse procedure to evaluate the zinc ion affinities considers a fully Co(II)-loaded protein and follows the Co(II) displacement by zinc via UV-Vis spectroscopy [6,15,26–30].

Cobalt(II), being a $d7$ ion used as a probe, can substitute the native metal into both structural and catalytic metal-binding sites of the examined proteins. Co(II) and Zn(II) are nearly the same size [31] (ionic radius of 0.58 Å and 0.60 Å, resp.), and many zinc-binding sites have been shown to be metal substitutable [32–35]. In some cases, enzymes with a catalytic zinc site have been shown to have similar or even higher enzymatic activity when Co(II) substitutes native Zn(II) [3].

Upon Co(II) coordination of ligands, a splitting of the energy levels of d-orbital electrons occurs. The Co(II)-ligand system absorbs light at specific wavelengths owing to the so-called d-d transitions, that is, the excitation and relaxation of the d-orbital electrons [36]. The nature and number of coordinating ligands together with the overall coordination geometry of the system dictate the wavelengths and the intensities at which this absorption occurs [37, 38]: an intense band ($\varepsilon > 300\,M^{-1}\,cm^{-1}$) at 625 ± 50 nm is diagnostic of a tetrahedral coordination and a weak band ($\varepsilon \leq 30\,M^{-1}\,cm^{-1}$) at 525 ± 50 nm reveals an octahedral complex. An intermediate band ($50 \leq \varepsilon \leq 250\,M^{-1}\,cm^{-1}$) indicates a penta-coordination [26].

Co(II) gives absorption bands also at different wavelengths: due to the $S^- \rightarrow$ Co(II) ligand-to-metal charge transfer (LMCT), an intense absorption band in the near UV, between 316 and 340 nm, can be observed. This band is very useful as the magnitude of the extinction coefficient at 320 nm permits to infer the number of S^--Co(II) bonds as each bond contributes to ε by about 900–1200 $M^{-1}\,cm^{-1}$ [39, 40]. Summarizing, while the ε at ~320 nm can count the number of S^- involved in the coordination, the ε at ~600 nm is utilized to detect the coordination geometry and to hypothesize the nature of the other ligands [26].

Here, exploiting the capability of the prokaryotic zinc finger to use different combinations of residues to properly coordinate the structural metal ion [18, 25, 41], we describe the effect of Co(II) binding on the larger prokaryotic zinc finger domain Ros87 and on one of its mutant Ros87_C27D.

2. Materials and Methods

2.1. Protein Expression and Purification. All the proteins used were expressed and purified as previously reported [42]. Only freshly prepared samples were used in all experiments. Briefly, the $_{pet}$Ros56-142 (Ros87) and $_{pet}$Ros56-142_C82D (Ros87_C27D) proteins were produced as follows: ^{15}N labeling for NMR experiments was achieved by growing the cells at 37°C in a modified minimal medium containing $^{15}NH_4Cl$ as the sole nitrogen source, while for UV-Vis and circular dichroism experiments, the proteins were expressed in LB medium. In both cases, the protein expression was induced for ~2.0 h with 1.0 mM IPTG.

The cells were then harvested, suspended in 20 mM Na_2HPO_4 (pH 6.8) buffer, and lysed by sonication. The crude cell extracts were purified by centrifugation, and the supernatant was applied to a Mono S HR 5/5 cation exchange chromatography column (Amersham Biosciences). The pooled fractions containing the proteins were applied to a HiLoad 26/60 Superdex 75 (Amersham Biosciences) gel filtration chromatography column.

2.2. UV-Vis Spectroscopy. The native zinc ion was removed obtaining apoRos87 and apoRos87_C27D by acidifying to pH 2.5 the protein solutions in the presence of 150 μM TCEP using HCl 0.1 M and dialyzing against 10 mM Tris, 150 μM TCEP, pH 2.5. The pH was finally readjusted to 6.5, and it has been strictly controlled throughout the experiments. UV-Vis spectra for the Co(II) addition experiments to Ros87 and to apoRos87_C27D were recorded in 10 mM Tris, 20 μM TCEP, pH 6.5, on a Shimadzu UV-1800 spectrophotometer in the range of 200–800 nm at room temperature. The apoprotein solution (4 μM in the case of Ros87 and 3 μM in the case of Ros87_C27D) has been titrated with aliquots corresponding each to an increase of 0.4 μM of final Co(II) concentration in solution for each step. 0.1 mM CoCl$_2$ solution was used up to 1.6 Co(II)/protein ratio. Each experiment has been repeated at least three times obtaining comparable results. Protein concentrations were obtained using absorption at 280 nm at pH 2.5.

2.3. NMR Spectroscopy. NMR samples contained 150 μM of proteins in 10 mM Tris and 150 μM TCEP at pH 6.5 in the presence of 1.4 equivalents of CoCl$_2$ and 90% H_2O/10% ^2H_2O. All the HSQC spectra were recorded at 298 K on a Bruker Avance III HD 600 MHz equipped with cryoprobe at the Department of Environmental, Biological and Pharmaceutical Science and Technology, University of Campania-Luigi Vanvitelli (Caserta, Italy). ^1H and ^{15}N chemical shifts were calibrated indirectly by using TMS as external references. All NMR spectroscopy data were processed with the TopSpin 3.5 software (Bruker) and analyzed by using the computer-aided resonance assignment [43] (CARA) software (downloaded from cara.nmr.ch).

2.4. Circular Dichroism. Circular dichroism experiments were collected using a JASCO J-815 CD spectropolarimeter

Ros87 : AVNVEKQKPAVSVRKSVQDDHIVCLECGGSFKSLKRHLTTHHSMTPEEYREKWDLPVDYPMVAPAYAEARSRLAKEMGLGQRRKANR

(a)

(b)

(c)

(d)

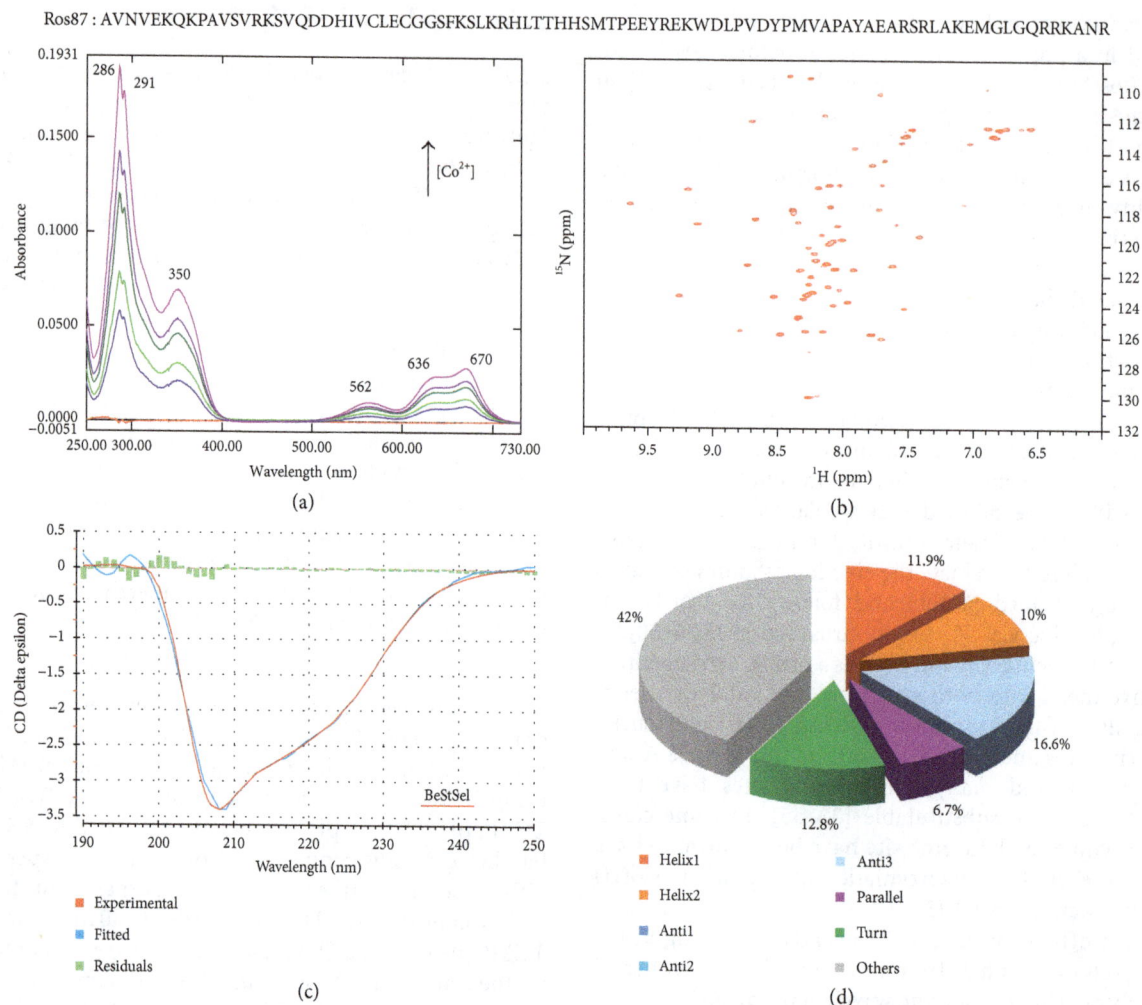

FIGURE 1: (Top) Ros87 amino acid sequence; (a) UV-Vis spectra of Ros87 titration with CoCl$_2$; (b) the ^{1}H-^{15}N HSQC spectrum of Ros87 in the presence of 1.4 equivalents of Co(II); (c) the experimental CD spectrum of Ros87 (red) overlaid to the fitted CD data (blue) by the server BeStSel; the green histogram indicates the deviations; (d) secondary structure content calculated from the CD data by the server BeStSel.

equipped with Peltier temperature control. Data were collected in the 200–260 nm wavelength range using a quartz cuvette with a 1 cm pathlength, with a data pitch of 1 nm, a band width of 1 nm, and a scanning speed of 50 nm/min. All CD samples contained ~15 μM of proteins in 10 mM Tris and 150 μM TCEP at pH 6.5. A fresh solution of CoCl$_2$ 5.0 mM has been used to reach a final [Co^{2+}]/[protein] ratio of 1.4. All the spectra were acquired in duplicates and were subtracted from the buffer contribution. Spectra deconvolution has been performed using the server BeStSel [44].

3. Results and Discussion

The UV-Vis spectra of the titration of apo-Ros87 (i.e., the unfolded prokaryotic zinc finger Ros87 with no native Zn(II) bound) and apo-Ros87_C27D (i.e., Ros87 with the second coordinating cysteine mutated in aspartate) with CoCl$_2$ are shown in Figures 1(a) and 2(a).

In the case of Co(II)-Ros87, the ε value in the near UV (at ~320 nm) that reflects the number of thiolate groups coordinated is 1950 M^{-1} cm^{-1} at 350 nm, indicating that the protein uses two thiol groups to coordinate with Co(II) ion. On the other hand, the ε value for Co(II)-Ros87_C27D is 1020 M^{-1} cm^{-1} at 345 nm, indicating the involvement of one thiol group in Co(II) coordination. In both cases, the lack of changes in the shape of the spectrum and in the wavelength of the transition during the titration permits to exclude the formation of complexes with different protein/Co(II) ratios (i.e., 2 : 1, 3 : 1, or more) formed at low Co(II) concentrations [28]. This UV-Vis behaviour was previously independently seen on the same proteins in HEPES buffer [25].

Intense absorption bands around 589–670 nm are also observed for both proteins. These results indicate that Ros87 coordinates the Co(II) with a tetrahedral geometry. Also, Co(II)-Ros87_C27D exhibits an intense d-d absorption band centered at about 589 nm with the ε value of 380 M^{-1} cm^{-1} indicating also in this case a tetrahedral geometry.

Ros87_C27D: AVNVEKQKPAVSVRKSVQDDHIVCLEDGGSFKSLKRHLTTHHSMTPEEYREKWDLPVDYPMVAPAYAEARSRLAKEMGLGQRRKANR

(a)

(b)

(c)

(d)

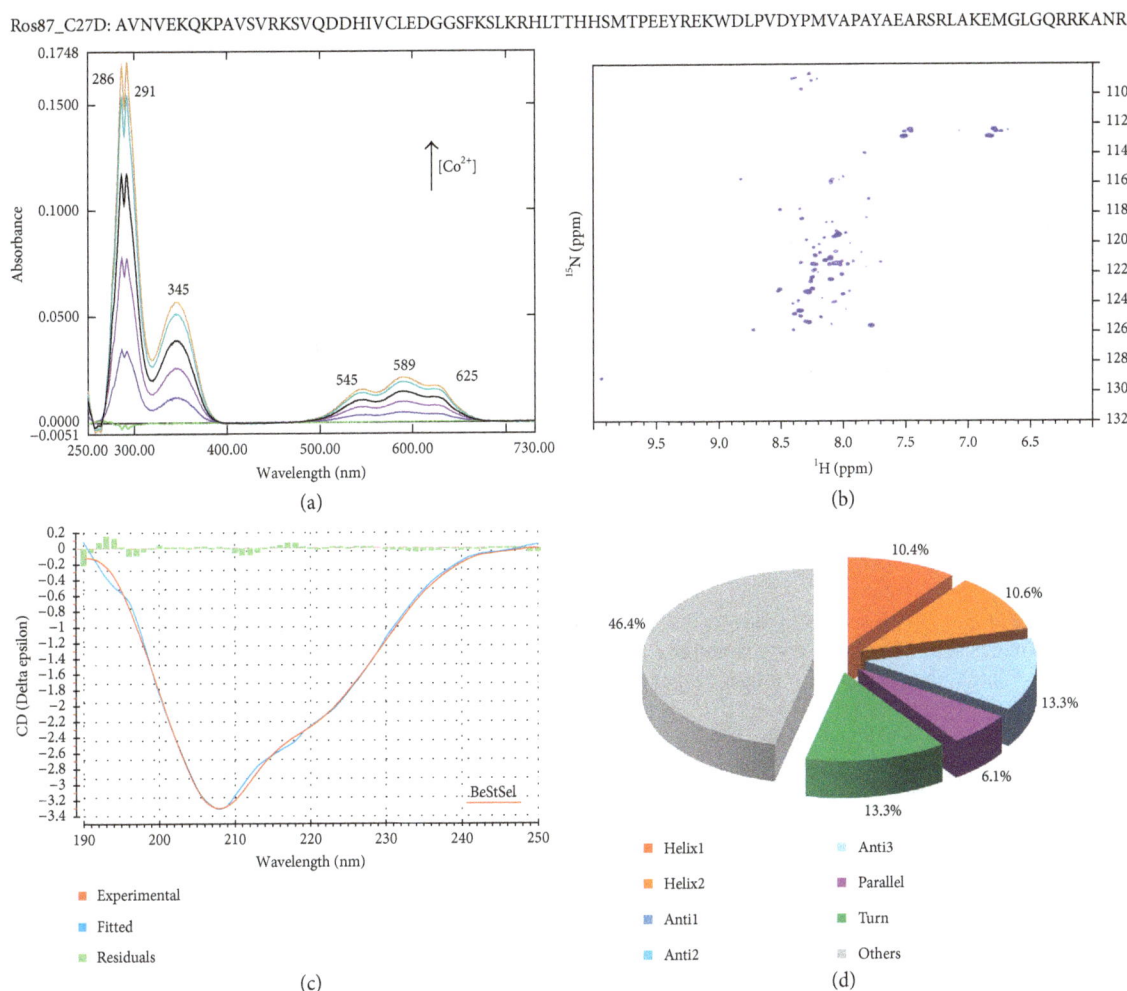

FIGURE 2: (Top) Ros87_C27D amino acid sequence; (a) UV-Vis spectra of Ros87 titration with CoCl$_2$; (b) the ^{1}H-^{15}N HSQC spectrum of Ros87 in the presence of 1.4 equivalents of Co(II); (c) the experimental CD spectrum of Ros87_C27D (red) overlaid to the fitted CD data (blue) by the server BeStSel; the green histogram indicates the deviations; (d) secondary structure content calculated from the CD data by the server BeStSel.

Accordingly, Figures 1(b) and 2(b) show the two ^{1}H-^{15}N HSQC spectra of Ros87 and Ros87_C27D, respectively, in the presence of 1.4 equivalents of Co(II) ion. Both spectra show a combination of intense and discrete signals in both proton and nitrogen dimensions indicating the interaction of Ros87 and Ros87_C27D with the paramagnetic Co(II), which gives rise in both cases to folded conformations with stable tertiary structures (Co(II)-Ros87 and Co(II)-Ros87_C27D). Importantly, the two spectra show a meaningful overlap with the holo-Ros87 spectra (data not shown) in the regions not influenced by the paramagnetism of Co(II), thus suggesting for the cobalt-loaded proteins a structure very similar to the zinc-loaded proteins.

Accordingly, the CD spectra indicate that also the secondary structure content of both proteins appears to be well conserved in the Co(II)-loaded structures with respect to the zinc-loaded conformations (Figures 1(c) and 2(c)). In fact, both CD spectra are characteristic of well-structured proteins containing both α-helical and β-sheet secondary structure. We estimated from the CD data the protein secondary structure

for the two proteins using the server BeStSel (Figures 1(d) and 2(d)). This server fits the CD experimental curve by linearly combining fixed basis components to get the percentage of the eight secondary structural elements [44]. The data indicate that Co(II)-Ros87 and Co(II)-Ros87_C27D structures have a content of secondary structure similar to that of the Ros87-calculated structure (PDB code 2JSP) and Ros87_C27D computational model [25] as determined using the software MOLMOL [45] and DSSP [46, 47].

The data reported here overall indicate for the Co(II) complexation a tetrahedral coordination geometry similar to that of the native zinc and that the replacement of the zinc ion by the Co(II) does not drastically perturb the structural properties of the prokaryotic zinc finger domain.

Interestingly, the comparison of the UV-Vis spectra of Ros87_C27D with those reported in literature for zinc fingers with Cys$_2$His$_2$, Cys$_3$His, and Cys$_4$ coordination outlines a blue shift of the d-d transition bands of the protein that uses a single cysteine to coordinate the metal ion [48] (Figure 2). This shift is in agreement with what has been

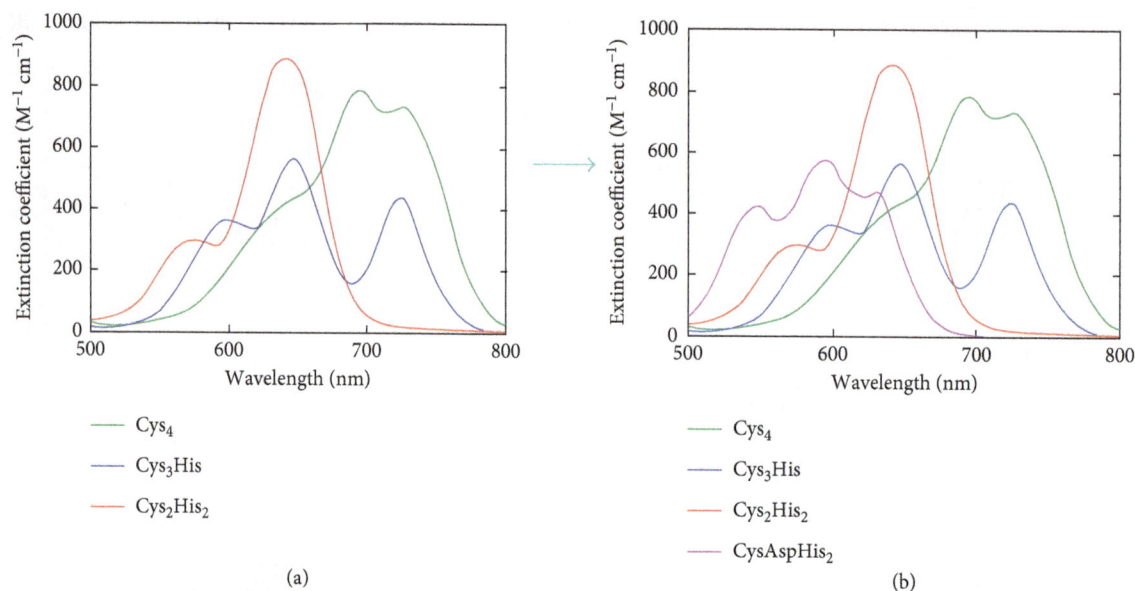

FIGURE 3: (a) Scheme of the UV-Vis spectra in the 500–800 nm range for Co(II) tetrahedral coordination by different ZFs: four Cys (green line), three Cys and one His (blue line), and two Cys and two His (red line) [37, 49]. (b) Introducing the single-cysteine coordination sphere in the scheme (violet line). The shape of the transition pattern is extremely sensitive to the structure [50].

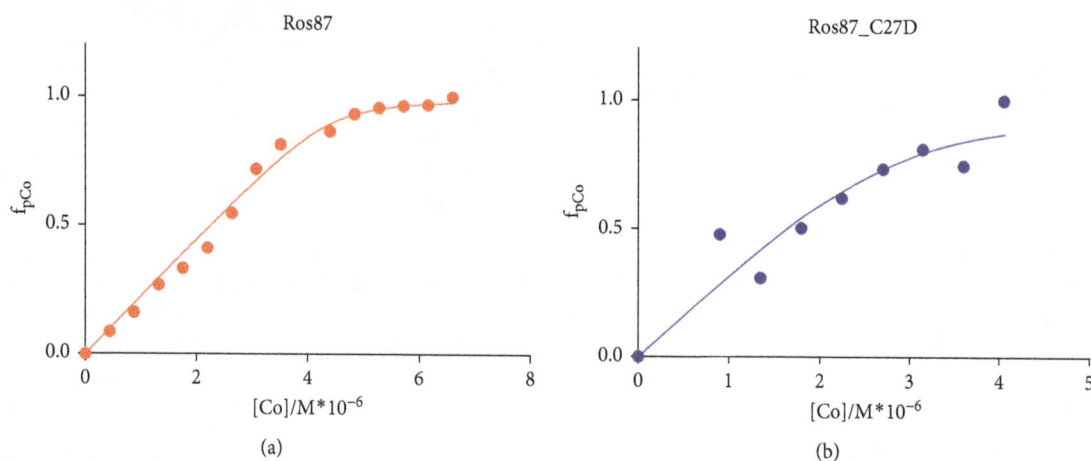

FIGURE 4: (a) Titration of Ros87 with CoCl$_2$ monitored by recording the absorbance at 352 nm. The absorbance is plotted against cobalt concentration. (b) Titration of Ros87_C27D with CoCl$_2$ monitored by recording the absorbance at 346 nm. The absorbance is plotted against cobalt concentration.

reported by Krizek et al. [10], who describe increasing shifts of the d-d transition to higher energies as the number of coordinating cysteines decreases. UV-Vis spectra of zinc finger metal sites containing only one cysteine have been rarely reported [48]. In the eukaryotic Cys$_2$His$_2$ ZF, the substitution of the second cysteine may result in some cases (i.e., the substitution with an aspartate or with a glutamate [48]) in coordination geometries different than the native tetrahedral coordination demonstrated by weak d-d absorption bands. Here, we found an intense band at 589 nm that, together with NMR and CD data, indicates a tetrahedral coordination of the metal ion with a resulting blue shift of the d-d absorption bands. We therefore propose that the scheme of the spectra of tetrahedral coordination of Co(II)

in zinc fingers with different numbers of cysteines and histidines (Figure 3(a)) [37, 49] can be implemented (Figure 3(b)) with our results.

We also determined the affinities of two proteins for Co(II), through direct titrations in Tris buffer at pH 6.5 which shows that complexes definitively form when the Co(II)/protein molar ratio was equal to 1.4. Using the 1 : 1 model to fit the UV data (Figure 4) [30], we obtain a lower limit for the β constant of 5.59 ($\pm$1.97) $\times 10^{-8}$ for Ros87 and 2.35 ($\pm$0.92) $\times 10^{-7}$ for Ros87_C27D.

The successive titration of the Co(II)-loaded proteins with Zn(II) induces a progressive reduction of both bands; the disappearance upon addition of a twofold excess of Zn(II) ion compared with Co(II) indicates that a Co(II) ion was substituted with the spectroscopically inert Zn(II) ion.

4. Conclusions

In this article, we report the spectroscopic and structural characterization of the Co(II)-substituted forms of the prokaryotic zinc finger Ros87, a native zinc protein, and of its mutant Ros87_C27D in which the second coordinating cysteine is mutated to aspartate. UV-Vis spectra of zinc finger sites containing only one cysteine, neither regarding zinc, nor other metals of interest, have been rarely reported [48]. In the case of the eukaryotic Cys_2His_2 zinc finger, the substitution of the second coordinating cysteine may result in some cases (i.e., the substitution with an aspartate or with a glutamic acid [48]) in a coordination geometry different than the native tetrahedral coordination demonstrated by weak d-d absorption bands; when a histidine substitutes the cysteine, the coordination remains tetrahedral. Here, we show that, in the prokaryotic domain, the substitution of the native zinc with cobalt mutation does not profoundly affect the structure of the domain and that the substitution of the second ligand amino acid with an aspartate gives rise to an intense band at 589 nm that indicates how this substitution does not markedly change the tetrahedral coordination geometry of the metal ion. We also show how the presence of a single cysteine in the coordination sphere of the protein implies strong d-d absorption bands in the UV-Vis spectra, blue-shifted with respect to the two cysteines coordination.

Differently from what happens for the small eukaryotic domain, our data outline how in the case of larger proteins like Ros87_C27D, other elements composing the structure (e.g., large hydrophobic cores) play a determinant role in determining the geometry of the coordination sphere. Overall, the UV-Vis spectroscopy confirms to be an excellent and extremely sensitive tool to determine the number and geometry of ligands in structural metal sites.

Acknowledgments

Financial support was provided by Ministero dell'Istruzione, dell'Università e della Ricerca Grant no. 20157WZM8A.

References

[1] A. L. Lehninger, D. L. Nelson, and M. M. Cox, *Principles of Biochemistry: With an Extended Discussion of Oxygen-Binding Proteins*, Worth Publisher, New York, NY, USA, 1993.

[2] G. Malgieri and G. Grasso, "The clearance of misfolded proteins in neurodegenerative diseases by zinc metalloproteases: an inorganic perspective," *Coordination Chemistry Reviews*, vol. 260, pp. 139–155, 2014.

[3] S. J. Lippard and J. M. Berg, *Principles of Bioinorganic Chemistry*, University Science Books, Mill Valley, CA, USA, 1994.

[4] R. H. Holm, P. Kennepohl, and E. I. Solomon, "Structural and functional aspects of metal sites in biology," *Chemical Reviews*, vol. 96, no. 7, pp. 2239–2314, 1996.

[5] A. Travaglia, D. La Mendola, A. Magrì et al., "Zinc(II) interactions with brain-derived neurotrophic factor N-terminal peptide fragments: inorganic features and biological perspectives," *Inorganic Chemistry*, vol. 52, no. 19, pp. 11075–11083, 2013.

[6] O. Sénèque and J. M. Latour, "Coordination properties of zinc finger peptides revisited: ligand competition studies reveal higher affinities for zinc and cobalt," *Journal of the American Chemical Society*, vol. 132, no. 50, pp. 17760–17774, 2010.

[7] T. Kochańczyk, A. Drozd, and A. Krężel, "Relationship between the architecture of zinc coordination and zinc binding affinity in proteins–insights into zinc regulation," *Metallomics*, vol. 7, no. 2, pp. 244–257, 2015.

[8] A. Miłoch and A. Krężel, "Metal binding properties of the zinc finger metallome–insights into variations in stability," *Metallomics*, vol. 6, no. 11, pp. 2015–2024, 2014.

[9] E. Kopera, T. Schwerdtle, A. Hartwig, and W. Bal, "Co(II) and Cd(II) substitute for Zn(II) in the zinc finger derived from the DNA repair protein XPA, demonstrating a variety of potential mechanisms of toxicity," *Chemical Research in Toxicology*, vol. 17, no. 11, pp. 1452–1458, 2004.

[10] B. A. Krizek, D. L. Merkle, and J. M. Berg, "Ligand variation and metal ion binding specificity in zinc finger peptides," *Inorganic Chemistry*, vol. 32, no. 6, pp. 937–940, 1993.

[11] J. C. Payne, B. W. Rous, A. L. Tenderholt, and H. A. Godwin, "Spectroscopic determination of the binding affinity of zinc to the DNA-binding domains of nuclear hormone receptors," *Biochemistry*, vol. 42, no. 48, pp. 14214–14224, 2003.

[12] J. H. Laity, B. M. Lee, and P. E. Wright, "Zinc finger proteins: new insights into structural and functional diversity," *Current Opinion in Structural Biology*, vol. 11, no. 1, pp. 39–46, 2001.

[13] S. S. Krishna, I. Majumdar, and N. V. Grishin, "Structural classification of zinc fingers: survey and summary," *Nucleic Acids Research*, vol. 31, no. 2, pp. 532–550, 2003.

[14] G. Malgieri, M. Palmieri, L. Russo, R. Fattorusso, P. V. Pedone, and C. Isernia, "The prokaryotic zinc-finger: structure, function and comparison with the eukaryotic counterpart," *Federation of European Biochemical Societies Journal*, vol. 282, no. 23, pp. 4480–4496, 2015.

[15] A. D. Frankel, J. M. Berg, and C. O. Pabo, "Metal-dependent folding of a single zinc finger from transcription factor IIIA," *Proceedings of the National Academy of Sciences*, vol. 84, no. 14, pp. 4841–4845, 1987.

[16] E. S. Lander, L. M. Linton, B. Birren et al., "Initial sequencing and analysis of the human genome," *Nature*, vol. 409, pp. 860–921, 2001.

[17] A. Klug, "The discovery of zinc fingers and their applications in gene regulation and genome manipulation," *Annual Review of Biochemistry*, vol. 79, no. 1, pp. 213–231, 2010.

[18] G. Malgieri, L. Russo, S. Esposito et al., "The prokaryotic Cys_2His_2 zinc-finger adopts a novel fold as revealed by the NMR structure of agrobacterium tumefaciens Ros DNA-binding domain," *Proceedings of the National Academy of Sciences*, vol. 104, no. 44, pp. 17341–17346, 2007.

[19] I. Baglivo, L. Russo, S. Esposito et al., "The structural role of the zinc ion can be dispensable in prokaryotic zinc-finger domains," *Proceedings of the National Academy of Sciences*, vol. 106, no. 17, pp. 6933–6938, 2009.

[20] C. Fumeaux, S. K. Radhakrishnan, S. Ardissone et al., "Cell cycle transition from S-phase to G1 in caulobacter is mediated

by ancestral virulence regulators," *Nature Communications*, vol. 5, p. 4081, 2014.

[21] G. Panis, S. R. Murray, and P. H. Viollier, "Versatility of global transcriptional regulators in alpha-Proteobacteria: from essential cell cycle control to ancillary functions," *Federation of European Microbiological Societies Microbiology Reviews*, vol. 39, no. 1, pp. 120–133, 2015.

[22] D. Moreira and F. Rodríguez-Valera, "A mitochondrial origin for eukaryotic C2H2 zinc finger regulators?" *Trends in Microbiology*, vol. 8, no. 10, pp. 448–450, 2000.

[23] E. Moreno, E. Stackebrandt, M. Dorsch, J. Wolters, M. Busch, and H. Mayer, "Brucella abortus 16S rRNA and lipid a reveal a phylogenetic relationship with members of the alpha-2 subdivision of the class proteobacteria," *Journal of Bacteriology*, vol. 172, no. 7, pp. 3569–3576, 1990.

[24] I. Baglivo, M. Palmieri, A. Rivellino et al., "Molecular strategies to replace the structural metal site in the prokaryotic zinc finger domain," *Biochimica et Biophysica Acta (BBA)–Proteins and Proteomics*, vol. 1844, no. 3, pp. 497–504, 2014.

[25] G. D'Abrosca, L. Russo, M. Palmieri et al., "The (unusual) aspartic acid in the metal coordination sphere of the prokaryotic zinc finger domain," *Journal of Inorganic Biochemistry*, vol. 161, pp. 91–98, 2016.

[26] A. Nomura and Y. Sugiura, "Contribution of individual zinc ligands to metal binding and peptide folding of zinc finger peptides," *Inorganic Chemistry*, vol. 41, no. 14, pp. 3693–3698, 2002.

[27] W. Kou, H. S. Kolla, A. Ortiz-Acevedo, D. C. Haines, M. Junker, and G. R. Dieckmann, "Modulation of zinc- and cobalt-binding affinities through changes in the stability of the zinc ribbon protein l36," *Journal of Biological Inorganic Chemistry*, vol. 10, no. 2, pp. 167–180, 2005.

[28] S. F. Michael, V. J. Kilfoil, M. H. Schmidt, B. T. Amann, and J. M. Berg, "Metal binding and folding properties of a minimalist Cys_2His_2 zinc finger peptide," *Proceedings of the National Academy of Sciences*, vol. 89, no. 11, pp. 4796–4800, 1992.

[29] M. C. Posewitz and D. E. Wilcox, "Properties of the Sp1 zinc finger 3 peptide: coordination chemistry, redox reactions, and metal binding competition with metallothionein," *Chemical Research in Toxicology*, vol. 8, no. 8, pp. 1020–1028, 1995.

[30] C. Isernia, E. Bucci, M. Leone et al., "NMR structure of the single qalggh zinc finger domain from the arabidopsis thaliana superman protein," *Chembiochem*, vol. 4, no. 2-3, pp. 171–180, 2003.

[31] I. Bertini, H. B. Gray, E. I. Stiefel, and J. S. Valentine, *Biological Inorganic Chemistry: Structure and Reactivity*, University Science Books, Sausalito, CA, USA, 2007.

[32] G. Malgieri, L. Zaccaro, M. Leone et al., "Zinc to cadmium replacement in the A. thaliana SUPERMAN Cys_2 His_2 zinc finger induces structural rearrangements of typical DNA base determinant positions," *Biopolymers*, vol. 95, no. 11, pp. 801–810, 2011.

[33] G. Malgieri, M. Palmieri, S. Esposito et al., "Zinc to cadmium replacement in the prokaryotic zinc-finger domain," *Metallomics*, vol. 6, no. 1, pp. 96–104, 2014.

[34] R. Gessmann, C. Kyvelidou, M. Papadovasilaki, and K. Petratos, "The crystal structure of cobalt-substituted pseudoazurin from *Alcaligenes faecalis*," *Biopolymers*, vol. 95, no. 3, pp. 202–207, 2011.

[35] C. J. Walsby, D. Krepkiy, D. H. Petering, and B. M. Hoffman, "Cobalt-substituted zinc finger 3 of transcription factor IIIA: interactions with cognate DNA detected by (31)P ENDOR spectroscopy," *Journal of the American Chemical Society*, vol. 125, no. 25, pp. 7502–7503, 2003.

[36] M. T. Worthington, B. T. Amann, D. Nathans, and J. M. Berg, "Metal binding properties and secondary structure of the zinc-binding domain of Nup475," *Proceedings of the National Academy of Sciences*, vol. 93, no. 24, pp. 13754–13759, 1996.

[37] Y. Shi, R. D. Beger, and J. M. Berg, "Metal binding properties of single amino acid deletion mutants of zinc finger peptides: studies using cobalt(II) as a spectroscopic probe," *Biophysical Journal*, vol. 64, no. 3, pp. 749–753, 1993.

[38] I. Bertini and C. Luchinat, "High spin cobalt(II) as a probe for the investigation of metalloproteins," *Advances in Inorganic Biochemistry*, vol. 6, pp. 71–111, 1984.

[39] S. W. May and J-Y. Kuo, "Preparation and properties of cobalt (II) rubredoxin," *Biochemistry*, vol. 17, no. 16, pp. 3333–3338, 1978.

[40] M. Vasák, J. H. Kägi, B. Holmquist, and B. L. Vallee, "Spectral studies of cobalt(II)- and nickel(II)-metallothionein," *Biochemistry*, vol. 20, no. 23, pp. 6659–6664, 1981.

[41] M. Palmieri, L. Russo, G. Malgieri et al., "Deciphering the zinc coordination properties of the prokaryotic zinc finger domain: the solution structure characterization of Ros87 H42A functional mutant," *Journal of Inorganic Biochemistry*, vol. 131, pp. 30–36, 2014.

[42] S. Esposito, I. Baglivo, G. Malgieri et al., "A novel type of zinc finger DNA binding domain in the agrobacterium tumefaciens transcriptional regulator Ros," *Biochemistry*, vol. 45, no. 34, pp. 10394–10405, 2006.

[43] C. Bartels, T. H. Xia, M. Billeter, P. Güntert, and K. Wüthrich, "The program XEASY for computer-supported NMR spectral analysis of biological macromolecules," *Journal of Biomolecular NMR*, vol. 6, no. 1, pp. 1–10, 1995.

[44] A. Micsonai, F. Wien, L. Kernya et al., "Accurate secondary structure prediction and fold recognition for circular dichroism spectroscopy," *Proceedings of the National Academy of Sciences*, vol. 112, no. 24, pp. E3095–E3103, 2015.

[45] R. Koradi, M. Billeter, and K. Wüthrich, "MOLMOL: a program for display and analysis of macromolecular structures," *Journal of Molecular Graphics*, vol. 14, pp. 51–55, 1996.

[46] W. G. Touw, C. Baakman, J. Black et al., "A series of PDB-related databanks for everyday needs," *Nucleic Acids Research*, vol. 43, pp. D364–D368, 2015.

[47] W. Kabsch and C. Sander, "Dictionary of protein secondary structure: pattern recognition of hydrogen-bonded and geometrical features," *Biopolymers*, vol. 22, no. 12, pp. 2577–2637, 1983.

[48] M. Imanishi, K. Matsumura, S. Tsuji et al., "Zn(II) binding and DNA binding properties of ligand-substituted CXHH-type zinc finger proteins," *Biochemistry*, vol. 51, no. 16, pp. 3342–3348, 2012.

[49] J. L. Michalek, A. N. Besold, and S. L. Michel, "Cysteine and histidine shuffling: mixing and matching cysteine and histidine residues in zinc finger proteins to afford different folds and function," *Dalton Transactions*, vol. 40, no. 47, pp. 12619–12632, 2011.

[50] O. Sénèque, E. Bonnet, F. L. Joumas, and J. M. Latour, "Cooperative metal binding and helical folding in model peptides of treble-clef zinc fingers," *Chemistry-A European Journal*, vol. 15, no. 19, pp. 4798–4810, 2009.

7

Cadmium Removal from Contaminated Water using Polyelectrolyte-Coated Industrial Waste Fly Ash

Fatai A. Olabemiwo,[1] **Bassam S. Tawabini,**[1] **Faheemuddin Patel,**[2] **Tajudeen A. Oyehan,**[1] **Mazen Khaled,**[3] **and Tahar Laoui**[2]

[1]*Geosciences Department, College of Petroleum & Geosciences, King Fahd University of Petroleum & Minerals (KFUPM), Dhahran 31261, Saudi Arabia*
[2]*Mechanical Engineering Department, KFUPM, Dhahran 31261, Saudi Arabia*
[3]*Chemistry Department, KFUPM, Dhahran 31261, Saudi Arabia*

Correspondence should be addressed to Bassam S. Tawabini; bassamst@kfupm.edu.sa and Tahar Laoui; tlaoui@kfupm.edu.sa

Academic Editor: Viktor Kochkodan

Fly ash (FA) is a major industrial waste generated from power stations that add extra cost for proper disposal. Recent research efforts have consequently focused on developing ways to make use of FA in environmentally sound applications. This study, therefore, investigates the potential ability of raw fly ash (RFA) and polyelectrolyte-coated fly ash (PEFA) to remove cadmium (Cd) from polluted water. Using layer-by-layer approach, functionalized fly ash was coated with 20 layers from 0.03% (v/v) of cationic poly(diallyldimethylammonium chloride) (PDADMAC) and anionic polystyrene sulfonate (PSS) solutions. Both surface morphology and chemical composition of the adsorbent (PEFA) were characterized using Field-Emission Scanning Electron Microscope (FE-SEM), X-Ray Diffraction (XRD), Fourier-Transform Infrared (FTIR), and X-Ray Fluorescence (XRF) techniques. The effects of pH, adsorbent dosage, contact time, initial contaminant concentration, and mixing rate of the adsorption of Cd were also studied in batch mode experiments. Results of the study revealed that a 4.0 g/L dosage of PEFA removed around 99% of 2.0 mg/L of Cd in 15 min at 150 rpm compared to only 27% Cd removal achieved by RFA under the same conditions. Results also showed that adsorption by PEFA followed both Langmuir and Freundlich models with correlation coefficients of 98% and 99%, respectively.

1. Introduction

In today's world, one issue of major concern is water pollution as the quality of water available for consumption greatly affects the health and wellbeing of humans and animals. Factors like industrialization, agricultural activities, urbanization, and population increase among others are likely reasons for water quality depreciation [1, 2]. The constant discharge of different pollutants such as organic compounds and heavy metals into the environment is causing growing concern to the entire world. Unlike most organic contaminants, heavy metals are mainly problematic because they accumulate in the tissues of living organisms and do not biodegrade, thereby leading to countless threats to the ecological environments and wellbeing of humans at large [3]. Majorly known heavy metals primarily consist of cadmium, chromium, mercury, lead, cobalt, nickel, and so forth; these metallic ions are toxic and pose severe effects on human health.

Cadmium (Cd) which is a deadly heavy metal of work-related and environmental worry has been recognized as a substance that is teratogenic and carcinogenic to human. The allowable limit for Cd in drinking water is set at 3.0 ppb by World Health Organization (WHO) [4]. If ingested beyond the limit, it would affect the kidney or probably damage it. Common ways via which Cd get leached to the environment include industrial processes like smelting, alloy manufacturing, and pesticide and anthropogenic activities such as improper disposal of cigarette, unused paints, fertilizers, and Ni/Cd batteries [5]. Therefore, the removal of this heavy metal from contaminated water has become a task of paramount importance. For that, numerous methods such as membrane separation, ion exchange, coagulation, softening, solvent extraction, and adsorption have been employed [6]. Some of these techniques are effective but are not widely applicable to different pollutants and also generate chemical waste. Application of these techniques relies on cadmium concentration and associated costs [7]. Mahvi & Bazrafshan (2007) applied electrocoagulation to remediate cadmium using Al electrode. Simulated wastewater of different concentrations of cadmium was filled in a tank and their removal was measured at different pH (3, 7, and 10) and at electric potential range of 20, 30, and 40 volts. Their investigation showed that initial pH was lower than the final pH value [8]. Numerous studies have used the adsorption mechanism for heavy metals removal using activated carbon owing to its very good adsorption features but with comparatively high operating cost [9, 10]. Therefore, the need to develop low-cost adsorbents for heavy metals removal from aqueous solution has greatly increased. Adsorbents such as *Setaria* grass [11], sawdust [12], zeolite [13], clay [9], biomass [14, 15], and fly ash [6, 16–19] have been used.

Fly ash (FA) is one of the major wastes from power stations that cannot be cheaply disposed of. Recent research efforts have consequently focused on developing ways to make use of FA in applications that are friendly to the environment. Apart from its limited applications in cement and concrete industries, fly ash alternative use/reuse in environmental study takes advantage of its reasonable adsorptive property for some water pollutants. Al-Khaldi et al. conducted a comparative study on Cd adsorption using activated carbon, CNT, CNF, and fly ash. They found out that, at pH 7 in 120 min with 50 mg and 150 rpm, percent removal of 95, 27, 34, and 38% was attained for fly ash, CNT, CNF, and activated carbon, respectively [7]. However, the efficiency of the FA for removing water pollutants was limited in a few previous studies and, therefore, there is a need to improve its adsorption efficiency [18]. One way to achieve this objective could be to coat the surface with polyelectrolytes which enhanced the adsorptive capacity of several adsorbent materials [18]. Literature search showed that no work has been conducted to assess the efficiency of FA to remove Cd from water after being coated with layers of polyelectrolytes which is the main aim of this work. As fly ash is cheaper compared to other adsorbent materials, any improvement in the efficiency of polyelectrolyte-coated fly ash in the removal of heavy metal ions from water gives it an advantage over other adsorbent materials.

Polyelectrolytes are charged organic polymers, which are soluble in water and are formed from monomers of different kinds. These polymers could be cationic or anionic depending on the functional ionic group. They are used in many applications such as water purification and paper production [20]. Examples of polyelectrolytes include poly(diallyldimethyl-ammonium chloride) (PDADMAC), polystyrene sulfonate (PSS), and polyethylenimine (PEI). Studies had been made to modify adsorbents with polyelectrolytes. Zhang et al. [21] modified multiwalled CNT with PDADMAC for chromium adsorption of which 32% removal was achieved at pH 6. Huang et al. [22] successfully applied silica-coated Fe_3O_4

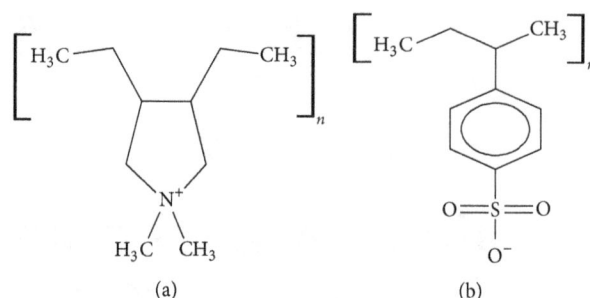

FIGURE 1: Chemical structures of (a) poly(diallyldimethylammonium chloride (PDADMAC) and (b) polystyrene sulfonate (PSS).

functionalized with c-mercaptopropyltrimethoxysilane for extraction of Cu^{2+}, Hg^{2+}, Cd^{2+}, and Pb^{2+} in a varied pH range and even in the presence of foreign ions acting as interferents such as Al^{3+}, Fe^{3+}, and Cl^-. Stanton et al. [23] showed that alternating polyelectrolyte deposition on porous supports can yield nanofiltration membranes allowing high water flux along with selective ion transport by using pairs of poly(styrene sulfonate)/poly(allylamine hydrochloride) on porous alumina.

The aim of this study is to explore the potential efficiency of fly ash to remove Cd ions from contaminated water and to evaluate the effect of acid treatment and polyelectrolyte coating of fly ash on removal efficiency. Moreover, the effects of experimental parameters such as adsorbent dose, contact time, pH, mixing rate, initial concentration, and temperature on the Cd ion removal efficiency were also deduced. Thermodynamic parameters like enthalpy, entropy, and Gibbs free energy were also investigated. The acid treatment of fly ash was done using HNO_3 and the polyelectrolyte coating of fly ash is done by layer-by-layer (LBL) deposition of two electrolytes, namely, PDADMAC and PSS, solutions on acid-treated fly ash (AFA). The Cd ion removal efficiency of adsorbents was measured using batch adsorption experiments. The sorption kinetics of Cd on the adsorbents were investigated using Langmuir and Freundlich isotherm models. Figure 1 shows the chemical structures of PSS and PDADMAC.

2. Materials and Methods

2.1. Chemicals/Stock Solution. All chemicals and solvents used were of analytical grade. PDADMAC (Mw: 200,000–350,000 kg/mol.) and PSS (Mw: 70,000 kg/mol.) were commercially acquired and used. Deionized (DI) water was generated in real time from Milli-Q Ultrapure water system (Millipore). Working standard solutions were prepared from stock Cadmium ICP Standard Solution supplied by ULTRA Scientific (USA) by serial progressive dilutions with deionized water. The prepared solutions were stirred for 30 mins with a magnetic stirrer to ensure homogeneity. The pH of the solutions was adjusted using either 0.1 M HNO_3 or 0.1 M NaOH solution. Buffer solutions were added as required in order to keep constant pH during the experiment.

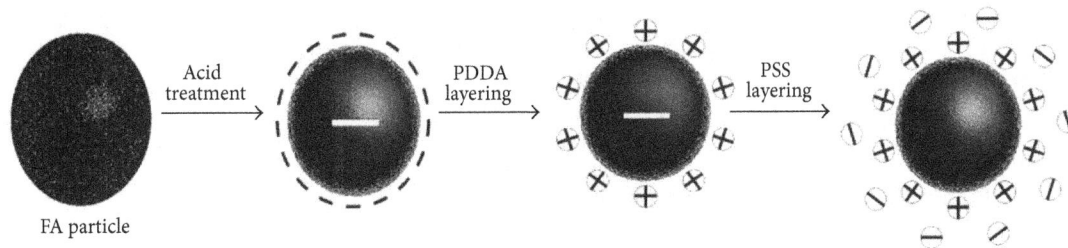

FIGURE 2: Schematic illustration of acid treatment and polyelectrolytes coating of fly ash (FA).

2.2. Adsorbent Preparation. The fly ash used in this study was obtained from a local power plant in the Eastern Province of Saudi Arabia. In this plant, raw fly ash is generated from the combustion of heavy liquid fuel and collected by electrostatic precipitation technique. This raw fly ash (RFA) was processed further to produce acid-treated fly ash (AFA) and polyelectrolyte-coated fly ash (PEFA).

2.2.1. Acid Treatment of Fly Ash. 150 g of fly ash materials was soaked in 1000 mL DI water and stirred for 2 h. After stirring, the mixture was allowed to settle for 10 min before the water was decanted and the procedure was repeated 3 times which gives a slurry phase, which was subsequently dried in the oven at 80°C temperature for 12 h and stored until used for the batch treatment experiments. 100 g of washed fly ash was soaked in 300 mL of 1 M HNO_3. The mixture was refluxed at 110°C temperature for 24 h. The acid was allowed to evaporate at 60°C, after which the reaction mixture was diluted with 500 mL DI water until the pH of the filtrate becomes neutral. The residue was then dried in the oven at 105°C for 72 h [24–29].

2.2.2. Layer-by-Layer Deposition (Polyelectrolyte Coating of Fly Ash). Polyelectrolyte coating of fly ash (PEFA) was prepared by coating the acid-treated fly ash (AFA) with polyelectrolyte (PE) using modified procedure of layer-by-layer method described by Li et al. [30]. Succinctly, the solutions used were prepared by dissolving 3 mL of poly(diallyldimethylammonium chloride) (PDADMAC) or polystyrene sulfonate (PSS) in 1000 mL of water; the solutions were stirred with a stirrer to ensure a homogenous mixture. The layering then followed Li et al.'s [30] procedure but without the addition of NaCl to ensure the formation of thinly coated PE layers as illustrated in Figure 2. The procedure was repeated until the desired number of layers was attained, that is (PDADMAC/PSS-FA)$_n$, where n could be 1, 2, 3, 4, 5, . . . , 20.

2.3. Characterization of FA Adsorbents. Characterization of adsorbents (RFA, AFA, and PEFA) surface morphology was conducted to understand elemental, mineralogical, and functional group composition. Scanning electron microscopy (SEM) micrographs were documented using FESEM (JSM-5900LV) fitted with an energy disperse X-ray spectroscopy (EDX) detector model X-max. Functional groups were determined using a Perkin-Elmer 16 FPC FTIR spectrometer with

the aid of KBr pellets and spectra were generated in the region of 600–4000 cm^{-1} wavenumber. Thermogravimetric analysis (TGA) was carried out using thermal analyzer (STA 449 F3 Jupiter) by Netzsch, Germany. The analysis was conducted in air at a distinct temperature ramped at 10°C per min to 900°C [6]. Phase analysis of the adsorbents was evaluated using D8 ADVANCE X-ray Diffractometer manufactured by BRUKER (Germany).

2.4. Batch Adsorption Studies. Batch mode adsorption studies were conducted at room temperature in 100 mL Erlenmeyer flasks covered with aluminium foil to avoid contamination. Effects of pH, contact time, adsorbent dosage, mixing rate, initial concentration, and temperature were investigated. Analysis of initial and final concentration of Cd ions was conducted using Optima 8000® ICP-OES Spectrometer (Perkin-Elmer, USA). The percent removal, as well as adsorption capacity of metal ions, was calculated with the following equations:

$$\% \text{ removal} = \frac{C_i - C_e}{C_i} \times 100$$

$$\text{adsorption capacity}, q_e \, (\text{mg/g}) = \frac{C_i - C_e}{M_s} \times V, \tag{1}$$

where C_i is the metal ion initial concentration in solution (mg/L), C_e is the final concentration of adsorbate ion in solution (mg/L), V is the total volume of solution (L), and M_s is adsorbent dosage.

Mean values of 5 replicates were used for data analysis to ensure reproducibility; relative standard deviation (RSD) was in the range of ±3–5%. The precision of the standard solution for analysis was better than 3%.

2.5. Adsorption Isotherm Model. The descriptions of adsorption behaviors are usually provided by mathematical models known as the adsorption isotherm models [6]. The distribution of adsorbate molecules between the liquid phase and a solid phase at equilibrium state can be indicated by the adsorption isotherm [25]. In this study, Langmuir and Freundlich isotherm models were employed to assess the adsorption behavior of polyelectrolyte-coated fly ash (PEFA) for Cd ion removal in an aqueous medium. Langmuir isotherm model explains the monolayer adsorption, suggesting that adsorbent materials have finite capacity, considered as the equilibrium state beyond which no further adsorption takes place [31]. The existence of specific homogeneous sites within the adsorbent at which adsorption occurs is the main assumption of this model. The Freundlich isotherm model also explains the adsorptive behavior of the adsorbent material. Adsorption on a heterogeneous surface with the interaction between adsorbate molecules is the main application of this model. The Langmuir and Freundlich isotherms are expressed by the following equation:

$$Q_e = \frac{Q_{max} K_L C_e}{1 + K_L C_e}. \tag{2}$$

The above equation can be linearized to

$$\frac{1}{Q_e} = \frac{1}{Q_{max} K_L C_e} + \frac{1}{Q_{max}}. \tag{3}$$

From (2), C_e is the equilibrium of Cd concentration (mg/L); Q_e is the amount of Cd (mg) adsorbed per gram of the adsorbent at equilibrium (mg/g); Q_{max} is the theoretical maximum adsorption capacity (mg/g); and K_L is the Langmuir isotherm constant (L/mg). A linear plot of $1/Q_e$ versus $1/C_e$ can be used to obtain the values of Q_{max} and K_L from slope and intercept, respectively.

$$Q_e = K_f C_e^{1/n}. \tag{4}$$

The above equation can be linearized to

$$\ln Q_e = \ln K_f + \frac{1}{n} \ln C_e. \tag{5}$$

From the equation above, K_f is the Freundlich adsorption constant related to the adsorption capacity [(mg/g) (L/mg)], while the remaining parameters (Q_e and C_e) were described above. A linear plot of $\ln Q_e$ versus $\ln C_e$ be used to obtain the values of K_f and n from intercept and slope, respectively.

2.6. Kinetic Modelling Studies. The adsorption of Cd (II) was analyzed using different kinetic models like pseudo-first-order model, pseudo-second-order model, and Weber intraparticle diffusion expressed in the following equations:

$$\log \frac{(q_e - q_t)}{q_e} = \frac{-K_L t}{2.303} \tag{6}$$

$$\frac{1}{(q_e - q_t)} = \frac{1}{q_e + Kt} \tag{7}$$

$$\frac{t}{q_t} = \frac{1}{(2K_s q_e^2)} + \frac{t}{q_e} \tag{8}$$

$$q_t = K_{id} t^{1/2} + C. \tag{9}$$

In the equations above, q_e and q_t are amounts of Cd adsorbed (mg/g) at equilibrium and at a given time, t (min), respectively. K_L is the pseudo-first-order rate constant sorption (min^{-1}). K_s and K are pseudo-second-order and second-order adsorption rate constants (g·mg^{-1}·min^{-1}). K_{id}, $t^{1/2}$, and C are intraparticle diffusion rate constant (mg/g·min^{-1}),

square root of time (min)$^{1/2}$, and intercept, respectively. The constants (K_L, K_s, and K) can be determined from the slopes of linear plots of $\log(q_e - q_t)$ against t, t/q_t against t, and $1/(q_e - q_t)$ against t, where q_e can be determined from the intercept data of pseudo-second-order and second-order rate equations.

3. Results and Discussions

3.1. Material Characterization

3.1.1. Surface Morphology. Surface morphology of raw fly ash (RFA), acid-treated FA (AFA), and polyelectrolyte-coated FA (PEFA) was examined with the aid of Field-Emission Scanning Electron Microscopy (FESEM) and energy disperse X-ray spectrometry (EDX). The surface morphologies of the RFA, AFA, and PEFA are presented in Figures 3, 4, and 5, respectively.

Figure 3(a) shows that the RFA has sizes that range from 50 to 500 microns. Elemental composition revealed by EDX spectra in Figure 3(b) shows that carbon (C) has 72%, oxygen has 16.6%, and the remaining elements, silicon (Si), copper (Cu), vanadium (V), aluminium (Al), and sulphur (S), were found to have 0.2, 6.0, 0.7, 2.0, and 2.8% composition, respectively.

When RFA was treated with nitric acid (HNO$_3$), most of the heavy metals impurities present in the as-received raw fly ash were removed as shown in the EDX spectrum in Figure 4(b). Moreover, it was observed that more pores were visible as a result of the treatment with nitric acid as shown in Figure 4(a) compared to raw fly ash shown in Figure 3(a). The spectrum in Figure 4(b) reveals that the carbon content increased from 72 to 92% and also reveals an increase in the silicon content from 0.2 to 0.4.

Figure 5(a) shows the morphology of polyelectrolyte-coated fly ash (PEFA) along with its elemental composition in Figure 5(b). The SEM image shows that a thin pore linen was coated with PDADMAC-PSS and the EDX spectrum shows an increase in the sulphur content (2.77–5.03%) of the fly ash upon coating with polyelectrolyte which might be due to the component of the polymer that has a polystyrene sulfonate compound (i.e., PSS) in its composition.

3.1.2. Elemental Analysis by X-Ray Fluorescence (XRF) Analysis. XRF analysis was carried out to determine the actual elemental composition of the RFA, AFA, and PEFA adsorbents. The results, as summarised in Table 1, identified the presence of some trace metals like vanadium, manganese, iron, nickel, zinc, and molybdenum with their percentage compositions. It was observed that RFA has no silicon content but has high sulphur content of 51% composition which could be attributed to the fact that the fly ash used in this study is an oil fly ash, received from local power plants operating on liquid fuel. This type of fly ash is usually characterized by low silicon and aluminium contents [32, 33]. As could be inferred from the EDX spectrum, the fly ash has a high carbonaceous content which is not commonly found at that rate in coal fly ash with high silicon and aluminium contents [32, 34]. After treatment with acid (AFA), the fly ash trace metal content was

(a)

Full scale 1073 cts cursor: 13.366 (1 cts)

(b)

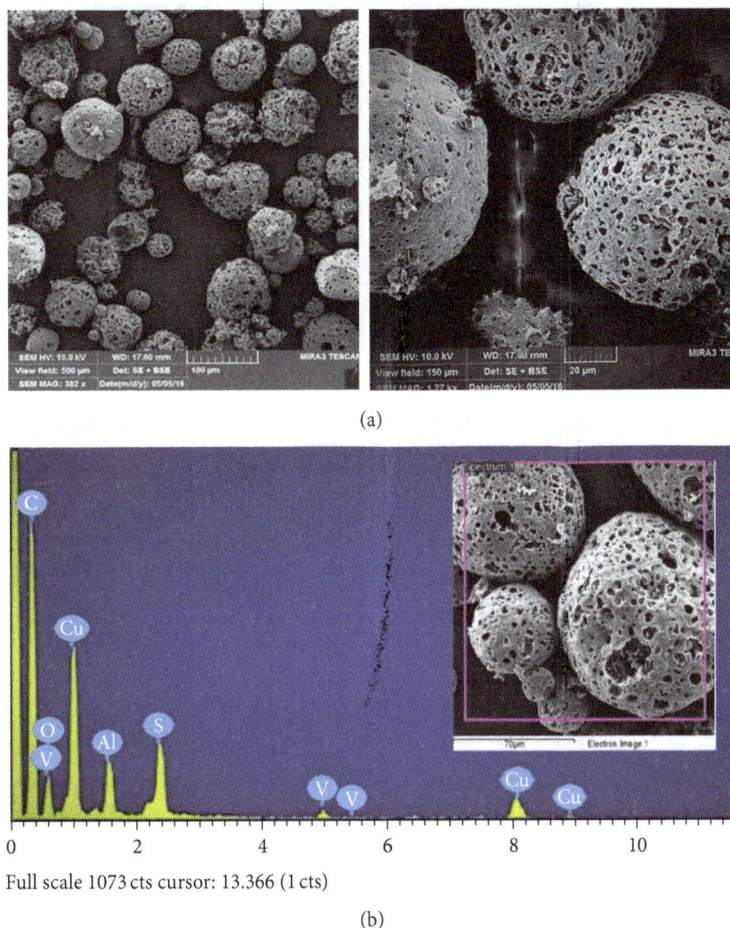

FIGURE 3: Raw FA (a) SEM micrographs (view field: 500 and 150 μm; voltage: 10 kV; resolution: 382x and 1.27 kx); (b) EDX spectrum.

reduced to a nonsignificant value, whereas sulphur content increased from 51% to 86% as shown in Table 1. In PEFA, trace metals were either absent or not present in detectable quantity but sulphur content increased to 92.5% which might be a result of additional sulphate group present in the polymer used for coating.

3.1.3. FTIR (Fourier-Transform Infrared) Spectroscopy Analysis. FTIR technique was used to ascertain the functional groups present in RFA, AFA, and PEFA surface. The samples were scanned from 500 to 4000 cm^{-1} and the intensity of peaks in the IR spectra was observed. Figure 6 shows FTIR spectra for RFA, AFA, and PEFA. The raw fly ash shows a mildly prominent peak at 604 cm^{-1} as a result of the naturally occurring C-S bond [35]. There was a prominent peak at 1367 cm^{-1} as a result of skeletal vibration of a C-C bond [26]. A peak was observed at 1628 cm^{-1}, which indicates the presence of C=C functional group of an alkene [28, 36]. A sharp peak at 1711 cm^{-1} represents C=O in ester group as noted by Shawabkeh et al. [27]. A broad trough was observed at 3436 cm^{-1} as a result of O-H stretching of alcoholic groups [27, 37]. However, after treatment with an acid (HNO_3), peaks were only seen at lower and higher region of the spectra; this might be a result of the bond breaking due to reactions

TABLE 1: Elemental composition of RFA, AFA, and PEFA as revealed by XRF.

Atomic number	Elements	RFA	AFA	PEFA
14	Silicon (Si)	0	0.76	0.00
15	Phosphorus (P)	1.09	0.66	1.18
16	Sulphur (S)	51.44	86.25	92.50
20	Calcium (Ca)	1.91	2.16	2.23
23	Vanadium (V)	20.22	5.10	1.09
25	Manganese (Mn)	0.1	0.00	0.00
26	Iron (Fe)	11.34	2.10	1.29
28	Nickel (Ni)	13.46	2.95	1.70
30	Zinc (Zn)	0.42	0.00	0.00
42	Molybdenum (Mo)	0.02	0.006	0.005
	Loss on Ignition (LOI)	0.009	0.014	0.005
	Total	*100*	*100*	*100*

between the acid and fly ash particles. After coating the AFA with polyelectrolytes (PDADMAC and PSS), a more prominent and sharp peak was observed at 607 cm^{-1} of PEFA which is evident of the presence of more C-S functional groups. The peak of C=C reappeared at 1635 cm^{-1} [29,

(a)

Full scale 585 cts cursor: −0.088 (7 cts)

(b)

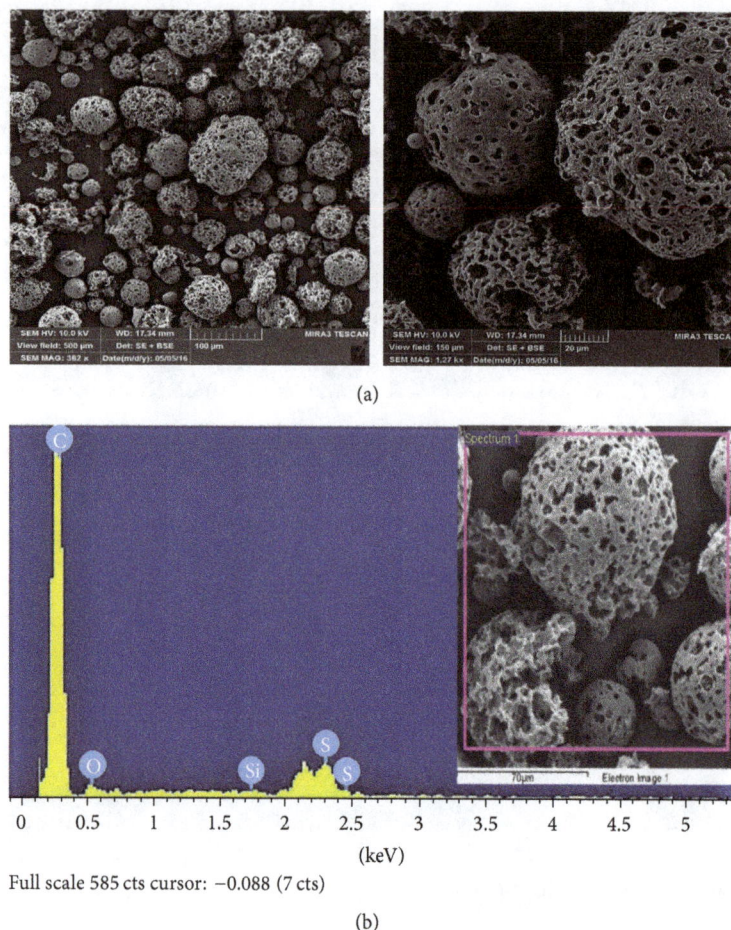

FIGURE 4: Acid-treated FA (a) SEM micrographs (view field: 500 and 150 μm; voltage: 10 kV; resolution: 382x and 1.27 kx); (b) EDX spectrum.

38]. Additionally, there was C-N peak at 2386 cm^{-1}, which indicates the presence of the polyelectrolytes PDADAMAC on the fly ash [39, 40]. The peaks observed at 3442 and 3451 cm^{-1} of the RFA and AFA spectra, respectively, were also observed at 3454 cm^{-1} in PEFA indicating the presence of carboxylic acid O-H functional groups.

3.1.4. Phase Identification by X-Ray Diffraction (XRD). The mineralogical compositions of RFA, AFA, and PEFA were studied using X-ray diffractometer. Overall, the XRD spectra shown in Figure 7 show the presence of carbon, sulphur, α-quartz (low quartz content), β-quartz (high quartz content), and cristobalite. A prominent peak was observed at 21.6° 2θ; this confirms the presence of highly concentrated carbon. The amorphous phase between 22° and 28° 2θ contains sulphur and quartz, respectively. A small peak of cristobalite at 31.4° 2θ was observed. Carbon was very prominent through the prepared adsorbents (RFA, AFA, and PEFA). One significant observation was the presence of β-quartz at 27.4° 2θ in both AFA and PEFA, respectively. No significant peaks were observed after 40° 2θ, indicating the presence of amorphous carbon. The XRD pattern of this material can be attributed to that of carbon black or oil fly ash which are both amorphous. Also, the crystalline structure of oil fly ash is known to consist

of carbon and metallic sulphur in the amorphous state [41]. Hence, the fly ash used to prepare PEFA in this study can be referred to as oil fly ash as confirmed from the XRD spectrum.

3.1.5. Thermogravimetric Analysis (TGA). Thermogravimetric analysis was performed to measure the thermal stability and purity of adsorbents. Figure 8 displays the thermograms of RFA, AFA, and PEFA. All samples analyzed exhibit similar curves and do not contain adsorbed water. Due to volatilization/decomposition of organic or inorganic substances, a 2% weight loss was observed between 100 and 470°C in RFA. Dramatic weight losses of 84% for RFA between 470 and 600°C, 80% for AFA at temperature range of 580–630°C, and 97% for PEFA between 580 and 670°C can be attributed to the phenomenon of gas generation (CO_2 and CO) upon pyrolysis [42]. Among the three adsorbents, AFA seems to be more thermally stable than the rest with a residual of approximately 5%. Other samples burn off almost completely before the maximum set temperature of 900°C.

3.2. Removal of Cadmium

3.2.1. Effect of pH. Generally, metal adsorption consists of a multifaceted mechanism of ion exchange, metal chelating

(a)

(keV)

Full scale 3270 cts cursor: −0.254 (0 cts)

(b)

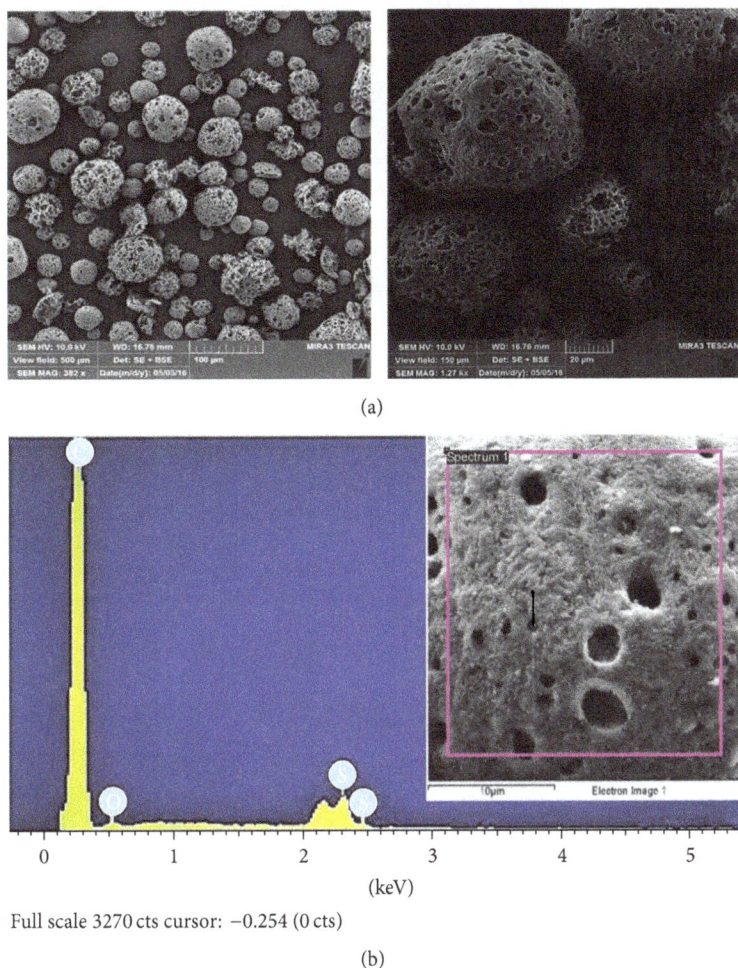

FIGURE 5: Polyelectrolyte-coated FA (a) SEM micrographs (view field: 500 and 150 μm; voltage: 10 kV; resolution: 382x and 1.27 kx); (b) EDX spectrum.

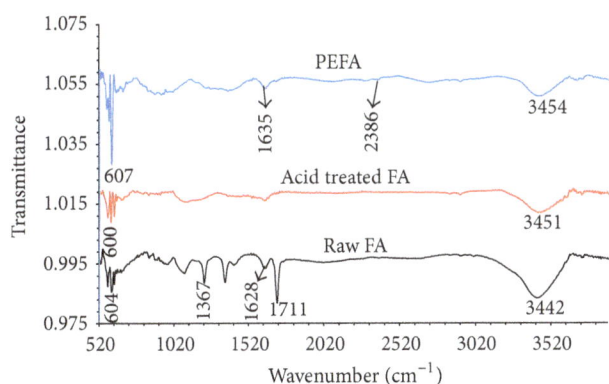

FIGURE 6: IR spectra of RFA, AFA, and PEFA.

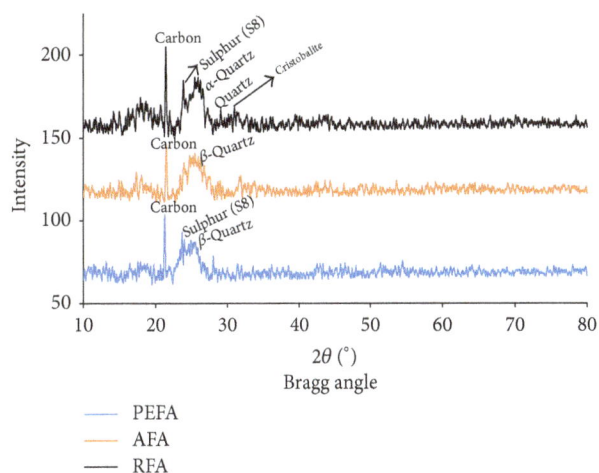

FIGURE 7: XRD spectra of RFA, AFA, and PEFA.

with numerous anionic functional groups, physical forces sorption, and trapping of ions in the interior sphere of adsorbents architectural network [9]. Different forms of Cd species occur in deionized water as Cd^{2+}, $Cd(OH)_{2(s)}$, and $Cd(OH)^{+}$ [43]. pH was a leading factor affecting Cd (II) ion removal under the investigated conditions. Nonetheless, Cd^{2+} often exists as a complex $[Cd(H_2O)_6]^{2+}$ at low pH and also as prevailing species [44]. With a specific focus on PEFA,

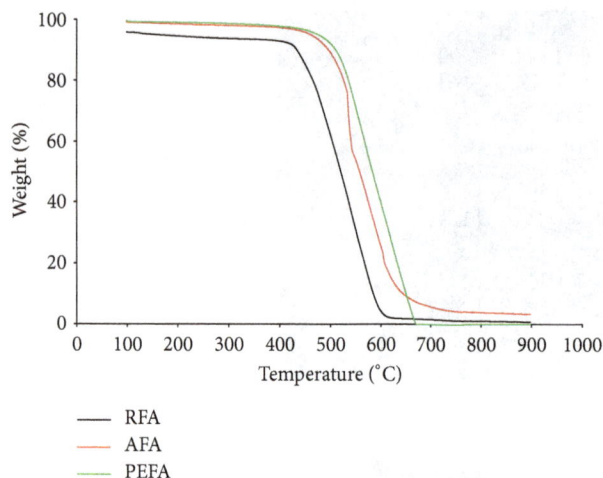

FIGURE 8: Thermogravimetric (TG) measurements of RFA, AFA, and PEFA.

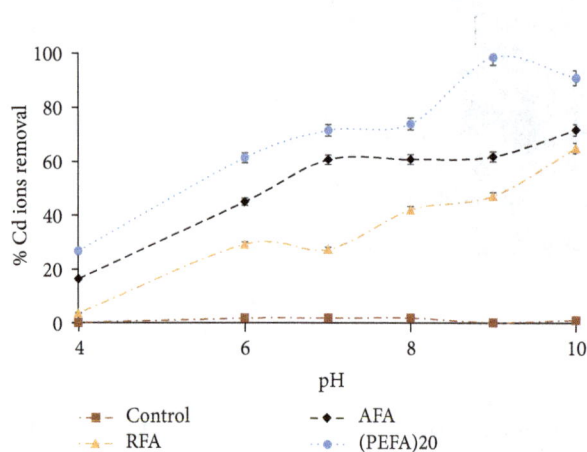

FIGURE 9: The influence of pH on the removal efficiency of Cd (II) ions on FA based adsorbents (RFA, AFA, and PEFA) as a function of 2 mg/L metal ion concentration, 4 g/L adsorbent dosage, 50 mL volume of aqueous solution, mixing rate of 150 rpm, 15 min contact time, and 298 K temperature.

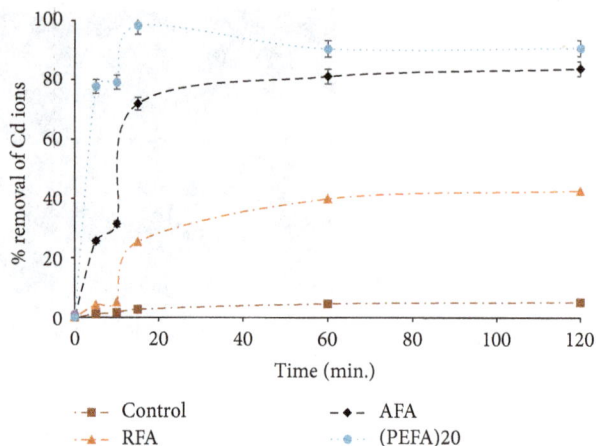

FIGURE 10: The influence of contact time on the removal efficiency of Cd (II) ions on FA based adsorbents (RFA, AFA, and PEFA) as a function of pH value of 9, 2 mg/L metal ion concentration, 4 g/L adsorbent dosage, mixing rate of 150 rpm, 50 mL volume of aqueous solution, and 298 K temperature.

ions in the internal sphere of the structural arrangement of the adsorbents [45].

3.2.2. Effect of Contact Time. Contact time is the time required for equilibrium to be attained in the process of adsorption when no substantial variations are detected in adsorptive concentration after a definite period of time [38]. It hinges on the surface features of the adsorbent in question. To find the optimum contact time for Cd (II) ions uptake, varying contact times from 5 min to 2 h were studied from aqueous solutions of 2 mg/L Cd (II) ions concentration, adsorbent mass of 4 g/L, pH value of 9, mixing rate of 150 rpm, and 298 K temperature. The results obtained indicated that at first there was rapid adsorption of Cd (II) ions for PEFA with 98% removal and a gradual decrease to attain equilibrium in 2 h as shown in Figure 10. Initial fast adsorption for this adsorbent might be a result of rich active sites on the adsorbent surface which become filled up with increasing time and turn out to be saturated [9, 38, 46]. Moreover, the decline in the removal efficiency could be attributed to the presence of metal impurities (V, Mn, Fe, Ni, Mo, and Zn) as revealed by XRF in Table 1, which might have occupied the active site needed for sorption. For this study, optimum contact time was chosen to be 15 min as maximum Cd ions removal was reached at this time. Percent removal for RFA and AFA was 25 and 72%, respectively, at the chosen optimum contact time.

3.2.3. Effect of Adsorbent Dosage. The mass of adsorbent has an effect on the active site available for binding of Cd (II) ions in aqueous solution [25, 46]. In this study, batch mode experiments were conducted by applying varying quantities of RFA, AFA, and PEFA from 1 to 6 g/L at pH value of 9, 2 mg/L metal ion concentration, 150 rpm mixing rate, 15 min contact time, 50 mL volume aqueous solution, and 298 K temperature. As illustrated in Figure 11, Cd (II) ion

the adsorption of Cd (II) ions by RFA, AFA, and PEFA was investigated at pH 4–10 to fix the optimum pH removal. Figure 9 illustrates an increase in Cd (II) removal efficiency with increased pH in aqueous solution with other parameters fixed at 2 mg/L of metal ion concentration, 4 g/L of adsorbent dosage, 50 mL volume of aqueous solution, mixing rate of 150 rpm, contact time of 15 min, and temperature of 298 K. Maximum sorption of Cd ion was attained at pH 9 due to the fact that in acidic medium Cd (II) ion sorption is low as a result of available large number of hydrogen ions (H$^+$) which outcompete Cd ions for active sites. However, as the pH increases, the number of positively charged ions available for active sites reduces with a rise in negatively charged ion for binding [38]. Moreover, the sudden increase and decrease in the removal efficiency as observed in Figure 9 suggest an elaborate process of exchanging ions, sorption driven by physical forces, metal chelation, and trapping of

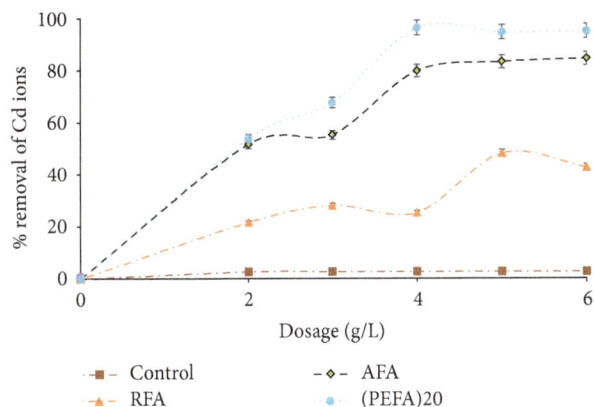

FIGURE 11: The influence of Adsorbent dose on the removal efficiency of Cd (II) ions on FA based adsorbents (RFA, AFA, and PEFA) as a function of pH value of 9, 2 mg/L metal ion concentration, 15 min contact time, mixing rate of 150 rpm, 50 mL volume of aqueous solution, and 298 K temperature.

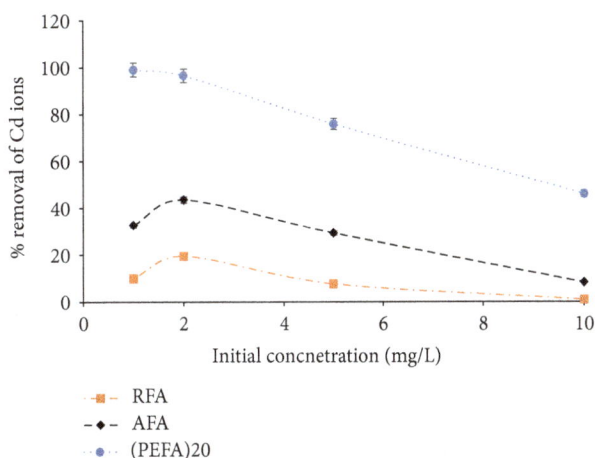

FIGURE 12: The influence of mixing rate on the removal efficiency of Cd (II) ions on FA based adsorbents (RFA, AFA, and PEFA) as a function of pH value of 9, 2 mg/L metal ion concentration, contact time of 15 min, 4 g/L adsorbent dosage, 50 mL volume of aqueous solution, and 298 K temperature.

sorption rises with an increase in dose of adsorbents up till 4 g/L and there was little or no significant adsorption for remaining dosage. Sorption increase with an increase in dose of adsorbent could be attributed to surface area increase, the rise in the exchange site ability of the ion, and an increase in active sites [18, 19, 38, 47]. PEFA reaches optimum at 4 g/L unlike RFA and AFA with 5 g/L and 6 g/L as well as removal efficiency of 48 and 84%, respectively. Incomplete adsorbent aggregation which leads to a decline in Cd ion uptake active surface area may be the reason for the drop in removal efficiency at higher concentration for RFA. 4 g/L adsorbent dose was used for other investigations.

3.2.4. Effect of Mixing Rate.
The mixing rate ensures that Cd (II) ions are transferred to the active sites by supporting the contact between Cd ions in aqueous solution and adsorbent binding sites [47]. The optimum removal of Cd (II) at pH value of 9 was used to investigate the effect of mixing rate on the adsorption of Cd (II) ion for RFA, AFA, and PEFA. Figure 12 indicates that the removal of Cd ion increases with mixing rate increase from 50 to 150 rpm. Maximum removal of over 96% was achieved for PEFA, 77% for AFA, and 27% for RFA at 150 rpm with 2 mg/L metal ion concentration, 50 mL volume of aqueous solution, 4 g/L dose of adsorbents, pH value of 9, contact time of 15 min, and 298 K temperature. Afterwards, there was no significant removal achieved above this mixing rate under similar conditions. This observation could be ascribed to improved interaction between the sorption-active sites and Cd ions in aqueous solution with an increase mixing rate [25]. The value of 150 rpm was chosen as optimum mixing rate.

3.2.5. Effect of Initial Concentrations.
Investigating the initial concentration of metal ion is essential in the sorption studies because water and wastewaters contamination does have diverse metal ion concentrations; hence, knowledge of its influence is required for an elaborate sorption

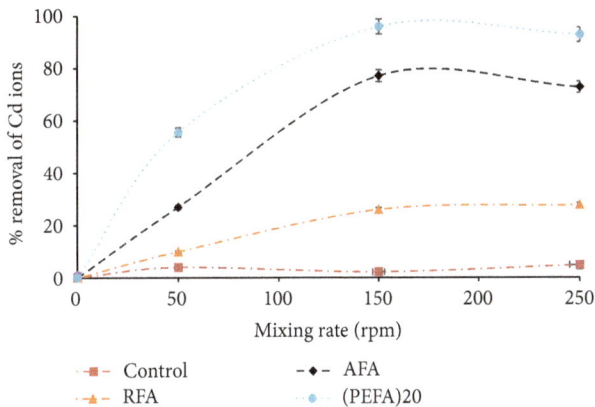

FIGURE 13: The influence of initial concentration on the removal efficiency of Cd (II) ions on FA based adsorbents (RFA, AFA, and PEFA) as a function of pH value of 9, 150 rpm mixing rate, contact time of 15 min, 4 g/L adsorbent dosage, 50 mL volume of aqueous solution, and 298 K temperature.

investigation [9]. The effect of Cd ion concentration in aqueous solution on its sorption by RFA, AFA, and PEFA was conducted with 4 g/L dose of adsorbent, pH value of 9, 150 rpm mixing rate, 15 min contact time, and 298 K temperature. Initial Cd ions concentrations investigated were varied from 1 to 10 mg/L and their effects on the removal efficiency were established. In Figure 13, it was observed that increasing the initial concentration of Cd (II) ions in solution could cause a decline in the removal efficiency of RFA, AFA, and PEFA. This can be ascribed to bulky quantities of Cd (II) ion with inadequate active sites on the surface of the adsorbents which resulted in increased concentration of Cd (II) ion in the greater part of the aqueous solution and as a result decreasing Cd ion removal efficiency [9, 25].

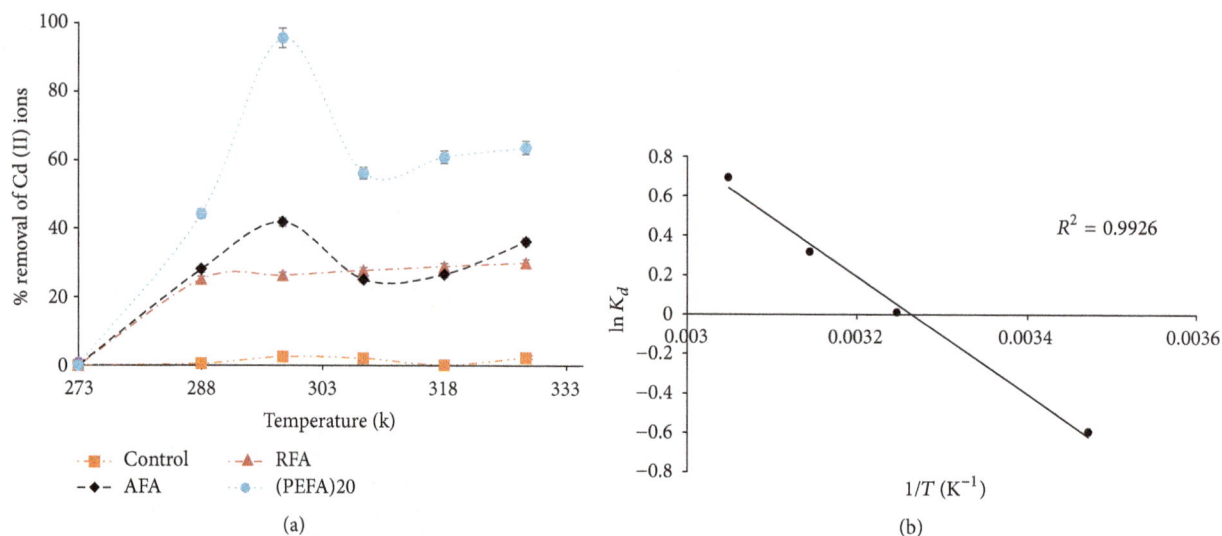

FIGURE 14: (a) The influence of temperature on the removal efficiency of Cd (II) ions on FA based adsorbents (RFA, AFA, and PEFA) as a function of pH value of 9, 2 mg/L metal ion concentration, 150 rpm mixing rate, contact time of 15 min, 4 g/L adsorbent dosage, and 50 mL volume of aqueous solution. (b) Van Hoff plot for Cd ion sorption at pH value of 9, 2 mg/L metal ion concentration, 150 rpm mixing rate, PEFA, contact time of 15 min, 4 g/L adsorbent dosage, and 50 mL volume of aqueous solution.

3.2.6. Effect of Temperature.

Figure 14(a) illustrates Cd ion sorption on RFA, AFA, and PEFA at different temperatures. It can be deduced from the graph that an initial rise in temperature brings about a sharp increase in Cd (II) sorption from 273 to 298 K [38, 48]. This observation could be attributed to the fact that more chemical sites were present as temperature rises from 288 to 298 K to surface component dissociation on PEFA. This also suggests that the adsorption mechanism of Cd (II) ion on PEFA could be chemical sorption in addition to physical sorption as observed for RFA in which sorption increases with an increase in temperature. After a drop in removal efficiency at 308 K, there was a steady increase up till 328 k, suggesting that a high temperature might be a favorable factor in the sorption process as well and indicating that the adsorption is endothermic [9, 38].

To assess the feasibility and spontaneity of sorption process, thermodynamic parameters like $\Delta G°$ (free energy change), $\Delta H°$ (enthalpy change), and $\Delta S°$ (entropy change) were determined as shown in Table 2. Gibbs free energy change of sorption was calculated from the following equation:

$$\Delta G° = -RT \ln K_d, \quad (10)$$

where R is 8.314 J/mol·K, T (K) is the absolute temperature, and K_d is the distribution coefficient expressed as $K_d = q_e/C_e$, where q_e is the amount of Cd ion adsorbed at equilibrium and C_e is the concentration of Cd ion in aqueous solution at equilibrium.

$$\ln K_d = -\frac{\Delta G°}{RT} \quad (11)$$

$$\ln K_d = -\frac{\Delta H°}{RT} + \frac{\Delta S°}{R}. \quad (12)$$

TABLE 2: Thermodynamic parameters for Cd (II) ions adsorption by polyelectrolyte-coated fly ash (PEFA).

T (K)	K_d	$\Delta G°$ (KJ/mol)	$\Delta H°$ (KJ/mol)	$\Delta S°$ (J/mol/K)
288	0.55306	1.41818	24.80814	80.98418
298	4.12617	−3.5116		
308	1.01149	−0.0293		
318	1.37544	−0.8428		
328	2.00362	−1.8951		

Equation (12) is known as the Van Hoff equation; the values of $\Delta H°$ and $\Delta S°$ were calculated from slope and intercept of the plot of $\ln K_d$ against T^{-1} (K^{-1}) as indicated in Figure 14(b).

3.3. Isotherm and Kinetic Studies

3.3.1. Langmuir and Freundlich Isotherm Models.

In order to determine the maximum sorption capacities of PEFA, data gotten at equilibrium for sorption experiment were employed. Figures 15(a) and 15(b) illustrate Langmuir and Freundlich isotherm models for Cd (II) at optimum pH (9). The maximum sorption capacity and adsorption intensity values were calculated from the slope and intercept of the plots between $1/q_e$ and $1/C_e$ for Langmuir as in q_m and K_L [31] and between $\ln q_e$ and $\ln C_e$ for Freundlich as in K_f and n [49], respectively. Table 3 shows the correlation coefficient values (R^2) for both Langmuir and Freundlich as well as other parameters. This implies that both models fitted well for the experimental data. Nonetheless, the important features of Langmuir parameters can be applied to further forecast the interaction between the adsorbate and adsorbent

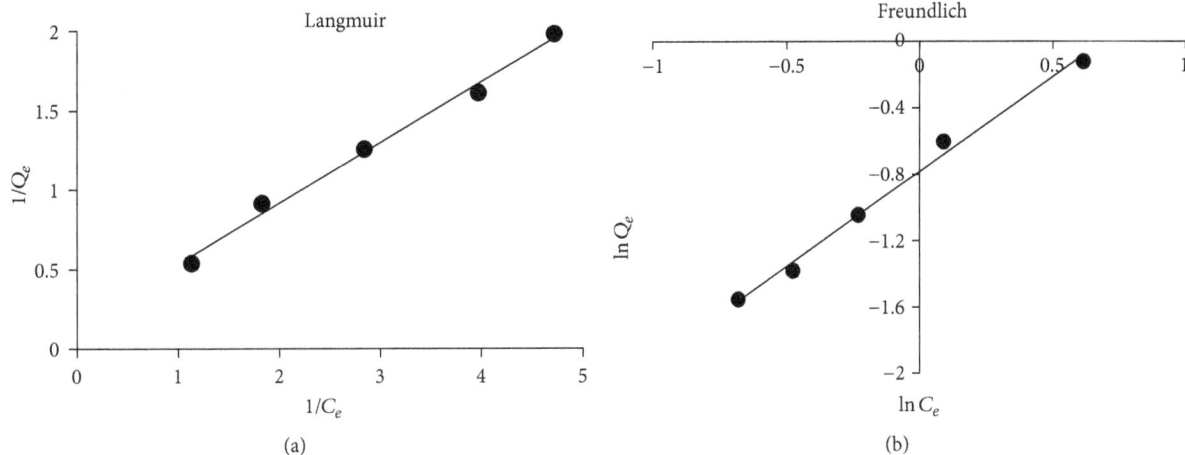

FIGURE 15: (a) Langmuir adsorption isotherm model. (b) Freundlich adsorption isotherm model.

TABLE 3: Langmuir and Freundlich constants for Cd (II) ion uptake.

	Langmuir constants					Freundlich constants		
	R^2	K_L	Q_{max}	R_L	θ	R^2	K_f	$1/n$
PEFA	0.9918	0.4101	6.3939	0.4714	0.5286	0.9924	0.4568	0.878

TABLE 4: Sorption kinetics parameters for Cd (II) ion adsorption by PEFA.

	Pseudo-1st-order			Pseudo-2nd-order			Intraparticle diffusion		
	q_e	K_1	R^2	q_e	K_2	R^2	K_i	C	R^2
PEFA	148.9	0.0619	0.9802	0.6052	4.2983	0.999	0.0046	0.8413	0.8365

with the aid of dimensionless separation parameters (R_L) as indicated in the following equation:

$$R_L = \frac{1}{1 + K_L C_i},\qquad(13)$$

where K_L is Langmuir constant and C_i is Cd (II) ions initial concentration. R_L value gives essential information on sorption nature. R_L value for this study as shown in Table 3 indicates a favorable adsorption process ($R_L < 1$) for 2 mg/L Cd (II) ion concentration [9, 38, 49]. Adsorption of PEFA can also be explained in terms of surface area coverage in contrast to initial concentration of Cd ion [38]. Langmuir model for surface area coverage of adsorbent surface can be illustrated with aid of the following equation:

$$K_L C_i = \frac{\theta}{1 - \theta},\qquad(14)$$

where θ is the surface area coverage of adsorbent surface as indicated in Table 3.

3.3.2. Kinetic Studies of Adsorption. As indicated in (7), (8), and (9), kinetic studies of sorption data were evaluated by different kinetic models like pseudo-1st-order model, pseudo-2nd-order model, and intraparticle diffusion [43, 49, 50]. Sorption of cadmium ions was supervised at different period of time. Sorption of Cd (II) ions was calculated

from data acquired. To determine the appropriate kinetic model, Cd (II) ion adsorption was plotted against time. These data were fitted into pseudo-1st-order, pseudo-2nd-order, and Weber intraparticle diffusion equations [43]. Table 4 shows that values of q_e and K_i were calculated from K_i (Slope) and $\ln q_e$ (intercept) of plot $\ln(q_e - q_t)$ versus t. The correlation coefficient value ($R^2 = 0.9802$) for pseudo-1st-order model was lower than that of pseudo-2nd-order model. This could be linked to the fact that sorption kinetics take place chemically and involve forces of valency via ions sharing or electron exchange between adsorbent and the adsorbed ions on PEFA [46, 51]. Values of q_e and K_2 were calculated from q_e and $1/q_e$ (slope) and $1/K_2 q_e^2$ (intercept) of the plot. The correlation ($R^2 = 0.9999$) for pseudo-2nd-order model was very strong, pointing towards the fact that sorption of cadmium ions occurred on a monolayer mode, with the assumption that the rate limiting factor could be chemical sorption [38]. This indicated that the cadmium ions were chemically bonded to definite active sites on the surface of PEFA. Weber and Morris' intraparticle diffusion equation was also plotted for q_t against $t^{1/2}$ [50]. Values of K_i and C were calculated from K_i (slope) and C (intercept) as shown in Table 4. Its correlation value ($R^2 = 0.8365$) was the lowest and the plot intercept did not pass through the origin pointing towards some control of boundary layers and suggesting that intraparticle pore diffusion is not the only rate limiting factor [38]. The intraparticle diffusion equation plot highlights multilinearity, indicating a three-stage process. The initial sharper part is linked to the diffusion of Cd (II) ions via the solution to the external surface of PEFA or boundary layer diffusion of solid molecules [38]. The second part gives description of ion phase, where intraparticle diffusion is a rate limiting factor. The third part is ascribed to the final equilibrium phase. Nonetheless, the intercept of the plot (not shown) fails to pass through the origin, which may be attributed to the difference in the rate of mass transfer in the initial and final phases of sorption [52].

4. Conclusion

This study has demonstrated that polyelectrolyte-coated fly ash (PEFA) performed as an excellent adsorbent for Cd (II) ion in aqueous solution. Adsorption of Cd (II) on PEFA surface was dependent on the dosage of adsorbent, pH of the aqueous solution, contact time, Cd (II) initial concentration, and temperature. Optimum conditions for Cd ions removal were found to be at an adsorbent dose of 4 g/L, pH value of 9, 15 min contact time, mixing rate of 150 rpm, 2 mg/L Cd initial concentration, and 298 k temperature. The maximum sorption capacity of PEFA was achieved at 6.40 mg/g with the experimental data fitting well to both Langmuir and Freundlich isotherm models and following pseudo-2nd-order kinetics. The investigation of thermodynamic parameters suggested that the adsorption of Cd (II) ions interaction with PEFA was endothermic and spontaneous and was increasing disorderliness of solute solution interface. This research highlights that fly ash material, a hazardous industrial waste, has a great potential in water treatment application.

Acknowledgments

The authors would like to acknowledge the support provided by King Abdulaziz City for Science and Technology (KACST) through Science & Technology Unit at King Fahd University of Petroleum & Minerals (KFUPM) for funding this work through Project no. *13-ADV161-04* as part of the National Science Technology and Innovation Plan (NSTIP).

References

[1] M. N. Chong, B. Jin, C. W. K. Chow, and C. Saint, "Recent developments in photocatalytic water treatment technology: a review," *Water Research*, vol. 44, no. 10, pp. 2997–3027, 2010.

[2] G.-M. Zeng, X. Li, J.-H. Huang et al., "Micellar-enhanced ultra-filtration of cadmium and methylene blue in synthetic wastewater using SDS," *Journal of Hazardous Materials*, vol. 185, no. 2-3, pp. 1304–1310, 2011.

[3] S. O. Lesmana, N. Febriana, F. E. Soetaredjo, J. Sunarso, and S. Ismadji, "Studies on potential applications of biomass for the separation of heavy metals from water and wastewater," *Biochemical Engineering Journal*, vol. 44, no. 1, pp. 19–41, 2009.

[4] WHO, "Guidelines for the safe use of wastewater, excreta and greywater," *World Health Organization*, vol. I, p. 95, 2006.

[5] K. Rao, M. Mohapatra, S. Anand, and P. Venkateswarlu, "Review on cadmium removal from aqueous solutions," *International Journal of Engineering, Science and Technology*, vol. 2, no. 7, pp. 81–103, 2010.

[6] F. A. Al-Khaldi, B. Abu-Sharkh, A. M. Abulkibash, and M. A. Atieh, "Cadmium removal by activated carbon, carbon nanotubes, carbon nanofibers, and carbon fly ash: a comparative study," *Desalination and Water Treatment*, vol. 53, no. 5, pp. 1417–1429, 2015.

[7] R. Han, H. Li, Y. Li, J. Zhang, H. Xiao, and J. Shi, "Biosorption of copper and lead ions by waste beer yeast," *Journal of Hazardous Materials*, vol. 137, no. 3, pp. 1569–1576, 2006.

[8] A. H. Mahvi and E. Bazrafshan, "Removal of Cadmium from Industrial effluents by electrocoagulation process using aluminum electrodes," vol. 2, no. 1, pp. 34–39, 2007.

[9] K. G. Akpomie, F. A. Dawodu, and K. O. Adebowale, "Mechanism on the sorption of heavy metals from binary-solution by a low cost montmorillonite and its desorption potential," *Alexandria Engineering Journal*, vol. 54, no. 3, pp. 757–767, 2015.

[10] M. Imamoglu and O. Tekir, "Removal of copper (II) and lead (II) ions from aqueous solutions by adsorption on activated carbon from a new precursor hazelnut husks," *Desalination*, vol. 228, no. 1–3, pp. 108–113, 2008.

[11] T. Vidhyadevi, A. Murugesan, S. S. Kalaivani et al., "Optimization of the process parameters for the removal of reactive yellow dye by the low cost *Setaria verticillata* carbon using response surface methodology: thermodynamic, kinetic, and equilibrium studies," *Environmental Progress & Sustainable Energy*, vol. 33, no. 3, pp. 855–865, 2014.

[12] S. Larous, A.-H. Meniai, and M. Bencheikh Lehocine, "Experimental study of the removal of copper from aqueous solutions by adsorption using sawdust," *Desalination*, vol. 185, no. 1-3, pp. 483–490, 2005.

[13] E. Erdem, N. Karapinar, and R. Donat, "The removal of heavy metal cations by natural zeolites," *Journal of Colloid and Interface Science*, vol. 280, no. 2, pp. 309–314, 2004.

[14] W. E. Marshall and E. T. Champagne, "Agricultural byproducts as adsorbents for metal ions in laboratory prepared solutions and in manufacturing wastewater," *Journal of Environmental Science and Health. Part A: Environmental Science and Engineering and Toxicology*, vol. 30, no. 2, pp. 241–261, 1995.

[15] H. Aydin, Y. Bulut, and Ç. Yerlikaya, "Removal of copper (II) from aqueous solution by adsorption onto low-cost adsorbents," *Journal of Environmental Management*, vol. 87, no. 1, pp. 37–45, 2008.

[16] A. Papandreou, C. J. Stournaras, and D. Panias, "Copper and cadmium adsorption on pellets made from fired coal fly ash," *Journal of Hazardous Materials*, vol. 148, no. 3, pp. 538–547, 2007.

[17] M. Visa and A. Duta, "Cadmium and nickel removal from wastewater using modified fly ash?: Thermodynamic and kinetic study," *Sci. Stu. and Res*, vol. IX, no. 1, pp. 73–82, 2008.

[18] G. Gupta and N. Torres, "Use of fly ash in reducing toxicity of and heavy metals in wastewater effluent," *Journal of Hazardous Materials*, vol. 57, no. 1-3, pp. 243–248, 1998.

[19] F. A. Olabemiwo, "The Potential Capacity of Polyelectrolyte coated Carbon Fly Ash in Removing Cadmium from Contaminated water," *King Fahd University of Petroleum and Minerals*, 2017.

[20] B. Bolto and J. Gregory, "Organic polyelectrolytes in water treatment," *Water Research*, vol. 41, no. 11, pp. 2301–2324, 2007.

[21] X. Zhang, M. Chen, Y. Yu, T. Yang, and J. Wang, "Polyelectrolyte-modified multi-walled carbon nanotubes for the adsorption of chromium(vi)," *Analytical Methods*, vol. 3, no. 2, pp. 457–462, 2011.

[22] D.-L. Huang, G.-M. Zeng, C.-L. Feng et al., "Degradation of lead-contaminated lignocellulosic waste by Phanerochaete chrysosporium and the reduction of lead toxicity," *Environmental Science and Technology*, vol. 42, no. 13, pp. 4946–4951, 2008.

[23] B. W. Stanton, J. J. Harris, M. D. Miller, and M. L. Bruening, "Ultrathin, multilayered polyelectrolyte films as nanofiltration membranes," *Langmuir*, vol. 19, no. 17, pp. 7038–7042, 2003.

[24] I. Y. El-Sherif and N. A. Fathy, "Modification of adsorptive properties of bagasse fly ash for uptaking cadmium from aqueous solution," *Environmental Research, Engineering and Management*, vol. 2, no. 2, pp. 19–28, 2013.

[25] H. A. Asmaly, Ihsanullah, B. Abussaud et al., "Adsorption of phenol on aluminum oxide impregnated fly ash," *Desalination and Water Treatment*, vol. 57, no. 15, pp. 6801–6808, 2016.

[26] M. Anwar Parvez, H. I. Al-Abdul Wahhab, R. A. Shawabkeh, and I. A. Hussein, "Asphalt modification using acid treated waste oil fly ash," *Construction and Building Materials*, vol. 70, pp. 201–209, 2014.

[27] R. Shawabkeh, M. J. Khan, A. A. Al-Juhani, H. I. Al-Abdul Wahhab, and I. A. Hussein, "Enhancement of surface properties of oil fly ash by chemical treatment," *Applied Surface Science*, vol. 258, no. 5, pp. 1643–1650, 2011.

[28] R. Shawabkeh, O. Khashman, and S. Tarawneh, "Synthesis of activated carbon from spent lubricating oil and application for adsorption of cadmium and lead ions from aqueous solution," *Combined and Hybrid Adsorbents, J.M. Loure*, pp. 195–200, 2006.

[29] F. A. Abuilaiwi, T. Laoui, M. Al-Harthi, and M. A. Atieh, "Modification and functionalization of multiwalled carbon nanotube (MWCNT) via fischer esterification," *The Arabian Journal for Science and Engineering*, vol. 35, no. 1, pp. 37–48, 2010.

[30] Q. Li, B. Wang, C. Li, J. Pang, and J. Zhai, "Synthesis of fly-ash cenospheres coated with polypyrrole using a layer-by-layer method," *Journal of Physics D: Applied Physics*, vol. 44, no. 44, Article ID 445301, 2011.

[31] I. Langmuir, "The adsorption of gases on plane surfaces of glass, mica and platinum," *The Journal of the American Chemical Society*, vol. 40, no. 9, pp. 1361–1403, 1918.

[32] Y. S. Al-Degs, A. Ghrir, H. Khoury, G. M. Walker, M. Sunjuk, and M. A. Al-Ghouti, "Characterization and utilization of fly ash of heavy fuel oil generated in power stations," *Fuel Processing Technology*, vol. 123, pp. 41–46, 2014.

[33] S. Vitolo, M. Seggiani, S. Filippi, and C. Brocchini, "Recovery of vanadium from heavy oil and Orimulsion fly ashes," *Hydrometallurgy*, vol. 57, no. 2, pp. 141–149, 2000.

[34] M. Pires and X. Querol, "Characterization of Candiota (South Brazil) coal and combustion by-product," *International Journal of Coal Geology*, vol. 60, no. 1, pp. 57–72, 2004.

[35] J. Coates, "Interpretation of infrared spectra, a practical approach interpretation of infrared spectra, a practical approach," *Encycl. Anal. Chem*, pp. 10815–10837, 2000.

[36] R. Shawabkeh, M. J. Khan, A. A. Al-Juhani, H. I. Al-Abdul Wahhab, and I. A. Hussein, "Enhancement of surface properties of waste oil fly ash by chemical treatment," pp. 1–8, Proceedings of the 3rd International Conference on Industrial and hazardous waste management (CRETE'12), 2012.

[37] Yaofa Jiang, E. R. Elswick, and M. Mastalerz, "Progression in sulfur isotopic compositions from coal to fly ash: Examples from single-source combustion in Indiana," *International Journal of Coal Geology*, vol. 73, no. 3-4, pp. 273–284, 2008.

[38] V. Rathod, H. Pansare, S. A. Bhalerao, and S. D. Maind, "Adsorption and Desorption Studies of Cadmium (II) ions from aqueous solutions onto Tur pod (Cajanus cajan)," *Int. J. Adv. Chem. Res*, vol. 4, no. 5, pp. 30–38, 2015.

[39] B. Wang, Q. Li, J. Kang, J. Pang, W. Wang, and J. Zhai, "Preparation and characterization of polypyrrole coating on fly ash cenospheres: role of the organosilane treatment," *Journal of Physics D: Applied Physics*, vol. 44, no. 41, Article ID 415301, 2011.

[40] C. Guo, L. Zhou, and J. Lv, "Effects of expandable graphite and modified ammonium polyphosphate on the flame-retardant and mechanical properties of wood flour-polypropylene composites," *Polymers and Polymer Composites*, vol. 21, no. 7, pp. 449–456, 2013.

[41] W. T. Kwon, D. H. Kim, and Y. P. Kim, "Characterization of heavy oil fly ash generated from a power plant," *AZo J. Mater. online*, pp. 1–10, 2005.

[42] J. Zou, Y. Dai, X. Wang et al., "Structure and adsorption properties of sewage sludge-derived carbon with removal of inorganic impurities and high porosity," *Bioresource Technology*, vol. 142, pp. 209–217, 2013.

[43] R. Leyva-Ramos, J. R. Rangel-Mendez, J. Mendoza-Barron, L. Fuentes-Rubio, and R. M. Guerrero-Coronado, "Adsorption of cadmium(II) from aqueous solution onto activated carbon," *Water Sci. Technol*, vol. 35, no. 7, pp. 205–211, 1997.

[44] Ihsanullah, F. A. Al-Khaldi, B. Abusharkh et al., "Adsorptive removal of cadmium(II) ions from liquid phase using acid modified carbon-based adsorbents," *Journal of Molecular Liquids*, vol. 204, pp. 255–263, 2015.

[45] G. H. Pino, L. M. Souza De Mesquita, M. L. Torem, and G. A. S. Pinto, "Biosorption of cadmium by green coconut shell powder," *Minerals Engineering*, vol. 19, no. 5, pp. 380–387, 2006.

[46] K. Akpomie and F. Dawodu, "Physicochemical analysis of automobile effluent before and after treatment with an alkaline-activated montmorillonite," *Journal of Taibah University for Science*, vol. 9, no. 4, pp. 465–476, 2015.

[47] M. A. Atieh, "Removal of phenol from water different types of carbon – a comparative analysis," *Procedia - Soc. Behav. Sci.*, vol. 10, pp. 136–141, 2014.

[48] G. C. Catena and F. V. Bright, "Thermodynamic study on the effects of β-cyclodextrin inclusion with anilinonaphthalenesulfonates," *Analytical Chemistry*, vol. 61, no. 8, pp. 905–909, 1989.

[49] G. McKay, Y. S. Ho, and J. C. Y. Ng, "Biosorption of copper from waste waters: A review," *Separation and Purification Methods*, vol. 28, no. 1, pp. 87–125, 1999.

[50] W. J. Weber, *Physicochemical processes for water quality control*, Wiley-Interscience, New York, NY, USA, 1972.

[51] C. Septhum, S. Rattanaphani, J. B. Bremner, and V. Rattanaphani, "An adsorption study of Al(III) ions onto chitosan," *Journal of Hazardous Materials*, vol. 148, no. 1-2, pp. 185–191, 2007.

[52] G. Prasad, K. K. Pandey, and V. N. Singh, "Mixed adsorbents for Cu(II) removal from aqueous solutions," *Environmental Technology Letters*, vol. 7, no. 1-12, pp. 547–554, 1986.

8

Adsorption Characteristics of Bixin on Acid- and Alkali-Treated Kaolinite in Aprotic Solvents

Winda Rahmalia,[1,2] **Jean-François Fabre,**[1] **Thamrin Usman,**[2] **and Zéphirin Mouloungui** ⓘ[1,3]

[1]*Université de Toulouse, INP-ENSIACET, Laboratoire de Chimie Agro-industrielle (LCA), 4 Allée Monso, 31030 Toulouse, France*
[2]*Department of Chemistry, Mathematics and Natural Science, Tanjungpura University, Jl. Ahmad Yani, Pontianak 78124, West Kalimantan, Indonesia*
[3]*INRA, UMR 1010 CAI, 31030 Toulouse, France*

Correspondence should be addressed to Zéphirin Mouloungui; zephirin.mouloungui@ensiacet.fr

Academic Editor: Spyros P. Perlepes

The adsorption of bixin in aprotic solvents onto acid- and alkali-treated kaolinite was investigated. Kaolinite was treated three times, for 6 h each, with 8 M HCl or 5 M KOH. The adsorbents were characterized by XRD, FT-IR, EDS, and BET-N_2. The effects of contact time and dye concentration on adsorption capacity and kinetics, electronic transition of bixin before and after adsorption, and also mechanism of bixin-kaolinite adsorption were investigated. Dye adsorption followed pseudo-second order kinetics and was faster in acetone than in dimethyl carbonate. The best adsorption results were obtained for KOH-treated kaolinite. In both of the solvents, the adsorption isotherm followed the Langmuir model and adsorption capacity was higher in dimethyl carbonate ($q_m = 0.43$ mg/g) than in acetone (0.29 mg/g). The adsorption capacity and kinetics of KOH-treated kaolinite ($q_m = 0.43$ mg/g, $k_2 = 3.27$ g/mg·min) were better than those of HCl-treated kaolinite ($q_m = 0.21$ mg/g, $k_2 = 0.25$ g/mg·min) and natural kaolinite ($q_m = 0.18$ mg/g, $k_2 = 0.32$ g/mg·min). There are shift in the band position of maximum intensity of bixin after adsorption on this adsorbent. Adsorption in this system seemed to be based essentially on chemisorption due to the electrostatic interaction of bixin with the strong basic and reducing sites of kaolinite.

1. Introduction

Bixin (methyl hydrogen 9′-*cis*-6,6′-diapocarotene-6,6′-dioate, $C_{25}H_{30}O_4$) is a carotenoid dye extracted from the seeds of the tropical shrub annatto (*Bixa orellana* L.) [1]. It is widely used in industry, cosmetics, and pharmaceutical products and as a food colouring and textile dye [2–5]. Bixin has photoactive properties and was recently explored as a sensitizing dye in solar cells [6–8] and for photodynamic therapy [9]. The potential uses of bixin in these applications are based on the conjugated double bond which can absorb energy in the visible region (400–500 nm), yielding colours in the yellow, orange, and red range. However, as other carotenoids, this double bond renders bixin unstable to light, temperature, and oxygen exposure [10, 11].

The poor heat and light stability of carotenoids in vitro is problematic when trying to construct photofunctional materials,

such as photosensitized semiconductors and nonlinear optical materials [12]. There have been reports of efforts to increase the stability of bixin to render it suitable for a broader range of applications, by incorporating this molecule into the surface or interlayer space of clay minerals. Kohno et al. [13] showed that annatto dye/organo-montmorillonites were more photostable than pure annatto dye because the layered structure of the montmorillonite protected the dye molecules from external oxygen. Rahmalia [14] reported the immobilization of bixin on natural kaolinite. The resulting product had slower degradation kinetics in acetone than pure bixin. Furthermore, solar cells sensitized with bixin immobilized on acid-activated kaolinite [7] had a higher energy conversion efficiency than cells sensitized with pure bixin [6].

The incorporation of various organic compounds into clays and clay minerals has been reported, due to the surface properties of these minerals, such as their adsorption capacities, surface charges, their large surface area, charge density, types of exchangeable cations, hydroxyl groups on the edges, silanol groups of crystalline defects or broken surfaces, and Lewis and Brönsted acidity [15–17]. Kaolinite ($Al_2Si_2O_5(OH)_4$) is a relatively inexpensive clay mineral that is highly effective as a carrier material. This behaviour is governed by the extent and nature of the external surface, which can be modified by appropriate treatment techniques [18–20]. Activation with acid or alkali has been widely studied as a chemical treatment for improving the surface characteristics of natural kaolinite in terms of its interactions with adsorbate [21–23]. Kaolinite is also hydrophobic and could therefore easily adsorb hydrophobic organic molecules, such as bixin [24–27]. The use of an activation process without heating for the ultimate enhancement of energy economics for bixin was investigated.

Clay minerals have been used as solid matrices to enhance the stability of bixin, but no systematic study has described the mechanism of bixin adsorption onto kaolinite. Rapid and efficient dye adsorption is important for industrial applications. The isotherm and adsorption kinetics of bixin were therefore determined on acid- and alkali-treated kaolinite. Untreated kaolinite was also tested for the purpose of comparison. Acetone and dimethyl carbonate were used as solvents because bixin was highly soluble in both these aprotic solvents. We previously reported that the transition energy and molar attenuation coefficient of bixin in dimethyl carbonate were similar to those in acetone [28]. Dimethyl carbonate would therefore be an appropriate alternative to volatile organic solvents like the nonpolar aprotic solvent acetone. Dimethyl carbonate is widely used as dialkyl carbonate with many applications in novel green chemistry [29]. It is a valuable chemical for industrial chemical engineering, due also to its low toxicity [30]. This study identified appropriate ketone and carbonate solvents for use in applications of bixin. Such applications, including the synthesis of a new sensitizing kaolinite-bixin dye, have become important in several fields.

2. Materials and Methods

2.1. Materials. Bixin crystals containing 88.11% *cis*-bixin and 11.75% di-*cis*-bixin and an unknown compound (0.14%) were obtained by the extraction and purification processes described by Rahmalia et al. [31]. Kaolinite ($Al_2O_7Si_2.2H_2O$), CAS Number 1318-74-7, was supplied by Sigma-Aldrich (Germany), together with analytical grade hydrochloric acid (HCl, 37%) and potassium hydroxide (KOH), and the HPLC grade solvents acetone (≥99.5%) and dimethyl carbonate (99%) were supplied by Sigma-Aldrich (Germany).

2.2. Methods

2.2.1. Preparation of Acid- and Alkali-Treated Kaolinite. Activation was achieved by adding 10 g of kaolinite to 100 mL 8 M HCl or 100 mL 5 M KOH separately. The mixtures were incubated at room temperature for 6 h, with constant shaking (300 rpm). The suspension was filtered, and the residue was washed with distilled water until neutral

and then dried in an oven at 103°C for 24 h. This process was repeated three times to optimize activation. The final products obtained are referred to as KA for HCl-treated kaolinite and KB for KOH-treated kaolinite. The untreated sample is referred to as KN.

2.2.2. Characterization of the Adsorbent. X-ray diffraction (XRD) patterns for the samples were obtained with a Bruker X-ray diffractometer (CuKα, $\lambda = 1.54$ Å, step scan size of 0.02° and a count time of 0.5 seconds, 25°C). Samples were prepared by allowing the particles of kaolinite to settle in water to obtain 2 μm particles. They were then treated with ethylene glycol and prepared as oriented mounts on a glass slide.

Fourier transform infrared spectroscopy (FT-IR) was carried out on a Shimadzu FT-IR spectrophotometer, over a spectral region of 400–4000 cm^{-1}, with a resolution of 1 cm^{-1}, and samples were evaluated in powder form mixed with KBr powder.

The BET surface area of the samples was determined by the multipoint N_2 adsorption-desorption method at the temperature of liquid nitrogen (−196°C), with an ASAP 2010 (Micrometrics) instrument.

Changes in elemental composition after treatment were assessed with an Edax Ametek high-resolution energy dispersive X-ray spectroscopy (EDS) detector.

2.2.3. Adsorption Experiment. Stock solutions of bixin (20 mg/L) were prepared in acetone and dimethyl carbonate separately. Solutions of the required concentration (3–18 mg/L) were prepared by diluting the stock solution. Adsorbent (0.05 g) was then added to 5 mL bixin solution (3–18 mg/L). The mixtures were incubated at room temperature (~22°C), with shaking at 300 rpm. The samples were withdrawn after 4 h (predetermined equilibrium time), and small aliquots of the supernatant were removed and diluted to an appropriate concentration if required. The absorption spectrum was determined immediately with a Shimadzu UV-1800 UV-Vis spectrophotometer. The bixin concentration of the solutions was determined with a UV spectrophotometer calibrated at 457 nm for acetone and 456 nm for dimethyl carbonate [28]. For the contact time studies, the residual concentration of the 5 mL bixin solution (±10 mg/L) with kaolinite (0.1 g) was determined at various time points, from 5 to 360 min. The experiments were carried out in triplicate.

3. Results and Discussion

3.1. Characterization of Adsorbents. The XRD patterns of KN (Supporting Information Figure S1) showed two intense diffraction reflections at 2θ values of 12.3 and 24.9°, less intense reflections at 2θ values of 23.2 and 26.6°, and a hump at 2θ values of 19.8–21.5°, associated with kaolinite (PDF 00-058-2001). The diffraction reflections of orthoclase were found at 2θ values of 15.4, 21.0, 25.7, 27.5, and 30.1° (PDF 00-022-1212), whereas the diffraction reflections of muscovite were found at a 2θ value of 18.0° (PDF 00-058-2036). After treatment with KOH, the reflection width and intensity of kaolinite decreased at 2θ values of 12.3 and

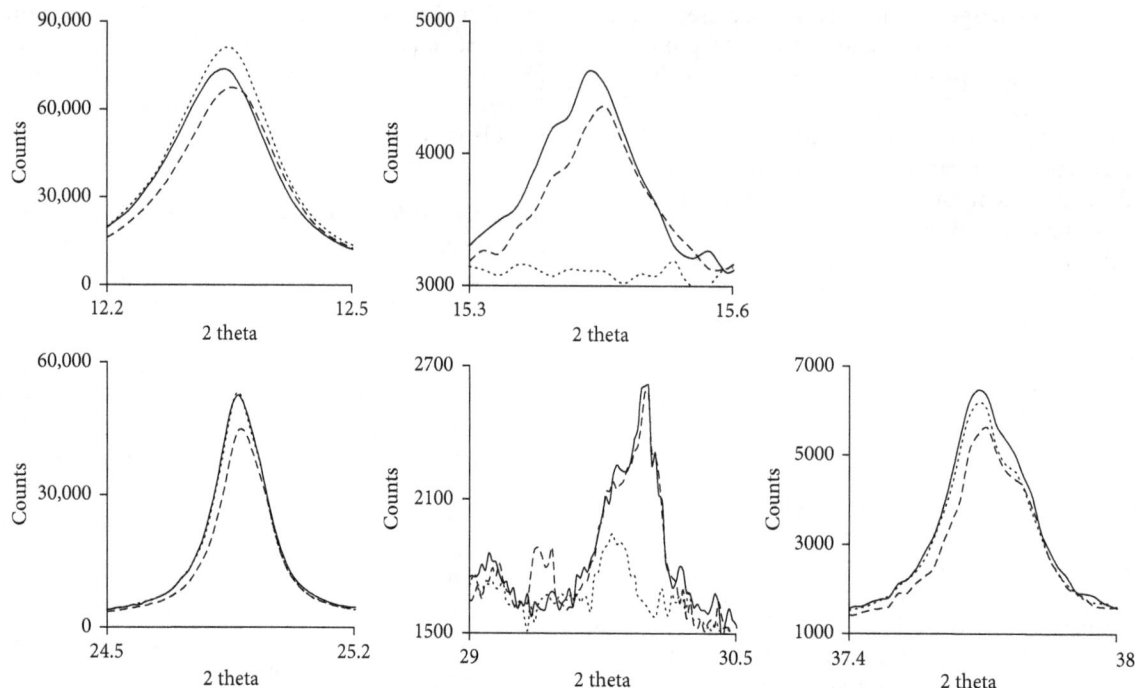

FIGURE 1: X-ray diffraction (XRD) patterns of KN (—), KA (....), and KB (----).

24.9°. This decrease was attributed to a minor structural disorder resulting from alkali treatment, which affects the crystalline nature of the clay [32].

A diffractogram for KA showed no significant difference with respect to KN, but 2θ values of 15.4 and 30.1° were unobservable, and the reflection increased at a 2θ of 12.3° (Figure 1). This finding may reflect the greater resistance to acid attack of the structure of kaolinite than that of orthoclase. The resistance of clay minerals to acid attack depends strongly on their crystallinity, with more regular crystals associated with greater resistance to acid attack [33]. It may also be due to the elimination of mineral impurities by acid leaching. The higher peak intensity may reflect the presence of larger crystallites or a decrease in mean lattice strain [32, 34].

The FT-IR spectra of KN, KA, and KB (Figure 2) revealed bands at 3696, 3669, 3652, and 3619 cm^{-1} corresponding to the stretching of inner-surface hydroxyl groups, at 3443 cm^{-1} corresponding to stretching of the hydroxyl group of water, 1631 cm^{-1} corresponding to the O–H deformation of water, 1114 cm^{-1} corresponding to Si–O stretching (longitudinal mode), 1030 and 1006 cm^{-1} corresponding to in-plane Si–O stretching, 937 cm^{-1} corresponding to the OH deformation of inner-surface hydroxyl groups, 911 cm^{-1} corresponding to Al–OH deformation, 794 cm^{-1} corresponding to Si–O vibration, 755 and 696 cm^{-1} corresponding to Si–O perpendicular vibrations, 536 cm^{-1} corresponding to Al–O–Si deformation, 468 cm^{-1} corresponding to Si–O–Si deformation, and 428 cm^{-1} corresponding to Si–O deformation [35, 36]. The FT-IR spectra patterns of KN, KA, and KB showed no significant differences between kaolinite before and after treatment, indicating an absence of significant change in

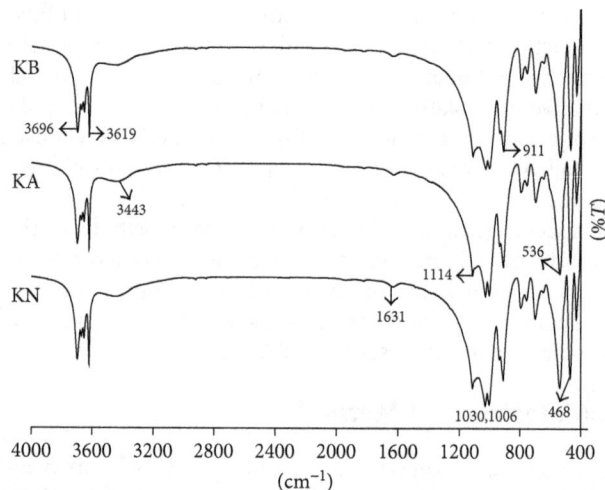

FIGURE 2: FT-IR spectra of KN, KA, and KB.

the kaolinite samples. Infrared absorption spectra, which displayed a fairly sharp absorption band at wave numbers around 911 cm^{-1} for both samples, showed an absence of change in the composition of the octahedral Al atoms following treatment with acid and alkali in the experimental conditions used.

Changes in elemental composition were investigated by EDS (Table 1). The samples of kaolinite tested were rich in silicon and aluminium. Their composition in terms of impurities, such as Na, Mg, K, and Fe, depended on the type of reagent used. Si and Al contents only slightly decreased after treatment. This shows the resistance of natural kaolinite minerals to acid and alkali attack.

TABLE 1: EDS analysis data.

Element	wt. %		
	KN	KA	KB
O	56.9	59.0	56.4
Na	0.42	0.37	0.33
Mg	0.15	0.06	0.37
Al	18.4	17.9	18.1
Si	23.1	21.3	22.7
K	0.79	0.55	0.93
Fe	0.22	0.34	0.33
Si/Al	1.26	1.19	1.25

TABLE 2: Nitrogen sorption isotherm data.

Sample	BET specific surface area, a_s BET (m^2g^{-1})	Total pore volume (10^{-2} cm^3g^{-1})	Mean pore diameter, d_p (nm)
KN	7.65	3.62	18.9
KA	7.28	3.74	20.6
KB	8.16	2.85	14.0

The results of nitrogen sorption isotherm analysis are summarised in Table 2. The surface area, pore volume, and mean diameter of the untreated kaolinite sample were 7.65 m^2/g, 3.62 × 10^{-2} cm^3/g, and 18.9 nm, respectively, indicating that the porosity of the original kaolinite was low. Pore volume and mean diameter increased after acid treatment, possibly due to the dissolution of metal ions present in the kaolinite and the rearrangement of its crystal structure, as a result of a reaction between the acid and the clay mineral. Treatment with 5 M alkali at room temperature has been shown to increase specific surface area but to decrease total pore volume and mean diameter. Belver et al. [21] reported an increase in specific surface area, presumably because of the disaggregation/separation of kaolinite particles. The decrease in specific surface area observed for KA may be due to an increase in crystallinity, as indicated by XRD.

3.2. Electronic Transition of Bixin.
The absorption spectrum of the supernatant of bixin in acetone and dimethyl carbonate was in the visible region, with peaks at 457 nm and 456 nm, respectively (Figure 3), associated with the 0-1 vibration band position, consistent with the finding of Rahmalia et al. [28]. After adsorption onto KA, the maximum intensity (λmax) of the bixin spectrum was still in the 0-1 band position. However, after adsorption onto KN and KB, the maximum wavelength shifted to shorter wavelengths, associated with a band position of 0–2 for KN and of 0–3 for KB, and intensity decreased. Schoonheydt and Johnston [37] reported that the absorption maxima of dye molecules in nonpolar solvents adsorbed onto the surface of clay minerals could shift to shorter wavelengths because the solvent-molecule interaction was stronger in the ground state than in the excited state. Yariv and Cross [38] suggested that the absorption band of the adsorbed dye displayed a blue shift due to interactions between the π-electrons of the dye and the hybridised orbitals of the surface oxygen atoms, leading to a stabilisation of the π-orbitals and a destabilisation of the π*-orbitals.

3.3. Effect of Contact Time.
The effects of contact time on the amount of bixin adsorbed onto kaolinite were investigated (Figure 4). Kaolinites adsorb bixin with different efficiencies,

and bixin was rapid and strong during the initial period of contact, between 5 and 60 minutes. During this period, the tendency towards adsorption was high, and the slope of the adsorption curve was steep. This early phase of steep increase was followed by a phase of slow increase between 120 and 360 minutes. During this period, the slope of the adsorption curve gradually flattened out, and the bixin adsorption gradually decreased eventually reaching zero. This corresponded to equilibrium being reached due to the saturation of adsorption sites.

The single, smooth, and continuous nature of the curves suggested that the bixin might cover the kaolinite as a monolayer. The percentage dye adsorption was highest on KB, consistent with the BET specific surface area analysis, which indicated that adsorption was most likely to occur on the external surface of kaolinite. It took 180 minutes to reach equilibrium for bixin in acetone with KN as the adsorbent and 240 minutes for the same mixture but with KA or KB as the adsorbent. It took 240 minutes to reach equilibrium for bixin in dimethyl carbonate, with KN or KB as the adsorbent, and 300 minutes for the same mixture but with KA as the adsorbent. This phenomenon is influenced by the surface properties of the adsorbent and the chemical and physical constants of the solvents.

3.4. Effect of Initial Dye Concentration.
The effect of initial dye concentration on equilibrium adsorption was investigated at different initial bixin concentrations. Initial bixin concentration affected the amount of bixin adsorbed at equilibrium (Figure 5). At low initial bixin concentrations, the adsorption capacity of KN, KA, and KB increased with initial bixin concentration. It therefore seems likely that an increase in adsorption with initial dye concentration leads to an increase in mass gradient between the solution and adsorbent, thereby driving the transfer of additional dye molecules from the bulk solution to the particle surface [39].

3.5. Adsorption Isotherm.
Adsorption properties and equilibrium parameters, commonly known as adsorption isotherms, describe the interaction of the adsorbate with the adsorbents, improving understanding of the nature of the interaction. Isotherms provide information about the optimal use of adsorbents. When optimizing the design of an adsorption system, it is essential to establish the most appropriate correlation for the equilibrium curve. Several isotherm equations are available for analysis of experimental sorption equilibrium parameters. However, Langmuir and Freundlich models are the most widely

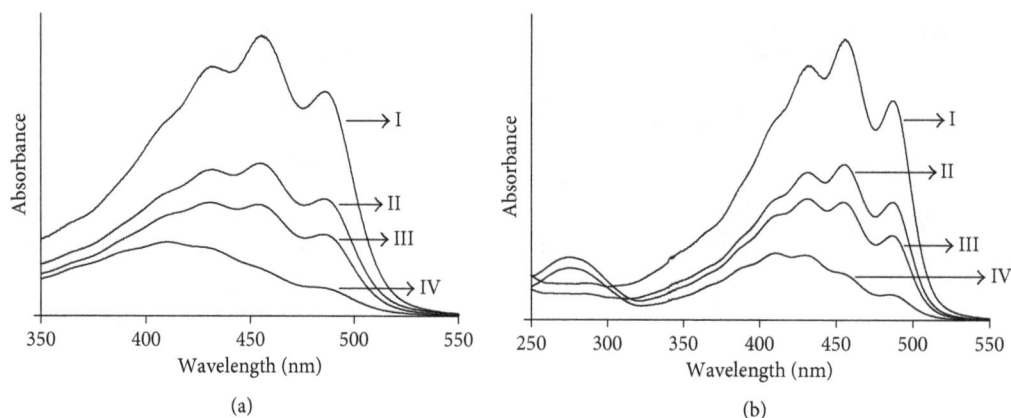

FIGURE 3: Absorption spectra of bixin in acetone (a) and dimethyl carbonate (b) before adsorption (I) and supernatant solution of bixin after adsorption on KN (III), KA (II), and KB (IV).

used type of isotherm [15, 23–25, 33, 40]. These models were used to explain the interaction between bixin and kaolinite in this study. They are the best models for explaining adsorption trends and are based on the rationale that the adsorbents become saturated with adsorbate after sufficiently long contact times.

The Freundlich isotherm describes the nonspecific adsorption of a heterogeneous system and reversible adsorption. The linear form of the Freundlich equation (1) is expressed as follows [41]:

$$\log q_e = \log K_F + \frac{1}{n} \log C_e, \qquad (1)$$

where $1/n$ is a combined measurement of the relative magnitude and diversity of energies associated with a particular sorption process.

In the Langmuir model, the mass of solute adsorbed per unit mass of adsorbent increases linearly with solute concentration at low surface coverage, approaching an asymptote as the adsorption sites become saturated. Equation (2) is based on three important assumptions: (1) the energy of adsorption is identical for all sites and is independent of surface coverage, (2) adsorption occurs only at localised sites, with no interaction between adjoining adsorbed molecules, and (3) the sorption maximum represents monolayer coverage. The linear form of the Langmuir equation (2) can be expressed as follows [42]:

$$\frac{C_e}{q_e} = \frac{1}{K_L \cdot q_m} + \left(\frac{1}{q_m}\right) C_e. \qquad (2)$$

We calculated the values of the parameters of the Freundlich and Langmuir model (Table 3). The equilibrium data were not consistent with the Freundlich equation for all adsorbents, in either of the two solvents. The poor fit of this model was demonstrated by the very low correlation coefficient ($r^2 < 0.95$). The values of $1/n < 1$ indicates a nonlinear adsorption of the Freundlich model and corresponds to a Langmuir-type isotherm curve, in which marginal sorption energy decreases with increasing surface concentration. The Langmuir equation

gave a better fit, with $r^2 > 0.95$, indicating a homogeneous active site and the coverage of the adsorbent surface with a monolayer of bixin. Based on q_m values, bixin adsorption to KB was more favourable than its adsorption to KA and KN.

The adsorption capacity of the adsorbents appeared to increase with specific surface area. The capacity of bixin to adsorb to adsorbents may reflect the extent to which the kaolinite was able to swell. The physical swelling of the kaolinite probably depended on the bulk properties of the intervening solvent molecules. In nonpolar solvents, increase in dielectric constant function ($R(\varepsilon)$) is associated with decreases in the volume of the kaolinite, due to lower levels of physical swelling [43]. Adsorption capacity was therefore greater when dimethyl carbonate ($R(\varepsilon) = 0.412$) was used as a solvent, because this solvent has lower dielectric constants than acetone ($R(\varepsilon) = 0.872$). The dimethyl carbonate also has several conformations (at least 3) against acetone which has only one conformation (Figure 6), causing the possibility of DMC molecules in the highly aggregated solvate. Two structure conformations of DMC (Figure 6b) are favourable between DMC and KB which oxygen of DMC interact with metal atom of Si and Al of KB [44–46].

3.6. Adsorption Kinetics. Lagergren's pseudo-first order and pseudo-second order models were used to investigate the dynamics of bixin adsorption onto kaolinite. The pseudo-first order model assumes that the rate of change of solute uptake over time is directly proportional to the difference in saturation concentration and the amount of solid uptake over time. In most cases, the adsorption reaction involves diffusion across a boundary (3) [47]. The adsorption process with chemisorptions controls the rate, according to the pseudo-second order model (4) [48].

$$\log (q_e - q_t) = \log q_e - k_1 \cdot \frac{t}{(2.303)}, \qquad (3)$$

$$\frac{t}{q_t} = \frac{1}{(k_2 q_e^2)} + \left(\frac{1}{q_e}\right) \cdot t. \qquad (4)$$

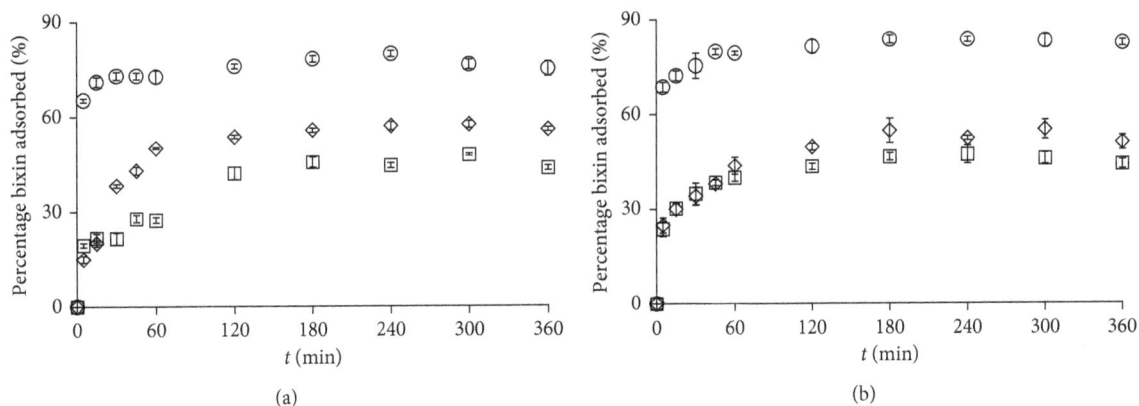

FIGURE 4: Effect of contact time on adsorption of bixin using KN (◇), KA (□), and KB (○) in acetone (a) and dimethyl carbonate (b).

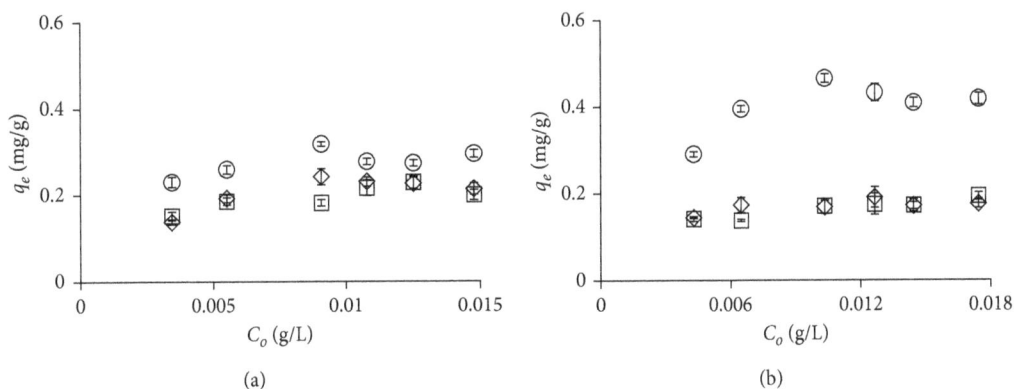

FIGURE 5: Effect of initial dye concentration on adsorption of bixin using KN (◇), KA (□), and KB (○) in acetone (a) and dimethyl carbonate (b).

k_1 and k_2 were calculated from the intercept of the corresponding plots of log $(q_e - q_t)$ against t and t/q_t against t. They are shown in Table 4, along with the values for the correlation coefficients, q_{e1} and q_{e2} (calc.) and q_e (exp.). The correlation coefficient values for the pseudo-second order rate equation were higher than those for the pseudo-first order rate equation (Table 4). The r^2 values for the plots were in the range 0.65–0.97 after application of the pseudo-first order model, but the calculated q_{e1} values obtained with this model did not give reasonable values because they were lower than the experimental q_e values. The q_{e2} and q_e values were very similar for the pseudo-second order model.

The adsorption process on all adsorbents in both solvents was found to follow the pseudo-second order kinetic model. These results suggest that chemisorption predominated in the adsorption occurring in this work [17]. The best results were obtained for KOH-treated kaolinite. Adsorption kinetics was faster with acetone ($k_2 = 3.27$) than with dimethyl carbonate ($k_2 = 1.08$) as the solvent. The smaller size of acetone molecules than of dimethyl carbonate molecules and the lower viscosity of acetone (0.295 cP) than of dimethyl carbonate (0.585 cP) may facilitate the diffusion of bixin into the interlayer region of kaolinite.

3.7. Mechanism of Bixin-Kaolinite Adsorption.

The FT-IR spectra of bixin dye and of bixin-KB following adsorption for different times were obtained (Figure 7). The spectrum of bixin dye may be assigned as follows: the –O–H stretching vibration at 3420 cm^{-1}, the H–C–H bending vibration at 2957, 2917, and 2850 cm^{-1}, the C=O ester group at 1731 cm^{-1}, the O–H bending vibration at 1620 cm^{-1}, the alkene C=C stretching at 1469 cm^{-1}, C–H bending of the methyl groups at 1378, C=O stretching at 1220 cm^{-1}, symmetric and asymmetric vibrations of the C–O–C ester group at 1180 cm^{-1}, and the methylene rocking vibration of cis-carotenoid at 720 cm^{-1} [49].

Figure 7 shows significant frequency modifications of the absorption bands from bixin in low frequency areas between 3400 and 3700 cm^{-1} and in high frequency areas between 1600 and 1750 cm^{-1}. Functional groups COOH and COOR from bixin strongly absorb at strong basic and reducing sites of KN through electrostatic interactions. The frequency of bands at 3420 cm^{-1} disappears in favour of frequency of bands at 3619 cm^{-1}. The same phenomenon was observed in high frequency areas; the frequency of bands at 1731 cm^{-1} disappears in favour of frequency of bands at 1620 cm^{-1}. This indicated strong interactions between two types of metal

TABLE 3: Adsorption isotherm parameters for the adsorption of bixin on kaolinite.

Solvent	Adsorbent	Freundlich			Langmuir		
		K_F	$1/n$	r^2	K_L	q_m	r^2
Acetone	KN	0.13	0.25	0.7043	1.16	0.24	0.9747
	KA	0.14	0.17	0.7107	1.22	0.22	0.9691
	KB	0.23	0.10	0.6005	4.00	0.29	0.9893
Dimethyl carbonate	KN	0.14	0.10	0.5600	2.03	0.18	0.9915
	KA	0.11	0.19	0.8362	0.48	0.21	0.9807
	KB	0.32	0.14	0.5701	5.00	0.43	0.9927

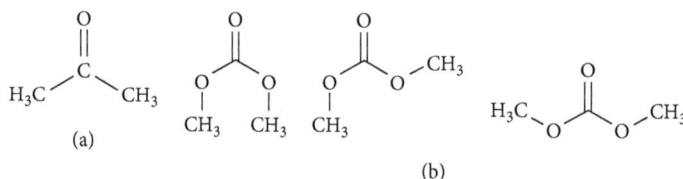

FIGURE 6: Structure conformations of acetone (a) and dimethyl carbonate (b).

TABLE 4: Adsorption kinetics parameters for the adsorption of bixin on kaolinite.

Solvent	Adsorbent	q_e exp	Pseudo-first order			Pseudo-second order			
			q_{e1}	$k_1 (10^{-2})$	r^2	q_{e2}	$h (10^{-2})$	k_2	r^2
Acetone	KN	0.25	0.21	2.53	0.9697	0.26	2.07	0.32	0.9951
	KA	0.21	0.16	1.17	0.8969	0.21	1.07	0.25	0.9823
	KB	0.35	0.09	1.70	0.6928	0.33	35.7	3.27	0.9956
Dimethyl carbonate	KN	0.29	0.20	1.77	0.9300	0.28	2.84	0.35	0.9947
	KA	0.25	0.15	1.77	0.9549	0.24	4.05	0.68	0.9962
	KB	0.45	0.12	1.31	0.6456	0.44	20.5	1.08	0.9991

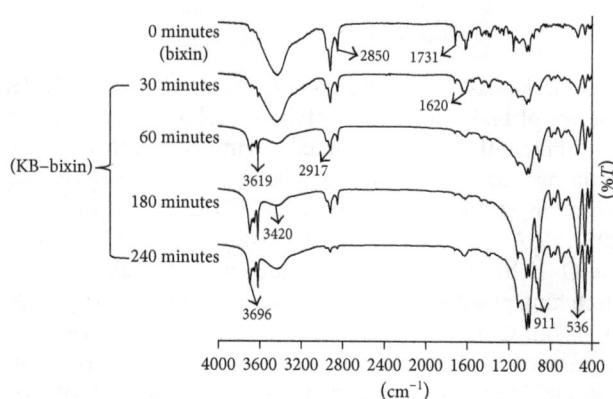

FIGURE 7: FT-IR spectra of bixin dye and bixin-KB obtained from the processes of adsorption in different times.

carboxylate groups, which result from a part of the interaction between the carboxylic group of the bixin and the other part of the carboxyester group of bixin, respectively, with Si and Al.

4. Conclusion

The adsorption characteristics of bixin onto kaolinite, especially for constructing photofunctional materials based on kaolinite-bixin organoclay, have been investigated. This adsorption is more dependent on the specific surface area of the adsorbent. The adsorption capacity of the kaolinite was considerably improved by an increase in the surface specific area. Alkali treatment (BET specific surface area = $8.16 \, \text{m}^2\text{g}^{-1}$) was therefore more suitable than acid treatment for increasing the capacity of kaolinite to adsorb organic molecules, such as bixin. Selection of the most appropriate aprotic solvent also increased the efficiency of bixin absorption onto kaolinite. Based on UV-visible spectroscopy data, the solvent-molecule interaction was stronger in the ground state than in the excited state. The adsorption isotherm was of the Langmuir-type and was higher in acetone than in dimethyl carbonate. Dye adsorption followed pseudo-second order kinetics and was faster in dimethyl carbonate complex solvate than in acetone. Adsorption in this system appears to be mostly due to chemisorption mediated by the electrostatic interaction of bixin with the strong basic and reducing sites of kaolinite. Finally, dimethyl carbonate has potential as a good solvent with no compound organic volatile for increasing bixin adsorption onto kaolinite.

Nomenclature and Units

q_e: Amount of dye adsorbed per unit of adsorbent at equilibrium (mg/g)

C_o: Initial concentration of the dye solution (mg/L)

C_e: Concentration of the dye solution at adsorption equilibrium (mg/L)

K_F: Freundlich adsorption isotherm constant $(\text{mg}^{1-1/n}\text{L}^{1/n}\text{g})$

N: Freundlich adsorption isotherm constant

K_L: Langmuir constant (L/mg)

q_m: Maximum adsorption resulting in monolayer coverage on the adsorbent surface (mg/g)

q_t: Amount of dye adsorbed per unit mass adsorbent at any time t (mg/g)

k_1:   Pseudo-first order adsorption rate constant (1/min)

k_2:   Pseudo-second order adsorption rate constant (g/mg·min)

H: Initial adsorption rate at any time approaching 0 (mg/g·min)

$R(\varepsilon)$:   Dielectric constant function.

Acknowledgments

This study received financial support from the Direktorat Jenderal Pendidikan Tinggi. The authors thank Cedric Charvillat (*Centre Inter-universitaire de Recherche et d'Ingénierie des Matériaux*), Gwénaëlle Raimbeaux (*Laboratoire de Genie Chimique*), and Deni Ferdian (University of Indonesia) for assistance with the analysis of kaolinite samples by XRD, BET-N_2, and EDS, respectively.

References

[1] G. Giuliano, C. Rosati, and P. M. Bramley, "To dye or not to dye: biochemistry of annatto unveiled," *Trends in Biotechnology*, vol. 21, no. 12, pp. 513–516, 2003.

[2] C. R. Cardarelli, M. de Toledo Benassi, and A. Z. Mercadante, "Characterization of different annatto extracts based on antioxidant and colour properties," *LWT-Food Science and Technology*, vol. 41, no. 9, pp. 1689–1693, 2008.

[3] D. Das, "Dyeing of wool and silk with *Bixa orellana*," *Indian Journal of Fibre and Textile Research*, vol. 32, pp. 366–372, 2007.

[4] G. C. dos Santos, L. M. Mendonça, G. A. Antonucci, A. C. dos Santos, L. M. G. Antunes, and M. de Lourdes Pires Bianchi, "Protective effect of bixin on cisplatin-induced genotoxicity in PC12 cells," *Food and Chemical Toxicology*, vol. 50, no. 2, pp. 335–340, 2012.

[5] A. Venugopalan, P. Giridhar, and G. A. Ravishankar, "Food, ethanobotanical and difersified applications of *Bixa orellana* L.: a scope for its improvement through biotechnological mediation," *Indian Journal of Fundamental and Applied Life Science*, vol. 1, pp. 9–31, 2011.

[6] N. M. Gómez-Ortíz, I. A. Vázquez-Maldonado, A. R. Pérez-Espadas, G. J. Mena-Rejón, J. A. Azamar-Barrios, and G. Oskam, "Dye-sensitized solar cells with natural dyes extracted from achiote seeds," *Solar Energy Materials and Solar Cells*, vol. 94, no. 1, pp. 40–44, 2010.

[7] A. Hiendro, F. Hadary, W. Rahmalia, and N. Wahyuni, "Enhanced performance of bixin sensitized TiO$_2$ solar cells with activated kaolinite," *International Journal of Engineering Research*, vol. 4, pp. 40–44, 2012.

[8] T. Ruiz-Anchondo, N. Flores-Holguín, and D. Glossman-Mitnik, "Natural carotenoids as nanomaterial precursors for molecular photovoltaics: a computational DFT study," *Molecules*, vol. 15, no. 7, pp. 4490–4510, 2010.

[9] M. Olson and N. D. Allen, "Natural photodynamic agents and their use," U.S. Patent 60, vol. 982, p. 419, 2010.

[10] M. A. Montenegro, A. d. O. Rios, A. Z. Mercadante, M. A. Nazareno, and C. D. Borsarelli, "Model studies on the photosensitized isomerization of bixin," *Journal of Agricultural and Food Chemistry*, vol. 52, no. 2, pp. 367–373, 2004.

[11] A. d. O. Rios, C. D. Borsarelli, and A. Z. Mercadante, "Thermal degradation kinetics of bixin in an aqueous model system," *Journal of Agricultural and Food Chemistry*, vol. 53, no. 6, pp. 2307–2311, 2005.

[12] T. Keinonen, "Investigation of grating recording in betacarotene-polystyrene film," *Journal of Modern Optics*, vol. 44, no. 3, pp. 555–562, 1997.

[13] Y. Kohno, M. Inagawa, S. Ikoma et al., "Stabilization of a hydrophobic natural dye by intercalation into organo-montmorillonite," *Applied Clay Science*, vol. 54, no. 3-4, pp. 202–205, 2011.

[14] W. Rahmalia, "Synthesis of kaolinite-bixin organoclay and their photostability test," Ph.D. thesis, Magister of Biology, Christian University of Satya Wacana, Salatiga, Indonesia, 2009.

[15] M. B. Abdul Rahman, S. M. Tajudin, M. Z. Hussein, R. N. Z. R. Abdul Rahman, A. B. Salleh, and M. Basri, "Application of natural kaolin as support for the immobilization of lipase from *Candida rugosa* as biocatalsyt for effective esterification," *Applied Clay Science*, vol. 29, no. 2, pp. 111–116, 2005.

[16] C. R. Reddy, Y. S. Bhat, G. Nagendrappa, and B. S. Jai Prakash, "Brønsted and Lewis acidity of modified montmorillonite clay catalysts determined by FT-IR spectroscopy," *Catalysis Today*, vol. 141, no. 1-2, pp. 157–160, 2009.

[17] V. Vimonses, S. Lei, B. Jin, C. W. K. Chow, and C. Saint, "Adsorption of congo red by three Australian kaolins," *Applied Clay Science*, vol. 43, no. 3-4, pp. 465–472, 2009.

[18] R. L. Frost, É. Makó, J. Kristóf, and J. T. Kloprogge, "Modification of kaolinite surface through mechanochemical treatment-a mid-IR and near-IR spectroscopic study," *Spectrochimica Acta Part A: Molecular and Biomolecular Spectroscopy*, vol. 58, no. 13, pp. 2849–2859, 2002.

[19] R. L. Frost, J. Kristóf, E. Makó, and E. Horváth, "A DRIFT spectroscopic study of potassium acetate intercalated mechanochemically activated kaolinite," *Spectrochimica Acta Part A: Molecular and Biomolecular Spectroscopy*, vol. 59, no. 6, pp. 1183–1194, 2003.

[20] T. A. Khan, E. A. Khan, and Shahjahan, "Removal of basic dyes from aqueous solution by adsorption onto binary iron-manganese oxide coated kaolinite: nonlinear isotherm and kinetics modeling," *Applied Clay Science*, vol. 107, pp. 70–77, 2015.

[21] C. Belver, M. A. Bañares Muñoz, and M. A. Vicente, "Chemical activation of a kaolinite under acid and alkaline conditions," *Chemistry of Materials*, vol. 14, no. 5, pp. 2033–2043, 2002.

[22] F. Slaty, H. Khoury, J. Wastiels, and H. Rahier, "Characterization of alkali activated kaolinitic clay," *Applied Clay Science*, vol. 75-76, pp. 120–125, 2013.

[23] Y. Hai, X. Li, H. Wu, S. Zhao, W. Deligeer, and S. Asuha, "Modification of acid-activated kaolinite with TiO$_2$ and its use for the removal of azo dyes," *Applied Clay Science*, vol. 114, pp. 558–567, 2015.

[24] Y. H. Chen and D. L. Lu, "CO$_2$ capture by kaolinite and its adsorption mechanism," *Applied Clay Science*, vol. 104, pp. 221–228, 2015.

[25] F. El Berrichi and S. Zen, "Removal of anionic dyes from aqueous solutions using local activated kaolins as adsorbers," in *Proceedings of 2014 International Conference on Power Systems, Energy, Environment*, pp. 191–196, Interlaken, Switzerland, February 2014.

[26] J. Fafard, O. Lyubimova, S. R. Stoyanov et al., "Adsorption of indole on kaolinite in nonaqueous media: organoclay preparation and characterization, and 3D-RISM-KH molecular theory of solvation investigation," *Journal of Physical Chemistry C*, vol. 117, no. 36, pp. 18556–18566, 2013.

[27] T. A. Khan, S. Dahiya, and I. Ali, "Use of kaolinite as adsorbent: equilibrium, dynamics and thermodynamic studies on the adsorption of Rhodamine B from aqueous solution," *Applied Clay Science*, vol. 69, pp. 58–66, 2012.

[28] W. Rahmalia, J. F. Fabre, T. Usman, and Z. Mouloungui, "Aprotic solvents effect on the UV-visible absorption spectra of bixin," *Spectrochimica Acta Part A: Molecular and Biomolecular Spectroscopy*, vol. 131, pp. 455–460, 2014.

[29] P. Tundo and M. Selva, "The chemistry of dimethyl carbonate," *Accounts of Chemical Research*, vol. 35, no. 9, pp. 706–716, 2002.

[30] B. A. V. Santos, V. M. T. M. Silva, J. M. Loureiro, and A. E. Rodrigues, "Adsorption of H$_2$O and dimethyl carbonate at high pressure over zeolite 3A in fixed bed column," *Industrial & Engineering Chemistry Research*, vol. 53, no. 6, pp. 2473–2483, 2014.

[31] W. Rahmalia, J. F. Fabre, and Z. Mouloungui, "Effect of cyclohexane/acetone ratio on bixin extraction yield by accelerated solvent extraction method," *Procedia Chemistry*, vol. 14, pp. 455–464, 2015.

[32] S. Kumar, A. K. Panda, and R. K. Singh, "Preparation and characterization of acid and alkaline treated kaolin clay," *Bulletin of Chemical Reaction Engineering & Catalysis*, vol. 8, no. 1, pp. 61–69, 2013.

[33] S. Sunardi, Y. Arryanto, and S. Sutarno, "Adsorption of giberellic acid onto natural kaolin from Tatakan, South Kalimantan," *Indonesian Journal of Chemistry*, vol. 9, no. 3, pp. 373–379, 2010.

[34] R. Delhez, Th. H. Keijser, and E. J. Mittemeijer, "Determination of crystallite size and lattice distortions through X-ray diffraction line profile analysis," *Fresenius' Zeitschrift fuer Analytische Chemie*, vol. 312, no. 1, pp. 1–16, 1982.

[35] P. A. Schroeder, "Infrared spectroscopy in clay science: in CMS workshop lectures," in *Teaching Clay Science*, A. Rule and S. Guggenheim, Eds., pp. 181–206, The Clay Mineral Society, Aurora, CO, USA, 2002.

[36] M. A. Qtaitat and I. N. Al-Trawneh, "Characterization of kaolinite of the Baten El-Ghoul region/south Jordan by infrared spectroscopy," *Spectrochimica Acta Part A: Molecular and Biomolecular Spectroscopy*, vol. 61, no. 7, pp. 1519–1523, 2005.

[37] R. A. Schoonheydt and C. T. Johnston, "Surface and interface chemistry of clay minerals," in *Developments in Clay Science*, Chapter 3, pp. 87–113, Elsevier, Amsterdam, Netherlands, 2006.

[38] S. Yariv and H. Cross, *Organo-Clay Complexes and Interactions*, CRC Press, Boca Raton, FL, USA, 2001.

[39] N. Sarier and Ç. Güler, "β-carotene adsorption on acid-activated montmorillonite," *Journal of the American Oil Chemists Society*, vol. 65, no. 5, pp. 776–779, 1988.

[40] L. F. M. Ismail, H. B. Sallam, S. A. Abo Farha, A. M. Gamal, and G. E. A. Mahmoud, "Adsorption behaviour of direct yellow 50 onto cotton fiber: equilibrium, kinetic and thermodynamic profile," *Spectrochimica Acta Part A: Molecular and Biomolecular Spectroscopy*, vol. 131, pp. 657–666, 2014.

[41] H. Freundlich, "Adsorption in solution," *Zeitschrift für Physikalische Chemie*, vol. 57, pp. 384–470, 1906.

[42] I. Langmuir, "The constitution and fundamental properties of solids and liquids. Part I. Solids," *Journal of the American Chemical Society*, vol. 38, no. 11, pp. 2221–2295, 1916.

[43] S. M. Rao, "Mechanism controlling the volume change behavior of kaolinite," *Clays and Clay Minerals*, vol. 33, no. 4, pp. 323–328, 1985.

[44] R. Boussessi, S. Guizani, M. L. S. Senent, and N. Jaïdane, "Theoretical characterization of dimethyl carbonate at low temperatures," *Journal of Physical Chemistry A*, vol. 119, no. 17, pp. 4057–4064, 2015.

[45] T. Afroz, D. M. Seo, S. D. Han, P. D. Boyle, and W. A. Henderson, "Structural interactions within lithium salt solvates: acyclic carbonates and esters," *Journal of Physical Chemistry C*, vol. 119, no. 13, pp. 7022–7027, 2015.

[46] B. P. Kar, N. Ramanathan, K. Sundararajan, and K. S. Viswanathan, "Conformations of dimethyl carbonate and its complexes with water: a matrix isolation infrared and ab initio study," *Journal of Molecular Structure*, vol. 1024, pp. 84–93, 2012.

[47] S. Lagergren, "About the theory of so-called adsorption of soluble substances," *Kungliga Svenska Vetenskapsakademiens, Handlingar, Band*, vol. 24, pp. 1–39, 1898.

[48] Y. Ho and G. McKay, "Pseudo-second order model for sorption processes," *Process Biochemistry*, vol. 34, no. 5, pp. 451–465, 1999.

[49] T. Lóránd, P. Molnár, J. Deli, and G. Tóth, "FT-IR study of some seco- and apocarotenoids," *Journal of Biochemical and Biophysical Methods*, vol. 53, no. 1–3, pp. 251–258, 2002.

9

Synthesis, Characterization, and BSA-Binding Studies of Novel Sulfonated Zinc-Triazine Complexes

Nalin Abeydeera,[1] **Inoka C. Perera,**[2] **and Theshini Perera** ⓘ[1]

[1]*Department of Chemistry, University of Sri Jayewardenepura, Nugegoda, Sri Lanka*
[2]*Department of Zoology and Environmental Science, University of Colombo, Colombo, Sri Lanka*

Correspondence should be addressed to Theshini Perera; theshi@sjp.ac.lk

Academic Editor: Francesco Paolo Fanizzi

Four Zn(II) complexes containing a pyridyl triazine core ($L1 = 3$-(2-pyridyl)-5,6-di(2-furyl)-1,2,4-triazine-$5',5''$-disulfonic acid disodium salt and $L2 = 3$-(2-pyridyl)-5,6-diphenyl-1,2,4-triazine-$4',4''$-disulfonic acid sodium salt) were synthesized, and their chemical formulas were finalized as $[Zn(L1)Cl_2]\cdot 5H_2O\cdot ZnCl_2$ (**1**), $[Zn(L1)_2Cl_2]\cdot 4H_2O\cdot 2CH_3OH$ (**2**), $[Zn(L2)Cl_2]\cdot 3H_2O\cdot CH_3OH$ (**3**), and $[Zn(L2)_2Cl_2]$ (**4**). The synthesized complexes are water soluble, making them good candidates for biological applications. All four complexes have been characterized by elemental analysis and ^1H NMR, IR, and UV-Vis spectroscopy. The IR stretching frequency of N=N and C=N bonds of complexes **1–4** have shifted to lower frequencies in comparison with free ligands, and a bathochromic shift was observed in UV-Vis spectra of all four complexes. The binding studies of ligands and complexes **1–4** with bovine serum albumin (BSA) resulted binding constants (K_b) of $3.09 \times 10^4\,M^{-1}$, $12.30 \times 10^4\,M^{-1}$, and $16.84 \times 10^4\,M^{-1}$ for ferene, complex **1**, and complex **2**, respectively, indicating potent serum distribution via albumins.

1. Introduction

The potential use of zinc complexes as antidiabetic insulin mimetics [1], antimicrobial [2], and anticancer agents [3] have garnered a renewed interest in such complexes among other applications, such as serving as tumor photo sensitizers [4], radioprotective agents [5], and antidandruff agents (Zn pyrithione-ZPT) [6]. Our interest in sulfa drug moieties has been fueled by the fact that they possess a wide range of pharmaceutical applications [7]. Of particular interest are 1,2,4-triazine derivatives because they have been reported to possess many biological activities such as kinase inhibition [8], antihypertensivity [9], antimicrobial [10], anticancer [11], anti-HIV [12, 13], and anti-inflammatory activities [8].

Novel polyanionic sulfonated aromatic synthetic platinum chelates were preliminarily evaluated for their HIV-1 virucidal activity, due to the presence of sulfonated aromatic groups and metals in the most active members [14, 15]. Furthermore, we recently reported that rhenium tricarbonyl complexes of ferene and ferrozine have exhibited the potential to be used as biological imaging agents [16]. (Chemical structures of ferene and ferrozine are illustrated in Figure 1.) However, to the best of our knowledge, no reports exist of zinc complexes of sulfonated 1,2,4-triazine derivatives.

It is noteworthy that although zinc complexes bearing the 5,6-diphenyl-3, 2-pyridyl-1 2,4-triazine ligand were first synthesized more than a decade ago [17, 18], biological studies have been reported for only one such complex, albeit only recently [19]. Therefore, our goal has been to synthesize zinc complexes bearing sulfonated pyridyl triazine derivatives and to assess their interaction with biological targets.

Being the most abundant protein in blood, serum albumin maintains the colloid osmotic pressure, while playing a major role in transport and sustained release of many biomolecules such as steroids, fatty acids, and hormones. Serum albumin also serves as a carrier protein for drug molecules [20]. Small molecule interaction with serum

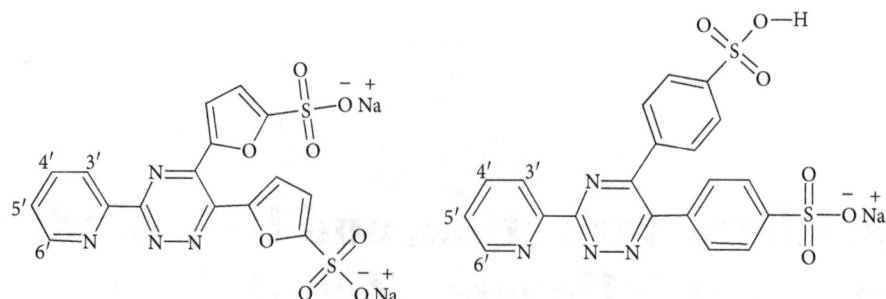

FIGURE 1: Chemical structures of 3-(2-pyridyl)-5,6-di(2-furyl)-1,2,4-triazine-5',5''-disulfonic acid disodium salt (ferene: L1) (left) and 3-(2-pyridyl)-5,6-diphenyl-1,2,4-triazine-4',4''-disulfonic acid sodium salt (ferrozine: L2) (right).

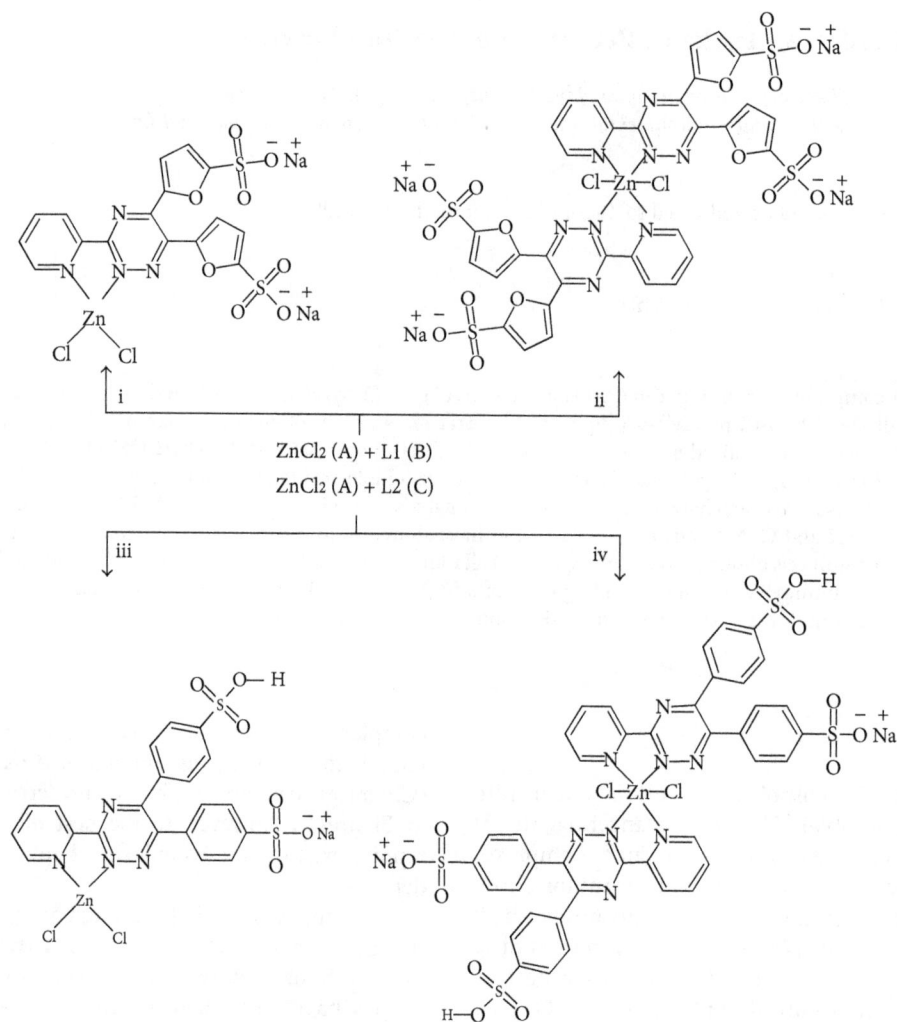

Conditions: (i) A : B = 1, (ii) A : B = 1 : 2, (iii) A : C = 1 : 1, (iv) A : C = 1 : 2

FIGURE 2: Synthetic routes for the preparation ML1Cl$_2$ (complex **1**) (i), M(L1)$_2$Cl$_2$ (complex **2**) (ii), ML2Cl$_2$ (complex **3**) (iii), and M(L2)$_2$Cl$_2$ (complex **4**) (iv) complexes. An associated molecule of ZnCl$_2$ in complex **1** and solvent molecules in complexes **1–4** have been omitted for clarity.

albumin is thus exploited in pharmaceutical research, where affinity to albumins is indicative of drugs with high serum distribution [21]. Through the interaction between bovine serum albumin (BSA), an analog of human serum albumin,

and the novel compounds, we seek to investigate their pharmacokinetic associations.

Thus, we report here the synthesis and characterization of four novel metal complexes of the type ML$_n$Cl$_2$ (Figure 2) (where $M = Zn^{2+}$, $L = 3$-(2-pyridyl)-5,6-di(2-furyl)-1,2,4-triazine-$5',5''$-disulfonic acid disodium salt/3-(2-pyridyl)-5,6-diphenyl-1,2,4-triazine-$4',4''$-disulfonic acid sodium salt and $n = 1/2$) and their BSA-binding studies.

2. Experimental

2.1. Materials and Methods. All chemicals (zinc chloride (3-(2-pyridyl)-5,6-di(2-furyl)-1,2,4-triazine-$5',5''$-disulfonic acid disodium salt (ferene/L1) and 3-(2-pyridyl)-5,6-diphenyl-1,2,4-triazine-$4',4''$-disulfonic acid sodium salt (ferrozine/L2)), methanol, diethyl ether, ethanol, bovine serum albumin (BSA), tris-HCl buffer (tris(hydroxymethyl)-aminomethane), sodium chloride (NaCl), and analytical grade water) were obtained from Sigma-Aldrich. All the solvents and chemicals were of analytical grade and were used as received, without further purification.

2.2. NMR Measurements. ^{1}H NMR spectra were recorded in D_2O on a Bruker 400 MHz spectrometer. Peak positions are relative to tetramethylsilane (TMS) as reference. All NMR data were processed with TopSpin 3.2 and Mestre-C software.

2.3. Elemental Analysis. CHNS elemental analysis was performed by PerkinElmer 2400 Series II CHNS/O Elemental analyzer at Atlantic Microlab, USA.

2.4. Melting Point Determination. Melting points were manually determined in open capillaries.

2.5. UV-Visible Spectroscopy. Electronic spectra for ligand and metal complex were obtained on Spectro UV-Vis auto version 3.10, UV-2602 spectrophotometer. The spectral range was 200–800 nm. Spectra were obtained in methanol with baseline correction. Spectral data were processed with UV WIN software.

2.6. FTIR Analysis. FTIR spectra were recorded on a Thermo Scientific NICOLET iS10 spectrophotometer. ATR spectra were obtained within the 4000–600 cm^{-1} spectral range. Spectral data were processed with OMNIC software.

2.6.1. Preparation of [Zn(L1)Cl$_2$]·5H$_2$O·ZnCl$_2$ (1). A solution of ferene (0.1 mmol, 0.0494 g) in methanol (4.0 cm^3) was added to zinc chloride (0.1 mmol, 0.0140 g) in methanol (1.0 cm^3). Then, the resulting mixture was stirred for 3 hours at 60–70°C [1]. The yellow precipitate which was obtained was collected by filtration, washed with ethanol and diethyl ether, and dried. C$_{16}$H$_8$Cl$_2$N$_4$Na$_2$O$_8$S$_2$Zn·5H$_2$O·ZnCl$_2$, yield 0.0517 g, 78%. Anal. calc. for C$_{16}$H$_8$Cl$_2$N$_4$Na$_2$O$_8$S$_2$Zn·5H$_2$O·ZnCl$_2$: C, 22.54; H, 1.92; N, 6.68; S, 7.64. Found: C, 22.13; H, 1.89; N, 6.57; S, 7.70%; melting point: >315°C; UV-Vis (MeoH) (λ_{max} (nm) (ε M^{-1}·cm^{-1})): 210, 242, 331, 364; FTIR (ATR) (cm^{-1}): 1504 (v(N=N)),1582 (v(C=N)); ^{1}H NMR (D$_2$O, δ ppm) 8.87 (d, H$_6$), 8.79 (d, H$_3$), 8.28 (t, H$_4$), 7.87 (t, H$_5$), 7.09–7.46 (m, 4H, furyl H).

2.6.2. Preparation of [Zn(L1)$_2$Cl$_2$]·4H$_2$O·2CH$_3$OH (2). A solution of ferene (0.2 mmol, 0.0988 g) in methanol (4.0 cm^3) was added to zinc chloride (0.1 mmol, 0.0140 g) in methanol (1.0 cm^3) [2]. Then, the resulting mixture was stirred for 6 hours at 60–70°C. The yellow precipitate which was obtained was collected by filtration, washed with ethanol and diethyl ether, and dried. C$_{32}$H$_{16}$Cl$_2$N$_8$Na$_4$O$_{16}$S$_4$Zn·4H$_2$O·2CH$_3$OH, yield 0.0689 g, 61% based on zinc chloride. Anal. calc. for C$_{32}$H$_{16}$Cl$_2$N$_8$Na$_4$O$_{16}$S$_4$Zn·4H$_2$O·2CH$_3$OH: C, 32.38; H, 2.56; N, 9.26; S, 10.57. Found: C, 32.5; H, 2.67; N, 9.63; S, 10.83%; melting point: >315°C; UV-Vis (MeoH) (λ_{max} (nm) (ε M^{-1}·cm^{-1})): 209, 244, 337, 370; FTIR (ATR) (cm^{-1}): 1511 (v(N=N)),1586 (v(C=N)); ^{1}H NMR (D$_2$O, δ ppm) 8.86 (d, H$_6$), 8.76 (d, H$_3$), 8.27 (t, H$_4$), 7.84 (t, H$_5$), 7.15–7.41 (m, 8H, furyl H).

2.6.3. Preparation of [Zn(L2)Cl$_2$]·3H$_2$O·CH$_3$OH (3). A solution of ferrozine (0.1 mmol, 0.0492 g) in methanol (4.0 cm^3) was added to zinc chloride (0.1 mmol, 0.0140 g) in methanol (1.0 cm^3) [3]. Then, the resulting mixture was stirred for 3 hours at 60–70°C. The light yellow precipitate which was obtained was collected by filtration, washed with ethanol and diethyl ether, and dried. C$_{20}$H$_{13}$Cl$_2$N$_4$NaO$_6$S$_2$Zn·3H$_2$O·CH$_3$OH, yield 0.0433 g, 69% based on zinc chloride. Anal. calc. for C$_{20}$H$_{13}$Cl$_2$N$_4$NaO$_6$S$_2$Zn·3H$_2$O·CH$_3$OH: C, 35.28; H, 4.24; N, 7.84; S, 8.97. Found: C, 35.27; H, 4.36; N, 7.89; S, 9.13%; melting point: >315°C; UV-Vis (MeoH) (λ_{max} (nm) (ε M^{-1}·cm^{-1})): 208, 240, 291, 323; FTIR (ATR) (cm^{-1}): 1497 (v(N=N)), 1599 (v(C=N)); ^{1}H NMR (D$_2$O, δ ppm) 8.86 (d, H$_6$), 8.76 (d, H$_3$), 8.27 (t, H$_4$), 7.84 (t, H$_5$), 7.46–8.28 (m, 8H, phenyl H).

2.6.4. Preparation of [Zn(L2)$_2$Cl$_2$]·5H$_2$O·CH$_3$OH (4). A solution of ferrozine (0.2 mmol, 0.0984 g) in methanol (4.0 cm^3) was added to zinc chloride (0.1 mmol, 0.0140 g) in methanol (1.0 cm^3) [4]. Then, the resulting mixture was stirred for 6 hours at 60–70°C. The light yellow precipitate which was obtained was collected by filtration, washed with ethanol and diethyl ether, and dried. C$_{41}$H$_{40}$Cl$_2$N$_8$Na$_2$O$_{18}$S$_4$Zn·5H$_2$O·CH$_3$OH, yield: 0.0579 g, 52% based on zinc chloride. Anal. calc. for C$_{41}$H$_{40}$Cl$_2$N$_8$Na$_2$O$_{18}$S$_4$Zn·5H$_2$O·CH$_3$OH: C, 39.59; H, 3.24; N, 9.01; S, 10.32. Found: C, 38.93; H, 3.04; N, 8.84; S, 10.67%; melting point: >315°C; UV-Vis (MeoH) (λ_{max} (nm) (ε M^{-1}·cm^{-1})): 216, 242, 298, 336; FTIR (ATR) (cm^{-1}): 1498 (v(N=N)), 1596 (v(C=N)); ^{1}H NMR (D$_2$O, δ ppm) 9.05 (d, H$_6$), 8.99 (d, H$_3$), 8.67 (t, H$_4$), 8.67 (t, H$_5$), 7.50–8.31 (m, 16H, phenyl H).

2.7. BSA-Binding Assay. A 6 μM BSA solution was prepared in a buffer containing 5 mM tris-HCl/50 mM NaCl by continuous stirring for 1 hr at room temperature. A 1×10^{-3} M stock solution of complexes and ligand was prepared in distilled water.

Absorption titration was carried out by keeping BSA concentration constant (6 μM) and varying the concentrations of the complexes and ligand (010 μM). After 10 min incubation at room temperature, absorbance was measured for each solution in the range of 250–300 nm wavelengths, and λ_{max} was recorded at 280 nm. Then, the change of λ_{max}

TABLE 1: Comparison of UV-Vis data of ferene, ferrozine, and complexes **1–4**.

$\lambda_{Zn^{2+}}$ (nm)	λ_{ferene} (nm)	$\lambda_{complex\ 1}$ (nm)	$\lambda_{complex\ 2}$ (nm)	$\lambda_{ferrozine}$ (nm)	$\lambda_{complex\ 3}$ (nm)	$\lambda_{complex\ 4}$ (nm)
207	209	210	209	211	208	216
222	243	242	244	236	240	242
—	306	331	337	285	291	298
—	339	364	370	310	323	336

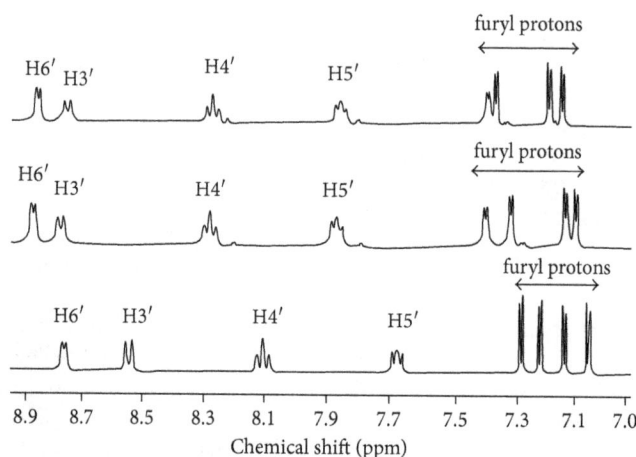

FIGURE 3: ^{1}H NMR spectrum of ferene (bottom) and complexes **1** (middle) and **2** (top).

was recorded for each solution. All absorbance measurements were triplicated and corrected for background absorbance by the compounds. The plot of $1/(A - A_0)$ (where A_0 is the initial absorbance of the free BSA at 280 nm and A is the absorbance of BSA in the presence of different concentrations of the complex) versus $1/[complex]$ is a linear curve, and the binding constant (K_b) can be obtained from the ratio of the intercept to slope [19].

3. Results and Discussion

3.1. Synthesis. In order to synthesize the metal complexes, zinc chloride and the relevant ligands in 1:1 and 1:2 ratios were used (Figure 2).

3.2. UV-Visible Spectroscopy. UV-visible spectra of ligands and complexes **1–4** recorded in methanol showed significant differences between the absorption peaks of ligands and their complexes (Table 1, Figures S1 and S2, Supporting Information). Since both ligands bear conjugated systems, π–π^* transition is possible. In all four complexes, the wavelengths have shifted towards the longer wavelength range (bathochromic shift) because of changes in the conjugated electron system due to formation of metal ligand bonds. These observations are in agreement with previously reported zinc pyridyl triazine derivatives [18] and copper pyridyl triazine derivatives [22], upon coordination of ligand to metal.

3.3. ^{1}H NMR Analysis. Complexes **1–4** were characterized using ^{1}H NMR spectroscopy in D$_2$O. All the peaks were

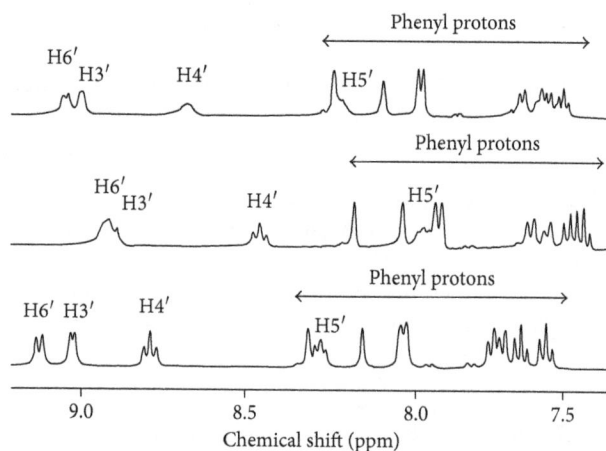

FIGURE 4: ^{1}H NMR spectrum of ferrozine (bottom) and complexes **3** (middle) and **4** (top).

assigned in comparison with related structures of both ligands (Figures 3 and 4).

The splitting pattern of the free ferene ligand can be observed for complexes **1** and **2**. However, due to the donation of electrons from the nitrogen in pyridine and triazine rings to the metal, the electron density of the ferene ligand is reduced, and thus protons of the metal complex should appear more downfield than the ferene ligand. Furthermore, the downfield shift, which will hence be denoted as $\Delta\delta$, of the H$_6$ signal is expected to be higher than that of other protons because it is closer to the pyridine N. In uncoordinated ferene ligand (L1), the pyridyl H$_6$ signal (8.74 ppm, Table 2) is its most downfield doublet consistent

TABLE 2: Selected ^{1}H NMR chemical shifts (ppm) of ferene, ferrozine, and complexes **1–4**.

Proton no.	H$_6$ (d)	H$_5$ (t)	H$_4$ (t)	H$_3$ (d)
Ferene	8.74	7.67	8.10	8.54
Ferrozine	9.15	8.27	8.81	9.05
Complex **1**	8.87	7.87	8.28	8.79
Complex **2**	8.86	7.84	8.27	8.76
Complex **3**	8.92	7.98	8.47	8.89
Complex **4**	9.05	8.21	8.67	8.99
$\Delta\delta$ (ppm) of complex **1**	(+) 0.13	(+) 0.20	(+) 0.18	(+) 0.25
$\Delta\delta$ (ppm) of complex **2**	(+) 0.12	(+) 0.17	(+) 0.17	(+) 0.22
$\Delta\delta$ (ppm) of complex **3**	(−) 0.23	(−) 0.29	(−) 0.34	(−) 0.16
$\Delta\delta$ (ppm) of complex **4**	(−) 0.10	(−) 0.06	(−) 0.14	(−) 0.06

TABLE 3: FTIR data of complexes **1–4**.

Ligand/complex	$\nu_{C=N}$ (cm^{-1})	$\nu_{N=N}$ (cm^{-1})
Ferene	1590	1510
Complex **1**	1582	1504
Complex **2**	1586	1511
Ferrozine	1608	1502
Complex **3**	1599	1497
Complex **4**	1596	1498

with its close proximity to the pyridyl nitrogen atom. In a spectrum of complex **1**, the H$_6$ signal appears further downfield (8.87 ppm, Figure 3) confirming metal-pyridine N bond formation. However, the observed change in shift of H$_6$ was small in both **1** ($\Delta\delta$; +0.13 ppm at 8.87 ppm) and **2** (($\Delta\delta$; +0.12 ppm at 8.86 ppm) in comparison with the change in shift observed for the H$_3$ proton (8.79 ppm and 8.76 ppm), which had the highest change in downfield shift (($\Delta\delta$; +0.25 and +0.22 ppm, resp.). The four doublets due to furyl ring protons (7.00–7.34 ppm range) also appear more downfield (7.15–7.45 ppm) upon metal bonding. Higher downfield shifts were observed for H$_3$ versus H$_6$ in ^{1}H NMR spectra of Zn(dppt)Cl$_2$·0.5H$_2$O and Zn(dppt)$_2$Cl$_2$·2H$_2$O (dppt = 5, 6 diphenyl-3-(2-pyridyl)-1,2,4- triazine) reported previously [18].

Although we expected similar observations for spectra of complexes **3** and **4**, unusual upfield shifts of the peaks attributed to H$_6$ (8.92 and 9.05 ppm), H$_5$ (7.98 and 8.21 ppm), H$_4$ (8.47 and 8.67 ppm), and H$_3$ (8.89 and 8.99 ppm) (Figure 4) were observed in complexes **3** and **4**, respectively, in comparison with that of the uncoordinated ferrozine ligand (H$_6$: 9.15 ppm, H$_5$: 8.27 ppm, H$_4$: 8.81 ppm, and H$_3$: 9.05 ppm). We attribute the observed upfield shift to possible π stacking of phenyl rings. Upfield shifts due to π stacking have been reported in previous studies on Zn(II) with mono- and dianionic pyrrole-2-imine complexes and zinc azadipyrromethene [23]. Although the ^{1}H NMR spectrum of ferrozine/L2 is comparatively more complicated due to the protons of the two phenyl rings, the pyridyl ring protons can be easily distinguished (Figure 4). Upfield shifts observed in complexes **3** and **4** were not observed in complexes **1** and **2**, which had furyl rings (Figure 3).

3.4. FTIR Analysis. Literature data have been used where relevant to get assignment of ligands [18]. The stretching frequency (ν) of the N=N bond in the triazine ring and C=N bond in the pyridine ring serves as important indicators of the formation of new metal ligand bonds [18]. In all four complexes, $\nu_{N=N}$ and $\nu_{C=N}$ have shifted to lower frequencies (Table 3) due to the formation of new metal ligand bonds which in turn lowers the strength of the N=N and C=N bonds. This observation allowed us to confirm that the complex was formed via the donation of a lone pair of

electrons each, from the triazine ring and from the pyridine ring, to zinc. For example, upon formation of complex **1**, $\nu_{C=N}$ (1590 cm^{-1}) and $\nu_{N=N}$ (1510 cm^{-1}) appear at 1582 cm^{-1} and 1504 cm^{-1} (Table 3), respectively, due to the change in chemical environment. FTIR spectra show broad peaks at around 3400 cm^{-1} region due to OH vibration.

3.5. Elemental Analysis. Empirical formulas related to experimental values of the complexes (Experimental) give the exact molecular formulas and experimental values are closer to the expected values. Some deviate from the theoretical values due to the residual solvent (methanol) and water molecules.

Elemental analysis data suggest that complex **1** exists as [Zn(L1)Cl$_2$]·5H$_2$O·ZnCl$_2$. The experimental data obtained from elemental analysis show a significant decrease in the carbon percentage of this complex than expected without the extra zinc ion, prompting us to include an associated zinc chloride molecule in **1**. However, the exact binding mode or type of the interaction of this extra ZnCl$_2$ molecule with ferene ligand cannot be explained with the obtained data. Complexes **2–4** have no such discrepancy, and the molecular formula were confirmed from elemental analysis to be [Zn(L1)$_2$Cl$_2$]·4H$_2$O·2CH$_3$OH, [Zn(L2)Cl$_2$]·3H$_2$O·CH$_3$OH, and [Zn(L2)$_2$Cl$_2$]·5H$_2$O·CH$_3$OH, respectively.

3.6. BSA-Binding Assay. Anjomshoa and coworkers have previously investigated bovine serum albumin- (BSA-) binding properties of Zn(dppt)$_2$Cl$_2$·2H$_2$O (dppt = 5,6-diphenyl-3,2-pyridyl-1 2,4-triazine) [19], which required an organic solvent such as dimethyl sulfoxide to be added to increase solubility. However, complexes **1–4** reported in this study are highly soluble in water which makes them compatible with biological systems.

Absorbance measurements at UV range are useful to identify the conformational changes in proteins. BSA has a maximum absorbance peak at 280 nm. Analyzing the absorbance spectra of BSA upon addition of ferene, complex **1**, and complex **2** (Figure 5) clearly shows that the peak intensity at 280 nm increases upon addition of the compounds, indicating association with BSA causes a change in the polarity of the microenvironment of tryptophan and tyrosine residues in a concentration dependent manner.

With increasing concentrations of ferene ligand and complexes **1** and **2**, the maximum absorbance was increased. Using the graph of $1/(A - A_0)$ against $1/[C]$, binding constants were calculated (Figure 6) as 3.09×10^4 M^{-1},

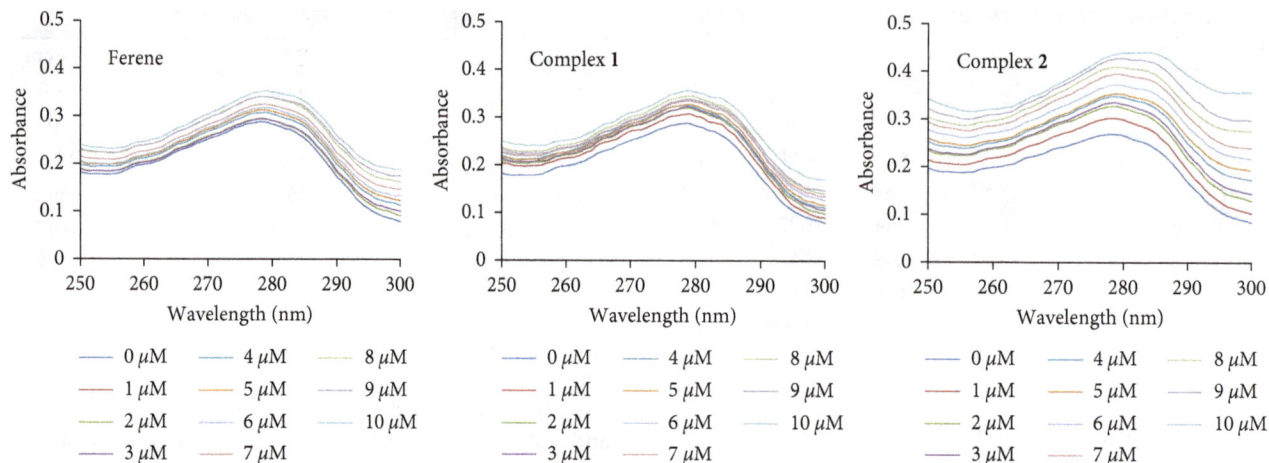

FIGURE 5: Absorbance spectra of BSA in the presence of different concentrations of ferene, complex **1**, and complex **2**.

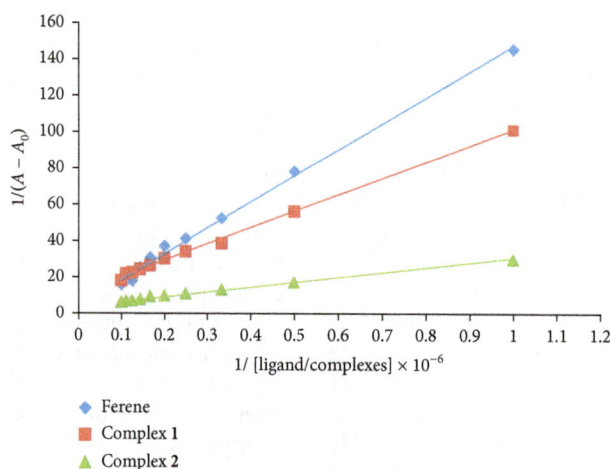

FIGURE 6: The graph of $1/(A - A_0)$ versus $1/[\text{ligand/complexes}]$ for ferene ligand ($R^2 = 0.9955$), complex **1** ($R^2 = 0.9951$), and complex **2** ($R^2 = 0.9955$).

$12.30 \times 10^4 \, \text{M}^{-1}$, and $16.84 \times 10^4 \, \text{M}^{-1}$ for ferene, complex **1**, and complex **2**, respectively. These values are within the range of 10^4–$10^6 \, \text{M}^{-1}$ as expected from a good BSA carrier activity *in vivo* [19]. Such changes upon BSA addition were not observed in ferrozine and complexes **3** and **4**.

4. Conclusions

In this study, we have described the synthesis and characterization of four novel zinc complexes by spectroscopic methods. According to the UV-Vis spectra, a bathochromic shift has been observed for all four complexes. FTIR data provide evidence that Zn-N bonds are formed via N atoms of the triazine and pyridine rings. Accordingly, the stretching frequency of N=N and C=N bonds of all four complexes have shifted to the low frequency range in comparison with the free ligands. Elemental analysis was used to determine the empirical formula of complexes **1–4**.

It is our belief that the scaffolds reported herein may provide a novel platform for drug designing. We have demonstrated that such systems possess high affinity to serum albumin, indicating their potential to be distributed in serum [24]. Sulfonated groups have aided to increase water solubility of the complexes, where aqueous solubility predicted by ChemAxon (https://disco.chemaxon.com) have also supported that the new complexes are soluble in water under biologically relevant pH ranges (data not shown). *In vivo* testing is warranted to explore the implications of such properties on cellular metabolism to delineate the function of the novel triazine complexes in biological systems.

Acknowledgments

Financial assistance by Grant no. ASP/01/RE/SCI/2015/19 of the University of Sri Jayewardenepura is gratefully acknowledged. The authors thank Professor Luigi Marzilli and Ms. Kokila Ranasinghe of Louisiana State University for obtaining NMR data and for useful discussions. Ms. Taniya Darshani is acknowledged for determination of melting points.

References

[1] S. Fujimoto, H. Yasui, and Y. Yoshikawa, "Development of a novel antidiabetic zinc complex with an organoselenium ligand at the lowest dosage in KK-A(y) mice," *Journal of Inorganic Biochemistry*, vol. 121, pp. 10–15, 2013.

[2] N. Poulter, M. Donaldson, G. Mulley et al., "Plasma deposited metal Schiff-base compounds as antimicrobials," *New Journal of Chemistry*, vol. 35, no. 7, pp. 1477–1484, 2011.

[3] S. Anbu, S. Kamalraj, B. Varghese, J. Muthumary, and M. Kandaswamy, "A series of oxyimine-based macrocyclic dinuclear zinc(II) complexes enhances phosphate ester hydrolysis, DNA binding, DNA hydrolysis, and lactate dehydrogenase inhibition and induces apoptosis," *Inorganic Chemistry*, vol. 51, no. 10, pp. 5580–5592, 2012.

[4] Q. Huang, Z. Pan, P. Wang, Z. Chen, X. Zhang, and H. Xu, "Zinc(II) and copper(II) complexes of β-substituted hydroxylporphyrins as tumor photosensitizers," *Bioorganic and Medicinal Chemistry Letters*, vol. 16, no. 11, pp. 3030–3033, 2006.

[5] S. Emami, S. J. Hosseinimehr, S. M. Taghdisi, and S. Akhlaghpoor, "Kojic acid and its manganese and zinc complexes as potential radioprotective agents," *Bioorganic and Medicinal Chemistry Letters*, vol. 17, no. 1, pp. 45–48, 2007.

[6] K. V. Thomas, "Determination of the antifouling agent zinc pyrithione in water samples by copper chelate formation and high-performance liquid chromatography–atmospheric pressure chemical ionisation mass spectrometry," *Journal of Chromatography A*, vol. 833, no. 1, pp. 105–109, 1999.

[7] M. Aly, A. Gobouri, S. A. Hafez, and H. Saad, "Synthesis, reactions, and biological activity of some triazine derivatives containing sulfa drug moieties," *Russian Journal of Bioorganic Chemistry*, vol. 41, no. 4, pp. 437–450, 2015.

[8] J. T. Hunt, T. Mitt, R. Borzilleri et al., "Discovery of the pyrrolo[2,1-f][1,2,4] triazine nucleus as a new kinase inhibitor template," *Journal of Medicinal Chemistry*, vol. 47, no. 16, pp. 4054–4059, 2004.

[9] F. Krauth, H.-M. Dahse, H.-H. Rüttinger, and P. Frohberg, "Synthesis and characterization of novel 1,2,4-triazine derivatives with antiproliferative activity," *Bioorganic and Medicinal Chemistry*, vol. 18, no. 5, pp. 1816–1821, 2010.

[10] S. K. Pandey, A. Singh, and A. Singh, "Antimicrobial studies of some novel quinazolinones fused with [1,2,4]-triazole, [1,2,4]-triazine and [1,2,4,5]-tetrazine rings," *European Journal of Medicinal Chemistry*, vol. 44, no. 3, pp. 1188–1197, 2009.

[11] Z. El-Gendy, J. Morsy, H. Allimony, W. Ali, and R. Abdel-Rahman, "Synthesis of heterobicyclic nitrogen systems bearing the 1,2,4-triazine moiety as anti-HIV and anticancer drugs, Part III," *Die Pharmazie*, vol. 56, no. 5, pp. 376–383, 2001.

[12] R. Abdel-Rahman, "Chemistry of uncondensed 1,2,4-triazines. Part IV. Synthesis and chemistry of bioactive 3-amino-1,2,4-triazines and related compounds–an overview," *Die Pharmazie*, vol. 56, no. 4, pp. 275–286, 2001.

[13] Y.-Z. Xiong, F.-E. Chen, J. Balzarini, E. De Clercq, and C. Pannecouque, "Non-nucleoside HIV-1 reverse transcriptase inhibitors. Part 11: structural modulations of diaryltriazines with potent anti-HIV activity," *European Journal of Medicinal Chemistry*, vol. 43, no. 6, pp. 1230–1236, 2008.

[14] A. Vzorov, D. Bhattacharyya, L. Marzilli, and R. Compans, "Prevention of HIV-1 infection by platinum triazines," *Antiviral Research*, vol. 65, no. 2, pp. 57–67, 2005.

[15] V. Maheshwari, D. Bhattacharyya, F. R. Fronczek, P. A. Marzilli, and L. G. Marzilli, "Chemistry of HIV-1 virucidal Pt complexes having neglected bidentate sp^2 N-donor carrier ligands with linked triazine and pyridine rings. synthesis, NMR spectral features, structure, and reaction with guanosine," *Inorganic Chemistry*, vol. 45, no. 18, pp. 7182–7190, 2006.

[16] K. Ranasinghe, S. Handunnetti, I. C. Perera, and T. Perera, "Synthesis and characterization of novel rhenium (I) complexes towards potential biological imaging applications," *Chemistry Central Journal*, vol. 10, no. 1, p. 71, 2016.

[17] F. Marandi, A. A. Soudi, A. Morsali, and R. Kempe, "Zinc(II) and cadmium(II) complexes of the 3-(2-Pyridyl)-5,6-diphenyl-1,2,4-triazine (PDPT) ligand, structural studies of [Zn (PDPT) 2Cl (ClO$_4$)] and [Cd (PDPT)$_2$ (NO$_3$)(ClO$_4$)]," *Zeitschrift für anorganische und allgemeine Chemie*, vol. 631, no. 15, pp. 3070–3073, 2005.

[18] V. Béreau and J. Marrot, "Coordination studies of 5,6-diphenyl-3-(2-pyridyl)-1,2,4-triazine towards Zn^{2+} cation. Synthesis and characterization by X-ray diffraction and spectroscopic methods," *Comptes Rendus Chimie*, vol. 8, no. 6-7, pp. 1087–1092, 2005.

[19] M. Anjomshoa, S.J. Fatemi, M. Torkzadeh-Mahani, and H. Hadadzadeh, "DNA-and BSA-binding studies and anticancer activity against human breast cancer cells (MCF-7) of the zinc (II) complex coordinated by 5,6-diphenyl-3-(2-pyridyl)-1,2,4-triazine," *Spectrochimica Acta Part A: Molecular and Biomolecular Spectroscopy*, vol. 127, pp. 511–520, 2014.

[20] O. A. Chaves, V. A. da Silva, C. M. R. Sant'Anna et al., "Binding studies of lophirone B with bovine serum albumin (BSA): combination of spectroscopic and molecular docking techniques," *Journal of Molecular Structure*, vol. 1128, pp. 606–611, 2017.

[21] M. E. Salive, J. Cornoni-Huntley, C. L. Phillips et al., "Serum albumin in older persons: relationship with age and health status," *Journal of Clinical Epidemiology*, vol. 45, no. 3, pp. 213–221, 1992.

[22] B. Machura, A. Świtlicka, J. Kruszynski, J. Mroziński, J. Kłak, and J. Kusz, "Coordination studies of 5,6-diphenyl-3-(2-pyridyl)-1,2,4-triazine towards Cu^{2+} cation. X-ray studies, spectroscopic characterization and DFT calculations," *Polyhedron*, vol. 27, no. 13, pp. 2959–2967, 2008.

[23] T. S. Teets, D. V. Partyka, J. B. Updegraff III, and T. G. Gray, "Homoleptic, four-coordinate azadipyrromethene complexes of d^{10} zinc and mercury," *Inorganic Chemistry*, vol. 47, no. 7, pp. 2338–2346, 2008.

[24] F. Kratz, "Albumin as a drug carrier: design of prodrugs, drug conjugates and nanoparticles," *Journal of Controlled Release*, vol. 132, no. 3, pp. 171–183, 2008.

10

Synthesis, Characterization, Cytotoxic Activity, and Interactions with CT-DNA and BSA of Cationic Ruthenium(II) Complexes Containing Dppm and Quinoline Carboxylates

Edinaldo N. da Silva,[1] Paulo A. B. da Silva,[1] Angélica E. Graminha,[2]
Pollyanna F. de Oliveira,[3] Jaqueline L. Damasceno,[3] Denise C. Tavares,[3]
Alzir A. Batista,[2] and Gustavo Von Poelhsitz[1]

[1]Instituto de Química, Universidade Federal de Uberlândia, 38400-902 Uberlândia, MG, Brazil
[2]Departamento de Química, Universidade Federal de São Carlos, 13565-905 São Carlos, SP, Brazil
[3]Universidade de Franca, 14404-600 Franca, SP, Brazil

Correspondence should be addressed to Gustavo Von Poelhsitz; gustavopoelhsitz@ufu.br

Academic Editor: Claudio Pettinari

The complexes cis-[Ru(quin)(dppm)$_2$]PF$_6$ and cis-[Ru(kynu)(dppm)$_2$]PF$_6$ (quin = quinaldate; kynu = kynurenate; dppm = bis(diphenylphosphino)methane) were prepared and characterized by elemental analysis, electronic, FTIR, ^{1}H, and ^{31}P{^{1}H} NMR spectroscopies. Characterization data were consistent with a cis arrangement for the dppm ligands and a bidentate coordination through carboxylate oxygens of the quin and kynu anions. These complexes were not able to intercalate CT-DNA as shown by circular dichroism spectroscopy. On the other hand, bovine serum albumin (BSA) binding constants and thermodynamic parameters suggest spontaneous interactions with this protein by hydrogen bonds and van der Waals forces. Cytotoxicity assays were carried out on a panel of human cancer cell lines including HepG2, MCF-7, and MO59J and one normal cell line GM07492A. In general, the new ruthenium(II) complexes displayed a moderate to high cytotoxicity in all the assayed cell lines with IC$_{50}$ ranging from 10.1 to 36 μM and were more cytotoxic than the precursor cis-[RuCl$_2$(dppm)$_2$]. The cis-[Ru(quin)(dppm)$_2$]PF$_6$ were two to three times more active than the reference metallodrug cisplatin in the MCF-7 and MO59J cell lines.

1. Introduction

The disseminated use of cisplatin and other platinum based metallodrugs as chemotherapeutic agents against ovarian, bladder, and testicular cancers, among others, is still a key aspect for the development of the medicinal inorganic chemistry [1–4]. In the search for coordination compounds active against tumors and less toxic than cisplatin, ruthenium compounds emerge as the most promising with biological features including mechanism of action, toxicity, and biodistribution which are very different from those of classical platinum compounds and might therefore be active against resistant human cancers [3, 5–8]. In the last years three ruthenium(III) complexes entered clinical trials: NAMI-A - [ImH][trans-RuCl$_4$(DMSO)(Im)], KP1019 - [InH][trans-RuCl$_4$(In)$_2$], and NKP3019 - Na[trans-RuCl$_4$(In)$_2$] (In = indazole) [7, 9–12]. For recent developments on the anticancer activity of these ruthenium(III) complexes see cited references.

Previous work from our group displayed biological results from the diphosphonic ruthenium(II) precursors cis-[RuCl$_2$(P-P)$_2$], P-P = dppm or dppe, and its derivatives with 2-pyridinecarboxylic acid anion (pic), the cis-[Ru(pic)(P-P)$_2$]PF$_6$, with an -N,O chelation for the pic ligand [13, 14]. The antimycobacterial activity against MTB H$_{37}$Rv indicated a MIC value of 26.6 μM for the precursor and a much higher activity for the cis-[Ru(pic)(dppm)$_2$]PF$_6$ with a MIC value of 0.69 μM [13]. Some additional studies performed with the analogous cis-[Ru(pic)(dppe)$_2$]PF$_6$ revealed a high antibacterial activity against *S. aureus*, *C. albicans*, and *M. smegmatis* with MIC in the range 0.3 to 5.3 μM [14]. In the assay of acute oral toxicity this complex belongs to class 5 (a substance with LD$_{50}$ greater than 2000 and less than 5000 mg·kg^{-1} body weight), indicating a relatively low acute toxicity [14]. Species containing bidentate carboxylates such as cis-[Ru(dicl)(dppm)$_2$]PF$_6$, cis-[Ru(ibu)(dppm)$_2$]PF$_6$ and cis-[Ru(prop)(dppe)$_2$]PF$_6$ were studied and presented moderate to high cytotoxic activity against human cancer cell lines [15, 16].

Due to this background of promising biological results in complexes containing the cis-[Ru(P-P)$_2$] unit our current strategy consists in evaluating other derivatives with different chelating moiety replacing the chlorido ligands in the search for new cytotoxic agents against tumor cells. In the current work quinoline carboxylates were chosen as coligands in order to generate cationic complexes and also to explore the possible different coordination modes of the ligands.

In this work the synthesis and characterization of two new derivatives with formula cis-[Ru(quin)(dppm)$_2$]PF$_6$ (**1**) and cis-[Ru(kynu)(dppm)$_2$]PF$_6$ (**2**) are reported. The interaction of these complexes with CT-DNA and BSA was investigated by circular dichroism (CD) and fluorescence spectroscopies, respectively. Besides, preliminary in vitro tests of cytotoxic activities against a variety of human cell lines are presented and discussed.

2. Experimental

2.1. Chemicals.
Solvents were purified by standard methods. All chemicals used were of reagent grade or comparable purity. The RuCl$_3$·3H$_2$O and the ligands 1,1-bis(diphenylphosphino)methane (dppm), quinaldic acid, and kynurenic acid were used as received from Aldrich. The cis-[RuCl$_2$(dppm)$_2$] precursor complex was prepared according to the literature method [17].

2.2. Instrumentation.
Elemental analyses were performed on a Perkin Elmer 2400 Series II CHNS/O microanalyser. Molar conductivities of freshly prepared 1.0×10^{-3} mol·dm^{-3} methanol solutions were measured using a Digimed DM-31 conductivity meter. IR spectra were recorded on a Shimadzu FTIR-Prestige 21 spectrophotometer, using KBr pellets. UV-vis spectroscopy was recorded on a Femto model 800 XI spectrophotometer using cuvettes of 1 cm path length. ^{1}H and ^{31}P{^{1}H} NMR experiments were performed on a Bruker Avance III HD 400 MHz (9.4 T) at 298 K. Spectra were recorded in CDCl$_3$ with TMS and 85% H$_3$PO$_4$ as external references, respectively, for ^{1}H and ^{31}P{^{1}H}.

2.3. Synthesis.
The precursor cis-[RuCl$_2$(dppm)$_2$] (0.103 mmol; 100 mg) was solubilized in 100 mL of methanol and the mixture was heated during 20 minutes. Quinaldic acid (0.103 mmol; 17.8 mg) or kynurenic acid (0.103 mmol; 19.5 mg), respectively, for synthesis of complexes **1** and **2**, was solubilized in 10 mL of methanol and deprotonated with triethylamine (0.132 mmol; 0.014 ml) and this solution was

added dropwise on the precursor solution. After this processes, the mixture was stirred and refluxed for 48 h. The final solution was concentrated to ca. 5 mL and an aqueous solution of NH$_4$PF$_6$ (0.150 mmol; 24.4 mg) was added for the precipitation of a yellow solid. The solid was filtered off and washed with water (3 × 5 mL) and diethyl ether (3 × 5 mL) and dried under reduced pressure.

2.3.1. cis-[Ru(quin)(dppm)$_2$]PF$_6$ (1).
Yield: 79.1 mg (63%). Anal. Calcd for C$_{60}$H$_{50}$F$_6$NO$_2$P$_5$Ru: exptl (calc) C, 60.97 (60.71); H, 4.93 (4.63); N, 1.10 (1.18). ^{1}H NMR (CDCl$_3$): δ 4.11, 4.73 (m × 2, 2 × 2H; PCH$_2$P), 6.26 (m, 4H; C$_6$H$_5$), 6.99–7.86 (m, 39H; C$_6$H$_5$ + quin), 7.91 (d, J_{HH} = 8.1 Hz, 1H; quin), 8.22 (d, J_{HH} = 8.5 Hz, 1H; quin), 8.31 (d, J_{HH} = 8.5 Hz, 1H; quin) ppm. ^{31}P{^{1}H} NMR (161.73 MHz - CDCl$_3$): δ 9.70 (t, 2P, J_{PP} = 39 Hz); −12.2 (t, 2P, J_{PP} = 39 Hz); −144.7 (sept, 1P, J_{PF} = 711 Hz). UV-vis (CH$_2$Cl$_2$, 5.0×10^{-5} M): λ/nm (ε/M^{-1} cm^{-1}) 256sh (4.33×10^4), 325 (8.00×10^3). Molar conductivity [Λ_M/(S·cm^2·mol^{-1})] in methanol: 79.0 (range for a 1 : 1 electrolyte: 80–115) [18].

2.3.2. cis-[Ru(kynu)(dppm)$_2$]PF$_6$ (2).
Yield: 73.0 mg (57%). Anal. Calcd for C$_{60}$H$_{50}$F$_6$NO$_3$P$_5$Ru: exptl (calc) C, 59.87 (59.91); H, 4.68 (4.39); N, 1.14 (1.16). ^{1}H NMR (CDCl$_3$): δ 4.14, 4.73 (m × 2, 2 × 2H; PCH$_2$P), 6.26 (m, 4H; C$_6$H$_5$), 6.80 (s, 1H; kynu), 6.96–7.77 (m, 39H; C$_6$H$_5$ + kynu), 8.37 (d, J_{HH} = 7.7 Hz, 1H; kynu), 8.62 (s, 1H; OH-kynu) ppm. ^{31}P{^{1}H} NMR (CDCl$_3$ - 161.73 MHz): δ (ppm) 9.94 (t, 2P, J_{PP} = 39 Hz); −12.1 (t, 2P, J_{PP} = 39 Hz); −144.7 (sept, 1P, J_{PF} = 711 Hz). UV-vis (CH$_2$Cl$_2$, 5.0×10^{-5} M): λ/nm (ε/M^{-1} cm^{-1}) 249sh (3.11×10^4), 315sh (5.22×10^3), 329 (7.10×10^3), 344 (7.4×10^3), 363 (4.01×10^3). Molar conductivity [Λ_M/(S·cm^2·mol^{-1})] in methanol: 102.0.

2.4. Interactions Studies

2.4.1. CT-DNA.
Measurements involving CT-DNA (calf thymus from Sigma-Aldrich) were carried out in a Trizma buffer (4.5 mM Trizma HCl, 0.5 mM Trizma base, and 50 mM NaCl, pH 7.4). The DNA concentration per nucleotide was determined by absorption spectrophotometric analysis using a molar absorption coefficient of 6600.0 mol^{-1}·L·cm^{-1} at 260.0 nm [19].

2.4.2. Circular Dichroism (CD) Experiments.
CD spectra were recorded on a spectropolarimeter JASCO J-720 between 540 and 240 nm in a continuous scanning mode (200 nm·min^{-1}). The final data are expressed in molar ellipticity (millidegrees). All of the CD spectra were generated and represented averages of three scans. Stock solutions of each complex were freshly prepared in DMSO prior to use. An appropriate volume of each solution was added to the samples of a freshly prepared solution of CT-DNA (50 μM) in Trizma buffer to achieve molar ratios ranging from 0.05 to 0.5 DNA·drug^{-1}. The samples were incubated at 37°C for 18 h.

2.4.3. BSA-Binding Experiments.
The protein interaction with complexes **1** and **2** was examined in 96-well plates used for fluorescence assays on a fluorimeter Synergy H1. BSA

($2.0\,\mu\text{mol·L}^{-1}$) was prepared in Trizma buffer at pH = 7.4 (4.5 mM Trizma HCl, 0.5 mM Trizma base and 50 mM NaCl).

The inner-filter effect for ruthenium complexes was corrected by using

$$F_{\text{corr}} = F_{\text{obs}} e^{(A_{\text{em}} + A_{\text{ex}})/2}, \tag{1}$$

where F_{corr} and F_{obs} are the corrected and measured fluorescence intensity of protein, respectively. A_{em} and A_{ex} are the absorption values of the system at the excitation wavelength and emission wavelength of the complex, respectively.

Complexes were dissolved in sterile DMSO. For fluorescence measurements, the BSA concentration was kept constant in all samples, while the complex concentration was increased from 3.125 to 150 μM, and quenching of the emission intensity of the tryptophan residues of BSA at 320 nm (excitation wavelength 280 nm) was monitored at different temperatures (300 and 310 K). The experiments were carried out in triplicate and analyzed using the classical Stern-Volmer equation

$$\frac{F_0}{F} = 1 + k_q \tau_o [Q] = 1 + K_{\text{sv}} [Q], \tag{2}$$

where F_0 and F are the fluorescence intensities in the absence and presence of quencher, respectively, $[Q]$ is the quencher concentration, and K_{sv} is the Stern-Volmer quenching constant, which can be written as

$$K_q = \frac{K_{\text{sv}}}{\tau_o}, \tag{3}$$

where K_q is the biomolecular quenching rate constant and τ_o is the average lifetime of the fluorophore in the absence of quencher (6.2×10^{-9} s) [20]. Therefore, (2) was applied to determine K_{sv} by linear regression of a plot of F_0/F versus $[Q]$.

The binding constant (K_b) and number of complexes bound to BSA (n) were determined by plotting the double log graph of the fluorescence data using

$$\log \frac{(F_0 - F)}{F} = \log K_b + n \log [Q]. \tag{4}$$

The thermodynamic parameters were calculated from the van't Hoff equation:

$$\ln K_b = \frac{\Delta H^\circ}{RT} + \frac{\Delta S^\circ}{R}, \tag{5}$$

where K_b is analogous to the Stern-Volmer quenching constant, K_{sv} is at the corresponding temperature (the temperatures used were 300 and 310 K), and R is the gas constant, from which the ΔH and ΔS of the reaction can be determined from the linear relationship between $\ln K_b$ and the reciprocal absolute temperature. Furthermore, the change in free energy (ΔG) was calculated from the following equation:

$$\Delta G^\circ = -RT \ln K_b = \Delta H^\circ - T\Delta S^\circ. \tag{6}$$

2.5. Human Cell Lines and Culture Conditions. Cells from the 4th through to the 12th passage were used. The different cell lines were maintained as monolayers in plastic culture flasks ($25\,\text{cm}^2$) containing HAM-F10 plus DMEM (1 : 1; Sigma-Aldrich) or only DMEM depending on the cell line, supplemented with 10% fetal bovine serum (Nutricell) and $2.38\,\text{mg·mL}^{-1}$ Hepes (Sigma-Aldrich) at 37°C in a humidified 5% CO_2 atmosphere. Antibiotics ($0.01\,\text{mg·mL}^{-1}$ streptomycin and $0.005\,\text{mg·mL}^{-1}$ penicillin; Sigma-Aldrich) were added to the medium to prevent bacterial growth.

2.6. Cell Viability Assay. The screening for cytotoxic activity of cell lines was assessed using the Colorimetric Assay In Vitro Toxicology-XTT Kit (Roche Diagnostics). For the experiments, 1×10^4 cells were seeded into microplates with 100 μL of culture medium (1 : 1 HAM F10 + DMEM or alone DMEM) supplemented with 10% fetal bovine serum containing concentrations of complexes ranging from 12.5 to $1600\,\mu\text{g·mL}^{-1}$. Negative (no treatment), solvent (0.02% DMSO), and positive (25% DMSO) controls were included. Positive controls comprising cisplatin (Sigma-Aldrich, ≥98% purity) were included. After incubation to 36.5°C for 24 h, the culture medium was removed. Cells were washed with 100 μL of PBS for removal of the treatments, after which they were exposed to 100 μL of culture medium HAM-F10 without phenol red. Then, 25 μL of XTT was added and incubated at 36.5°C for 17 h. The absorbance of the samples was determined using a multiplate reader (ELISA-Tecan-SW Magellan versus 5.03 STD 2P) at a wavelength of 450 nm and a reference length of 620 nm.

2.7. Statistical Analysis. Cytotoxicity was assessed using the IC_{50} response parameter (50% cell growth inhibition) calculated with the GraphPad Prism program, plotting cell survival against the respective concentrations of the treatments. One-way ANOVA was used for the comparison of means ($P < 0.05$). The selectivity index was calculated by dividing the IC_{50} value of the isolated compounds on GM07492-A cells by the IC_{50} value determined for human cancer cells.

3. Results and Discussion

3.1. Synthesis. The simple reaction of the quinoline carboxylic acids with ruthenium(II) diphosphine precursor complex *cis*-[RuCl$_2$(dppm)$_2$] resulted in the products *cis*-[Ru(quin)(dppm)$_2$]PF$_6$ (**1**) and *cis*-[Ru(kynu)(dppm)$_2$]PF$_6$ (**2**), Figure 1, by simple chlorido exchange under mild conditions.

3.2. Spectroscopical Characterization. The diamagnetic and monoelectrolytes compounds **1-2** exhibited satisfactory microanalytical (C, H, and N) data. The ^{1}H NMR spectra showed two broad signals due to the CH$_2$ group of the dppm ligand close to 4.10 and 4.70 ppm [21]. The hydrogens of the phenyl groups (H$_o$, H$_m$ and H$_p$) were observed as several multiplets between 6.26 and 7.86 ppm [21]. For complex **1** three doublets of the quin ligand were observed at 7.91, 8.22, and 8.31 ppm while for complex **2** one doublet at 8.37 ppm and two singlet signals at 6.80 and 8.62 ppm

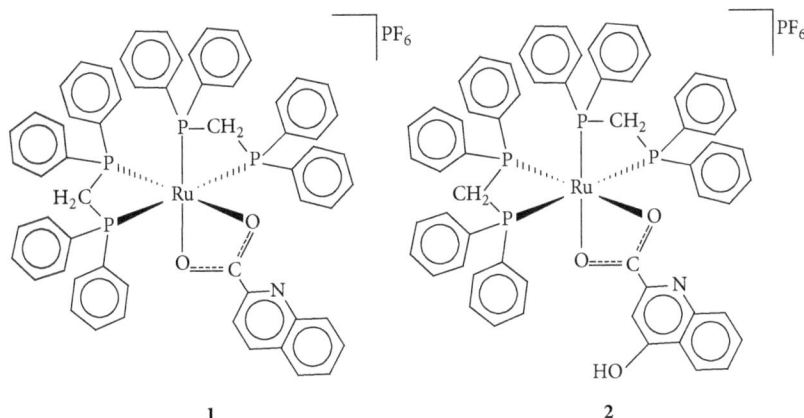

FIGURE 1: Structures of the ruthenium(II) compounds obtained in this work.

could be assigned to the kynu ligand [22]. Other hydrogens of the quin and kynu ligands were obscured by the signals corresponding to the hydrogens of the phenyl groups. The total number of hydrogens and proportion between the ligands dppm and quin or kynu was confirmed by the integral values for both complexes. In the $^{31}P\{^1H\}$ NMR spectra these complexes displayed a pair of triplets resonance signals, corresponding to two *trans*positioned phosphorus atoms and two phosphorus atoms *trans*positioned to oxygen atoms from carboxylate groups of the quinaldate and kynurenate anions, respectively. Triplets are observed at 9.70 and −12.2 ppm and at 9.94 and −12.1 ppm, respectively, for **1** and **2**, with the splitting pattern typical of an AX pattern [16]. These signals are downfield shifted when compared with the triplets signals for the cis-$[RuCl_2(dppm)_2]$ that are observed at 0.64 and −25.3 ppm with J_{PP} = 36 Hz [17].

The IR spectrum displayed the typical asymmetric (v_{asym}) and symmetric (v_{sym}) carboxylate stretching frequencies at 1523 and 1455 cm^{-1} (Δ = 68 cm^{-1}) and 1516 and 1454 (Δ = 62 cm^{-1}), respectively, for **1** and **2,** confirming the presence of the quinoline carboxylate ligands coordinated in the chelating mode through the carboxylate to the metal center [23, 24]. The characteristic P-F stretch of the PF_6^- counterion was seen at 840 and 557 cm^{-1} [23].

3.3. Circular Dichroism (CD) Experiments.

The CD spectral technique is very sensitive for diagnosing changes in the secondary structure of DNA, resulting from drug-DNA interactions [25]. A typical CD spectrum of CT-DNA shows a maximum at 275 nm, due to the base-stacking and a minimum at 248 nm attributed to the right-handed helicity, characteristic of the B conformation [26]. Thus simple groove binding and electrostatic interaction of small molecules show less or no perturbation on the base-stacking and helicity bands, while intercalation enhances the intensities of both the bands stabilizing the right-handed B conformation of CT-DNA as observed for the classical intercalator methylene blue [27]. To determine if the ruthenium(II) complexes cause changes in DNA, CD spectra of CT-DNA with increasing concentrations of **1** and **2** were acquired, up to molar ratio drug·DNA^{-1} (Ri) = 0.4. As shown in Figure 2 significant

changes were not observed indicating that these compounds were not able to intercalate DNA [28].

3.4. BSA-Binding Experiments.

Serum albumin is the most abundant protein in plasma and is involved in the transport of metal ions and metal complexes with drugs through the blood stream. Binding to these proteins to complexes may lead to loss or conformational change in the protein subunit and provide paths for drug transportation. Bovine serum albumin, BSA (containing two tryptophans, Trp-134 and Trp-212) is the most extensively studied serum albumin, due to its structural homology with human serum albumin, HAS (one Trp-214). The BSA solution exhibits a strong fluorescence emission with a peak at 340 nm, due to the tryptophan residues, when excited at 280 nm [29]. Fluorescence quenching of BSA can occur by different mechanisms, usually classified as either dynamic or static quenching, which can be distinguished by their differing on temperature, viscosity, and lifetime measurements [30]. A dynamic quenching refers to the collisional process between the fluorophore and the quencher (in this case, ruthenium complexes) during the transient existence of the excited state. Dynamic quenching depends on diffusion, as higher temperatures result in high diffusion coefficient, and consequently, the constant quenching must also increase. In contrast, for the static quenching, an increase in temperature results in lower extinction values of the constants due to a fluorophore and quencher complex formation in the ground state [31].

The Stern-Volmer equation (2) has been used to understand the nature of the quenching mechanism of BSA in the presence of complexes **1** and **2** at different temperatures [31]. Figure 3 shows the quenching of the BSA in the presence of the different concentrations of the complexes **1** and **2**.

In general, these complexes showed no significant variation K_{sv} with increasing temperature. However, the results of K_q have values greater than the maximum possible for a dynamic mechanism (2.0 × 10^{10} L·mol^{-1}·s^{-1}) [32] and in both cases were of the order 10^{13} M^{-1}·s^{-1} (Table 1), which is 1000-fold higher than the maximum value possible for diffusion controlled quenching of various kinds of quencher to biopolymer. It is suggested that the suppression of BSA

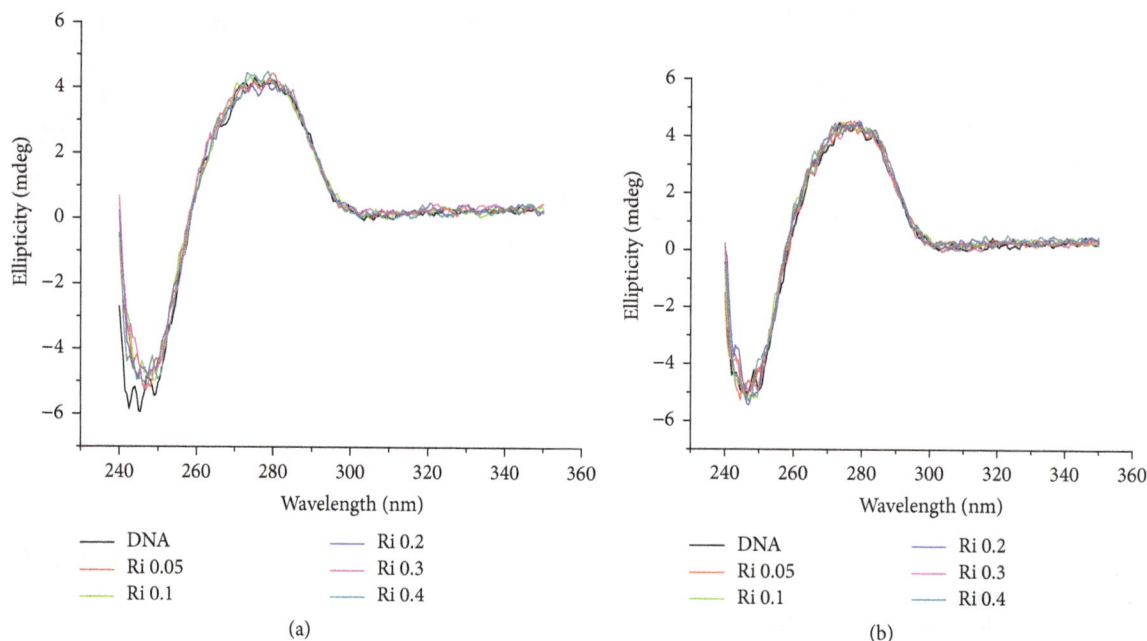

FIGURE 2: Circular dichroism (CD) spectra of CT-DNA incubated 18 h with complexes **1** (a) and **2** (b) at different [complex]/[DNA] ratios at 37°C.

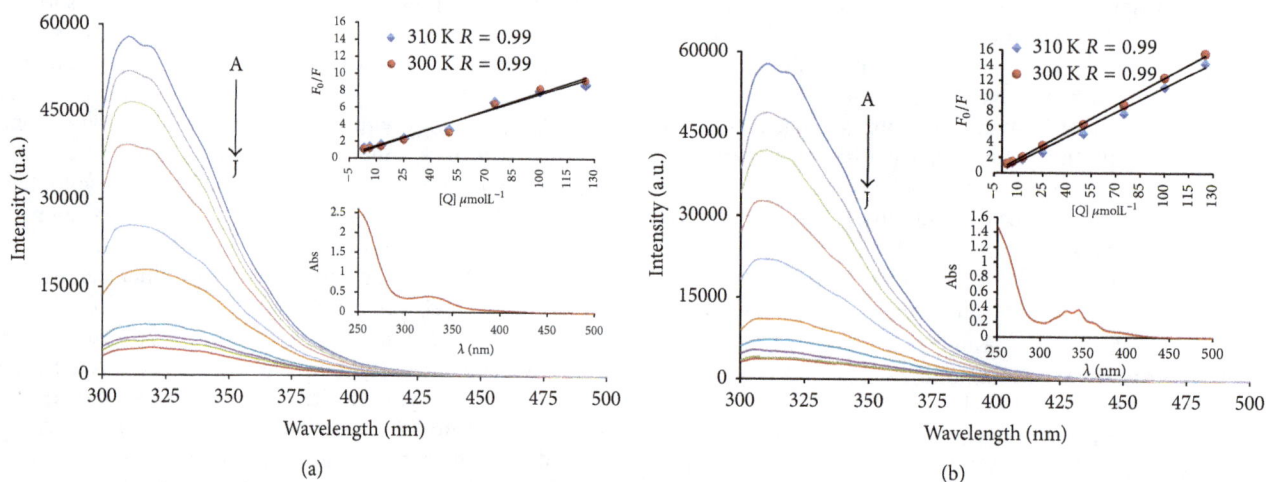

FIGURE 3: Fluorescence emission spectra of the BSA (2.5 μM λ_{ex} 280 nm) at different concentrations of complexes **1** (a) and **2** (b) at 300 K. Inset: Stern-Volmer plots showing tryptophan quenching in BSA at 300 and 310 K and UV-vis spectra of complexes **1** and **2**.

with these complexes is a static quenching mechanism. The decreasing of K_q with increasing temperature is in accordance with K_{sv} dependence on temperature.

In both cases, the obtained values of n indicate that the proportion between BSA-complex is equal 1:1, indicating that there is only one binding site in the BSA for each ruthenium complex, similar or equal to those reported before for other metal complexes [33–35]. Furthermore, values of K_b confirmed the moderated interaction force between complex-BSA [36–38] and the temperature is not significant. Thus, these complexes can be stored and carried by protein in the body. The interaction forces between drugs and biomolecules may include van der Waals interaction, hydrogen bonds, and electrostatic and hydrophobic interactions. The thermodynamic parameters ΔG (free energy change), ΔH (enthalpy change), and ΔS (entropy change) were calculated to evaluate the intermolecular forces involving the molecules of complex and BSA. The values for $\Delta H > 0$ and $\Delta S > 0$ imply the involvement of hydrophobic forces in protein binding, $\Delta H < 0$ and $\Delta S < 0$ correspond to van der Waals and hydrogen bonding interactions, and $\Delta H < 0$ and $\Delta S > 0$ suggest an electrostatic force [39]. Thermodynamic parameters (ΔH and ΔS) were calculated from the van't Hoff plots, (5); ΔG was estimated from (6). All the results are shown in Table 1.

TABLE 1: Stern-Volmer quenching constant (K_{sv}, L·mol^{-1}), biomolecular quenching rate constant (K_q, L·mol^{-1}·s^{-1}), binding constant (K_b, L·mol^{-1}), the number of binding sites (n), and ΔG^0 (KJ·mol^{-1}), ΔH^0 (KJ·mol^{-1}), and ΔS^0 (J·mol^{-1}·K) values for the complex-BSA system at different temperatures.

		K_{sv} ($\times 10^5$)	K_q ($\times 10^{13}$)	K_b ($\times 10^5$)	n	ΔG^0	ΔH^0	ΔS^0
1	300	0.77	1.10	4.77	1.2	−32.61	−87.78	−183.91
	310	0.75	1.10	1.53	1.0	−30.77		
2	300	1.24	1.79	3.25	1.1	−31.65	−50.16	−61.70
	310	1.00	1.50	1.70	1.0	−31.04		

TABLE 2: Inhibitory activity of ruthenium complexes and cisplatin against normal and tumor cell lines for 24 h incubation, expressed as IC$_{50}$, μg·mL^{-1} (μM) and selectivity index (SI).

| | Cell line | | | | | | |
	HepG2[a]	MCF-7[b]	MO59J[c]	GM07492A[d]	SI^1	SI^2	SI^3
cis-[RuCl$_2$(dppm)$_2$]	102 ± 26 (108 ± 26)	180 ± 13 (191 ± 13)	126 ± 10 (134 ± 10)	62 ± 4 (66 ± 4)	0.61	0.34	0.49
1	11.8 ± 0.7 (10.1 ± 0.6)	14.2 ± 0.6 (12.1 ± 0.5)	13 ± 2 (11 ± 2)	11 ± 1 (10 ± 1)	0.93	0.77	0.85
2	31 ± 6 (26 ± 5)	162 ± 14 (135 ± 11)	44 ± 3 (36 ± 2)	45 ± 1 (37 ± 1)	1.45	0.28	1.02
cisplatin[e]	1.9 ± 0.2 (6.3 ± 0.7)	10 ± 1 (34 ± 4)	7 ± 1 (22 ± 4)	8 ± 1 (26 ± 3)	4.21	0.80	1.14

[a]Human hepatocellular carcinoma, [b]human breast adenocarcinoma, [c]human glioblastoma, [d]normal human lung fibroblasts, and [e]reference drug. SI^1 = IC$_{50}$ GM07492A/IC$_{50}$ HepG2; SI^2 = IC$_{50}$ GM07492A/IC$_{50}$ MCF-7; SI^3 = IC$_{50}$ GM07492A/IC$_{50}$ MO59J.

The negative values of free energy, ΔG, suggested that the interaction process was spontaneous; the values entropy ΔS and enthalpy ΔH negative indicated that the hydrogen bonds and van der Waals forces are the more important interactions in the reaction [38].

3.5. *Cytotoxicity Assays.* Ruthenium compounds 1 and 2, cis-[RuCl$_2$(dppm)$_2$] and cisplatin were evaluated for their capability of inhibiting tumor cell growth in vitro using three human cell lines, HepG2, MCF-7, and MO59J. One nontumorigenic cell line (GM07492A) was also assayed in the same conditions in order to verify the selectivity of these compounds. The resulting concentration-effect curves obtained with continuous exposure for 24 h are depicted in Figure 4. A more convenient comparison of the cytotoxic potency (expressed as IC$_{50}$ values) is listed in Table 2.

Overall sensitivities of the three tumor cell lines (Figures 4(a)–4(c)) and of the normal cell (Figure 4(d)) are more pronounced for complex 1, followed by complex 2, and the less active is the precursor cis-[RuCl$_2$(dppm)$_2$].

Complexes 1 and 2 have showed, in general, a moderate cytotoxicity against all the human tumor cell lines assayed with IC$_{50}$ values ranging from 10.1 to 36 μM, except for complex 2 in MCF-7 cells that showed a very low cytotoxicity as shown in Table 2. Complex 1 displayed higher activity than 2 in all the cell lines assayed with IC$_{50}$ close to 10 μM. This nonselective activity of complex 1 was not observed for complex 2 that was almost inactive in MCF-7 cells. The selectivity index (SI) was very close to 1 (or smaller) for all the cell lines assayed indicating a lack of selectivity for both complexes.

In the same experimental conditions the precursor complex cis-[RuCl$_2$(dppm)$_2$] was less active than the complexes 1 and 2 by factors ranging from 1.4 to 15.8. A similar increased in activity was also observed against the normal cell line GM07492A. This lack of selectivity was also observed for the cis-[RuCl$_2$(dppm)$_2$] for all the cell lines assayed and for cisplatin in the MCF-7 and MO59J cell lines. These data clearly indicate that exchanging two chlorido ligands by a bidentate quinoline carboxylate group turns the unity cis-[Ru(dppm)$_2$] complex more cytotoxic, probably due to its higher solubility and disponibility in the culture medium. Interestingly, the presence of one OH-group in complex 2 led to a significant decrease in cytotoxic activity, indicating that this may be a way to modulate the cytotoxic potency of this type of complex.

4. Conclusion

In this investigation two new ruthenium(II) complexes containing dppm and the quinaldate and kynurenate anions with formula cis-[Ru(quin)(dppm)$_2$]PF$_6$ and cis-[Ru(kynu)(dppm)$_2$]PF$_6$ were synthesized and characterized by elemental analysis and spectroscopic methods. Characterization data are in agreement with a cis geometry and a chelated coordination, through the carboxylate group, for the quinaldate and kynurenate ligands. Utilizing circular dichroism spectroscopy showed that these complexes lack the ability to intercalate DNA. On the other hand BSA-binding constants and thermodynamic parameters suggest spontaneous interactions with this protein by hydrogen bonds and van der

(a)

(b)

(c)

(d)

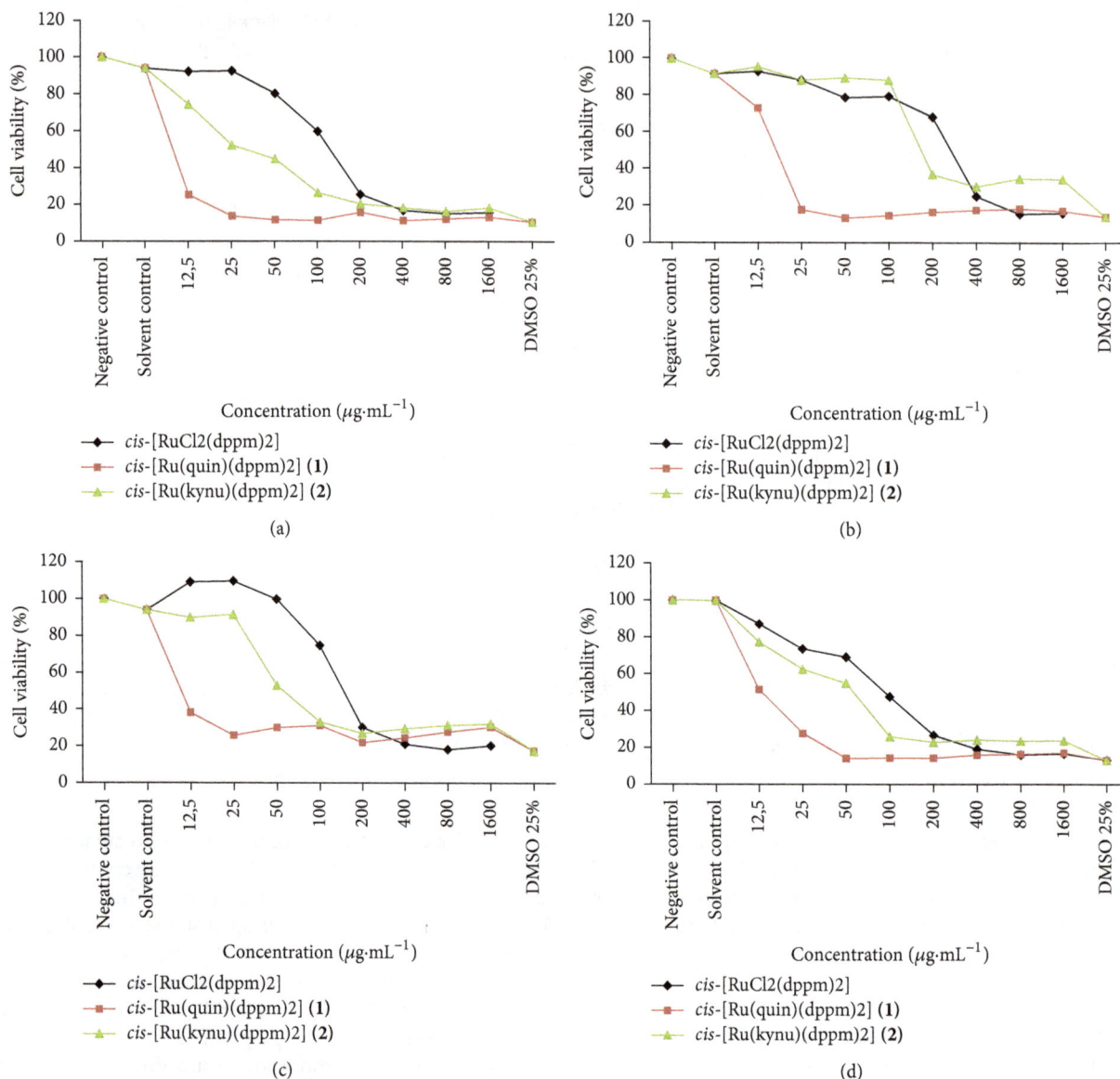

FIGURE 4: In vitro cytotoxicity activities of ruthenium complexes in HepG2 (a), MCF-7 (b), MO59J (c), and GM07492A (d) cells after exposure for 24 h. The concentration-effect curves were determined using XTT assay. Each data point is the mean ± standard error obtained from three independent experiments.

Waals forces. The in vitro cytotoxicity activity assays indicate a moderate to high cytotoxicity against a panel of human tumor cell lines; however, these complexes lacks selectivity. Interestingly, modulation of cytotoxic potency can probably be done by exchanging the substituents of the quinolone carboxylate ligand and will be explored in future work.

Acknowledgments

This work was supported by Fundação de Amparo à Pesquisa de Minas Gerais (FAPEMIG) (Grant APQ-04010-10), CAPES, and CNPq. Gustavo Von Poelhsitz is thankful to the *Grupo de Materiais Inorgânicos do Triângulo*-GMIT research group supported by FAPEMIG (APQ-00330-14) and Rede Mineira de Química (RQ-MG) supported by FAPEMIG (Project: CEX - RED-00010-14).

References

[1] S. Dasari and P. B. Tchounwou, "Cisplatin in cancer therapy: molecular mechanisms of action," *European Journal of Pharmacology*, vol. 740, pp. 364–378, 2014.

[2] N. J. Wheate, S. Walker, G. E. Craig, and R. Oun, "The status of platinum anticancer drugs in the clinic and in clinical trials," *Dalton Transactions*, vol. 39, no. 35, pp. 8113–8127, 2010.

[3] K. D. Mjos and C. Orvig, "Metallodrugs in medicinal inorganic chemistry," *Chemical Reviews*, vol. 114, no. 8, pp. 4540–4563, 2014.

[4] S. Gómez-Ruiz, D. Maksimović-Ivanić, S. Mijatović, and G. N. Kaluđerović, "On the discovery, biological effects, and use of cisplatin and metallocenes in anticancer chemotherapy," *Bioinorganic Chemistry and Applications*, vol. 2012, Article ID 140284, 14 pages, 2012.

[5] C. G. Hartinger, S. Zorbas-Seifried, M. A. Jakupec, B. Kynast, H. Zorbas, and B. K. Keppler, "From bench to bedside - preclinical and early clinical development of the anticancer agent indazolium trans-[tetrachlorobis(1H-indazole)ruthenate(III)] (KP1019 or FFC14A)," *Journal of Inorganic Biochemistry*, vol. 100, no. 5-6, pp. 891–904, 2006.

[6] W. H. Ang, A. Casini, G. Sava, and P. J. Dyson, "Organometallic ruthenium-based antitumor compounds with novel modes of action," *Journal of Organometallic Chemistry*, vol. 696, no. 5, pp. 989–998, 2011.

[7] A. Bergamo, C. Gaiddon, J. H. Schellens, J. H. Beijnen, and G. Sava, "Approaching tumour therapy beyond platinum drugs: status of the art and perspectives of ruthenium drug candidates," *Journal of Inorganic Biochemistry*, vol. 106, no. 1, pp. 90–99, 2012.

[8] V. Moreno, J. Lorenzo, F. X. Aviles et al., "Studies of the antiproliferative activity of ruthenium (II) Cyclopentadienyl-derived complexes with nitrogen coordinated ligands," *Bioinorganic Chemistry and Applications*, vol. 2010, Article ID 936834, 11 pages, 2010.

[9] S. Pillozzi, L. Gasparoli, M. Stefanini et al., "NAMI-A is highly cytotoxic toward leukaemia cell lines: Evidence of inhibition of KCa 3.1 channels," *Dalton Transactions*, vol. 43, no. 32, pp. 12150–12155, 2014.

[10] R. Trondl, P. Heffeter, C. R. Kowol, M. A. Jakupec, W. Berger, and B. K. Keppler, "NKP-1339, the first ruthenium-based anticancer drug on the edge to clinical application," *Chemical Science*, vol. 5, no. 8, pp. 2925–2932, 2014.

[11] C. G. Hartinger, M. A. Jakupec, S. Zorbas-Seifried et al., "KP1019, a new redox-active anticancer agent—preclinical development and results of a clinical phase I study in tumor patients," *Chemistry and Biodiversity*, vol. 5, no. 10, pp. 2140–2155, 2008.

[12] M. Oszajca, E. Kuliś, G. Stochel, and M. Brindell, "Interaction of the NAMI-A complex with nitric oxide under physiological conditions," *New Journal of Chemistry*, vol. 38, no. 8, pp. 3386–3394, 2014.

[13] F. R. Pavan, G. V. Poelhsitz, F. B. do Nascimento et al., "Ruthenium (II) phosphine/picolinate complexes as antimycobacterial agents," *European Journal of Medicinal Chemistry*, vol. 45, no. 2, pp. 598–601, 2010.

[14] F. R. Pavan, G. V. Poelhsitz, L. V. P. da Cunha et al., "In vitro and in vivo activities of ruthenium(II) phosphine/diimine/picolinate complexes (SCAR) against Mycobacterium tuberculosis," *PLoS ONE*, vol. 8, no. 5, Article ID e64242, 2013.

[15] T. M. P. Pagoto, L. L. G. Sobrinho, A. E. Graminha et al., "A ruthenium(II) complex with the propionate ion: Synthesis, characterization and cytotoxic activity," *Comptes Rendus Chimie*, vol. 18, no. 12, pp. 1313–1319, 2015.

[16] J. C. S. Lopes, J. L. Damasceno, P. F. Oliveira et al., "Ruthenium(II) Complexes containing anti-inflammatory drugs as ligands: synthesis, characterization and in vitro cytotoxicity activities on cancer cell lines," *Journal of the Brazilian Chemical Society*, vol. 26, no. 9, pp. 1838–1847, 2015.

[17] B. P. Sullivan and T. J. Meyer, "Comparisons of the physical and chemical properties of isomeric pairs. 2. Photochemical, thermal, and electrochemical cis-trans isomerizations of M(Ph2PCH2PPh2)2Cl2 (M = RuII, OsII)," *Inorganic Chemistry*, vol. 21, no. 3, pp. 1037–1040, 1982.

[18] W. J. Geary, "The use of conductivity measurements in organic solvents for the characterisation of coordination compounds," *Coordination Chemistry Reviews*, vol. 7, no. 1, pp. 81–122, 1971.

[19] E. Fredericq, A. Oth, and F. Fontaine, "The ultraviolet spectrum of deoxyribonucleic acids and their constituents," *Journal of Molecular Biology*, vol. 3, no. 1, pp. 11–17, 1961.

[20] E. Gratton, N. Silva, G. Mei, N. Rosato, I. Savini, and A. Finazzi-Agro, "Fluorescence lifetime distribution of folded and unfolded proteins," *International Journal of Quantum Chemistry*, vol. 42, no. 5, pp. 1479–1489, 1992.

[21] J. A. Robson, F. Gonzàlez De Rivera, K. A. Jantan et al., "Bifunctional chalcogen linkers for the stepwise generation of multimetallic assemblies and functionalized nanoparticles," *Inorganic Chemistry*, vol. 55, no. 24, pp. 12982–12996, 2016.

[22] M. C. Barral, R. Jiménez-Aparicio, E. C. Royer et al., "Synthesis and crystal structure of a ruthenium complex containing two monodentate DPPM ligands (DPPM = bis(diphenylphosphino)methane)," *Inorganica Chimica Acta*, vol. 209, no. 1, pp. 105–109, 1993.

[23] K. Nakamoto, *Infrared and Raman Spectra of Inorganic and Coordination Compounds*, Wiley-Interscience, New York, NY, USA, 5th edition, 1997.

[24] W. Lewandowski, M. Kalinowska, and H. Lewandowska, "The influence of metals on the electronic system of biologically important ligands. Spectroscopic study of benzoates, salicylates, nicotinates and isoorotates. Review," *Journal of Inorganic Biochemistry*, vol. 99, no. 7, pp. 1407–1423, 2005.

[25] P. Lincoln, E. Tuite, and B. Norden, "Short-circuiting the molecular wire: cooperative binding of Δ-[Ru(phen)2dppz]2+ and Δ-[Rh(phi)2bipy]3+ to DNA," *Journal of the American Chemical Society*, vol. 119, no. 6, pp. 1454–1455, 1997.

[26] V. I. Ivanov, L. E. Minchenkova, A. K. Schyolkina, and A. I. Poletayev, "Different conformations of double-stranded nucleic acid in solution as revealed by circular dichroism," *Biopolymers*, vol. 12, no. 1, pp. 89–110, 1973.

[27] B. Norden and F. Tjerneld, "Structure of methylene blue-DNA complexes studied by linear and circular dichroism spectroscopy," *Biopolymers*, vol. 21, no. 9, pp. 1713–1734, 1982.

[28] P. Uma Maheswari and M. Palaniandavar, "DNA binding and cleavage properties of certain tetrammine ruthenium(II) complexes of modified 1,10-phenanthrolines - Effect of hydrogen-bonding on DNA-binding affinity," *Journal of Inorganic Biochemistry*, vol. 98, no. 2, pp. 219–230, 2004.

[29] F. Dimiza, A. N. Papadopoulos, V. Tangoulis et al., "Biological evaluation of cobalt(II) complexes with non-steroidal anti-inflammatory drug naproxen," *Journal of Inorganic Biochemistry*, vol. 107, no. 1, pp. 54–64, 2012.

[30] L. Shang, Y. Wang, J. Jiang, and S. Dong, "PH-dependent protein conformational changes in albumin:Gold nanoparticle bioconjugates: a spectroscopic study," *Langmuir*, vol. 23, no. 5, pp. 2714–2721, 2007.

[31] M. Ganeshpandian, R. Loganathan, E. Suresh, A. Riyasdeen, M. A. Akbarsha, and M. Palaniandavar, "New ruthenium(II) arene complexes of anthracenyl-appended diazacycloalkanes: effect of ligand intercalation and hydrophobicity on DNA and protein binding and cleavage and cytotoxicity," *Dalton Transactions*, vol. 43, no. 3, pp. 1203–1219, 2014.

[32] X. Zhao, R. Liu, Z. Chi, Y. Teng, and P. Qin, "New insights into the behavior of bovine serum albumin adsorbed onto carbon

nanotubes: comprehensive spectroscopic studies," *Journal of Physical Chemistry B*, vol. 114, no. 16, pp. 5625–5631, 2010.

[33] R. S. Correa, K. M. Oliveira, H. Pérez et al., "cis-bis(N-benzoyl-N',N'-dibenzylthioureido)platinum(II): Synthesis, molecular structure and its interaction with human and bovine serum albumin," *Arabian Journal of Chemistry*, vol. 2015, 2015.

[34] R. S. Correa, K. M. De Oliveira, F. G. Delolo et al., "Ru(II)-based complexes with N-(acyl)-N*i*,N*i*-(disubstituted)thiourea ligands: synthesis, characterization, BSA- and DNA-binding studies of new cytotoxic agents against lung and prostate tumour cells," *Journal of Inorganic Biochemistry*, vol. 150, article 9708, pp. 63–71, 2015.

[35] L. Colina-Vegas, J. L. Dutra, W. Villarreal et al., "Ru(II)/clotrimazole/diphenylphosphine/bipyridine complexes: Interaction with DNA, BSA and biological potential against tumor cell lines and Mycobacterium tuberculosis," *Journal of Inorganic Biochemistry*, vol. 162, pp. 135–145, 2016.

[36] M. Mathew, S. Sreedhanya, P. Manoj, C. T. Aravindakumar, and U. K. Aravind, "Exploring the interaction of bisphenol-S with serum albumins: a better or worse alternative for bisphenol A?" *Journal of Physical Chemistry B*, vol. 118, no. 14, pp. 3832–3843, 2014.

[37] B. Ojha and G. Das, "The interaction of 5-(Alkoxy)naphthalen-1-amine with bovine serum albumin and Its effect on the conformation of protein," *Journal of Physical Chemistry B*, vol. 114, no. 11, pp. 3979–3986, 2010.

[38] S.-L. Zhang, G. L. V. Damu, L. Zhang, R.-X. Geng, and C.-H. Zhou, "Synthesis and biological evaluation of novel benzimidazole derivatives and their binding behavior with bovine serum albumin," *European Journal of Medicinal Chemistry*, vol. 55, pp. 164–175, 2012.

[39] P. D. Ross and S. Subramanian, "Thermodynamics of protein association reactions: forces contributing to stability," *Biochemistry*, vol. 20, no. 11, pp. 3096–3102, 1981.

11

Synthesis, Structural Analysis, and Biological Activities of Some Imidazolium Salts

Gühergül Uluçam ⑩ **and Murat Turkyilmaz**

Department of Chemistry, Trakya University, 22030 Edirne, Turkey

Correspondence should be addressed to Gühergül Uluçam; gulergul@trakya.edu.tr

Academic Editor: Rais A. Khan

Four newly synthesized imidazolium salts were characterized by nuclear magnetic resonance, vibrational spectra, and mass spectra. Then, the density functional theory calculations were performed to obtain the molecular configurations on which the theoretical nuclear magnetic resonance and infrared spectra were consequently obtained. The comparison of calculated spectra with the experimental spectra for each molecule leads to the conclusion that the theoretical results can be assumed to be a good approach to their molecular configurations. The in vitro biological activities of the salts on the selected bacteria and cancer cell lines were determined by using the broth dilution method according to Clinical and Laboratory Standards Institute guidelines. The 1,3-bis(2-hydroxyethyl) imidazolidinium bromide and 3-(2-ethoxy-2-oxoethly)-1-(3-aminopropyl)-1H-imidazol-3-ium bromide showed efficiency on *Bacillus cereus* ATCC 11778. The 3-bis(2-carboxyethyl)-4-methyl-1-H-imidazol-3-ium bromide was effective on HeLa while a similar effect was observed on Hep G2 with 3-(2-carboxyethyl)-1-(3-aminopropyl)-1H-imidazol-3-ium bromide.

1. Introduction

Imidazole rings are building blocks in amino acids [1] together with the fact that their membership in the development of new antifungal drugs [2, 3] and antibiotics [4, 5] are crucial. Its derivatives are widely used in other medicinal applications [6]. The pregnane derivatives with imidazole moiety and triazole moiety, for example, were tested on the prostate, breast, and lung cancer cell lines, and dose-effective proliferation of the cells was determined [7]. Similarly, the novel hybrid compounds of imidazole scaffold-based 2-benzylbenzofloran have been prepared and used in cytotoxic activity studies on various cancer cell lines [8].

As a five-membered aromatic ring containing two nonadjacent nitrogen atoms [9], imidazole is also subjected to various computational chemistry research beyond its biological applications. Its ability to capture CO_2 was determined in the investigations of the greenhouse effect compensation in the framework of van der Waals bonded host-guest relation [10]. The hydroxyl conductivity in polymembranes based on imidazole salts was simulated using radial distribution functions and found that the imidazole groups provide better conductivity than that of water and methanol [11]. Moreover, the specific imidazole derivatives exhibited good cross section values for two-photon absorption [12]. The detoxification of phosphotriesters by imidazole rings was clarified comparing the same effects with methylimidazoles depending on the methyl positioning [13].

N-Heterocyclic carbenes (NHCs) are the imidazole-based carbene groups which are isolated and crystallized by the deprotonation of imidazole salts [14]. Also, the imidazole salts naturally transform into NHCs over metal complex building reactions, as exampled on the synthesis and characterization of the silver-NHC complexes [15] and the iron-imidazole salts [16].

The constitution and functions of imidazole ligands in organometallic chemistry and inorganic chemistry have been widely studied, and these particular researches have been evaluated as a scientific competition field due to its importance in the related industry [17]. Also, they are set as alternative to usual ligands in the carbon-carbon coupling reactions of the pharmaceutical reagents [18]. The synthesis and spectroscopic characterizations of four new NHC ligands, namely, *1,3-bis(2-hydroxyethyl) imidazolidinium bromide* (L_A), *3-(2-ethoxy-2-oxoethly)-1-(3-aminopropyl)-1H-imidazol-3-ium bromide* (L_B), *1,3-bis(2-carboxyethyl)-4-methyl-1H-imidazol-3-ium bromide* (L_C), and *3-(2-carboxyethyl)-1-(3-aminopropyl)-1H-imidazol-3-ium bromide* (L_D) were exhibited in this study. Using their nuclear magnetic resonance (NMR) and infrared (IR) spectra, the molecular properties of the ligands were obtained. Also, the *in vitro* biological activities of the synthesized molecules were presented.

2. Materials and Methods

2.1. Instrumentation and Methods. The ^1H and ^{13}C NMR spectra of the compounds in deuterium oxide (D_2O) were recorded on Varian 300 MHz and Varian 75.5 MHz, respectively. The IR spectra by KBr pellets were recorded in the range 450–4000 cm^{-1} by a PerkinElmer BXII spectrometer. The mass spectra were acquired by the electron impact technique using a Thermo Finnegan Trace DSQ GC/MS. Elemental analyses for C, H, and N were realized on the dried samples using a PerkinElmer 2400 CHN analyzer. The absorbance measurements in determining the biological activities of the material were carried out in Thermo Scientific Multiskan Go multiplate spectrophotometer.

2.2. Synthesis

2.2.1. 1,3-Bis(2-hydroxyethyl) Imidazolidinium Bromide, L_A. Imidazole (10 mmol, 0.68 g) was dissolved in tetrahydrofuran (THF), and bromoethanol (22 mmol, 2.75 g) was added as the mixture was stirred for 20 hours. The completion of the reaction was monitored by thin-layer chromatography (TLC) in ethyl acetate/hexane (1 : 5) analyses, and the solid residue was filtered out with a sintered glass funnel. The solvent in the filtrate was evaporated using a rotary evaporator, and the product was dried in a vacuum desiccator. After that, the product was purified by column chromatography (ethyl acetate/hexane, 1 : 5). The best yield was obtained when the reaction was carried out at room temperature with a 1 : 2 mole ratio of the reagents. 1.54 g of the final product was obtained with 65% yield. It was in yellowish liquid form. The elemental analyses result for L_A with the chemical formula $C_7H_{13}BrN_2O_2$ are C, 35.46%; H, 5.53%; and N, 11.82%; found: C, 35.35%; H, 5.43%; and N, 11.73%. The mass spectroscopy reads (*m/z*) 158.22 $(M + H)^+$ which is consistent with the expected molecular weight.

2.2.2. 3-(2-Ethoxy-2-oxoethly)-1-(3-aminopropyl)-1H-imidazol-3-ium Bromide, L_B. 3-(1H-imidazol-1-yl)propan-1-amine (10 mmol, 1.27 g) was dissolved in THF and stirred at room temperature, and then, ethyl bromoacetate (11 mmol, 1.837 g) was added with a 1 : 1 mole ratio. The mixture was stirred for 15 hr. The completion of the reaction was monitored by thin layer chromatography (TLC) in ethyl acetate/hexane (1 : 5) analyses. The solvent was evaporated in a rotary evaporator, and the substance was kept under vacuum in a desiccator. After that, the product was purified by column chromatography (ethyl acetate/hexane, 1 : 5). It was pale brown oily liquid. 0.94 g of the final product was obtained with 32% yield. The elemental analyses result for L_B with the chemical formula $C_{10}H_{18}BrN_3O_2$ are C, 41.11%; H, 6.21%; and N, 14.38%; found: C, 39.93%; H, 6.35%; N, and 14.21%. The mass spectroscopy result reads (*m/z*) 213.12 $(M + H)^+$ which is consistent with the expected molecular weight.

2.2.3. 1,3-Bis(2-carboxyethyl)-4-methyl-1H-imidazol-3-ium Bromide, L_C. The procedure applied in the synthesis of L_A was also used for the synthesis of L_C by replacing imidazole and bromoethanol with 4-methylimidazole (10 mmol, 0.82 g) and 3-bromopropionic acid (22 mmol, 3.366 g), respectively. After that, the product was purified by column chromatography (ethyl acetate/hexane, 1 : 5). It was in white solid form. 1.48 g of the final product was obtained with 48% yield. The elemental analyses result for L_C with the chemical formula $C_{10}H_{15}BrN_2O_4$ are C, 39.10%; H, 4.92%; and N, 9.12%; found: C, 38.97%; H, 4.77%; and N, 19.22%. The mass spectroscopy result reads (*m/z*) 227.68 $(M + H)^+$ which is consistent with the expected molecular weight.

2.2.4. 3-(2-Carboxyethyl)-1-(3-aminopropyl)-1H-imidazol-3-ium Bromide, L_D. The procedure applied in the synthesis of L_B was also used for the synthesis of L_D by replacing ethyl bromoacetate with 3-bromopropanoic acid (11 mmol, 1.683 g), respectively. The product was purified by column chromatography (ethyl acetate/hexane, 1 : 5). It was pale brown oily liquid. 1.22 g of the final product was obtained with 44% yield. The elemental analyses result for L_D with the chemical formula $C_9H_{16}BrN_3O_3$ are C, 38.86%; H.5.80%; and N, 15.11%; found: C, 38.77%; H, 5.63%; and N, 15.23%. The mass spectroscopy result reads (*m/z*) 198.25 $(M + H)^+$ which is consistent with the expected molecular weight.

2.3. Biological Activities

2.3.1. Antibacterial Activity. Broth microdilution method as in the guidelines of Clinical Laboratory Standards Institute was applied to determine the antibacterial activities of the L_A, L_B, L_C, and L_D on the Gram-negative bacteria (*Escherichia coli* ATCC 25922, *Escherichia coli* 0157:H7, and *Salmonella typhimurium* ATCC 14028), Gram-positive bacteria (*Bacillus cereus* ATCC 11778, *Staphylococcus aureus* ATCC 25923, and *Listeria monocytogenes* ATCC 19115), and standard yeast *Candida albicans* ATCC 10231. The bacteria and yeast were obtained from the American Type Culture Collection. The incubations were done in Tryptic Soy Broth medium at 37°C for 24 hr in the scale of McFarland 0.5. The antibiotic controls were carried by gentamicin on *Bacillus cereus* ATCC 1177 and amphotericin-b on *Candida albicans* ATCC 10231, while ampicillin was used on the other bacteria samples. For sterilization purposes, the antibiotic solutions and the stock solution of the chemicals were filtered through a 0.45 µm sterile filter. The solvent in which the test compounds

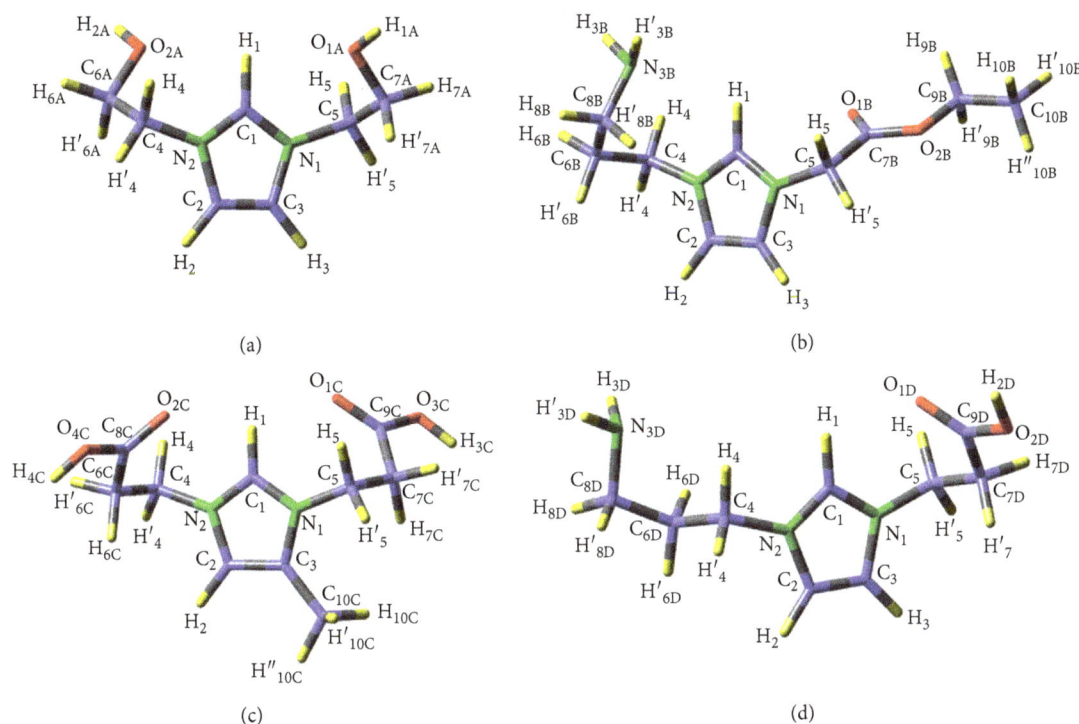

SCHEME 1: Optimized conformers of the synthesized imidazole molecules with their labeled and numbered atoms: (a) 1,3-bis(2-hydroxyethyl) imidazolidinium bromide, (b) 3-(2-ethoxy-2-oxoethly)-1-(3-aminopropyl)-1H-imidazol-3-ium bromide, (c) 1,3-bis(2-carboxyethyl)-4-methyl-1H-imidazol-3-ium bromide, and (d) 3-(2-carboxyethyl)-1-(3-aminopropyl)-1H-imidazol-3-ium bromide.

TABLE 1: Some selected geometrical parameters of the common properties of the investigated molecules.

	L_A	L_B	L_C	L_D	L
Selected bond lengths (Å)					
C_1-H_1	1.102	1.043	1.061	1.095	0.976
C_1-N_1	1.295	1.301	—	—	1.312 (7)
C_3-N_1	1.403	1.398	1.335	1.396	1.374 (8)
C_2-C_3	1.297	1.332	1.366	1.314	1.312 (10)
C_4-N_2	1.464	1.437	1.418	1.523	1.444 (8)
N_1-C_5	1.464	1.474	1.429	1.505	1.444 (8)
C_{7B}-O_{1B}	—	1.204	—	—	1.241 (8)
$N_{3B(D)}$-$H_{3B(D)}$	—	1.015	—	1.013	1.010
Selected bond angles (°)					
N_1-C_1-N_2	108.7	112.6	111.7	108.9	109.2 (6)
C_1-N_1-C_3	108.9	105.7	108.9	108.6	107.9 (5)
C_1-N_1-C_5	125.1	127.6	126.0	126.0	126.2 (5)
C_4-N_2-C_1	125.1	126.5	126.9	125.2	126.2 (5)
H_2-C_2-N_2	122.6	121.8	121.8	122.3	123.4
C_3-C_2-H_2	130.6	129.4	129.9	130.5	128.2
Selected dihedral angles (°)					
C_4-N_2-C_1-N_1	177.5	179.2	179.6	178.6	178.4 (5)
N_2-C_2-C_3-N_1	0.2	0.4	0.3	0.1	0.9 (8)
C_5-N_1-C_1-N_2	177.5	175.1	179.1	179.6	178.4 (5)
C_5-N_1-C_3-C_2	177.7	175.2	179.8	179.7	177.9 (6)

The data of the molecule showed by L in the last column reflect the X-ray diffraction measurements and calculated values on *1,3-bis(acetamide)imidazol-3-ium bromide* as given in [22]. The numbers in the paranthesis indicates the experimental error margins on the specific measurements. The labeled and numbered atoms in the first column are presented for each molecule in Scheme 1.

dissolved was dimethyl sulfoxide (DMSO) which did not show any inhibition effect on the bacteria. The pure L_A, L_B, L_C, and L_D solutions and the pure microorganism planted mediums were used as the sterility and the growth controls, respectively. The six different concentrations of the each compound were applied to the cells starting from $32\,\mu m$ and diluting to half each time. Each of the 96 microplates was planted in $150\,\mu l$ of Tryptic Soy Broth medium, $30\,\mu l$ of the

TABLE 2: Proton and carbon nuclear magnetic resonance spectral data of the molecules.

Assignment	Proton chemical shift (ppm)		Assignment	Carbon chemical shift (ppm)	
	Experimental	Theory		Experimental	Theory
1,3-Bis(2-hydroxyethyl) imidazolidinium bromide (L$_A$)					
s, 1H, NCHN	9.06	9.06	NCHN	136.33	137.66
d, 2H, HC=CH	7.70	7.32	HC=CH	122.60	123.06
t, 4H, N-CH$_2$	3.93	4.01	O-CH$_2$	59.75	61.68
t, 4H, O-CH$_2$	3.82	3.87	N-CH$_2$	51.98	52.59
s. 2H, OH	—	1.73	—	—	—
3-(2-Ethoxy-2-oxoethly)-1-(3-aminopropyl)-1H-imidazol-3-ium bromide (L$_B$)					
s, 1H, NCHN	7.79	7.73	CO	171.85	172.23
d, 1H, HC=CH	7.25	7.23	NCHN	136.78	137.10
d, 1H, HC=CH	7.02	7.00	HC=CH	125.13	124.74
s, 2H, N-CH$_2$ CO	4.23	4.22 (2)	HC=CH	120.73	121.34
q, 2H, C$\underline{H}_2$-CH$_3$	3.23	3.21 (3)	O-C$\underline{H}_2$-CH$_3$	60.78	57.36
t, 2H, N-C$\underline{H}_2$-C	2.95	2.96 (3)	N-CH$_2$	54.97	54.10
t, 2H, NH$_2$-C$\underline{H}_2$	2.55	2.54 (2)	N-C$\underline{H}_2$-CH$_2$-	44.97	38.64
m, 2H, C-C$\underline{H}_2$-C	2.20	2.18 (2)	NH$_2$-CH$_2$-	36.84	36.61
t, 3H, CH$_3$	2.02	2.04 (3)	C-C$\underline{H}_2$-C	28.72	31.52
NH$_2$	—	0.5 (3)	CH$_3$	13.64	17.70
1,3-Bis(2-carboxyethyl)-4-methyl-1H-imidazol-3-ium bromide (L$_C$)					
s, 1H, NCHN	8.84	8.83	COOH	173.03	170.3 (1)
s, 1H, $\underline{H}$C=C-CH$_3$	7.33	7.34	NCHN	133.30	132.96
s, OH	—	5.95 (2)	$\underline{C}$-CH$_3$	129.94	130.93
			$\underline{C}$=C-CH$_3$	116.01	119.67
t, 4H, N-CH2	3.64	3.62 (5)	N-CH$_2$	37.53	39.7 (9)
t, 4H, C$\underline{H}_2$-COOH	2.91	3.10 (2)	C$\underline{H}_2$-COOH	26.79	33.8 (5)
s, 3H, -CH3	2.38	2.37	CH$_3$	9.01	10.26
3-(2-Carboxyethyl)-1-(3-aminopropyl)-1H-imidazol-3-ium bromide (L$_D$)					
s, 1H, NCHN	9.11	9.11	COOH	170.21	177.74
d, 1H, HC=CH	7.72	7.71	NCHN	137.16	137.47
d, 1H, HC=CH	7.65	7.63	HC=CH	123.94	124.97
s, OH	—	6.39	HC=CH	121.50	124.69
t, 4H, N-CH$_2$	4.45	4.47 (1)	N-CH$_2$	60.01	60.29
t, 2H, COOH-C	3.33	3.34	N-CH$_2$	52.02	53.30
t, 2H, NH$_2$-CH$_2$	3.05	3.03	NH$_2$-CH$_2$	45.42	45.41
m, 2H, C-C$\underline{H}_2$-C	2.33	2.31	C-C$\underline{H}_2$-C	36.18	40.05
NH$_2$	—	0.7 (3)	COOH-CH$_2$	27.76	34.43

bacteria or the yeast culture, and $20\,\mu l$ of the chemical compound solution. All microplates were incubated at 37°C for 24 hr. The absorbance was measured at 600 nm.

2.3.2. Cytotoxicity.
3-(4,5-Dimethylthiazol-2-yl)-2,5-diphenyl tetrazolium bromide (MTT) assay was used to observe the cytotoxicity of synthesized compounds on human cervical cancer cell line (HeLa), human liver cancer cell line (Hep G2), and healthy mouse embryonic fibroblast cell line (MEF). All cell lines were provided with American Type Culture Collection (ATCC, Manassas, VA, USA). A certain population of the cells were incubated in Dulbecco's modified Eagle's Medium (DMEM) provided with Life Technologies GIBCO, Grand Island, NY, USA involving 1% penicillin streptomycin and 1% L-glutamine and HAMS F12 (1 : 1) broth medium at 37°C under 5% CO$_2$. The cells were planted in 96-multiwell plates with approximately equal numbers of 10^5, and they were allowed further incubation for 24 hr. The seven different doses of each compound were applied to the cancer cell lines and the MEF cells. The applied

doses were $400\,\mu m$, $200\,\mu m$, $100\,\mu m$, $50\,\mu m$, $25\,\mu m$, $12.5\,\mu m$, and $6.25\,\mu m$, and the dose application time was 24 hr. The surviving control for each cell was carried out against the cells not exposed to any dose. Then, $20\,\mu l$/plate of 5 mg/ml MTT solution was added into the each well and left to further incubation between 2 and 4 hr. The excess MTT solutions were removed from the wells, and $200\,\mu l$ of ultra-pure DMSO was added. The set was left in dark for 5 min before measuring the color intensities on a 492 nm spectrophotometer.

2.4. Computational Modeling.
All calculations were carried out in the framework of Gaussian 09 package. The molecules were optimized in their ground state using Becke3-Lee-Yang-Parr (B3LYP) exchange correlation functional method and 6-311G + (2d, p) basis set within density functional theory (DFT). Then, the NMR spectra and the IR spectra were calculated on the optimized geometries using the same method and the same basis set. The gauge-independent atomic orbital (GIAO) method was adopted to acquire theoretical ^{1}H and ^{13}C NMR shifts, which were

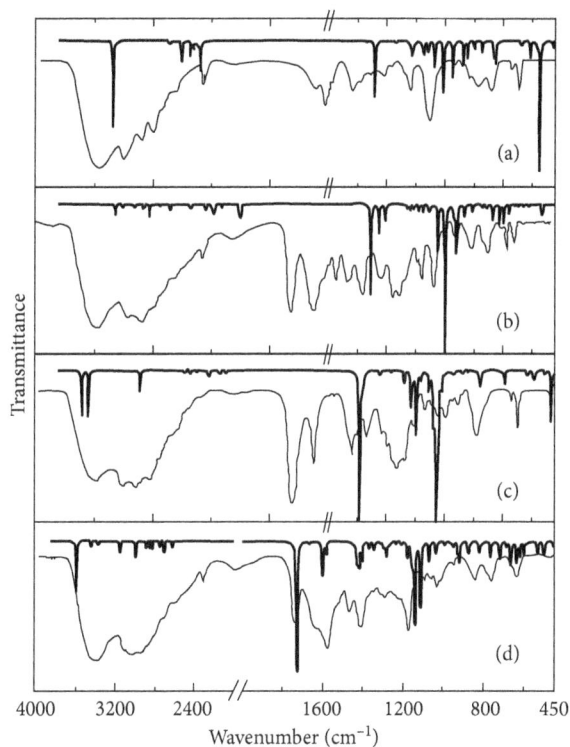

FIGURE 1: Calculated (bold lines) and experimental (pale lines) infrared spectra of the synthesized imidazole molecules: (a) *1,3-bis(2-hydroxyethyl) imidazolidinium bromide*, (b) *3-(2-ethoxy-2-oxoethly)-1-(3-aminopropyl)-1H-imidazol-3-ium bromide*, (c) *1,3-bis(2-carboxyethyl)-4-methyl-1H-imidazol-3-ium bromide*, and (d) *3-(2-carboxyethyl)-1-(3-aminopropyl)-1H-imidazol-3-ium bromide*.

converted to that of tetramethylsilane scale. The IR spectra were scaled by the factor 0.9613 due to the theoretical miscalculations [19–21]. The vibrational modes were assigned by observing the animation property of the frequency calculations provided with Gaussian 09 package.

3. Result and Discussion

NMR and IR spectra are reliable methods to elucidate an organic material or some of metal complexes. They can also be used to verify calculated molecular structure by comparing calculated spectra with corresponding experimental spectra and thus determining structural parameters and complete description of chemicals investigated.

3.1. Molecular Structures.
The NMR measurements were taken in dilute D_2O solution, while the IR spectra were recorded in the solid KBr pellet. Therefore, NMR and infrared spectra may reflect different molecular structures as the molecule surrounded by D_2O molecules in NMR spectra while the intermolecular interactions exist in the IR spectra, especially on -CH and -OH bonds [22]. On the theoretical calculation side, the crystalline phase calculations were excluded due to the single crystal form of the molecules obtained for X-ray analysis. The gas phase calculations were adopted considering differences with other phases with 2%

maximum error margin which were especially on the -CH and -OH bond lengths.

The process of the theoretical modeling of the molecules has initially been realized by using the potential energy surface scanning method for the selected dihedral angles of each molecule, thus obtaining the lowest energy conformers in the gas phase, and the resulting optimized conformers are given in Scheme 1. The imidazole rings except that of L_C with its CH_3 attachment and the CH_2 groups directly bonded to the nitrogen of the rings are common in the all molecules.

L_A has the point group symmetry C2 as the other molecules were found to have C1 symmetry. The lower point group symmetry of the L_B, L_C, and L_D in comparison with the C2 symmetry of L_A is because of the attachment of the different aliphatic chains to their imidazole rings. The calculated optimized energies and the dipole moments of L_A, L_D, L_B, and L_C are, respectively, −534.4, −667.3, −706.6, and −800.5 in units of a.u. and 2.11, 3.41, 4.54, and 7.78 in units of Debye.

The calculated parameters of the common properties of the molecules are presented in Table 1. Although the molecules differ from each other by their moieties bonded to the imidazole rings, the bond lengths, the bond angles, and the dihedral angles belong to the ring, and its immediate vicinity are in good agreement with each other and with the corresponding X-ray diffractometer (XRD) results of previously investigated similar molecule *1,3-bis(acetamide) imidazol-3-ium bromide* which was crystallized successfully [22]. The dihedral angles chosen on the imidazole ring are either about 0° or 180°, implying the aromatic structure as expected. The C-N bond lengths which bind the rings and the aliphatic moieties, that is, C_4-N_2 and N_1-C_5 are equal in L_A reflecting the symmetric structure of the molecule. However, these are different in L_B, L_C, and L_D because of the different moieties on the both side of their aromatic rings. Although the symmetry of L_C was broken by the C-CH_3 group instead of C-H in the ring, no drastic changes were observed in the concerning ring parameters. It can be inferred that the common geometrical parameters are consistent with each other with a priori theoretical confirmations before NMR and IR spectra clarifications of the calculated molecular structures.

3.2. Nuclear Magnetic Resonance Spectra.
The experimental and theoretical chemical shifts of ^1H NMR and ^{13}C NMR spectra of L_A, L_B, L_C, and L_D molecules are given in Table 2. The calculations were carried out in the gas phase considering their optimized geometries given in Scheme 1. The proton signals for -NH_2 and -OH in all were absent in the experimental ^1H NMR spectra because the solvent was D_2O concerning the solubility of the specimens which exchanged NH_2 and OH protons with deuterium. Although their detailed analysis left to the IR spectra in the following section, the calculated shifts δ 1.73 ppm of OH in L_A and δ 0.5 ppm and δ 0.7 ppm of NH_2 in L_B and L_D, respectively, are in the typical chemical shift range of R-NH_2 and R-OH groups.

The imidazole proton (NCHN, H_1) shifts of L_A, L_C, and L_D with 9.06 ppm, 8.84 ppm, and 9.11 ppm, respectively, are noticeably bigger than that of L_B with 7.79 ppm. That can be due to the intramolecular interactions of the NCHN

TABLE 3: The experimental and theoretical vibrational wave numbers for the infrared spectra of *1,3-bis(2-hydroxyethyl) imidazolidinium bromide* (L_A) with its symbolled and numbered atoms in Figure 1(a).

Vibrational assignments/vibrating atoms	Theory	Experiment
$\nu_s(OH)/(O_{1A}\text{-}H_{1A}, H_{2A}\text{-}O_{2A})$	3477	
$\nu_{as}(OH)/(O_{1A}\text{-}H_{1A}, H_{2A}\text{-}O_{2A})$	3477	3358
$\nu_{as}(CH_2)/(C_4\text{-}H_4, H'_4\text{-}C_4), (C_5\text{-}H_5, H'_5\text{-}C_5)$	3093	
$\nu_{as}(CH_2)/(C_{6A}\text{-}H_{6A}, H'_{6A}\text{-}C_{6A}), (C_{7A}\text{-}H_{7A}, H'_{7A}\text{-}C_{7A})$	3011	
$\nu(NCHN)/(C_1\text{-}H_1)$	2949	
$\nu_s(CH)/(C_2\text{-}H_2, H_3\text{-}C_3)$	2928	3114–2822
$\nu_s(CH_2)/(C_4\text{-}H_4, H'_4\text{-}C_4), (C_5\text{-}H_5, H'_5\text{-}C_5)$	2914	
$\nu_{as}(CH)/(C_2\text{-}H_2, H_3\text{-}C_3)$	2899	
$\nu_s(CH_2)/(C_{6A}\text{-}H_{6A}, H'_{6A}\text{-}C_{6A}), (C_{7A}\text{-}H_{7A}, H'_{7A}\text{-}C_{7A})$	2880	2808
$\nu_{as}(CN)/(C_1\text{-}N_1, N_2\text{-}C_1)$	1682	1630
$\nu_s(CN)/(C_1\text{-}N_1, N_2\text{-}C_1)$	1543	1592
$\delta_{sc}(CH_2)/(C_{6A}\text{-}H_{6A}, H'_{6A}\text{-}C_{6A}), (C_{7A}\text{-}H_{7A}, H'_{7A}\text{-}C_{7A})$	1452	
$\delta_{sc}(CH_2)/(C_4\text{-}H_4, H'_4\text{-}C_4), (C_5\text{-}H_5, H'_5\text{-}C_5)$	1428	1463
$\delta(OH)/(O_{1A}\text{-}H_{1A}), (O_{2A}\text{-}H_{2A})$	1411	
$\gamma_t(CH_2) + \delta_r(CH) + \delta(NCHN)/(C_4\text{-}H_4, H'_4\text{-}C_4), (C_2\text{-}H_2, C_3\text{-}H_3), (C_1\text{-}H_1)$	1375	1373
$\gamma_w(CH_2) + \nu_s(CN)/(C_5\text{-}H_5, H'_5\text{-}C_5), (C_{7A}\text{-}H_{7A}, H'_{7A}\text{-}C_{7A}), (C_4\text{-}N_2, N_1\text{-}C_5)$	1355	
$\delta_r(CH)/(C_2\text{-}H_2, H_3\text{-}C_3)$	1345	1330
$\gamma_t(CH_2) + \delta(NCHN) + \delta_r(CH)/(C_{6A}\text{-}H_{6A}, H'_{6A}\text{-}C_{6A}), (C_1\text{-}H_1), (C_2\text{-}H_2, H_3\text{-}C_3)$	1323	
$\gamma_t(CH_2) + \delta(NCHN) + \delta_r(CH)/(C_4\text{-}H_4, H'_4\text{-}C_4), (C_1\text{-}H_1), (C_2\text{-}H_2, H_3\text{-}C_3)$	1313	1301
$\delta(NCHN) + \gamma_w(OH)/(C_1\text{-}H_1), (O_{1A}\text{-}H_{1A}, H_{2A}\text{-}O_{2A})$	1274	
$\gamma_t(CH_2) + \delta_r(OH)/(C_4\text{-}H_4, H'_4\text{-}C_4), (C_5\text{-}H_5, H'_5\text{-}C_5), (O_{1A}\text{-}H_{1A}, O_{2A}\text{-}H_{2A})$	1212	1265
$\gamma_t(CH_2) + \delta_r(OH)/(C_{6A}\text{-}H_{6A}, H'_{6A}\text{-}C_{6A}), (C_{7A}\text{-}H_{7A}, H'_{7A}\text{-}C_{7A}), (O_{1A}\text{-}H_{1A}, O_{2A}\text{-}H_{2A})$	1209	
$\delta_{sc}(CH)/(C_2\text{-}H_2, H_3\text{-}C_3)$	1149	
$\nu_{as}(CN)/(C_3\text{-}N_1, N_2\text{-}C_2) + (C_4\text{-}N_2, N_1\text{-}C_5)$	1143	1179
$\nu_{as}(CO) + \nu_{as}(CC)/(C_{6A}\text{-}O_{2A}, O_{1A}\text{-}C_{7A}), (C_{6A}\text{-}C_4, C_5\text{-}C_{7A})$	1077	
$\nu_s(CO) + \nu_s(CC)/(C_{6A}\text{-}O_{2A}, O_{1A}\text{-}C_{7A}), (C_{6A}\text{-}C_4, C_5\text{-}C_{7A})$	1076	1080
$\delta_r(CH_2) + \nu_s(CO)/(C_4\text{-}H_4, H'_4\text{-}C_4), (C_5\text{-}H_5, H'_5\text{-}C_5), (C_{6A}\text{-}O_{2A}, C_{7A}\text{-}O_{1A})$	1047	
$\delta_r(CH_2) + \nu_s(CN)/(C_{6A}\text{-}H_{6A}, H'_{6A}\text{-}C_{6A}), (C_{7A}\text{-}H_{7A}, H'_{7A}\text{-}C_{7A}), (C_3\text{-}N_1, N_2\text{-}C_2)$	997	
$\gamma(NCHN)/(C_1\text{-}H_1)$	946	956
$\nu_s(CC)/(C_{6A}\text{-}C_4, C_5\text{-}C_{7A})$	930	917
$\delta_r(CH_2) + \nu_{as}(CO)/(C_4\text{-}H_4, H'_4\text{-}C_4), (C_5\text{-}H_5, H'_5\text{-}C_5), (C_{6A}\text{-}O_{2A}, O_{1A}\text{-}C_{7A})$	867	
$\delta_r(CH_2)/(C_4\text{-}H_4, H'_4\text{-}C_4), (C_{6A}\text{-}H_{6A}, H'_{6A}\text{-}C_{6A})$	861	875
$\gamma_w(CH) + \gamma(NCHN)/(C_2\text{-}H_2, H_3\text{-}C_3), (C_1\text{-}H_1)$	853	840
$\nu_{as}(CN) + \delta_r(CH_2)/(N_2\text{-}C_4, N_1\text{-}C_5), (C_{6A}\text{-}H_{6A}, H'_{6A}\text{-}C_{6A}), (C_{7A}\text{-}H_{7A}, H'_{7A}\text{-}C_{7A})$	677	673
$\gamma(NCHN)/(C_1\text{-}H_1)$	617	639

ν, stretching; δ, in-plane bending; γ, out-of-plane bending; s, symmetric; as, asymmetric; sc, scissoring; r, rocking; t, twisting; w, wagging.

hydrogen with the close oxygen atoms since the oxygen reduces the electron density on the hydrogen depending on the distance between them, thus causing higher NMR shifts. Indeed, the $H_1\text{-}O_{1A}$ in L_A, the $H_1\text{-}O_{1C}$ in L_C, and the $H_1\text{-}O_{1D}$ in L_D are 2.61 Å, 2.42 Å, and 2.41 Å, respectively. These distances which were measured from the theoretical models of the molecules are quite longer than that 2.85 Å distance of the $H_1\text{-}O_{1B}$ measured in L_B which results in small shifts for the H_1 of L_B. This is supported by the fact that the other imidazole proton shifts of H_2 and H_3 are all in close values about 7.7 ppm for all molecules as no distinct intermolecular interactions possible on H_2 and H_3 in any of them. In addition, the NMR signals of CH_2 protons in the aliphatic chains of all molecules and the smallest proton shifts of CH_3 protons in L_B and L_C molecules are theoretically and experimentally in agreement with the expected NMR results. These NMR data are in accordance with the previous study on a similar imidazole salt [23, 24] in which the protons belong to the aliphatic chain found in 4.15 ppm–2.08 ppm, the carbons belong to the imidazole ring, and the aliphatic

chain is found in 139.2 ppm–123 ppm and 62.5 ppm–33.9 ppm as in this study.

The eight R-squared tests using the data in Table 2 provide least 99.8% agreement between the experimental and the theoretical ^1H and ^{13}C NMR for L_A, L_B, L_C, and L_D. The theoretical results for NMR are in very good agreement with the experimental results as well as the observation of the expected specific values. Thus, one can infer that the calculated atomic configurations of the all molecules are good estimations except the exclusion of the interchangeable hydrogens.

3.3. Infrared Spectra. In Figure 1, the experimental infrared spectra of L_A, L_B, L_C, and L_D in the 450–4000 cm^{-1} region are given against their IR spectra calculations. The detailed account of the IR spectra including the in-plane vibrations of the stretching, scissoring, and rocking and the out-of-plane vibrations of the wagging and twisting for each molecule is presented for L_A, L_B, L_C, and L_D in Tables 3–6, respectively.

TABLE 4: The experimental and theoretical vibrational wave numbers for the infrared spectra of 3-(2-ethoxy-2-oxoethly)-1-(3-aminopropyl)-1H-imidazol-3-ium bromide (L_B) with its symbolled and numbered atoms in Figure 1(b).

Vibrational assignments/vibrating atoms	Theory	Experiment
ν(NCHN)/(C$_1$-H$_1$)	3460	
ν_{as}(NH$_2$)/(N$_{3B}$-H$_{3B}$, H$'_{3B}$-N$_{3B}$)	3412	3400
ν_s(NH$_2$)/(N$_{3B}$-H$_{3B}$, H$'_{3B}$-N$_{3B}$)	3345	
ν_{as}(CH$_2$)/(C$_{9B}$-H$_{9B}$, H$'_{9B}$-C$_{9B}$)	3327	
ν_s(CH$_2$) + ν_s(CH$_2$)/(C$_1$-H$_5$, H$'_5$-C$_1$), (C$_{9B}$-H$_{9B}$, H$'_{9B}$-C$_{9B}$)	3270	
ν (CH$_2$)/(C$_4$-H$'_4$)	3263	
ν (CH$_2$)/(C$_{8B}$-H$_{8B}$)	3226	
ν_{as}(CH$_2$)/(C$_5$-H$_5$, H$'_5$-C$_5$)	3221	
ν_s(CH$_2$)/(C$_5$-H$_5$, H$'_5$-C$_5$)	3157	
ν(CH)/(C$_3$-H$_3$)	3087	3192–2720
ν(CH)/(C$_2$-H$_2$)	2946	
ν_{as}(CH$_2$)/(C$_{6B}$-H$_{6B}$, H$'_{6B}$-C$_{6B}$)	2843	
ν_s(CH$_2$)/(C$_{6B}$-H$_{6B}$, H$'_{6B}$-C$_{6B}$)	2795	
ν_{as}(CH$_3$)/(C$_{10B}$-H$_{10B}$, H$'_{10B}$-C$_{10B}$)	2784	
ν_{as}(CH$_3$)/(C$_{10B}$-H$_{10B}$, H$''_{10B}$-C$_{10B}$)	2783	
ν_s(CH$_3$)/(C$_{10B}$-H$_{10B}$, H$'_{10B}$-C$_{10B}$, H$''_{10B}$-C$_{10B}$)	2731	
ν(CH$_2$)/(C$_{8B}$-H$'_{8B}$)	2606	2618
ν(CO)/(C$_{7B}$-O$_{1B}$)	1707	1753
ν(CN) + δ(CH) + ν(C=C)/(C$_1$-N$_2$), (C$_2$-H$_2$), (C$_2$-C$_3$)	1652	
ν(CN) + ν(C=C)/(C$_1$-N$_1$), (C$_2$-C$_3$)	1619	1645
δ_{sc}(NH$_2$)/(C$_{3B}$-H$_{3B}$, H$'_{3B}$-C$_{3B}$)	1606	
ν_s(CN) + ν(NC)/(N$_2$-C$_1$,C$_1$-N$_1$), (N$_2$-C$_2$)	1494	1561
δ_{sc}(CH$_3$) + δ(CH$_3$)/(C$_{10B}$-H$_{10B}$, C$_{10B}$-H$''_{10B}$), (C$_{10B}$-H$'_{10B}$)	1464	1528
δ_{sc}(CH$_2$)/(C$_{8B}$-H$_{8B}$, H$'_{8B}$-C$_{8B}$)	1462	
δ_{sc}(CH$_3$) + δ(CH$_3$)/(C$_{10B}$-H$'_{10B}$, C$_{10B}$-H$''_{10B}$), (C$_{10B}$-H$_{10B}$)	1451	1473
δ_{sc}(CH$_2$)/(C$_{6B}$-H$_{6B}$, H$'_{6B}$-C$_{6B}$)	1441	
δ_{sc}(CH$_2$) + γ_w(CH$_3$)/(C$_{9B}$-H$_{9B}$, H$'_{9B}$-C$_{9B}$), (C$_{10B}$-H$_{10B}$, C$_{10B}$-H$'_{10B}$)	1415	1465
δ_{sc}(CH$_2$)/(C$_5$-H$_5$, H$'_5$-C$_5$)	1381	
γ_w(CH$_2$) + δ_{sc}(CH$_2$)/(C$_{8B}$-H$_{8B}$, H$'_{8B}$-C$_{8B}$), (C$_4$-H$_4$, H$'_4$-C$_4$)	1372	1458
γ_w(CH$_2$) + δ_{sc}(CH$_2$)/(C$_{6B}$-H$_{6B}$, H$'_{6B}$-C$_{6B}$), (C$_4$-H$_4$, H$'_4$-C$_4$)	1352	1394
ν_s(CN) + γ_t(CH$_2$)/(C$_4$-N$_2$, N$_1$-C$_5$), (C$_{6B}$-H$_{6B}$, H$'_{6B}$-C$_{6B}$), (C$_5$-H$_5$, H$'_5$-C$_5$)	1319	1300
γ_w(CH$_2$)/(C$_{9B}$-H$_{9B}$, H$'_{9B}$-C$_{9B}$)	1309	
γ_t(NH$_2$) + γ_t(CH$_2$)/(N$_{3B}$-H$_{3B}$, H$'_{3B}$-N$_{3B}$), (C$_{6B}$-H$_{6B}$, H$'_{6B}$-C$_{6B}$)	1296	1242
δ(NCHN) + δ_r(CH)/(C$_1$-H$_1$), (C$_2$-H$_2$, C$_3$-H$_3$)	1282	
ν_{as}(CO) + γ_w(CH$_2$)/(C$_{7B}$-O$_{2B}$, O$_{2B}$-C$_{9B}$), (C$_5$-H$_5$, H$'_5$-C$_5$), (C$_{9B}$-H$_{9B}$, H$'_{9B}$-C$_{9B}$)	1248	1214
ν_{as}(CO) + γ_w(CH$_3$)/(C$_{7B}$-O$_{2B}$, O$_{2B}$-C$_{9B}$), (C$_{10B}$-H$_{10B}$, C$_{10B}$-H$''_{10B}$)	1198	1183
ν_{as}(CC) + ν(CO) + ν_s(CN)/(C$_{8B}$-C$_{6B}$, C$_{6B}$-C$_4$), (C$_{3B}$-O$_{2B}$), (C$_4$-N$_2$, N$_1$-C$_5$)	1123	1118
ν_s(CO) + ν(CC) + δ(NCHN)/(C$_{7B}$-O$_{2B}$, O$_{2B}$-C$_{9B}$), (C$_{9B}$-C$_{10B}$), (C$_{7B}$-C$_5$), (C$_1$-H$_1$)	1069	1097
γ_w(NH$_2$) + δ_r(CH$_2$)/(N$_{3B}$-H$_{3B}$, H$'_{3B}$-N$_{3B}$), (C$_{8B}$-H$_{8B}$, H$'_{8B}$-C$_{8B}$)	1011	1039
γ_w(NH$_2$) + ν(CC)/(N$_{3B}$-H$_{3B}$, H$'_{3B}$-N$_{3B}$) + (C$_{6B}$-C$_4$)	869	931
ν(CC) + γ_w(CH$_3$) + ν(CO)/(C$_5$-C$_{7B}$), (C$_{10B}$-H$_{10B}$, C$_{10B}$-H$'_{10B}$), (C$_{7B}$-O$_{2B}$)	860	852
γ_w(NH$_2$) + ν(CC)/(N$_{3B}$-H$_{3B}$, H$'_{3B}$-N$_{3B}$), (C$_4$-C$_{6B}$)	824	787
γ(NCHN) + γ_w(CH)/(C$_1$-H$_1$), (C$_2$-H$_2$, C$_3$-H$_3$)	797	765
γ_w(CH) + γ_t(NH$_2$) + γ_t(CN)/(C$_2$-H$_2$, C$_3$-H$_3$), (N$_{3B}$-H$_{3B}$, H$'_{3B}$-N$_{3B}$), (N$_1$-C$_5$)	759	700
γ_t(NH$_2$) γ_t(CN) + γ(NCHN)/(N$_{3B}$-H$_{3B}$, H$'_{3B}$-N$_{3B}$), (N$_2$-C$_4$), (C$_1$-H$_1$)	722	652
γ_t(CH$_2$) + δ(CC)/(C$_{8B}$-H$_{8B}$, H$'_{8B}$-C$_{8B}$), (C$_{6B}$-H$_{6B}$, H$'_{6B}$-C$_{6B}$), (C$_5$-C$_{7B}$)	665	636
γ_t(CH$_2$) + δ(CC)/(C$_{6B}$-H$_{6B}$, H$'_{6B}$-C$_{6B}$) + (C$_{6B}$-C$_4$) + (C$_5$-C$_{7B}$)	626	592
γ_t(C=C)/(N$_2$-C$_2$, C$_3$-N$_1$)	603	
Γ(NCHN)/(C$_1$-H$_1$)	534	534

ν, stretching; δ, in-plane bending; γ, out-of-plane bending; s, symmetric; as, asymmetric; sc, scissoring; r, rocking; t, twisting; w, wagging.

The R-square test results exhibit 99.8%, 99.5%, 99.5%, and 99.7% agreement between the experimental and the theoretical IR spectra of L_A, L_B, L_C, and L_D, respectively.

The intermolecular interactions and the correlations of close frequency IR signals cause the -OH, -CH, -CH$_2$, -CH$_3$, and -NH$_2$ stretching vibrations to appear under broad peaks in the experimental spectra. These broad peaks are in very well-defined frequency region. The vibrations coming from the other functional groups of the imidazole rings and the aliphatic chains including the other vibrations of the hydrogenic groups in all modes appeared in 1750–450 cm^{-1} region.

The first vibration signals observed in the 1800–1700 cm^{-1} region of the experimental spectra of L_B, L_C, and L_D are distinct peaks arising from ν(C=O) stretching as their values

TABLE 5: The experimental and theoretical vibrational wave numbers for the infrared spectra of *1,3-bis(2-carboxyethyl)-4-methyl-1H-imidazol-3-ium bromide* (L_C) with its symbolled and numbered atoms in Figure 1(c).

Vibrational assignments/vibrating atoms	Theory	Experiment
$\nu(OH)/(O_{3C}\text{-}H_{3C})$	3685	3416
$\nu(OH)/(O_{4C}\text{-}H_{4C})$	3643	
$\nu(CH_2)/(C_5\text{-}H_5)$	3328	
$\nu(CH_2)/(C_4\text{-}H_4)$	3311	
$\nu(NCHN)/(C_1\text{-}H_1)$	3292	
$\nu(CH_3)/(C_{10}\text{-}H''_{10C})$	3046	
$\nu(CH_2)/(C_5\text{-}H'_5)$	2983	
$\nu(CH_2)/(C_4\text{-}H'_4)$	2949	3268–2677
$\nu_{as}(CH_3)/(C_{10C}\text{-}H_{10C}, H'_{10C}\text{-}C_{10C})$	2904	
$\nu_{as}(CH_2)/(C_{7C}\text{-}H_{7C}, H'_{7C}\text{-}C_{7C})$	2836	
$\nu_{as}(CH_2)/(C_{6C}\text{-}H_{6C}, H'_{6C}\text{-}C_{6C})$	2828	
$\nu_s(CH_2)/(C_{6C}\text{-}H_{6C}, H'_{6C}\text{-}C_{6C})$	2752	
$\nu_s(CH_2)/(C_{7C}\text{-}H_{7C}, H'_{7C}\text{-}C_{7C})$	2739	
$\nu(CH)/(C_2\text{-}H_2)$	2706	2663
$\nu_s(COOH) + \nu_s(OH)/(C_{9C}\text{-}O_{1C}, O_{2C}\text{-}C_{8C}), (O_{3C}\text{-}H_{3C}, H_{4C}\text{-}O_{4C})$	1785	
$\nu_{as}(COOH) + \nu_{as}(OH)/(C_{9C}\text{-}O_{1C}, O_{2C}\text{-}C_{8C}), (O_{3C}\text{-}H_{3C}, H_{4C}\text{-}O_{4C})$	1783	1774
$\nu_{as}(CN)/(C_1\text{-}N_1, N_2\text{-}C_1)$	1756	
$\nu_{as}(CN) + \nu(CC) + \nu(CN)/(C_1\text{-}N_1, N_2\text{-}C_5), (C_{6C}\text{-}C_4), (N_2\text{-}C_4)$	1679	1662
$\nu_s(CN) + \nu(C=C) + \nu(CC)/(C_1\text{-}N_1, N_2\text{-}C_1), (C_2\text{-}C_3), (C_3\text{-}C_{10C}), (C_4\text{-}N_2)$	1642	
$\gamma_t(CH_2) + \nu(CN)/(C_5\text{-}H_5, H'_5\text{-}C_5), (N_1\text{-}C_3)$	1510	1576
$\gamma_t(CH_2) + \nu(CN)/(C_4\text{-}H_4, H'_4\text{-}C_4), (N_2\text{-}C_2)$	1476	1561
$\delta_{sc}(CH_2) + \delta_{sc}(CH_3)/(C_4\text{-}H_4, H'_4\text{-}C_4), (C_{6C}\text{-}H_{6C}, H'_{6C}\text{-}C_{6C}), (C_{10C}\text{-}H_{10C}, C_{10C}\text{-}H''_{10C})$	1434	1468
$\gamma(CH_3)/(C_{10C}\text{-}H'_{10C})$	1398	1393
$\delta(CH) + \gamma_t(CH_2) + \delta(NCHN)/(C_2\text{-}H_2) + (C_4\text{-}H_4, H'_4\text{-}C_4) + (C_5\text{-}H_5, H'_5\text{-}C_5) + (C_1\text{-}H_1)$	1305	1317
$\delta(OH) + \nu(CO)/(O_{4C}\text{-}H_{4C}), (C_{8C}\text{-}O_{4C})$	1259	1282
$\delta(CH) + \nu(CC) + \nu(CN)/(C_1\text{-}H_1), (C_2\text{-}H_2), (C_3\text{-}C_{10C}), (N_2\text{-}C_4)$	1239	1235
$\delta(CH) + \nu(CC) + \nu_s(CN)/(C_1\text{-}H_1), (C_2\text{-}H_2), (C_3\text{-}C_{10C}), (N_2\text{-}C_4, C_5\text{-}N_1)$	1221	
$\gamma_t(CH_2)/(C_4\text{-}H_4, H'_4\text{-}C_4), (C_{6C}\text{-}H_{6C}, H'_{6C}\text{-}C_{6C})$	1191	1199
$\gamma_t(CH_2)/(C_5\text{-}H_5, H'_5\text{-}C_5), (C_{7C}\text{-}H_{7C}, H'_{7C}\text{-}C_{7C})$	1181	
$\gamma_w(CH_2) + \delta(OH)/(C_{6C}\text{-}H_{6C}, H'_{6C}\text{-}C_{6C}), (O_{4C}\text{-}H_{4C})$	1149	1157
$\gamma_w(CH_2) + \delta(OH)/(C_{7C}\text{-}H_{7C}, H'_{7C}\text{-}C_{7C}), (O_{3C}\text{-}H_{3C})$	1139	
$\nu(CC)/(C_5\text{-}C_{7C})$	1086	1099
$\nu(CC)/(C_4\text{-}C_{6C})$	1074	
$\delta(CH_3)/(C_{10}\text{-}H_{10C}, C_{10}\text{-}H'_{10C}, C_{10}\text{-}H''_{10C})$	1046	1031
$\gamma(CH_3) + \gamma(NCHN) + \gamma(CH)/(C_{10}\text{-}H_{10C}, C_{10}\text{-}H'_{10C}, C_{10}\text{-}H''_{10C}), (C_1\text{-}H_1), (C_2\text{-}H_2)$	953	990
$\gamma_w(CH_2) + \delta_r(CH_2)/(C_{6C}\text{-}H_{6C}, H'_{6C}\text{-}C_{6C}), (C_4\text{-}H_4, H'_4\text{-}C_4)$	940	931
$\nu(CC) + \delta_r(CH_2)/(C_3\text{-}C_{10}), (C_{7C}\text{-}C_{9C}), (C_4\text{-}H_4, H'_4\text{-}C_4), (C_{6C}\text{-}H_{6C}, H'_{6C}\text{-}C_{6C})$	807	836
$\gamma(NCHN)/(C_1\text{-}H_1)$	786	
$\gamma_t(C=C) + \delta(OH)/(N_2\text{-}C_2, C_3\text{-}N_1) + (O_{4C}\text{-}H_{4C})$	624	629
$\delta(OH)/(O_{4C}\text{-}H_{4C})$	470	462
$\delta(OH)/(O_{3C}\text{-}H_{3C})$	424	

ν, stretching; δ, in-plane bending; γ, out-of-plane bending; s, symmetric; as, asymmetric; sc, scissoring; r, rocking; t, twisting; w, wagging.

in agreement with the previously observed $\nu(C=O)$ stretching [25] and theoretically calculated values as reflected in Table 3. The imidazole $\nu(C=N)$ stretching of L_A appears as individual signals while they are coupled under broad peaks with L_B, L_C, and L_D molecules as seen about 1650 cm^{-1} while the frequencies for $\nu(C=N)$ stretching are in agreements with the corresponding data as given in [26]. The aliphatic $\nu(C\text{-}N)$ stretching in 1069–1235 cm^{-1} interval and the aliphatic $\nu(C\text{-}C)$ stretching in 917–1099 cm^{-1} are consistent with the previous corresponding measurements [27].

Beyond the consistency of the common imidazole properties of the molecules, we also give unique infrared signals of molecules in Table 3. C-O stretching belong to the H$_2$C-OH group in L_A, in-plane δ vibrations of CH$_3$ at the end of the aliphatic chain of L_B, and at the attachment of the imidazole ring of L_C are such unique vibrations. The calculated and observed wave numbers of these vibrations are consistent with the concerning previous studies [28, 29].

The molecules in consideration analyzed by infrared spectrum because of the lack of exchangeable protons of -OH and -NH$_2$ in the NMR spectra of the chemicals and the existence of these groups is proved in the molecules. Also, the IR spectra of the molecules reverifies the theoretically obtained configurations of them as they were first verified by the comparison the theoretical NMR spectra with that of the experimental NMR spectra.

3.4. Antibacterial and Cytotoxic Activities. The antibacterial tests of the synthesized four molecules were run on the six

TABLE 6: The experimental and theoretical vibrational wave numbers for the infrared spectra of *3-(2-carboxyethyl)-1-(3-aminopropyl)-1H-imidazol-3-ium bromide* (L_D) with its symbolled and numbered atoms in Figure 1(d).

Vibrational assignments/vibrating atoms	Theory	Experiment
$\nu(OH)/(O_{2D}\text{-}H_{2D})$	3590	3423
$\nu_{as}(NH_2)/(N_{3D}\text{-}H_{3D}, H'_{3D}\text{-}N_{3D})$	3446	
$\nu(CH_2)/(C_4\text{-}H_4)$	3381	
$\nu_s(NH_2)/(N_{3D}\text{-}H_{3D}, H'_{3D}\text{-}N_{3D})$	3370	3222–2666
$\nu(CH_2)/(C_5\text{-}H_5)$	3154	
$\nu(CH_2)/(C_{8D}\text{-}H_{8D})$	3152	
$\nu(NCHN)/(C_1\text{-}H_1)$	2995	
$\nu(CH_2)/(C_{6D}\text{-}H_{6D})$	2890	
$\nu_s(CH)/(C_2\text{-}H_2, H_3\text{-}C_3)$	2857	
$\nu_{as}(CH_2)/(C_{7D}\text{-}H_{7D}, H'_{7D}\text{-}C_{7D})$	2850	—
$\nu_{as}(CH) + \nu_s(CH_2)/(C_2\text{-}H_2, H_3\text{-}C_3), (C_5\text{-}H_5, H'_5\text{-}C_5)$	2828	
$\nu_s(CH_2)/(C_{7D}\text{-}H_{7D}, H'_{7D}\text{-}C_{7D})$	2789	
$\nu(CH_2)/(C_4\text{-}H'_4)$	2752	
$\nu(CH_2)/(C_{8D}\text{-}H'_{8D})$	2709	
$\nu(COOH)/(C_{9D}\text{-}O_{1D})$	1728	1745
$\nu_s(CN) + \nu(C=C)/(C_1\text{-}N_1, N_2\text{-}C_1), (C_2\text{-}C_3)$	1668	1642
$\nu_{as}(CN) + \delta_r(CH)/(C_1\text{-}N_1, N_2\text{-}C_1), (C_2\text{-}H_2, H_3\text{-}C_3)$	1601	1585
$\delta_{sc}(NH_2)/(N_{3D}\text{-}H_{3D}, H'_{3D}\text{-}N_{3D})$	1582	
$\nu_s(CN)/(C_1\text{-}N_1, N_2\text{-}C_1)$	1468	1470
$\delta_{sc}(CH_2)/(C_{6D}\text{-}H_{6D}, H'_{6D}\text{-}C_{6D})$	1454	
$\delta_{sc}(CH_2)/(C_{8D}\text{-}H_{8D}, H'_{8D}\text{-}C_{8D})$	1443	
$\delta_{sc}(CH_2) + \gamma_w(CH_2)/(C_5\text{-}H_5, H'_5\text{-}C_5), (C_{7D}\text{-}H_{7D}, H'_{7D}\text{-}C_{7D})$	1430	1419
$\delta_{sc}(CH_2)/(C_4\text{-}H_4, H'_4\text{-}C_4)$	1423	
$\gamma_w(CH_2) + \nu(CC) + \nu(CO)/(C_{7D}\text{-}H_{7D}, H'_{7D}\text{-}C_{7D}), (C_{7D}\text{-}C_{9D}) + (C_{9D}\text{-}O_{2D})$	1417	
$\delta_{sc}(CH_2)/(C_{7D}\text{-}H_{7D}, H'_{7D}\text{-}C_{7D})$	1404	1354
$\delta(NCHN) + \delta_r(CH)/(C_1\text{-}H_1), (C_2\text{-}H_2, H_3\text{-}C_3)$	1372	
$\gamma_w(CH_2) + \nu(CC)/(C_4\text{-}H_4, H'_4\text{-}C_4), (C_{6D}\text{-}H_{6D}, H'_{6D}\text{-}C_{6D}), (C_{6D}\text{-}C_{8D})$	1369	
$\gamma_w(CH_2) + \gamma_t(NH_2)/(C_4\text{-}H_4, H'_4\text{-}C_4), (C_{6D}\text{-}H_{6D}, H'_{6D}\text{-}C_{6D}), (N_{3D}\text{-}H_{3D}, H'_{3D}\text{-}N_{3D})$	1348	1318
$\gamma_w(CH_2)/(C_5\text{-}H_5, H'_5\text{-}C_5), (C_{7D}\text{-}H_{7D}, H'_{7D}\text{-}C_{7D})$	1344	
$\delta_r(CH) + \gamma_t(CH_2) + \gamma_w(CH_2)/(C_2\text{-}H_2, H_3\text{-}C_3), (C_5\text{-}H_5, H'_5\text{-}C_5), (C_{8D}\text{-}H_{8D}, H'_{8D}\text{-}C_{8D})$	1314	
$\delta(NCHN) + \delta_r(CH) + \delta(OH)/(C_1\text{-}H_1), (C_2\text{-}H_2, H_3\text{-}C_3), (O_{2D}\text{-}H_{2D})$	1298	1297
$\gamma_w(CH_2) + \delta(OH)/(C_{7D}\text{-}H_{7D}, H'_{7D}\text{-}C_{7D}), (O_{2D}\text{-}H_{2D})$	1283	
$\gamma_t(NH_2) + \gamma_t(CH_2)/(N_{3D}\text{-}H_{3D}, H'_{3D}\text{-}N_{3D}), (C_{6D}\text{-}H_{6D}, H'_{6D}\text{-}C_{6D})$	1266	1225
$\gamma_t(NH_2)/(N_{3D}\text{-}H_{3D}, H'_{3D}\text{-}N_{3D})$	1209	
$\delta_{sc}(CH) + \delta(CH) + \gamma_t(CH_2)/(C_2\text{-}H_2, H_3\text{-}C_3), (C_1\text{-}H_1), (C_4\text{-}H_4, H'_4\text{-}C_4)$	1178	1174
$\delta_{sc}(CH)/(C_2\text{-}H_2, H_3\text{-}C_3)$	1148	1119
$\delta(NCHN) + \gamma_w(CH_2) + \delta(OH)/(C_1\text{-}H_1), (C_{7D}\text{-}H_{7D}, H'_{7D}\text{-}C_{7D}), (O_{2D}\text{-}H_{2D})$	1139	
$\delta(NCHN) + \gamma_t(CH_2) + \gamma_t(NH_2)/(C_1\text{-}H_1), (C_5\text{-}H_5, H'_5\text{-}C_5), (N_{3D}\text{-}H_{3D}, H'_{3D}\text{-}N_{3D})$	1111	1098
$\gamma_t(NH_2) + \delta_{sc}(CH) + \nu_{as}(CN)/(N_{3D}\text{-}H_{3D}, H'_{3D}\text{-}N_{3D}), (C_2\text{-}H_2, H_3\text{-}C_3), (C_4\text{-}N_2, N_1\text{-}C_5)$	1070	1069
$\nu(CC)/(C_5\text{-}C_{7D})$	1037	1033
$\gamma_w(NH_2) + \nu(CC) + \nu(CN)/(N_{3D}\text{-}H_{3D}, H'_{3D}\text{-}N_{3D}), (C_{6D}\text{-}C_4), (C_4\text{-}N_2)$	1034	
$\gamma_t(CH)/(C_2\text{-}H_2, H_3\text{-}C_3)$	973	
$\gamma_w(CH_2) + \delta_r(CH_2) + \delta(OH)/(C_{7D}\text{-}H_{7D}, H'_{7D}\text{-}C_{7D}), (C_5\text{-}H_5, H'_5\text{-}C_5), (O_{2D}\text{-}H_{2D})$	960	976
$\gamma_t(NH_2) + \nu(CC)/(N_{3D}\text{-}H_{3D}, H'_{3D}\text{-}N_{3D}), (C_{6D}\text{-}C_{8D})$	948	
$\gamma(NCHN) + \gamma_w(CH)/(C_1\text{-}H_1), (C_2\text{-}H_2, H_3\text{-}C_3)$	918	954
$\gamma_t(NH_2) + \nu(CN) + \delta_r(CH_2)/(N_{3D}\text{-}H_{3D}, H'_{3D}\text{-}N_{3D}), (N_{3D}\text{-}C_{8D}), (C_{6D}\text{-}H_{6D}, H'_{6D}\text{-}C_{6D})$	818	846
$\nu_s(CC) + \nu(CO) + \delta(OH)/(C_5\text{-}C_{7D}, C_{7D}\text{-}C_{9D}), (C_{9D}\text{-}O_{2D}), (O_{2D}\text{-}H_{2D})$	801	767
$\gamma_t(C=C) + \delta(OH)/(C_2\text{-}C_3), (O_{2D}\text{-}H_{2D})$	663	637
$\delta(OH) + \delta_r(CH_2)/(O_{2D}\text{-}H_{2D}), (C_{7D}\text{-}H_{7D}, H'_{7D}\text{-}C_{7D})$	530	—

ν, stretching; δ, in-plane bending; γ, out-of-plane bending; s, symmetric; as, asymmetric; sc, scissoring; r, rocking; t, twisting; w, wagging.

different bacteria and yeast, as mentioned in Section 2.3.1. Figures 2(a) and 3(b) show the inhibitory effects of the various concentrations of the L_A and L_B on *Candida albicans* ATCC 10231 and *Bacillus cereus* ATCC 11778 (Gram positive) together with that of the antibiotic controls. Their effects on *Escherichia coli* O157:H7 (Gram negative), *Escherichia coli* ATCC 25922, *Salmonella typhimurium*

ATCC 14028, *Staphylococcus aureus* ATCC 25923, and *Listeria monocytogenes* ATCC 19115 are excluded because of their very weak efficacy in comparison with that of the antibiotic control. Although the absorbance measurements involve some statistical errors, L_A showed better inhibition than the antibiotic on *Bacillus cereus* ATCC 11778, while both chemicals are effective on the selected bacteria and

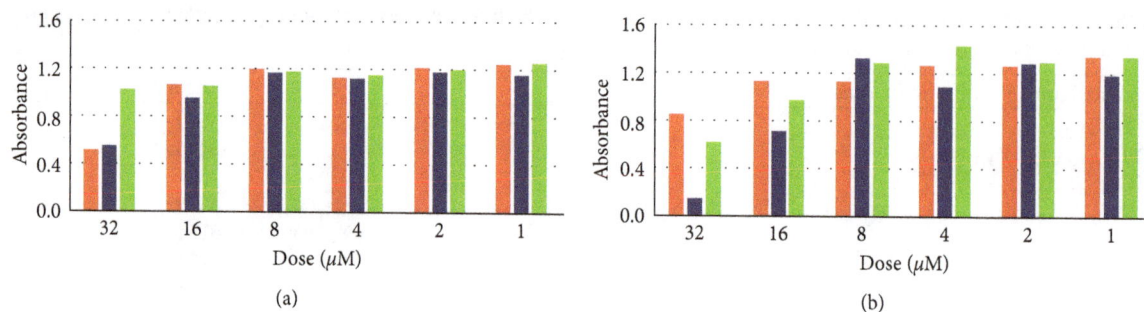

FIGURE 2: (a) The antibacterial activity on *Candida albicans*. The first bars on the left are for the commercial antibiotic inhibition for each dose. The second bars are for *1,3-bis(2-hydroxyethyl) imidazolidinium bromide* (L_A), and the third bars are for *3-(2-ethoxy-2-oxoethly)-1-(3-aminopropyl)-1H-imidazol-3-ium bromide* (L_B), (b) same as in Figure 2(a), but for the antibacterial activity on *Bacillus cereus*.

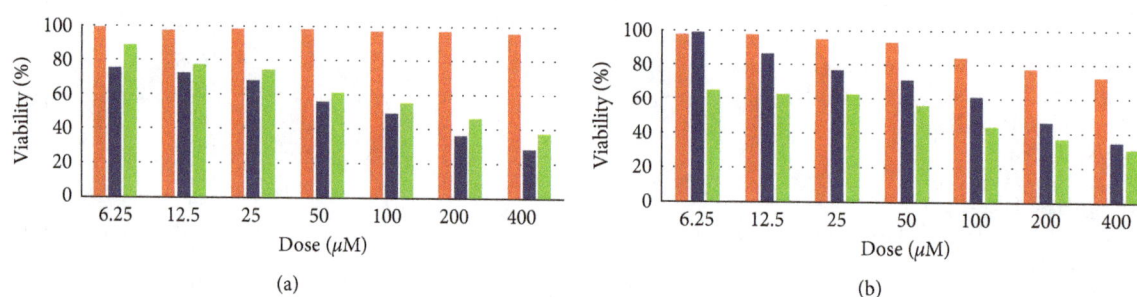

FIGURE 3: (a) The cytotoxicity of *1,3-bis(2-carboxyethyl)-4-methyl-1H-imidazol-3-ium bromide*. The first bars represent the effect on healthy mouse embryonic fibroblast cell line for each dose, as do the second bars on human cervical cancer cell line, and the third bars on human liver cancer cell line, (b) same as in Figure 3(a), but for the cytotoxicity of *3-(2-carboxyethyl)-1-(3-aminopropyl)-1H-imidazol-3-ium bromide*.

TABLE 7: Half inhibition concentrations of the molecules on the selected bacteria and cancer cell lines.

	Antibacterial activity IC$_{50}$ (μM)			Cytotoxic activity IC$_{50}$ (μM)		
	Escherichia coli O157:H7	*Candida albicans*	*Bacillus cereus*	HeLa	Hep G2	MEF
L_A	32	30	17	316	100	—
L_B	39	156	29	141	182	—
L_C	—	—	—	81	150	—
L_D	—	—	—	167	57	—
Antibiotic	10	30	56	—	—	—

L_A, 1,3-bis(2-hydroxyethyl) imidazolidinium bromide; L_B, 3-(2-ethoxy-2-oxoethly)-1-(3-aminopropyl)-1H-imidazol-3-ium bromide; L_C, 1,3-bis(2-car-boxyethyl)-4-methyl-1H-imidazol-3-ium bromide; L_D 3-(2-carboxyethyl)-1-(3-aminopropyl)-1H-imidazol-3-ium bromide; IC$_{50}$, half inhibitory concen-trations; HeLa, human cervical cancer cell line; Hep G2, human liver cancer cell line; MEF, healthy mouse embryonic fibroblast cell line.

yeast as much as the antibiotics. L_C and L_D show no no-ticeable inhibitory effects on the target bacteria, and thus, their absorbance values as function of their concentration were not given for the sake of brevity.

Figure 3(a) shows the cytotoxic activity of L_C on HeLa and Hep G2 cell lines against healthy MEF cell lines, as does Figure 3(b) for the cytotoxic activity of L_D. The cytotoxic activities of L_A and L_B are not exhibited as they showed lesser cytotoxic activities in comparison with L_C and L_D. The percentage cell viability of HeLa and Hep G2 cell lines was significantly reduced by L_C and L_D at the end of 24 hr ap-plication of the doses. In addition, L_C did not harm the healthy MEF cell lines for any dose so that its half inhibitory concentrations (IC$_{50}$) cannot be calculated. Meantime, L_D

showed some activity on the MEF cells with the high doses together with the fact that the activity was not as strong as it did on the cancer cell lines.

The antibacterial and cytotoxic effectiveness of the syn-thesized molecules are summarized in Table 7 by presenting their IC$_{50}$ concentrations in the units of μM on the bacteria together with the IC$_{50}$ values of the antibiotics and on the cancer cell lines. The lack of the IC$_{50}$ values of L_C and L_D for antibacterial activity in Table 7 indicates that the IC$_{50}$ values of them cannot be calculated due to their very weak effect on the bacteria sample within the dose range considered. L_A and L_B were, respectively, thrice and twice more effective on *Bacillus cereus* than the antibiotic (gentamicin) as the L_A equals the antibiotic (amphotericin-b) effect on *Candida albicans*.

The cell viability assay of the chemicals exhibits no harmful effect on the healthy MEF cell lines as their IC_{50} values cannot be calculated within the dose range considered. The L_C inhibition on HeLa and the L_D inhibition on Hep G2 are distinctive when they are compared with the inhibition of the other imidazole-based chemicals on different cancer cell lines in the concerning studies. The IC_{50} values of the L_C on HeLa and the L_D on Hep G2 were 81 μM and 57 μM, respectively. The similar imidazole compounds which have alkyl moieties were tested on the cancer cells different from the cells used in this study [8]. The IC_{50} values of *1-(benzofuran-2-yl (phenyl)methyl)-3-allyl-2-ethyl-1H-imidazol-3-ium bromide* and *1-(benzofuran-2-yl(phenyl)methyl)-3-butyl-2-ethyl-1H-imidazol-3-ium iodide* on leukemia (HL-60), lung carcinoma (A549), colon carcinoma (SW480), breast carcinoma (MCF-7), and myeloid liver carcinoma (SMMC-7721) cancer lines have been detected over 40 μM. Additionally, the IC_{50} activity of *3β-hydroxy-21-(1H-imidazol-1-yl)pregna-5,16-dien-20-one* on prostate cancer (PC-3), breast cancer (MCF7), and lung cancer (SK-LU- 1) were 20 μM, 19 μM, and 18 μM, respectively [7]. These results are quantitatively better than 81 μM and 57 μM on Hep G2 and HeLa. However, this is compensated by the fact that the L_C and L_D have not any harmful effect on the healthy MEF cell lines.

4. Conclusions

Novel imidazole salts, or *N*-heterocyclic carbene ligands, namely, *1,3-bis(2-hydroxyethyl) imidazolidinium bromide L_A, 3-(2-ethoxy-2-oxoethly)-1-(3-aminopropyl)-1H-imidazol-3-ium bromide L_B, 1,3-bis(2-carboxyethyl)-4-methyl-1H-imidazol-3-ium bromide L_C*, and *3-(2-carboxyethyl)-1-(3-aminopropyl)-1H-imidazol-3-ium bromide L_D* were synthesized, and they were preliminary confirmed by GC-MS and elemental analysis methods. Their molecular structures were theoretically determined, and they were confirmed by comparing calculated ^{1}H, ^{13}C NMR, and IR spectra with those of experimentally observed data. Also, the calculated structures were verified by the XRD results on a similar imidazole salt [22].

The antimicrobial and cytotoxic activities of the synthesized ligands on some specific bacteria and cancer cell lines were measured using spectrophotometric methods. It is seen that L_A showed better inhibition than the selected antibiotic on *Bacillus cereus* ATCC 11778 while it is effective on the selected bacteria and the yeast together with L_B. On the cytotoxicity evaluation, L_C showed considerable inhibition effect on HeLa, as does L_D on Hep G2. Although their IC_{50} doses are quite high in comparison with the similar chemicals in the literature, the cytotoxicity of L_C and L_D is affirmed by not causing harmful effect on the healthy MEF cells as much as they do on the cancel cell lines.

References

[1] X. Chen, Y. Zou, M. Han et al., "Solubilisation of myosin in a solution of low ionic strength L-histidine: significance of the imidazole ring," *Food Chemistry*, vol. 196, pp. 42–49, 2016.

[2] R. I. Benhamou, K. B. Steinbuch, and M. Fridman, "Anti-fungal imidazole-decorated cationic amphiphiles with markedly low hemolytic activity," *Chemistry*, vol. 22, no. 32, pp. 11148–11151, 2016.

[3] M. Y. Wani, A. Ahmad, R. A. Shiekh, K. J. Al-Ghamdi, and A. J. Sobral, "Imidazole clubbed 1,3,4-oxadiazole derivatives as potential antifungal agents," *Bioorganic and Medicinal Chemistry*, vol. 23, no. 15, pp. 4172–4180, 2015.

[4] K. Krowicki and J. W. Lown, "Synthesis of novel imidazole-containing DNA minor groove binding oligopeptides related to the antiviral antibiotic netropsin," *Journal of Organic Chemistry*, vol. 52, no. 16, pp. 3493–3501, 1987.

[5] W.-N. Lee, A. Y.-C. Lin, and X.-H. Wang, "The occurrence of quinolone and imidazole antibiotics in rivers in Central Taiwan," *Desalination and Water Treatment*, vol. 52, no. 4–6, pp. 1143–1152, 2013.

[6] G. Castanedo, Y. Liu, J. J. Crawford, and M. G. Braun, "Synthesis of fused imidazole-containing ring systems via dual oxidative amination of C(sp(3))-H bonds," *Journal of Organic Chemistry*, vol. 81, no. 18, pp. 8617–8624, 2016.

[7] A. V. Silva-Ortiz, E. Bratoeff, M. T. Ramirez-Apan et al., "Synthesis and biological activity of two pregnane derivatives with a triazole or imidazole ring at C-21," *Journal of Steroid Biochemistry and Molecular Biology*, vol. 159, pp. 8–18, 2016.

[8] X.-Q. Wang, L.-X. Liu, Y. Li et al., "Design, synthesis and biological evaluation of novel hybrid compounds of imidazole scaffold-based 2-benzylbenzofuran as potent anticancer agents," *European Journal of Medicinal Chemistry*, vol. 62, pp. 111–121, 2013.

[9] A. Mrozek, J. Karolak-Wojciechowska, and K. Kieć-Kononowicz, "Five-membered heterocycles. Part III. Aromaticity of 1,3-imidazole in 5+n hetero-bicyclic molecules," *Journal of Molecular Structure*, vol. 655, no. 3, pp. 397–403, 2003.

[10] R. Boussessi, S. Dalbouha, V. Timón, N. Komiha, N. Jaïdane, and M. L. Senent, "Stability of Van der Waals complexes of the greenhouse effect gases CH4 and SF6 with imidazole in gas mixtures containing CO_2," *Computational and Theoretical Chemistry*, vol. 1094, pp. 82–91, 2016.

[11] L. Feng, X. Zhang, C. Wang et al., "Effect of different imidazole group positions on the hydroxyl ion conductivity," *International Journal of Hydrogen Energy*, vol. 41, no. 36, pp. 16135–16141, 2016.

[12] G.-C. Zheng, Z.-B. Cai, Y.-L. Pan et al., "Synthesis and two-photon absorption properties of novel 2-substituted-4,5-diphenyl-1H-imidazoles," *Tetrahedron*, vol. 72, no. 22, pp. 2988–2996, 2016.

[13] R. B. Campos, L. R. A. Menezes, A. Barison, D. J. Tantillo, and E. S. Orth, "The importance of methyl positioning and tautomeric equilibria for imidazole nucleophilicity," *Chemistry–A European Journal*, vol. 22, no. 43, pp. 15521–15528, 2016.

[14] A. J. Arduengo, R. L. Harlow, and M. Kline, "A stable crystalline carbene," *Journal of the American Chemical Society*, vol. 113, no. 1, pp. 361–363, 1991.

[15] M. O. Karatas, B. Olgundeniz, S. Gunal, I. Ozdemir, B. Alici, and E. Cetinkaya, "Synthesis, characterization and antimicrobial activities of novel silver(I) complexes with coumarin substituted N-heterocyclic carbene ligands," *Bioorganic and Medicinal Chemistry*, vol. 24, no. 4, pp. 643–650, 2016.

[16] H. A. Özbek, P. S. Aktaş, J.-C. Daran, M. Oskay, F. Demirhan, and B. Çetinkaya, "Synthesis, structure, electrochemical and antimicrobial properties of N,N'-bis(ferrocenylmethyl)imidazolinium salts," *Inorganica Chimica Acta*, vol. 423, pp. 435–442, 2014.

[17] W. A. Herrmann, "N-heterocyclic carbenes: a new concept in organometallic catalysis," *Angewandte Chemie International Edition*, vol. 41, no. 8, pp. 1290–1309, 2002.

[18] N. Hadei, E. A. Kantchev, C. J. O'Brien, and M. G. Organ, "Electronic nature of N-heterocyclic carbene ligands: effect on the Suzuki reaction," *Organic Letters*, vol. 7, no. 10, pp. 1991–1994, 2005.

[19] J. S. Al-Otaibi and R. I. Al-Wabli, "Vibrational spectroscopic investigation (FT-IR and FT-Raman) using ab initio (HF) and DFT (B3LYP) calculations of 3-ethoxymethyl-1,4-dihy-droquinolin-4-one," *Spectrochimica Acta Part A: Molecular and Biomolecular Spectroscopy*, vol. 137, pp. 7–15, 2015.

[20] J. B. Foresman and A. E. Frisch, *Exploring Chemistry with Electronic Structure Methods*, Gaussian, Inc., Wallingford, CT, USA, 2nd edition, 1996.

[21] G. Uluçam, S. E. Okan, Ş. Aktaş, and G. P. Öğretmen, "Characterization of dinaphthosulfoxide molecule," *Journal of Molecular Structure*, vol. 1102, pp. 146–152, 2015.

[22] M. Turkyilmaz, G. Uluçam, Ş. Aktaş, and S. E. Okan, "Synthesis and characterization of new N-heterocyclic carbene ligands: 1,3-Bis(acetamide)imidazol-3-ium bromide and 3-(acetamide)-1-(3-aminopropyl)-1H-imidazol-3-ium bromide," *Journal of Molecular Structure*, vol. 1136, pp. 263–270, 2017.

[23] X. Mi, S. Luo, and J.-P. Cheng, "Ionic liquid-immobilized quinuclidine-catalyzed Morita–Baylis–Hillman reactions," *Journal of Organic Chemistry*, vol. 70, no. 6, pp. 2338–2341, 2005.

[24] X. Mi, S. Luo, H. Xu, L. Zhang, and J.-P. Cheng, "Hydroxyl ionic liquid (HIL)-immobilized quinuclidine for Baylis–Hillman catalysis: synergistic effect of ionic liquids as organocatalyst supports," *Tetrahedron*, vol. 62, no. 11, pp. 2537–2544, 2006.

[25] J. Sasaki, J. K. Lanyi, R. Needleman, T. Yoshizawa, and A. Maeda, "Complete identification of C:O stretching vibrational bands of protonated aspartic acid residues in the difference infrared spectra of M and N intermediates versus bacteriorhodopsin," *Biochemistry*, vol. 33, no. 11, pp. 3178–3184, 1994.

[26] L. Clougherty, J. Sousa, and G. Wyman, "Notes-C=N stretching frequency in infrared spectra of aromatic azomethines," *Journal of Organic Chemistry*, vol. 22, no. 4, p. 462, 1957.

[27] S. Gunasekaran, E. Sailatha, S. Seshadri, and S. Kumaresan, "FTIR, FT Raman spectra and molecular structural confirmation of isoniazid," *Indian Journal of Pure and Applied Physics*, vol. 47, no. 1, pp. 12–18, 2009.

[28] M. Schwanninger, J. C. Rodrigues, H. Pereira, and B. Hinterstoisser, "Effects of short-time vibratory ball milling on the shape of FT-IR spectra of wood and cellulose," *Vibrational Spectroscopy*, vol. 36, no. 1, pp. 23–40, 2004.

[29] J. S. Singh, "FT-IR and Raman spectra, ab initio and density functional computations of the vibrational spectra, molecular geometries and atomic charges of uracil and 5-methyluracil (thymine)," *Spectrochimica Acta Part A: Molecular and Biomolecular Spectroscopy*, vol. 137, pp. 625–640, 2015.

12

Synthesis of *N*-Tetradecyl-1,10-phenathrolinium-Based New Salts for Biological Applications

Atakilt Abebe ⓘ,[1] Minaleshewa Atlabachew,[1] Misganaw Liyew,[2] and Elsabet Ferede[2]

[1]*Chemistry Department, Science College, Bahir Dar University, Bahir Dar, Ethiopia*
[2]*Biology Department, Science College, Microbiology Research Laboratory, Bahir Dar University, Bahir Dar, Ethiopia*

Correspondence should be addressed to Atakilt Abebe; atakiltabebe1@gmail.com

Academic Editor: Giovanni Natile

New organic salts were synthesized by quaternizing 1,10-phenanthroline using 1-bromotetradecane. The first step yielded an organic salt of formula $[C_{26}H_{37}N_2]Br$. Anion exchange reaction using $Li[(CF_3SO_2)_2N]$ resulted in a more stable salt of formula $[C_{26}H_{37}N_2][(CF_3SO_2)_2N]$. The organic salts were investigated by spectrometry (^1H, ^{13}C, ^{19}F NMR, X-ray photoelectron spectroscopy (XPS), UV-Vis, and matrix-assisted laser desorption/ionization mass spectroscopy (MALDI MS), CHNSBr elemental analysis, and thermal analysis (TGA and DSC). The thermal characterization showed the melting and decomposition points of $[C_{26}H_{37}N_2][(CF_3SO_2)_2N]$ to be 48°C and 290°C, respectively, which indicates it is an ionic liquid with large liquidus range. The biological activities of the salts were investigated against two Gram-positive (*Staphylococcus aureus* and *Streptococcus pyogenes*) and two Gram-negative (*Escherichia coli* and *Klebsiella pneumoniae*) bacteria, and they are found to be active against all of them. They were compared with [Cu(1,10-phenanthroline)₂Cl]Cl. They are found more active against the Gram-negative bacteria. The salts demonstrated minimum inhibitory concentration as low as 50 μg/L. These results suggest the synthesized salts can be considered as a better alternative to certain transition metal complex drugs. This minimizes the concern of introducing metal ions into the organism.

1. Introduction

Nucleic acids play a central role in critical cellular processes including cell division and protein expression [1]. Therefore, they are very attractive targets for small molecule therapeutics [2]. Most biological processes are reliant upon molecular recognition and the reversible interactions of one set of molecules with another [3, 4]. Failure in these activities results in malfunctions in cell replication or gene expression. This could be controlled by modulating nucleic acid activity through use of sequence- or structure-specific drugs. Such compounds would have a direct and beneficial role in the treatment of major diseases [5]. Compounds that have the potential to be clinically useful are normally either intercalators, groove binders, or capable of external electrostatic interaction to DNA [6]. In this regard, aromatic molecules with rigid planar or approximately planar aromatic ring structure systems are the primary choices [7–9]. One such biologically active moiety is 1,10-phenanthroline. However, molecular 1,10-phenanthroline

removes metal ions from biological systems using its conveniently placed coordinating two nitrogen atoms. This is expressed by the inhibition of metalloenzymes removing metal ions required for catalytic activity [10, 11]. Therefore, one present strategy in the field of pharmacological research is coordinating 1,10-phenanthroline with transition metal ions to make its nitrogen atoms unavailable for denaturation [12–15]. Furthermore, this activity enhances the medicinal activity of transition metal complexes [16, 17]. Based on this, many transition metal complexes of 1,10-phenanthroline have been found exhibiting numerous biological activities such as antiviral [18, 19], anti-inflammatory, antitumor [20], anti-Candida [21], antimycobacterial [22], and antimicrobial [23] activity upon binding to DNA. For example, [bis(1,10-phenanthroline) L₂copper(II)] (L = *cis*-5-norbornene-endo-2, 3-dicarboxylic acid; 2,2′-bipyridine; dicyanamide, adenine, and thymine) is well known for its numerous biological activities such as anti-Candida [21], antimycobacterial, and antimicrobial [23] activity. Its activity is believed to emerge from an efficient DNA cleavage activity [24]. However, their rigid three-dimensional structures allow intercalation of the coordinated 1,10-phenanthroline only partially in the major groove of the DNA strand. The latter diminishes the activity of the complex [25, 26].

In this work, an alternative to the employment of transition metal complexes of 1,10-phenanthroline for transportation of 1,10-phenanthroline to the cellular target is described. This is achieved by quaternizing one of the nitrogen atoms in 1,10-phenanthroline using 1-bromo-tetradecane which produces N-tetradecyl-1,10-phenanthrolinium bromide organic salt. This results in an amphiphilic, serpent-like, flat head cation. The amphiphilicity is a consequence of the combined effects of the hydrophobic alkyl chain and the positive charge delocalized throughout the aromatic ring portion of 1,10-phenanthroline [27]. Moreover, its required flat structure is retained intact. This possibly results a complete intercalation in both the major and minor grooves of DNA. The bromide is exchanged with the lipophilic bis(trifluoromethylsulfonyl) imide, $[(CF_3SO_2)_2N]^-$, which compounds the penetration of the salt through the lipophilic cell membrane and cell wall which maximizes the biological activity of the salt. Furthermore, the biological activities of these organic salts will be compared with chloro-bis(1,10-phenanthroline)copper(II) chloride. In this comparison, the number of 1,10-phenanthroline moieties in both systems would be made equal. The synthesis of the organic salt from 1,10-phenanthroline is described in the experimental section. The synthesis of the copper complex was carried out following reported procedures [28].

2. Experimental

2.1. Materials and Methods. Chemicals used in this work include 1,10-phenanthroline monohydrate and lithium bis (trifluoromethanesulfonyl) imide obtained from Alfa Aesar. 1-Bromotetradecane, 1,4-dioxane, and all the solvents used were obtained from Sigma-Aldrich and used as received.

The prepared compounds were characterized by their ^{1}H NMR, ^{13}C NMR, and ^{19}F NMR spectra which were recorded on a 400 MHz Bruker 400 Ultrashield NMR with operating frequencies 270 MHz (^{1}H), 68 MHz (^{13}C), and 376 MHz (^{19}F). Chemical shifts (δ) are reported in parts per million (ppm) with reference to residual traces in the commercial deuterated solvent, dimethyl sulfoxide $(CD_3)_2SO$ (δ_H 2.54, (δ_C 40.45)), at ambient temperature. Coupling constants (J) are given in Hz. The electronic environment of the component atoms of the cation, $[C_{26}H_{37}N_2]^{1+}$, and the anion, bistrifluorosulfonyl imide $([(CF_3SO_2)_2N]^-)$, were investigated using ultra-high vacuum (UHV) characterization using X-ray photoelectron spectroscopy (XPS). We followed the method of Men et al. [29]. The X-ray photoelectron spectra were recorded using a Kratos Axis Ultra spectrometer employing a focused, monochromated Al Kα source ($h\nu = 1486.6$ eV), hybrid (magnetic/electrostatic) optics, hemispherical analyzer, and a multichannel plate and delay line detector (DLD) with an X-ray incident angle of 0° (relative to the surface normal). The spectrum was processed without charge correction. The information depth (ID) of these experiments may be defined as the depth, within the sample, from which 95% of the measured signal will originate. ID is assumed to vary mainly with

cos θ, where θ is the electron emission angle relative to the surface normal. ID = 7–9 nm and the data obtained may be considered as a representative of the bulk composition. X-ray gun power was set to 100 W. All spectra were recorded using an entrance aperture of $300 \times 700\,\mu$m with pass energy of 80 eV for survey spectra and 20 eV for high-resolution spectra. The instrument sensitivity was 7.5×10^5 counts$\cdot$s^{-1} while measuring the Ag 3d$_{5/2}$ photoemission peak for a clean Ag sample recorded at a pass energy of 20 eV and 450 W emission power. Ag 3d$_{5/2}$ full width at half maximum (FWHM) was 0.55 eV for the same instrument settings. Binding energy calibration was made using Au 4f$_{7/2}$ (83.96 eV), Ag 3d$_{5/2}$ (368.21 eV), and Cu 2p$_{3/2}$ (932.62 eV). The absolute error in the acquisition of binding energies is ±0.1 eV, as quoted by the instrument manufacturer (Kratos); consequently, any binding energy within 0.1 eV can be considered the same, within the experimental error. The absorption wave length in the UV-Vis was recorded using Cary 60, version 2.00, in the range 800 to 200 nm, with the UV-Vis scan rate 600 nm/min taking 0.01 mM solution. CHNS elemental analyses were performed with a Flash EA 1112 elemental analyzer (Thermo Quest) taking 15 mg sample. Bromide estimation was conducted taking 100 mg sample dissolved in 40 mL distilled water. Excess AgNO$_3$ solution was added for the formation of silver bromide (AgBr) precipitate. Then the cruddy white precipitate formed was filtered, dried in an oven, and the amount of bromide was calculated from the weight difference. Matrix-assisted laser desorption/ionization mass spectrometry (MALDI-MS) in the reflectron mode was performed in a Bruker Ultraflex III for the investigation of the cation and anions. The matrix used was *trans*-2[3-(4-*tert*-butylphenyl)-2-methyl-2-propenylidene] malononitrile, commonly referred to as DCTB. The sample solution and the DCTB in acetonitrile were mixed together to give about 10:1 excess of matrix. 0.5 μL of this mixture was spotted onto a stainless steel target plate and allowed to evaporate to dryness before introduction into the mass spectrometer. The laser operates at a wavelength of 337 nm. Calibration of the data was performed in the FlexControl software.

2.2. Synthesis of $[C_{26}H_{37}N_2]Br$. The coordinated water in 1,10-phenanthroline monohydrate was removed heating in an oven at 105°C for 2 h. 1-Bromotetradecane dissolved in 1,4-dioxane was added from a dropping funnel to a molar equivalent 1,4-dioxane solution of 1,10-phenanthroline in a 100 ml two-necked round-bottomed flask fitted to the reflux condenser and guarded from moisture using CaCl$_2$, being stirred in an oil bath at 50°C for 36 h. The completion of the reaction was followed up using thin-layer chromatography (TLC). Gray precipitate was obtained and washed thoroughly with 1,4-dioxane three times.

2.3. ^{1}H and ^{13}C of $C_{26}H_{37}N_2$Br. ^{1}H NMR (270 MHz, DMSO-d_6) δ ppm: 0.81 (s, 3H), 1.19 (s, 18H), 1.34 (br. s., 2H), 1.48 (br. s., 2H), 2.07 (s, 2H), 5.89 (br. s., 2H), 8.07 (s, 1H), 8.43 (s, 3H), 8.80 (s, 1H), 9.27 (s, 1H), 9.44 (dd, $J = 8.19$, 1.31 Hz, 1H), and 9.70 (dd, $J = 5.92$, 1.38 Hz, 1H).

^{13}C NMR (68 MHz, DMSO-d_6) δ ppm: 14.49 (s, 1C), 22.64 (s, 1C), 26.27 (s, 1C), 29.11 (s, 1C), 29.25 (s, 1C), 29.44 (s, 1C), 29.58 (s, 2C), 31.57 (s, 1C), 31.84 (s, 1C), 63.92 (s, 1C), 125.19 (s, 1C), 125.91 (s, 1C), 127.72 (s, 1C), 131.10 (s, 1C), 132.23 (s, 1C), 133.16 (s, 1C), 136.94 (s, 1C), 138.52 (s, 1C), 140.15 (s, 1C), 147.59 (s, 1C), 150.45 (s, 1C), and 151.46 (s, 1C).

Spectra are indicated in Supporting information 1(a) and 1(b).

2.4. Synthesis of $[C_{26}H_{37}N_2][(CF_3SO_2)_2N]$. The halide anion was exchanged with $(CF_3SO_2)_2N^-$ by dissolving $[C_{26}H_{37}N_2]$Br in deionized water in round-bottomed flask at 70°C to which slightly excess from equimolar amount solution of LiN(CF$_3$SO$_2$)$_2$ was added dropwise. A separate phase viscous liquid following the formation of white suspension was formed. The viscous liquid was decanted, washed thoroughly at 70°C, and dried in vacuum. The viscous mass was solidified after several days. The path of the synthesis is indicated in Scheme 1.

2.5. ^1H, ^{13}C, and ^{19}F NMR of $[C_{26}H_{37}N_2][(CF_3SO_2)_2N]$. ^{1}H NMR (270 MHz, DMSO-d_6) δ ppm: 0.83 (t, $J = 6.20$ Hz, 3H), 1.21 (s, 18H), 1.35 (d, $J = 6.47$ Hz, 2H), 1.52 (br. s., 2H), 2.06 (d, $J = 7.71$ Hz, 2H), 5.80–5.98 (m, 2H), 7.99–8.11 (m, 1H), 8.32–8.54 (m, 3H), 8.74–8.86 (m, 1H), 9.30 (dd, $J = 4.06$, 1.58 Hz, 1H), 9.39 (d, $J = 7.71$ Hz, 1H), and 9.62 (d, $J = 5.65$ Hz, 1H).

^{13}C NMR (68 MHz, DMSO-d_6) δ ppm: 14.47 (s, 1C), 22.63 (s, 1C), 26.28 (s, 1C), 28.43–29.99 (m, 4C), 31.41–32.69 (m, 2C), 64.01 (s, 1C), 112.92–127.15 (q, 1C), 125.14 (s, 1C), 125.22–126.21 (m, 1C), 126.92–127.35 (m, 1C), 127.77–128.34 (m, 1C), 130.47–131.60 (m, 1C), 132.03–132.74 (m, 1C), 133.20 (s, 1C), 136.71–137.71 (m, 1C), 138.13–139.12 (m, 1C), 140.19 (s, 1C), 147.58 (s, 1C), 150.19–151.05 (m, 1C), and 151.39 (s, 1C).

Spectra are indicated in Supporting information 1(c) and 1(d).

^{19}F NMR (376 MHz, DMSO-d_6) δ ppm: −78.75 (s, 1 F). Spectra are indicated in Supporting information 1(e).

2.6. Antibacterial Activity Testing. The organic salts were evaluated for *in vitro* antibacterial activities against strains of two Gram-positive (*S. aureus* and *S. pyogenes*) and two Gram-negative (*E. coli* and *K. pneumoniae*) bacteria. We followed the methods of Lawal et al. [30]. The bacterial strains were maintained in the appropriate blood agar base at 4°C. Antibiotic disc (gentamicin 10 μg) was used as a reference. The minimum inhibitory concentration (MIC) against each bacterium was determined by preparing ethanolic solutions of different concentrations of the salts by serial dilution (50 μg/mL, 75 μg/mL, 100 μg/mL, 125 μg/mL, 150 μg/mL, 175 μg/mL, and 200 μg/mL).

3. Results and Discussions

The salts are stable in air. They are soluble methanol, ethanol, acetonitrile, acetone, dichloromethane, and DMSO. The bromide salt dissolves in water as well. Elemental analyses value is in agreement with the assigned formulae. The elemental analysis values are given in Table 1.

3.1. ^1H, ^{13}C, and ^{19}F NMR Results. 1,10-Phenanthroline is a symmetric molecule that shows four types of protons and six types of carbon atoms in the aromatic region of its ^{1}H NMR and ^{13}C NMR spectra, respectively. However, following the quaternization reaction, it loses its symmetry evident from the appearance of eight types of protons and twelve types of carbons in this region. This new characteristic feature helps in the identification of the new salt. Moreover, the upfield appearance of fourteen types of protons and fourteen types of carbons and the number of alkyl protons of each type identified using their integration taking aromatic protons as references is a strong confirmation for the occurrence of quaternization (Supporting information 1(a)–1(d)). However, the possibility of di-quaternization of 1,10-phenanthroline is ruled out due to the steric hindrance of one of the nitrogen atoms after the first quaternization [31]. The successful anion exchange performed to get the intended salt is evident from the appearance of four new peaks in ^{13}C NMR at δ ppm 112.92–127.15 assignable to the carbon in $[(CF_3SO_2)_2N]^-$ (Supporting information 1(d)). It is quartet because of the coupling with the bonded fluorine atoms. Moreover, the single strong peak signaled in the ^{19}F NMR is an additional confirmation for the purity of the salt and the successful anion exchange reaction (Supporting information 1(e)).

3.2. XP Spectroscopy Results. X-ray photoelectron spectroscopy is very reliable in confirming purity. The survey spectrum (Figure 1(a)) signaled only those elements expected from $[C_{26}H_{37}N_2][(CF_3SO_2)_2N]$ in the appropriate percentage. This strongly supports the CHNS elemental analysis result.

3.3. Wide-Scan XPS Spectra: The Electronic Environments of the Component Atoms of $[C_{26}H_{37}N_2][(CF_3SO_2)_2N]$. In $[C_{26}H_{37}N_2][(CF_3SO_2)_2N]$, there are 28 C atoms whose type of electronic environment is differentiated into five groups as CF$_3$, C^{1+2}, C^3, C^4 + C^5, and C$_{Alkyl}$. Their corresponding binding energies are 293.00, 286.90, 286.30, 285.70, and 285.00 eV, respectively (Figure 1(b)). The N 1s XP spectrum contains three characteristic peaks. The peak at higher binding energy, 402.20 eV, is assigned to the alkylated nitrogen of the cation, N$_{Cation}$, while the peak at lower binding energy whose peak is nearly twice the former represents enveloped two nitrogens, the unalkylated nitrogen, N$'_{Cation}$ (399.52 eV), of the cation and that of the anion, $[(CF_3SO_2)_2N]^-$ (399.35 eV) (Figure 1(c)). Fluorine, oxygen, and sulfur of these compounds each show a single electronic environment (Figures 1(d)–1(f)). This is because of the obvious reason that the six fluorine atoms are chemically indistinguishable. This fact works the same for the four oxygen atoms and the two sulfur atoms. The sulfur signal is a doublet due to spin-orbit splitting.

3.4. Mass Spectra. The matrix-assisted laser desorption/ionization (MALDI MS) spectra of $[C_{29}H_{37}N_2][(CF_3SO_2)N]$ were recorded for both the cation and anion targeting the

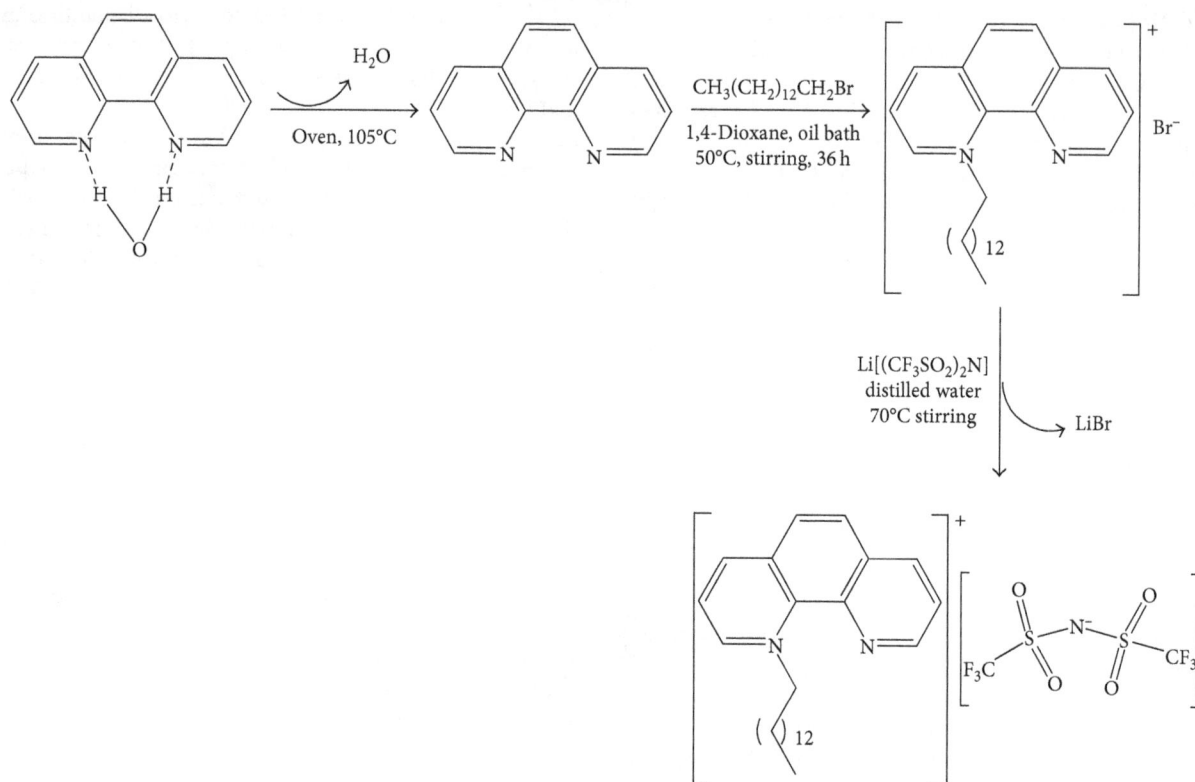

SCHEME 1: Synthesis path of the salts.

TABLE 1: CHNBrS elemental analysis measurements.

Compound	Elemental estimation Calculated (found) (%)				
	H	C	N	Br	S
[C_{26}H_{37}N_2]Br	8.10 (7.98)	68.27 (68.04)	6.13 (6.00)	17.51 (17.28)	—
[C_{26}H_{37}N_2][(CF_3SO_2)_2N]	5.67 (5.64)	51.14 (51.01)	6.39 (6.28)	—	9.74 (9.59)

cation and anion, respectively (Figure 2). The obtained molecular ion peaks for the cation appeared at m/z 377.2956 confirmed the quaternization of 1,10-phenanthroline with 1-bromotetradecane and the acquisition of the intended salt with the proposed formula (Figure 2(a)). Moreover, the recorded molecular ion peak for the anion appeared at m/e 279.9165 confirmed the successful anion exchange and isolation of pure salt (Figure 2(b)).

3.5. UV-Vis Spectra.

In addition to XPS, UV-Vis spectroscopy was employed to investigate the electronic environments of the final product and the starting of 1,10-phenanthroline to confirm the successful monoquaternization. The UV-Vis spectrum of 1,10-phenanthroline shows bands at 229 nm and 264 nm corresponding to $n \rightarrow \pi^*$ and $\pi \rightarrow \pi^*$ transitions, respectively. Following the quaternization, these bands appeared shifted to 215 nm and 274 nm, respectively (Figure 3). This is because, the quaternization involves the nonbonding electron in the bonding, thereby lowers the energy. The latter increases the energy gap between the nonbonding and the π^* orbitals. On the other hand, the quaternization develops a delocalized

positive charge in the ring system that increases the energy of the π orbitals. Subsequently, it decreases the energy gap between π and π^* orbitals.

3.6. Thermal Properties.

The response obtained from heating of the samples clearly demonstrated the influence of the anion on the nature of the salt produced. The cation is large in size over which the monopositive charge is highly delocalized over the entire aromatic portion. This makes it classified as a soft acid. The increase in the size of the counter mono-charged anion increases the extent of its softness [32]. In [C_{26}H_{37}N_2]Br, the soft cation is coupled with a relatively hard anion which creates incompatibility in their interaction that made the salt relatively unstable. Because of this, it showed decomposition without melting, starting around 129.72°C. This is the consequence of the localized negative charge on the bromide that easily attacks the cation which creates relative instability [33]. On the other hand, the coupling of the soft cation with a relatively softer base in [C_{26}H_{37}N_2][(CF_3SO_2)_2N] resulted in thermally stable salt. This is expressed by its higher decomposition point which

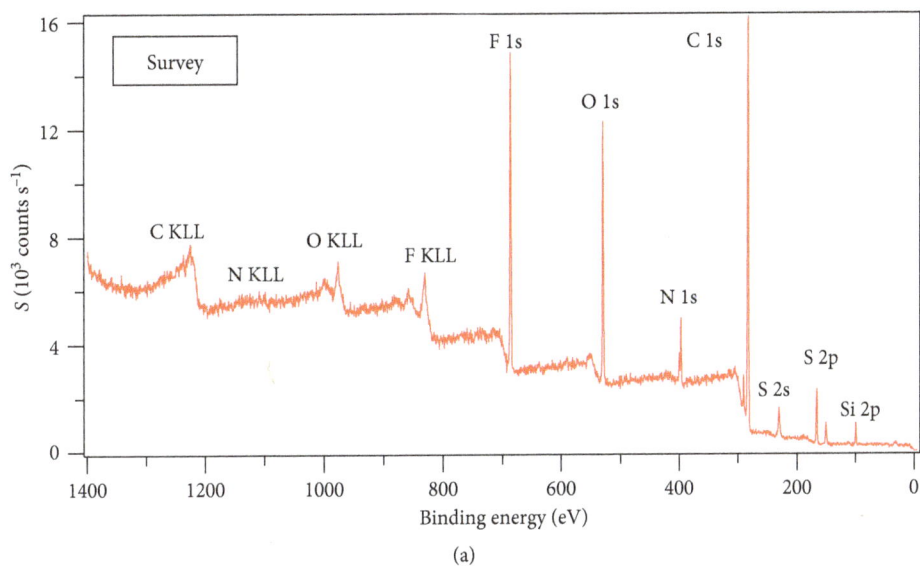

(a)

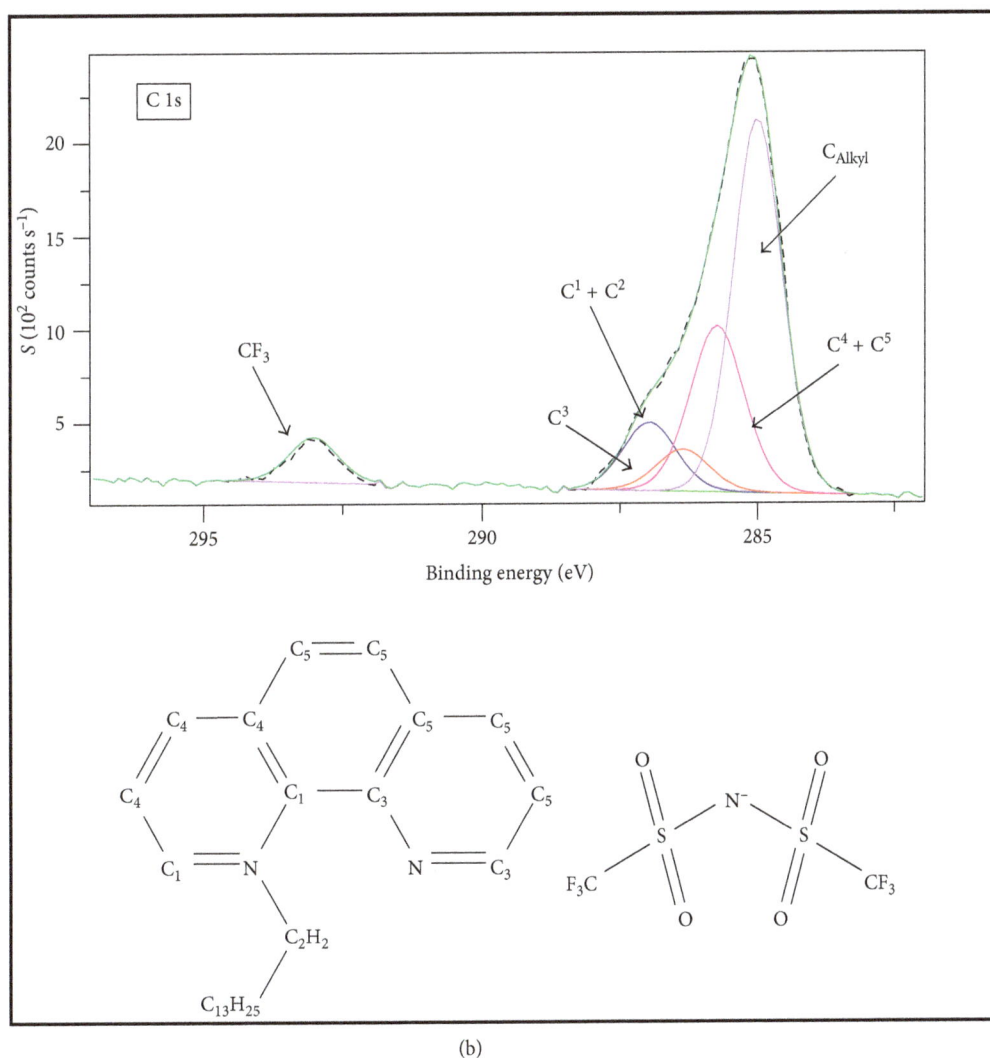

(b)

FIGURE 1: Continued.

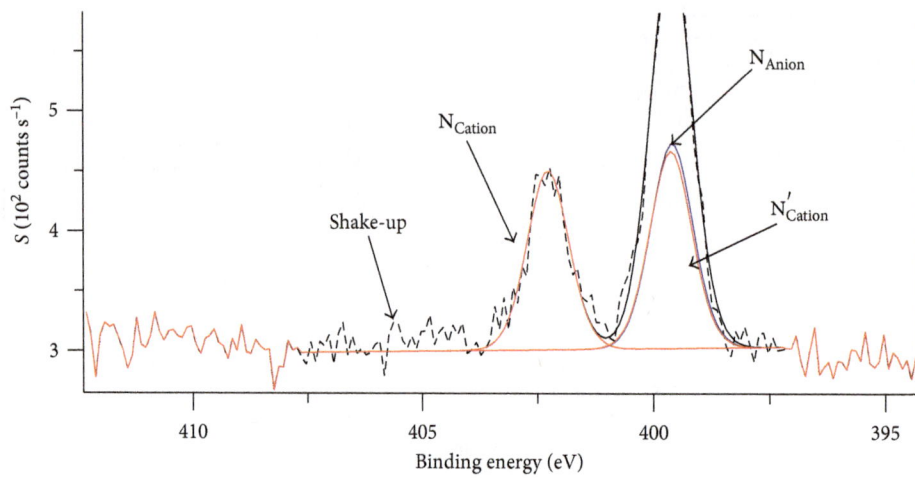

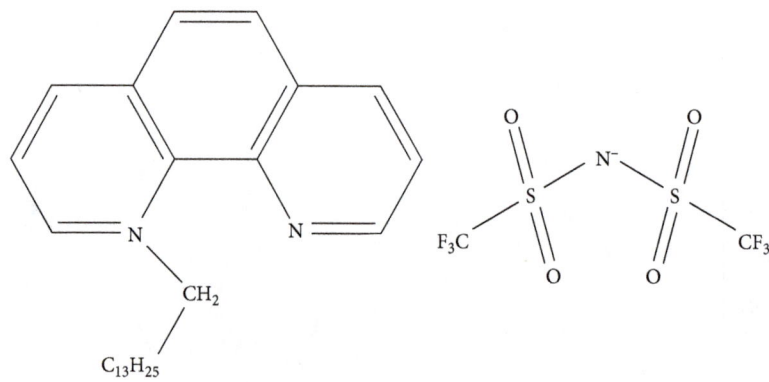

(c)

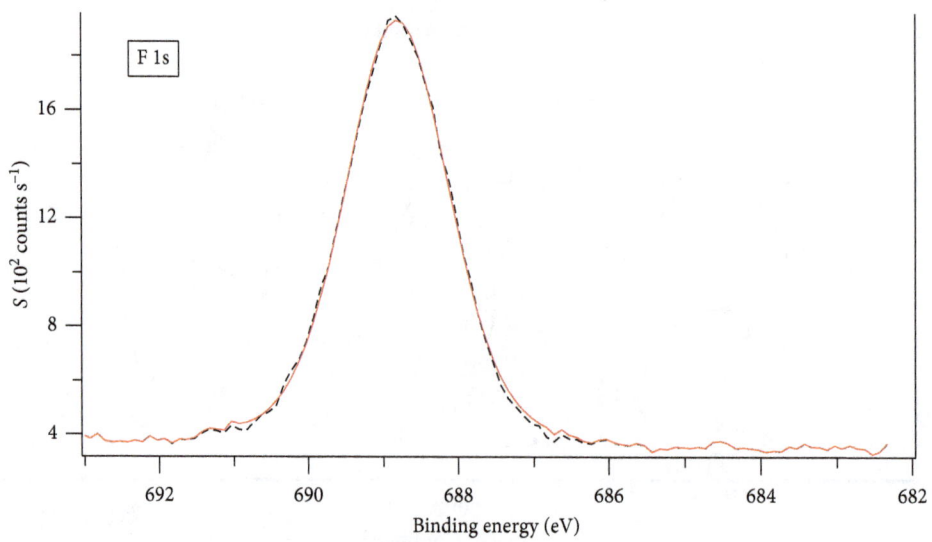

(d)

FIGURE 1: Continued.

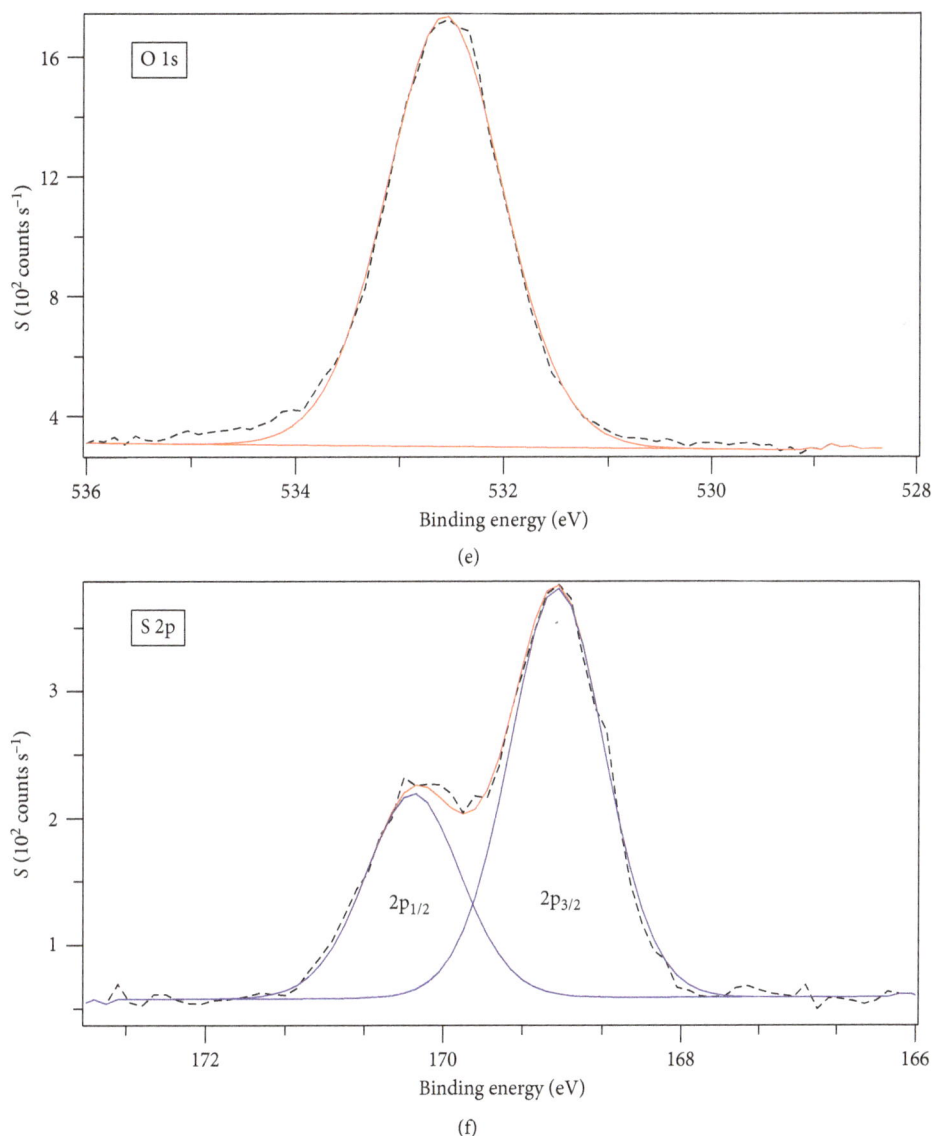

FIGURE 1: X-ray photoelectron spectrum recorded for (a) survey, (b) wide scan for C1s, (c) wide scan for N1s, (d) wide scan for F1s, (e) wide scan for O1s, and (f) wide scan for S2p of $[C_{26}H_{37}N_2][(CF_3SO_2)_2N]$.

started at around 244°C (Table 2 and Supporting information 2(b) and 3). Furthermore, the cation and anion are both unsymmetrical that they hinder the crystalline packing of the salt. This fact significantly reduced the melting point to 48.77°C. The latter property makes it to be classified as a new ionic liquid, and the wide gap between its melting and decomposition temperatures gives $[C_{26}H_{37}N_2][(CF_3SO_2)_2N]$, an attractive feature for potential applications such as electrochemistry [31, 34, 35] (Supporting information 2(b) and 3 and Table 2). Moreover, $[C_{26}H_{37}N_2][(CF_3SO_2)_2N]$ typically demonstrated substantial supercooling as its freezing point is significantly lower than the melting point (−11.28 to 48.77°C) (Supporting information 3 and Table 2).

3.7. Antibacterial Screening. The compounds were tested for their in vitro antimicrobial activity and were compared with the commercially available gentamicin. They were tested against two Gram-positive (S. aureus and S. pyogenes) and two Gram-negative (E. coli and K. pneumoniae) bacteria, and they all were found active against all the tested pathogens (Figure 4). The synthesized compounds, the metal complex, and gentamicin showed less activities than the molecular 1,10-phenanthroline against all the bacteria. The metal complex, $[Cu(Phen)_2Cl]Cl$, showed better activities than the organic salts and gentamicin against the Gram-positive bacteria (S. aureus and S. pyogenes). On the other hand, the synthesized organic salts demonstrated better activities than $[Cu(Phen)_2Cl]Cl$ against the Gram-negative (E. coli and K. pneumoniae) bacteria (Figure 4 and Table 3). Furthermore, $[C_{26}H_{37}N_2][(CF_3SO_2)_2N]$ is found better than its precursor, $[C_{26}H_{37}N_2]Br$. The former statement is encouraging because these compounds succeeded in reaching the target passing two barriers of the Gram-negative bacteria,

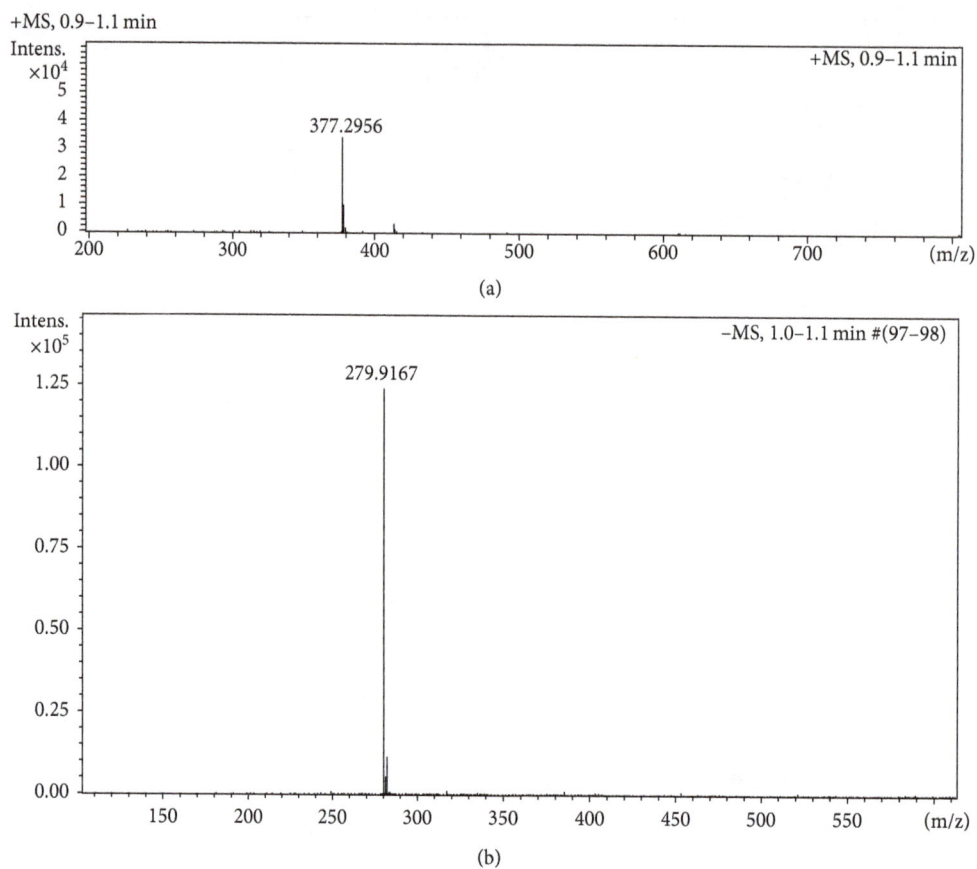

FIGURE 2: (a) MALDI MS+ and (b) MALDI MS for $[C_{26}H_{37}N_2][(CF_3SO_2)_2N]$.

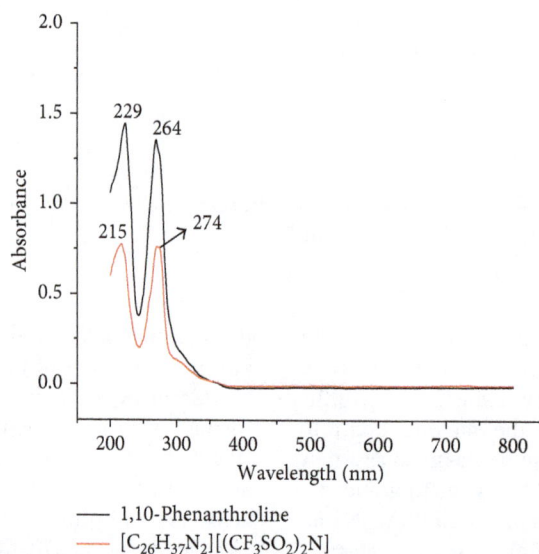

FIGURE 3: UV-Vis spectra of 1,10-phenanthroline and $[C_{26}H_{37}N_2][(CF_3SO_2)_2N]$.

TABLE 2: Starting point and onset temperatures with the weight loss curves and melting and crystallization temperatures of the salts.

Compound	Temperature (°C)				
	Start	Onset	Glass transition	Crystallization	Melting
$[C_{26}H_{37}N_2]Br$	129.72	157.41	—	—	—
$[C_{26}H_{37}N_2][(CF_3SO_2)_2N]$	244.41	290.31	−0.60	−11.28	48.77

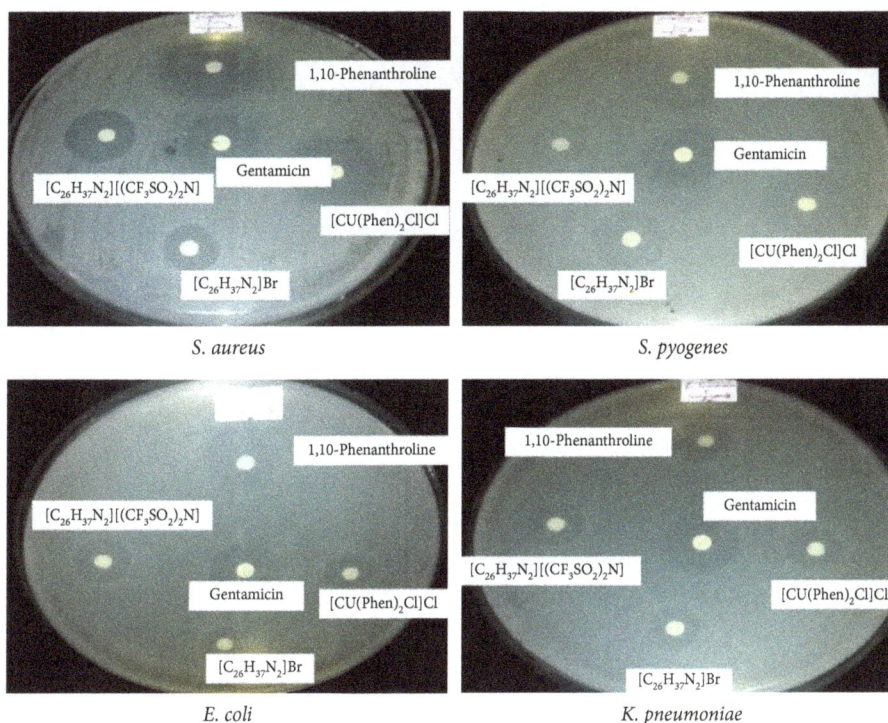

FIGURE 4: The inhibition observed by the actions of the salts.

TABLE 3: Antibacterial studies of the investigated compounds (inhibition zones).

| Compound | Inhibition zone (mm) | | | |
| | Gram-negative bacteria | | Gram-positive bacteria | |
	E. coli	K. pneumoniae	S. aureus	S. pyogenes
$[C_{26}H_{37} N_2]Br$	19.75	19.75	16.50	25.00
$[C_{26}H_{37} N_2][(CF_3SO_2)_2N]$	20.25	20.25	23.50	20.25
Gentamicin	31.25	25.00	26.50	28.25
$[Cu(Phen)_2Cl]Cl$	15.00	17.25	29.75	29.50
1,10-Phenanthroline	36.00	18.75	36.00	31.25

TABLE 4: MIC assays of the salts against four bacterial pathogens.

| Compound | Minimum concentration of microorganism growth (μg/mL) | | | |
| | Gram-negative bacteria | | Gram-positive bacteria | |
	E. coli	K. pneumoniae	S. aureus	S. pyogenes
$[C_{26}H_{37} N_2]Br$	50	50	50	50
$[C_{26}H_{37} N_2][(CF_3SO_2)_2N]$	50	75	75	75
$[Cu(Phen)_2Cl]Cl$	100	125	75	75

namely, the cell membrane and the cell wall of even the highly drug resistant K. pneumoniae. This is probably due to the very long and rod-like cation which contained relatively long lipophilic alkyl chain; its positive charge is highly delocalized throughout the aromatic ring portion; subsequently, it becomes amphiphilic which increased the cell permeability. Subsequently, the cell wall and cell membrane that surrounds the cell favors the passage of salts to reach to their target [36]. Moreover, significant biological activity differences are observed between $[C_{26}H_{37}N_2][(CF_3SO_2)_2N]$ and $[C_{26}H_{37}N_2]Br$. This is attributed to the better penetration of the former into the cellular target due to the lipophilicity of $[(CF_3SO_2)_2N]^-$. This is in very good agreement with the solubility experiment result that $[C_{26}H_{37}N_2][(CF_3SO_2)_2N]$ is soluble only in organic solvents. The minimum inhibitory concentration (MIC) values of the salts are summarized in Table 4. The result shows that even though the organic salts show less inhibition zones than the metal complex against Gram-positive bacteria, they are found inhibiting at such very small minimum concentrations (Table 4).

4. Conclusions

The purity and synthesis of the intended salt was confirmed from the data obtained using all characterization techniques employed here. The results demonstrated that the effect of the anion on the properties of the salt is significant. This was revealed on their thermal and antibacterial results. The result of the *in vitro* biological activity studies indicated that the organic salts are biologically active against all the tested pathogens. This result makes them classified as a wide-range antibacterial agent. These organic salts are found better than the copper complex in their activities against the Gram-negative bacteria. This can be considered promising news as alternative drugs providing the cytotoxicity issue is resolved. This can lighten the concern of introducing metal ions into organisms. The DSC result showed that $[C_{26}H_{37}N_2]$ $[(CF_3SO_2)_2N]$ is an ionic liquid with large liquidus range.

Acknowledgments

The authors express sincere thanks to Bahir Dar University for financial support.

References

[1] R. R. Breaker and G. F. Joyce, "The expanding view of RNA and DNA function," *Chemistry & Biology*, vol. 21, no. 9, pp. 1059–1065, 2014.

[2] T. A. Cooper, L. Wan, and G. Dreyfuss, "RNA and disease," *Cell*, vol. 136, no. 4, pp. 777–793, 2009.

[3] J. W. Lown, K. Krowicki, U. G. Bhat, A. Skorobogaty, B. Ward, and J. C. Dabrowiak, "Molecular recognition between oligopeptides and nucleic acids: novel imidazole-containing oligopeptides related to netropsin that exhibit altered DNA sequence specificity," *Biochemistry*, vol. 25, no. 23, pp. 7408–7416, 1986.

[4] M. H. Caruthers, "Deciphering the protein-DNA recognition code," *Accounts of Chemical Research*, vol. 13, no. 5, pp. 155–160, 1980.

[5] L. Liu, Y. Lin, and S. L. Gerson, "MGMT: a critical DNA repair gene target for chemotherapy resistance and for stem cell protection," in *DNA Repair in Cancer Therapy*, pp. 17–28, Elsevier Inc., New York, NY, USA, 2012.

[6] S. Neidle, D. G. Brown, T. C. Jenkins, C. A. Laughton, M. R. Sanderson, and J. V. Skelly, "Drug-DNA recognition: sequence-specificity of the DNA minor groove binder berenil," in *Molecular Basis of Specificity in Nucleic Acid-Drug Interactions*, pp. 43–57, Springer, Dordrecht, Netherlands, 1990.

[7] S. C. Zimmerman, "Molecular tweezers: synthetic receptors for π-sandwich complexation of aromatic substrates," *Bioorganic Chemistry Frontiers*, vol. 2, pp. 33–71, 1991.

[8] A. Tawani, A. Amanullah, A. Mishra, and A. Kumar, "Evidences for Piperine inhibiting cancer by targeting human G-quadruplex DNA sequences," *Scientific Reports*, vol. 6, no. 1, article 39239, 2016.

[9] S. Yadav, B. Narasimhan, S. M. Lim et al., "Synthesis, characterization, biological evaluation and molecular docking studies of 2-(1H-benzo [d] imidazol 2-ylthio)-N-(substituted 4-oxothiazolidin-3-yl) acetamides," *Chemistry Central Journal*, vol. 11, no. 1, p. 137, 2017.

[10] D. Schomburg and I. Schomburg, "N-acetyl-1-d-myo-inositol-2-amino-2-deoxy-α-d-glucopyranoside deacetylase 3.5.1.103," in *Class 3.4–6 Hydrolases, Lyases, Isomerases, Ligases: Springer Handbook of Enzymes*, D. Schomburg and I. Schomburg, Eds., vol. 10, pp. 247–254, Springer, Berlin, Heidelberg, Germany, 2013.

[11] A. Meister, Ed., *Advances in Enzymology and Related Areas of Molecular Biology*, vol. 56, John Wiley & Sons, Hoboken, NJ, USA, 2009, ISBN: 978-0-470-12380-5.

[12] A. Abebe and T. Hailemariam, "Synthesis and assessment of antibacterial activities of ruthenium (III) mixed ligand complexes containing 1, 10-phenanthroline and guanide," *Bioinorganic Chemistry and Applications*, vol. 2016, Article ID 3607924, 9 pages, 2016.

[13] R. Gup and B. Kırkan, "Synthesis and spectroscopic studies of mixed-ligand and polymeric dinuclear transition metal complexes with bis-acylhydrazone tetradentate ligands and 1, 10-phenanthroline," *Spectrochimica Acta Part A: Molecular and Biomolecular Spectroscopy*, vol. 64, no. 3, pp. 809–815, 2006.

[14] I. Turel, A. Golobič, A. Kljun, P. Samastur, U. Batista, and K. Sepčić, "New synthetic routes for the preparation of ruthenium-1, 10-phenanthroline complexes: tests of cytotoxic and antibacterial activity of selected ruthenium complexes," *Acta Chimica Slovenica*, vol. 62, no. 2, pp. 337–345, 2015.

[15] M. Cusumano, M. L. Di Pietro, and A. Giannetto, "DNA interaction of platinum (II) complexes with 1, 10-phenanthroline and extended phenanthrolines," *Inorganic Chemistry*, vol. 45, no. 1, pp. 230–235, 2006.

[16] M. R. Jones, D. Duncan, and T. Storr, "Ligand Design in Medicinal Inorganic Chemistry," in *Introduction to Ligand Design in Medicinal Inorganic Chemistry*, T. Storr, Ed., pp. 1–8, Wiley, Chichester, UK, 2014.

[17] T. Aiyelabola, E. Akinkunmi, E. Obuotor, I. Olawuni, D. Isabirye, and J. Jordaan, "Synthesis characterization and biological activities of coordination compounds of 4-hydroxy-3-nitro-2h-chromen-2-one and its aminoethanoic acid and pyrrolidine-2-carboxylic acid mixed ligand complexes," *Bioinorganic Chemistry and Applications*, vol. 2017, Article ID 6426747, 9 pages, 2017.

[18] N. Margiotta, F. P. Fanizzi, J. Kobe, and G. Natile, "Synthesis, characterisation and antiviral activity of platinum (II) complexes with 1, 10-phenanthrolines and the antiviral agents acyclovir and penciclovir," *European Journal of Inorganic Chemistry*, vol. 2001, no. 5, pp. 1303–1310, 2001.

[19] N. Margiotta, A. Bergamo, G. Sava, G. Padovano, E. De Clercq, and G. Natile, "Antiviral properties and cytotoxic activity of platinum (II) complexes with 1, 10-phenanthrolines and acyclovir or penciclovir," *Journal of Inorganic Biochemistry*, vol. 8, pp. 1385–1390, 2004.

[20] J. D. Ranford, P. J. Sadler, and D. A. Tocher, "Cytotoxicity and antiviral activity of transition-metal salicylato complexes and crystal structure of bis (diisopropylsalicylato)(1, 10-

phenanthroline) copper (II)," *Journal of the Chemical Society, Dalton Transactions*, vol. 22, pp. 3393–3399, 1993.

[21] M. Geraghty, V. Sheridan, M. McCann, M. Devereux, and V. McKee, "Synthesis and anti-Candida activity of copper (II) and manganese (II) carboxylate complexes: X-ray crystal structures of [Cu (sal)(bipy)]·C$_2$H$_5$OH·H$_2$O and [Cu (norb) (phen)$_2$]·6.5 H$_2$O (salH$_2$ = salicylic acid; norbH$_2$ = cis-5-norbornene-endo-2, 3-dicarboxylic acid; bipy= 2, 2′-bipyridine; phen = 1, 10-phenanthroline)," *Polyhedron*, vol. 18, no. 22, pp. 2931–2939, 1999.

[22] D. K. Saha, U. Sandbhor, K. Shirisha et al., "A novel mixed-ligand antimycobacterial dimeric copper complex of ciprofloxacin and phenanthroline," *Bioorganic & Medicinal Chemistry Letters*, vol. 14, no. 12, pp. 3027–3032, 2004.

[23] M. A. Zoroddu, S. Zanetti, R. Pogni, and R. Basosi, "An electron spin resonance study and antimicrobial activity of copper (II)-phenanthroline complexes," *Journal of Inorganic Biochemistry*, vol. 63, no. 4, pp. 291–300, 1996.

[24] C. H. Chen and D. S. Sigman, "Nuclease activity of 1, 10-phenanthroline-copper: sequence-specific targeting," *Proceedings of the National Academy of Sciences*, vol. 83, no. 19, pp. 7147–7151, 1986.

[25] I. Potočňák, M. Dunaj-Jurčo, D. Miklóš, M. Kabešová, and L. Jäger, "Bis (dicyanamide) bis (1, 10-phenanthroline) copper (II)," *Acta Crystallographica Section C: Crystal Structure Communications*, vol. 51, no. 4, pp. 600–602, 1995.

[26] J. K. Barton, A. Danishefsky, and J. Goldberg, "Tris (phenanthroline) ruthenium (II): stereoselectivity in binding to DNA," *Journal of the American Chemical Society*, vol. 106, no. 7, pp. 2172–2176, 1984.

[27] M. Bonchio, M. Carraro, G. Casella, V. Causin, F. Rastrelli, and G. Saielli, "Thermal behaviour and electrochemical properties of bis (trifluoromethanesulfonyl) amide and dodecatungstosilicate viologen dimers," *Physical Chemistry Chemical Physics*, vol. 14, pp. 2710–2717, 2012.

[28] A. Abebe, A. Sendek, S. Ayalew, and M. Kibret, "Copper (II) mixed-ligand complexes containing 1, 10-phenanthroline, adenine and thymine: Synthesis, characterization and antibacterial activities," *Chemistry International*, vol. 3, no. 3, pp. 230–239, 2017.

[29] S. Men, K.R. Lovelock, and P. Licence, "X-ray photoelectron spectroscopy as a probe of rhodium-ligand interaction in ionic liquids," *Chemical Physics Letters*, vol. 645, pp. 53–58, 2016.

[30] A. Lawal, A. S. Shodeinde, S. A. Amolegbe, S. E. Elaigwu, and M. T. Yunus-Issa, "Synthesis, characterization and antimicrobial activity of mixed transition metal complexes of salicylic acid with 1, 10-phenanthroline," *Journal of Applied Sciences and Environmental Management*, vol. 21, no. 3, pp. 568–573, 2017.

[31] I. J. Villar-Garcia, A. Abebe, and Y. Chebude, "1, 10-Phenanthrolinium ionic liquids exhibiting excellent solubility for metal complexes: potential solvents for biphasic and supported ionic liquid phase (SILP) catalysis," *Inorganic Chemistry Communications*, vol. 19, pp. 1–3, 2012.

[32] J. E. Huheey, E. A. Keiter, and R. L. Keiter, *Inorganic Chemistry: Principles of Structure and Reactivity*, College Publishers, New York, NY, USA, 4th edition, 1993.

[33] H. L. Ngo, K. LeCompte, L. Hargens, and A. B. McEwen, "Thermal properties of imidazolium ionic liquids," *Thermochimica Acta*, vol. 357-358, pp. 97–102, 2000.

[34] R. D. Rogers and K. R. Seddon, Eds., *Ionic Liquids: Industrial Applications for Green Chemistry*, American Chemical Society, Washington, DC, USA, 2002.

[35] T. Welton, "Room-temperature ionic liquids: solvents for synthesis and catalysis," *Chemical Reviews*, vol. 99, no. 8, pp. 2071–2084, 1999.

[36] L. A. Summers, "The Phenanthrolines," in *Advances in Heterocyclic Chemistry*, A. R. Katritzky and A. J. Boulton, Eds., vol. 22, pp. 1–69, Elsevier, Gainesville, FL, USA, 1978.

13

Study of Isothermal, Kinetic, and Thermodynamic Parameters for Adsorption of Cadmium: An Overview of Linear and Nonlinear Approach and Error Analysis

Fozia Batool [ID],[1] **Jamshed Akbar,**[1] **Shahid Iqbal,**[1] **Sobia Noreen** [ID],[1] and **Syed Nasir Abbas Bukhari** [ID][2]

[1]*Department of Chemistry, University of Sargodha, Sargodha 40100, Pakistan*
[2]*Department of Pharmaceutical Chemistry, College of Pharmacy, Jouf University, Aljouf, Sakaka 2014, Saudi Arabia*

Correspondence should be addressed to Fozia Batool; fozia.batool@uos.edu.pk and
Syed Nasir Abbas Bukhari; snab_hussaini@yahoo.com

Academic Editor: Zhe-Sheng Chen

Reports about presence and toxicity of Cd^{2+} in different chemical industrial effluents prompted the researchers to explore some economical, rapid, sensitive, and accurate methods for its determination and removal from aqueous systems. In continuation of series of investigations, adsorption of Cd^{2+} onto the stem of *Saccharum arundinaceum* is proposed in the present work. Optimization of parameters affecting sorption potential of Cd^{2+} including pH, contact time, temperature, sorbent dose, and concentration of sorbate was carried out to determine best suited conditions for maximum removal of sorbate. To understand the nature of sorption process, linear and nonlinear forms of five sorption isotherms including Freundlich and Langmuir models were employed. Feasibility and viability of sorption process were evaluated by calculating kinetics and thermodynamics of the process, while error analysis suggested best fitted sorption model on sorption data. Thermodynamic studies demonstrated exothermic nature of reaction, while kinetic studies suggested pseudo-second order of reaction.

1. Introduction

Environmental pollution should be taken into special consideration because it is a very serious matter affecting every type of organism at every level. The most adversely affected environmental resource is water [1]. As water is an essential element for the survival of living beings, it is very necessary to keep it pure and clean [2]. Quality of drinking water is of prime importance for mankind because waterborne diseases can decimate population of the whole area. These diseases arise due to toxic release of chemicals from industrial zones [3]. Particularly in industrial areas, these waterborne diseases are a great threat towards safety of water supplies. Other sources which may pollute water include domestic waste, pesticides run off from agricultural land, metal plating operations, and so on. Key contaminants present in water include heavy metals, chlorinated hydrocarbons, pathogens, detergents, pesticides, algal nutrients, trace organic compounds, dyes, and so on. These hazardous substances are of concern because of their ultimate effect on survival of human life [4]. Heavy metals like Cd, Zn, Ni, and Pb are present in relatively major amounts in industrial effluents and enter in rivers and oceans and ultimately pollute groundwater leading to adverse effects on aquatic life. Metals resist the process of biodegradability and hence remain in ecosystem, affecting food chain and human health [5]. In order to provide clean environment and healthy lifestyle to our coming generations, it is necessary to remove hazardous pollutants from the environment. In environmental restoration areas, conventional techniques are practiced to eradicate those pollutants from the environment which include chemical precipitation, evaporative method, electrolytic extraction, reverse osmosis, ion exchange, and electrochemical and membrane processes [6]. All these methods are costly and produce large amount of sludge which is difficult to dispose of. Use of biological materials for the removal of pollutants from aqueous media is considered superior to other methods in terms of cost effectiveness and simple design. It is a surface phenomenon, in which pollutants get accumulated on the surface of the adsorbent material. Binding nature is based on type of sorbent and sorbate, but mostly physisorption or chemisorption takes place [7]. Materials with ease in availability and low cost are preferred for the purpose. In this context, agrowastes are considered a significant material for adsorption. Binding capacity of these materials can be intensified by physical and chemical treatments and heat therapy [8].

To explore the appropriate adsorbent, it is necessary to establish equilibrium correlation of sorbent to predict behavior of sorbent under different experimental conditions. This equilibrium correlation is developed by using equilibrium isotherms. These isotherms express way of sorbent interaction with the surface of adsorbent, that is, whether it is monolayer or multilayer sorption [9]. Similarly, thermodynamic studies are of prime importance to predict whether the adsorption is spontaneous or not. Furthermore, it provides information about suitable temperature range for sorption and nature of sorbent and sorbate at equilibrium [10].

The aim of the present research was to explore *Saccharum arundinaceum* for adsorption of cadmium under different operating conditions including pH, contact time, initial concentration, and temperature. The application of linear and nonlinear forms of equilibrium isotherms was to determine appropriate isotherm for the purpose, and thermodynamic and kinetic studies were performed to check reaction nature of the adsorption phenomenon. Error analysis based on five different error functions was also performed.

2. Materials and Methods

2.1. Preparation of Adsorbent. On the basis of literature survey and indigenous availability of agrowaste materials, the stem of *Saccharum arundinaceum* (hardy sugar cane) was collected from different regions of Sargodha District, Pakistan. After collection, the sample was properly washed with deionized water to remove dust and surface impurities. The sorbent was initially dried in an open container at room temperature and later in an electric oven (Model, LEB-1-20) at 105°C for 24 h to remove all the moisture contents. The dried sorbent was ground, and appropriate particle size was separated by sieves and was stored for further analyses.

2.2. Chemicals. All the chemicals, reagents, and solvents used in the present work were of analytical reagent grade and purchased from Merck (Germany) or Sigma-Aldrich (Germany). Standard solutions were prepared, and successive dilutions were made with double-distilled water to make working solutions.

2.3. Pretreatment of Sorbents. *Saccharum arundinaceum* was pretreated with HCl (0.1 M) and NaOH (0.1 M) to evaluate the effects of acid and base treatments on pore size, that is, pore area, pore volume, and sorption capacity. For chemical treatment, the sorbent (20 g) was stirred for 4 h in 1 L solution of 0.1 M NaOH or HCl followed by filtration and extensive

washing with distilled water to remove any traces of acid/base. After that, the treated sorbent material was dried at 110°C and stored in airtight zipper bags at −4°C before further use.

2.4. Characterization of Sorbent. To determine different physical and chemical parameters affecting adsorption, it is necessary to characterize the sorbent. Therefore, physical and chemical characterization was done by scanning electron microscopy (SEM) and Fourier transform infrared spectroscopy (FTIR).

2.4.1. Scanning Electron Microscopy. Surface analysis was performed using scanning electron microscope JEOL model 2300. SEM provides information about surface area available for adsorption and morphology of sorbent [11]. Analysis of each sorbent was carried out in optimized conditions under argon atmosphere.

2.4.2. Fourier Transform Infrared Spectroscopy. Functional groups present in structure of sorbent were determined by Fourier transform infrared spectrophotometer (Model Shimadzu AIM-8800). These functional groups are responsible for adsorption of sorbate on the surface of sorbent, and their detection helps in determining the nature of binding interactions between sorbate and sorbent surface [12]. Diffused reflectance infrared technique (DRIFT) was used for analysis taking KBr as a background reagent.

2.5. Equilibrium Isotherms. In order to study adsorption pathway and equilibrium relationship between sorbent and sorbate, it is necessary to design proper adsorption isotherms. Isotherms predict the appropriate parameters and behavior of sorbent towards different sorption systems [13]. In this context, linear and nonlinear models are utilized using Microsoft Excel®2007 (equilibrium isotherms applied on the present work are given in Table 1s of supplementary data).

2.6. Error Functions. In order to determine best fitting of linear or nonlinear models onto adsorption data, it is necessary to calculate the error function [14]. These error functions include sum square error, hybrid functional error, average relative error, sum of absolute error, nonlinear chi-square, and so on (calculated error functions and their equations are present in Table 2s of supplementary data).

2.7. Thermodynamic Investigations. Thermodynamic investigations are another important parameter of adsorption studies. For thermodynamic studies, the adsorption experiment was carried out at different temperature conditions and calculated parameters included enthalpy (ΔH), entropy (ΔS), and Gibbs free energy (ΔG).

For this purpose, (1)–(3) were applied

$$\Delta G^{\circ} = \Delta H^{\circ} - T\Delta S^{\circ}, \tag{1}$$

$$\ln K_{\mathrm{C}} = -\frac{\Delta H}{RT} + \frac{\Delta S}{R}, \tag{2}$$

$$\Delta G = -RT \ln K_{\mathrm{C}}, \tag{3}$$

where R is the natural gas constant and K_C is the constant at equilibrium and is calculated as

$$K_C = \frac{C_e}{1 - C_e}. \tag{4}$$

where C_e is the concentration of sorbent at equilibrium condition.

2.8. Adsorption Kinetics.

In batch adsorption process, kinetic studies provide information about optimum conditions, mechanism of sorption, and possible rate controlling step. For this purpose, linear and nonlinear form of pseudo-first- and pseudo-second-order kinetics is applied on adsorption data [15]. In order to check the effect of contact time (10–70 min) on adsorption, initial concentration of 100 mg/L for cadmium was prepared and 100 ml of this sample was used for study. Sorbent (0.5 g) was added in this cadmium solution and applied for shaking at 150 rpm speed. After fixed interval of time, the sample was removed from flask and analyzed for cadmium concentration by atomic absorption spectrophotometer. The amount of cadmium adsorbed at different time intervals was calculated by employing the following formula:

$$Q_t = \frac{(Q_o - Q_e)}{W_{sorbent}} \times V, \tag{5}$$

where Q_t is the amount of cadmium adsorbed at any time t, Q_o and Q_e are initial and equilibrium concentrations, respectively. The volume of cadmium solution taken is represented by $V(L)$, and $W_{sorbent}$ is the amount of sorbent in g.

2.8.1. Pseudo-First-Order Kinetics.

In order to calculate pseudo-first-order kinetics for adsorption system, following equations were used:

$$\ln\left(Q_e - Q_t\right) = \ln\left(Q_e\right) - k_1 t \text{ linear form,}$$
$$Q_t = Q_e\left(1 - e^{-k_1 t}\right) \text{ nonlinear form,} \tag{6}$$

where Q_t is the amount adsorbed at time t, Q_e is the equilibrium amount, t is time in minutes, and k_1 is the rate constant.

2.8.2. Pseudo-Second-Order Kinetics.

For pseudo-second-order kinetics, linear and nonlinear forms were applied as follows:

$$\frac{t}{Q_t} = \frac{1}{k_2 Q_e^2} + \left(\frac{1}{Q_e}\right)t \text{ linear form,}$$
$$Q_t = \frac{k_2 Q_e^2 t}{1 + k_2 Q_e t} \text{ nonlinear form.} \tag{7}$$

3. Results and Discussion

3.1. Effect of Pretreatment.

Pretreatment has promising effect on adsorption potential of *Saccharum arundinaceum*. Results reveal that base-treated (97.5%) sorbent shows good efficiency for cadmium sorption as compared to raw

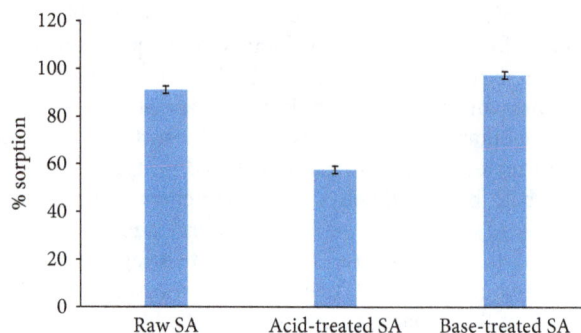

FIGURE 1: Effect of pretreatment on adsorption capacity of *Saccharum arundinaceum* (0.1 M HCl and NaOH treated sorbent, 60 minutes time, and 60 ppm initial concentration).

(91.15%) and acid-treated (57.6) sorbent as shown in Figure 1. Adsorption capacity depends upon functional groups present on the surface of sorbent and its microporous structure [16]. Increase in sorption capacity by base treatment can be attributed to hydroxyl groups created on the surface of adsorbent by base treatment or modification of cell wall components by base [17]. Decrease in adsorption after acid treatments was found as the binding sites available on the surface of biosorbent got destructed due to acid [18]. So, base-treated *Saccharum arundinaceum* was used for adsorption analysis.

3.2. Characterization of Sorbents.

Saccharum arundinaceum was characterized in terms of surface morphology and functional group analysis by scanning electron microscopy and Fourier transform infrared spectroscopy.

3.2.1. Scanning Electron Microscopy.

Three native and two treated sorbents (acid- and base-treated *Saccharum arundinaceum*) were analyzed through scanning electron microscope to study surface morphology. Large pore size available on the surfaces of native and base-treated sorbent was responsible for enhanced adsorption on the surface of these agrowaste materials. Results for SEM analysis are given in Figure 2. Hollow cavities appear in the structure of raw adsorbent, which were responsible for binding of sorbate onto the surface of sorbent. Acid treatment decreases these cavities by deforming surface of the sorbent, so adsorption decreases after acid treatment because surface becomes smooth and thin adsorption layer is formed on the sorbent surface. Raw and base-treated sorbent surface is found rough and cylindrical, which make possible multilayer and thick adsorption on the sorbent surface as compared to smooth surface. Results obtained in the SEM micrograph are in good agreement with reported data [19].

3.2.2. Fourier Transform Infrared Spectroscopy.

Fourier transform infrared spectrometer provides information about functional groups present on the surface of sorbent and makes possible attachment of sorbate [20]. FTIR spectra of sorbent obtained in the range of 4000–450 cm^{-1} wavenumber and major functional groups present in adsorbent

(a)

(b)

(c)

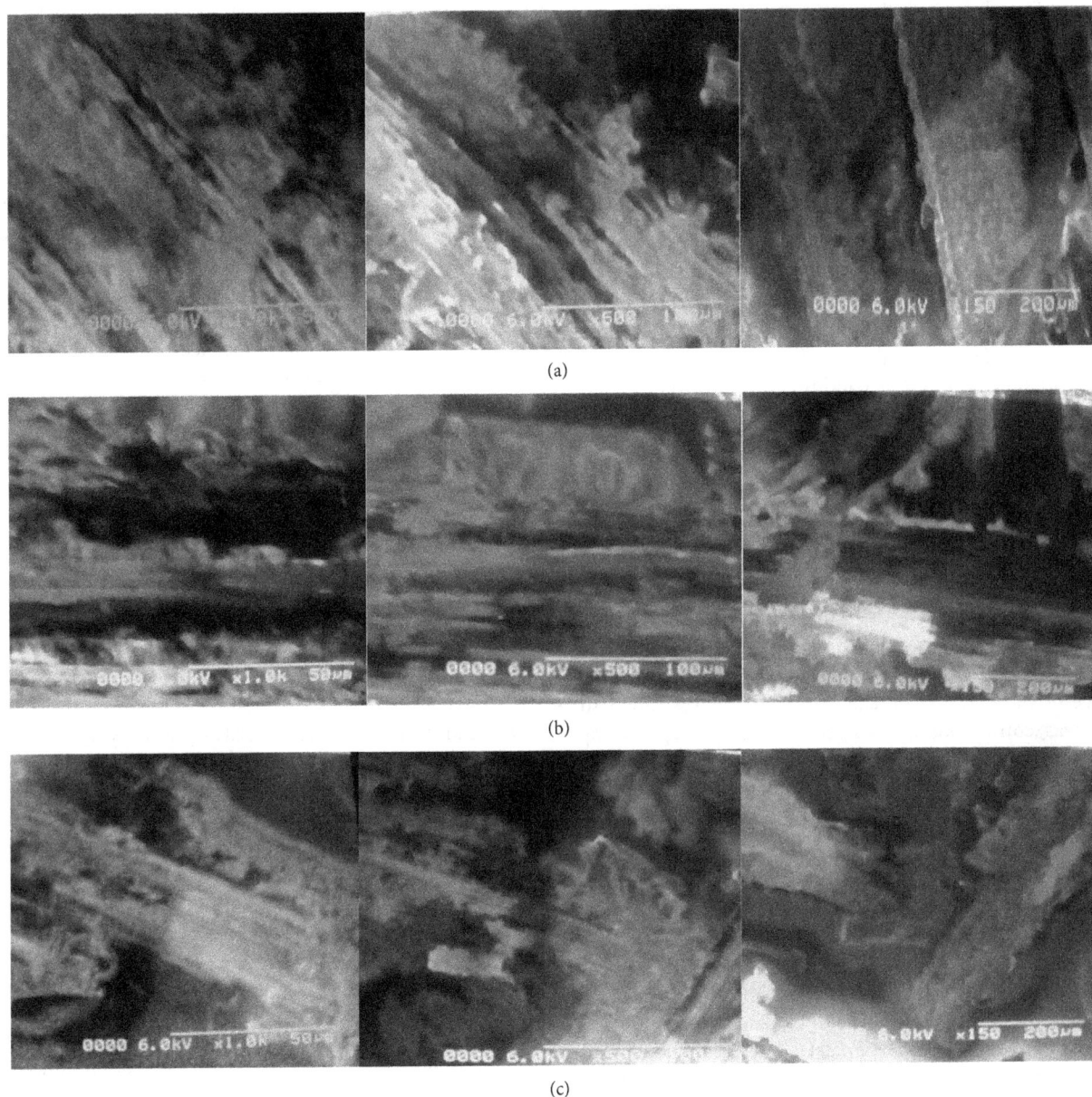

Figure 2: (a) SEM image of raw *Saccharum arundinaceum* at three different resolutions. (b) SEM image of base-treated *Saccharum arundinaceum* at three different resolutions. (c) SEM image of acid-treated *Saccharum arundinaceum* at three different resolutions.

Table 1: Identified functional groups present in *Saccharum arundinaceum* by FTIR spectroscopy.

Possible functional groups	Raw (cm^{-1})	Base treated (cm^{-1})	Acid treated (cm^{-1})
–OH stretching	3309–3751	3211–3400	
=CH			3294
C–H	2858–2918 (bifurcate)	2910	2922
C=O			
Secondary amide	1645	1608	1654.92
–NH			
C–O	1107	1222	
C–N	1051	1056	1043

are listed in Table 1. A broadband appears in the range of 3000–$3700\,cm^{-1}$ which was due to –OH stretching vibration of hydroxyl functional groups including hydrogen bonding because broadband of –OH group in this range is indication of hydrogen bonding present in the compound. This peak appears in raw and base-treated sorbent but disappears in case of acid treated due to reaction of –OH group with acid hydrogen. Stretching band of –CH appears in $2900–3000 \, cm^{-1}$ wavenumber range for all sorbents. The peak at $1750 \, cm^{-1}$ appears due to C=O and $1200 \, cm^{-1}$ due to C-O functional group. In some cases, –CN also appears at $1049 \, cm^{-1}$ value. Vibration due to secondary amide appears at $1645 \, cm^{-1}$ [21]. For adsorption purpose, significant role is played by –OH group and heteroatoms to attach sorbate on the surface.

3.3. Adsorption Study.

Adsorption study was performed by the batch adsorption method by varying different parameters including pH, contact time, and initial concentration of sorbate to find best suited conditions for the removal of cadmium from aqueous media.

3.3.1. Effect of Contact Time.

Contact time was varied from 10 to 100 minutes under neutral conditions with constant amount of sorbent (1 g), initial concentration (60 ppm), and shaking speed (150 rpm), and results obtained are shown in Figure 1s (Supplementary material).

Maximum adsorption was achieved at 60 minutes time interval and no significant increase found by further increase in time. Initially, excess of vacant places are available on the surface of sorbent, and uptake of metal ions was more, so there was continuous increase in adsorption capacity by increasing time slot from zero to 60 minutes. But, further increase could not cause sufficient change in adsorption of metals as vacant spaces are already filled, and equilibrium is achieved [22].

3.3.2. Effect of pH.

Initial pH of the adsorption system has significant role in adsorption of sorbate, as it affects the surface morphology of the sorbent and binding nature of sorbate. The range of pH selected was 2–10 with 1 g sorbent and 60 ppm initial concentration of sorbate. The result given in Figure 2s (Supplementary material) reveals the fact that adsorption capacity is quite low under acidic conditions. When pH is increased, it causes increase in adsorbed amount of sorbate on the surface of sorbent. At low pH value, metals have to compete with H^+ ions for adsorption on sorbent surface since H^+ ions are present in excess at that pH value. But when pH value is raised, it causes significant increase in adsorption due to attraction developed between negatively charged surfaces of sorbent by –OH groups and positively charged metal ions [23]. So, for cadmium, optimum pH range was found from 6 to 8; in this range, cadmium shows best adsorption behavior. When pH is further increased, there is decline in adsorption capacity due to formation of metal hydrides.

3.3.3. Effect of Initial Concentration.

Initial concentration of sorbate is another important parameter, which affects the adsorption phenomenon. For this purpose, initial concentration of cadmium was varied in range of 10–100 ppm by keeping all other parameters constant (Figure 3s Supplementary material).

Rapid increase in adsorption capacity was observed initially for adsorption of cadmium on *Saccharum arundinaceum* as vacant spaces were available on the surface of sorbent. So, rise in concentration also raised adsorption of sorbate on available sites [24]. Adsorbent readily occupies these adsorption sites, and adsorption capacity has positive influence of concentration in this range. Further increase in concentration from 60 to 100 ppm has no significant effect on adsorption phenomenon. Surface of adsorbent becomes saturated with sorbate, and after establishment of equilibria, increase in concentration has no significant influence on adsorption phenomenon. Previous studies also report that accommodation for sorbate decreases as concentration is very high due to unavailability of resident sites [25].

3.3.4. Effect of Temperature.

The effect of temperature on adsorption was studied by using temperature range 20, 30, 40, and 50°C with pH 6 (Figure 4s) (Supplementary material). Adsorption of cadmium onto *Saccharum arundinaceum* agrowaste was found to increase with increase in temperature. At high temperature, intraparticle diffusion increases and more adsorption sites are created which boost up adsorption phenomenon [26].

3.4. Equilibrium Isotherms.

Adsorption system can be designed by adsorption isotherms commonly known as equilibrium isotherms which represent the amount of solute adsorbed per unit weight of sorbent [27]. These isotherms use equilibrium concentration of sorbent at constant temperature. In order to remove effluents from the system, particular design is optimized to generate proper correlation for experimental data which is called adsorption isotherm. Researches proposed many isotherms in this regard which are based on the adsorption system including Langmuir, Freundlich, Redlich–Peterson, Temkin, and Elovich [28, 29]. Sorption was carried out by employing linear as well as nonlinear adsorption models by varying initial concentration from 10 to 100 ppm.

3.4.1. Freundlich Isotherm.

Freundlich adsorption isotherm was developed for the heterogeneous system, and it gives concept of multilayer adsorption on the surface of sorbent (Figure 5s Supplementary material).

Parameters calculated for Freundlich isotherm by employing its linear and nonlinear form are given in Table 4. Freundlich isotherm was obtained by plotting log Cad versus log Ce. K_F and n are constants obtained from intercept and slope, respectively. Freundlich adsorption capacity (K_F) is an indicator of a system, whether it is favorable for adsorption or not. Adsorption is considered promising if value of K_F is found in range of 1–20, and results reveal that in the present study, K_F was 9.5 and 12.2, respectively, for linear and nonlinear approaches of Freundlich adsorption isotherm. Similarly, adsorption intensity represented by n indicates fitness of model for adsorption purposes if value of n is above 1. Value of R^2 obtained from the plot is significant

TABLE 2: Linear and nonlinear parameters of isothermal models for sorption of Cd onto *Saccharum arundinaceum*

Models	Linear method	Nonlinear method
Freundlich		
K_F (mg/g) (L/mg)^n	9.4558	12.2001
N	2.42	3.0271
R^2	0.9446	
Langmuir		
Q_o (mg/g)	48.309	48.0821
b (L/mg)	0.1446	0.1461
R_L	0.408	0.406
R^2	0.9958	
Dubinin–Radushkevich		
k_{ad} (mol^2/kJ^2)	0.6543	3.1816
q_s (mg/g)	34.3533	39.667
R^2	0.7912	
Temkin		
B (J/mol)	264.39	161.8954
K_T	1.0227	0.38
R^2	0.9852	
Elovich		
Q_m (mg/g)	16.58	17.2006
K_E (L/mg)	1.1677	0.6631
R^2	0.9433	

(0.9446) representing good fitness of this model for adsorption of cadmium onto *Saccharum arundinaceum*.

3.4.2. Langmuir Isotherm. Langmuir adsorption isotherm is based on monolayer adsorption of metal ions on the surface of agrowastes, and energy of adsorption system is considered constant. In order to calculate Langmuir model, initial concentrations were changed from 10 to 100 ppm with 1 g sorbent amount and 1 h shaking time. Distribution of metal ions between liquid and solid surface was calculated by equations given in Table 1, employing linear and nonlinear forms of the model. Langmuir adsorption isotherm was obtained by plotting Ce/Cad versus Ce as shown in Figure 6s Supplementary material. R^2 value obtained for plot was found satisfactory showing fitness of model on the adsorption experiment. Q_o represents metal ion uptake per unit mass of adsorbent (mg/g) and b is Langmuir constant [15]. A dimensionless constant R_L is calculated by using Langmuir constant, and initial concentration represents model fitness for a particular system. If value of R_L falls between 0 and 1, the system is considered appropriate for adsorption purpose and Table 2 shows results which are in this range. Furthermore, experimental data and predicted results obtained for the present work were found in close correlation with low value of residual sum of square (0.006) making this model applicable for the present work.

3.4.3. Dubinin–Radushkevich Isotherm. Dubinin–Radushkevich isotherm was designed as an empirical model for adsorption of vapors onto solid surface. It is successfully applied for adsorption of heterogeneous system including solid and liquid. This model is considered more general than Langmuir because in its derivation homogenous surface and constant

sorption potential are not assumed [30]. The relationship given in Table 1s (supplementary data) was employed to relate ln Cad with ε^2, where ε^2 is Polanyi potential which is based on temperature, natural gas constant, and equilibrium concentration as given in the following equation:

$$\varepsilon = RT \ln\left(1 + \frac{1}{C_e}\right). \tag{8}$$

Slope of the plot gives value of k_{ad} and the intercept is q_s. The model showed good applicability on the adsorption system in nonlinear form with high value of R^2.

Dubinin–Radushkevich isotherm has found very promising applications for determination of nature of sorption, whether it is physical or chemical. For this purpose, k_{ad} obtained from the slope of the plot is used in the following equation:

$$E = \frac{1}{\sqrt{2k_{ad}}}. \tag{9}$$

The value of E calculated for the present research was 0.764 and suggests physical nature of sorption. Because the value of E below 8 kj/mol reflects physical sorption and 8–16 kj/mol reflects chemical sorption (Figure 7s Supplementary material).

3.4.4. Temkin Isotherm. Temkin adsorption isotherm discusses interaction of sorbent and sorbate, and the model is based on assumption that heat of adsorption will not remain constant. It decreases due to interaction between sorbent and sorbate during adsorption phenomenon [31]. Linear and nonlinear forms of Temkin model are given in Table 1s (Supplementary data). Equilibrium constant of binding K_T provides information about binding energy, and β expresses heat of adsorption for a particular adsorption experiment (Figure 8s Supplementary material). Linear form of Temkin model is found more suitable with high value of binding constant as given in Table 2. The model indicates the exothermic nature of adsorption reaction as $B > 0$ which is an indicator of heat release during the process [32].

3.4.5. Elovich Isotherm. According to Elovich model, mechanism of adsorption is based on chemical reactions which are responsible for adsorption. Plot of ln C_{ad}/C_e versus C_{ad} gives R^2 value close to unity. K_E and Q_m are obtained from intercept and slope of plot, respectively. K_E shows initial sorption rate and Q_m is adsorption constant. Initial sorption rate obtained from linear form of Elovich model is quite high (35,100.411) as compared to nonlinear form (11.7891), so making linear form adequate to describe adsorption of cadmium onto *Saccharum arundinaceum*. Furthermore, R^2 value (0.9033) for linear form is also high than nonlinear form (0.835) (Figure 9s-Supplementary material).

3.5. Error Analysis for Equilibrium Isotherms. In order to check the fit of adsorption model to experimental data, error functions are used [33]. In the present work, six error functions were applied on linear and nonlinear form of data by minimizing the error function in a range of concentration

TABLE 3: Error functions for optimization of equilibrium isotherms.

Error functions	R^2	ERRSQ/RSS	ARE	EABS	Chi-square (χ^2)
Linear approach					
Freundlich	0.8092	0.0261	−0.8917	−1.9864	3.7154
Langmuir	0.9958	0.0060	−1.2623	−0.0054	0.0106
Dubinin–Radushkevich	0.0523	17.6722	26.6963	9.2320	4.9631
Elovich	0.9432	1.8130	−196.878	−3.9160	10.0831
Temkin	1	18.9131	−0.0723	−0.0055	0.6727
Nonlinear approach					
Freundlich	0.9476	69.1613	−4.9554	−1.4878	40.3303
Langmuir	0.9893	14.0441	1.0363	0.6722	0.5029
Dubinin–Radushkevich	0.9148	166.7881	10.4851	9.4789	11.7308
Elovich	0.8354	1.0911	0.7977	5.3805	0.5322
Temkin	0.8359	766.4798	38.3613	49.2605	65.3952

used for analysis by employing solver add-in with Microsoft Excel 2010. Results for optimization of equilibrium isotherms by error analysis are given in Table 3. For meaningful results, a comparison of each error function for linear and nonlinear forms was made (Figure 10s Supplementary material).

For linear form of adsorption isotherms, a comparison of error functions reflects that Langmuir, Freundlich, and Elovich isotherms have good correlation with experimental values for the present adsorption study. These isotherms give low values for most of error functions. Applicability of these models for removal of cadmium ions from aqueous media is also studied by many researchers [34, 35].

R^2: Temkin > Langmuir > Elovich > Freundlich > Dubinin–Radushkevich

RSS: Temkin > Dubinin–Radushkevich > Elovich > Freundlich > Langmuir

ARE: Dubinin–Radushkevich > Temkin > Freundlich > Langmuir > Elovich

EABS: Dubinin–Radushkevich > Langmuir > Temkin > Freundlich > Elovich

Chi-square (χ^2): Elovich > Dubinin–Radushkevich > Freundlich > Temkin > Langmuir

Similar study was carried out by employing nonlinear form of adsorption models, and results are summarized below. Nonlinear form of Temkin isotherm was not found suitable for adsorption of cadmium onto *Saccharum arundinaceum* agrowaste because of high value of error functions. Freundlich and Elovich isotherms have been proved to be suitable models for this study with low value for error functions.

R^2: Langmuir > Freundlich > Dubinin–Radushkevich > Temkin > Elovich

RSS: Temkin > Dubinin–Radushkevich > Freundlich > Langmuir > Elovich

ARE: Temkin > Dubinin–Radushkevich > Langmuir > Elovich > Freundlich

EABS: Temkin > Dubinin–Radushkevich > Elovich > Langmuir > Freundlich

Chi-square (χ^2): Temkin > Freundlich > Dubinin–Radushkevich > Elovich > Langmuir

A comparison between linear and nonlinear approaches of each adsorption isotherm was also made to select the most appropriate form for adsorption study. Linear form of Freundlich adsorption isotherm was found superior over nonlinear form with small error functions in most of the cases.

R^2: Freundlich (linear approach) < Freundlich (nonlinear)

RSS: Freundlich (linear approach) < Freundlich (nonlinear)

ARE: Freundlich (linear approach) > Freundlich (nonlinear)

EABS: Freundlich (linear approach) < Freundlich (nonlinear)

Chi-square (χ^2): Freundlich (linear approach) < Freundlich (nonlinear)

For Langmuir adsorption isotherm, error function for nonlinear form was obtained high as compared with linear to exception of R^2.

R^2: Langmuir (linear approach) > Langmuir (nonlinear approach)

RSS: Langmuir (linear approach) < Langmuir (nonlinear approach)

ARE: Langmuir (linear approach) < Langmuir (nonlinear approach)

EABS: Langmuir (linear approach) < Langmuir (nonlinear approach)

Chi-square (χ^2): Langmuir (linear approach) < Langmuir (nonlinear approach)

Applicability of linear Dubinin–Radushkevich model was found better than nonlinear due to small value of error function (except ARE).

R^2: Dubinin–Radushkevich (linear approach) < Dubinin–Radushkevich (nonlinear)

RSS: Dubinin–Radushkevich (linear approach) < Dubinin–Radushkevich (nonlinear)

ARE: Dubinin–Radushkevich (linear approach)
> Dubinin–Radushkevich (nonlinear)

EABS: Dubinin–Radushkevich (linear approach)
< Dubinin–Radushkevich (nonlinear)

Chi-sq/χ^2: Dubinin–Radushkevich (linear approach)
< Dubinin–Radushkevich (nonlinear)

Linear form of Elovich isotherm, which is also based on multilayer sorption on the surface of sorbent, shows small value of ARE and EABS, but other error functions were found lower for nonlinear form.

R^2: Elovich (linear approach) > Elovich
(nonlinear approach)

RSS: Elovich (linear approach) > Elovich
(nonlinear approach)

ARE: Elovich (linear approach) < Elovich
(nonlinear approach)

EABS: Elovich (linear approach) < Elovich
(nonlinear approach)

Chi-square (χ^2): Elovich (linear approach)
> Elovich (nonlinear approach)

Linear approach for Temkin isotherm was found favorable for adsorption of cadmium ions onto *Saccharum arundinaceum* with low error function.

R^2: Temkin (linear approach) > Temkin
(nonlinear approach)

RSS: Temkin (linear approach) < Temkin
(nonlinear approach)

ARE: Temkin (linear approach) < Temkin
(nonlinear approach)

EABS: Temkin (linear approach) < Temkin
(nonlinear approach)

Chi-square (χ^2): Temkin (linear approach)
< Temkin (nonlinear approach)

3.6. Thermodynamic Studies. Effect of temperature on adsorption was studied by using temperature range 20, 30, 40, and 50°C with pH 6 and variable initial concentration (30–120 ppm). Adsorption of cadmium onto *Saccharum arundinaceum* agrowaste was found to increase with increase in temperature. At high temperature, intraparticle diffusion increases and more adsorption sites are created which boost up adsorption phenomenon. Results for thermodynamic study are given in Table 4:

$$\text{Log}\frac{\text{Cad}}{\text{Ce}} - \frac{\Delta H}{2.303RT} + \frac{\Delta S}{2.303R}. \tag{10}$$

Plot of log Cad/Ce versus $1/T$ was obtained with R-squared value 0.927. Slop and intercept provide value of $\Delta H°$ and $\Delta S°$, respectively, as shown in (10).

$\Delta G°$ was calculated by employing (1) given in Section 2 in temperature range 292–328 K. Results show a negative value for Gibb's free energy at all temperature ranges, and $\Delta G°$ increases with the increase in temperature. These negative values represent spontaneous nature as well as feasibility of

TABLE 4: Thermodynamic parameters for adsorption of cadmium onto *Saccharum arundinaceum*.

Parameters	Temperature (K)	Results
$\Delta G°$ (kJ/mol)	293	−591.24
	303	−612.34
	313	−633.44
	328	−665.09
$\Delta H°$ (kJ/mol)		26.99
$\Delta S°$ (kJ/mol·K)		2.11
Sorption energy (kJ/mol)		0.764

adsorption reaction [36]. Decrease in $\Delta G°$ with the increase in temperature reflects better sorption at elevated temperature. The positive value for change in enthalpy is due to endothermic nature of adsorption of cadmium. Enthalpy was also found positive because randomness in system increases due to solid-liquid interaction during adsorption phenomenon. Sorption energy calculated by Dubinin–Radushkevich model was found below 1 which is an indication of physical nature of cadmium sorption on the surface of sorbent. $E < 8$ kJ/mol is representative of physical sorption, and $E > 8$–16 kJ/mol is due to chemical sorption [37]. For adsorption of cadmium value of E is found below 8, so adsorption of cadmium occurred on the surface and no chemical bonding took place between sorbent and sorbate. Similar results for adsorption of cadmium onto agrowaste were reported in the literature [38].

3.7. Adsorption Kinetics. Adsorption kinetics has prime importance in describing solute uptake rate and time required for adsorption process. In the present work, kinetic study was performed at different time intervals for cadmium adsorption by employing linear and nonlinear forms of pseudo-first- and second-order kinetics. Results indicate that amount of cadmium adsorbed increases with the increase in time interval; however, this increase was sharp in the start of reaction and gradually magnitude of adsorption decreases down. Initially, plenty of active sites were available on the surface of sorbent, so sharp rise in adsorption occurred, but these sites got occupied with the passage of time, so magnitude of adsorption gradually decreases [39].

3.7.1. Pseudo-First-Order Kinetics. For pseudo-first-order kinetic model, log $(Q_e - Q_t)$ was plotted against time interval and value of k was obtained from slope of the line and Q_e from intercept. Initial sorption rate, h, was calculated by the following equation:

$$h = k_2 Q_e^2. \tag{11}$$

Poor correlation was obtained for linear form of the model with low value of R^2 (0.0918). Results indicate that adsorption of cadmium onto *Saccharum arundinaceum* does not follow pseudo-first-order kinetics. Nonlinear form of pseudo-first-order kinetics was obtained by using Microsoft Excel 2010 [40].

3.7.2. Pseudo-Second-Order Kinetics. Second-order kinetics is applicable on small amount of initial concentration for determination of initial sorption rate. Different linear and nonlinear forms of pseudo-second-order kinetics are given in Table 3s (Supplementary material). These four linear forms of pseudo-second-order kinetic models were applied on experimental data, and Figures 11s–15s (Supplementary material) show results for these models. Coefficient of determination (R^2), found for type 1, was quite high indicating the best fitting of this model on adsorption data of cadmium. Results obtained for pseudo-second-order kinetic are given in Table 4s (Supplementary material). Theoretical results obtained for amount of cadmium adsorbed at equilibrium are found best fitted with experimental data for pseudo-second-order kinetics. For nonlinear form of pseudo-second-order, a computer-based procedure was used in Microsoft Excel 2010 using solver add-in method as reported in the literature [41]. For sorption of cadmium onto *Saccharum arundinaceum* pseudo-second-order kinetic may describe the method of adsorption in quite appropriate way as compared to pseudo-first-order approach. Furthermore, nonlinear form of pseudo-second-order gives close results to experimental data.

3.8. Effect of Interfering Ions. Process of adsorption becomes complicated in case of multicomponent adsorption as many interactions of sorbent and sorbate are involved. Effect of interfering ions was measured by observing adsorption of one metal ion, and then, by addition of interfering ion change in adsorption capacity was noted. Interfering effect was calculated by employing the following equation:

$$\text{interfering capacity} = \frac{C_{\text{mix}}}{C}, \quad (12)$$

where C_{mix} is the % adsorption of mixture of two metal ions and C is the % adsorption of pure metal on selected sorbent. If the value of C_{mix}/C is found equal to 1, then there is no effect of interfering ions on adsorption phenomenon. However, if it is found less than 1, then adsorption capacity is found to be affected by addition of these interfering ions. To study interference, ions were divided into three categories as monovalent, bivalent, and trivalent ions based on valences of ions. One ion was selected from each class of ions to check the interfering effect on adsorption of cadmium ions. Metals are attached on the surface of sorbent through electrostatic forces, and competition of metals for sorbent place is mainly based on metal ion charge and its attraction towards functional groups present on the surface on adsorbent. Results for interference of metal ions are summarized in Table 5. Metals with high charge value were found to have maximum effect on adsorption of cadmium as compared with those with low charge. Anions have also found to affect the adsorption phenomenon of metal ions but their interference is comparatively very less than cations. Adsorption of anions on sorbent is dependent on charge present on the surface of sorbent. Since negatively charged hydroxyl groups are present on the surface of adsorbent, adsorption of anions is not as favored as cations [42].

TABLE 5: Effect of interfering cations and anions on % adsorption of cadmium.

Interfering cations		Effect on adsorption of Cd^{+2}
Monovalent	K^{+1}	0.958
Bivalent	Ca^{+2}	0.644
Trivalent	Cr^{+3}	0.491
Interfering anions	Cl^{-1}	0.945
	NO_3^{-1}	0.881
	SO_4^{-2}	0.907

4. Conclusion

Removal of cadmium was performed by employing the stem powder of *Saccharum arundinaceum*. In order to generate proper correlation for the removal of cadmium, five adsorption isotherms were applied on experimental data including Freundlich, Langmuir, Dubinin–Radushkevich, Elovich, and Temkin. Error analysis provides information about fitness of these models on experimental data. The model with minimum error was selected best for adsorption data. Order of equilibrium isotherms according to increasing RSS value was found as Temkin > Dubinin–Radushkevich > Elovich > Freundlich > Langmuir.

Linear form of Freundlich and Langmuir models was found best fitted with minimum value of error. Effect of temperature on cadmium adsorption was investigated by thermodynamic analysis, and it was found to increase with increase in temperature. Gibb's free energy ($\Delta G° = -612.34$ at 303 K) revealed spontaneous nature of sorbent-sorbate binding reaction, and it followed pseudo-second-order kinetics.

Supplementary Materials

Figures 1s–4s: effect of different parameters (contact time, pH, initial concentration of cadmium, and temperature) which were adjusted during adsorption for maximum removal of cadmium from aqueous media. Figures 5s–9s: graphical results of isothermal study. Data derived from these figures are present in Table 4 of the main manuscript. Figure 10s: thermodynamic studies. Results of Figures 11s–15s of kinetic studies are described in these figures. Linear and nonlinear forms of equilibrium isotherms are given in Tables 1s and 2s. Table 2s provides information about equations of error functions applied on the results. Tables 3s and 4s describe kinetic studies and formulas applied on work for calculation of pseudo-first- and second-order kinetics. (*Supplementary Materials*)

References

[1] E. Szabó, Z. Pap, G. Simon, A. Dombi, L. Baia, and K. Hernádi, "New insights on the simultaneous removal by adsorption on

organoclays of humic acid and phenol," *Water*, vol. 8, no. 1, p. 21, 2016.

[2] J. N. Halder and M. N. Islam, "Water pollution and its impact on the human health," *Journal of Environment and Human*, vol. 2, no. 1, pp. 36–46, 2015.

[3] T. Vaughan, C. W. Seo, and W. E. Marshall, "Removal of selected metal ions from aqueous solution using modified corncobs," *Bioresource Technology*, vol. 78, no. 2, pp. 133–139, 2001.

[4] P. Húmpola, H. Odetti, J. C. Moreno-Piraján, and L. Giraldo, "Activated carbons obtained from agro-industrial waste: textural analysis and adsorption environmental pollutants," *Adsorption*, vol. 22, no. 1, pp. 23–31, 2016.

[5] V. K. Gupta, M. Gupta, and S. Sharma, "Process development for the removal of lead and chromium from aqueous solutions using red mud—an aluminium industry waste," *Water Research*, vol. 35, no. 5, pp. 1125–1134, 2001.

[6] Z. Carmen and S. Daniela, "Textile organic dyes-characteristics, polluting effects and separation/elimination procedures from industrial effluents–a critical overview," in *Organic Pollutants Ten Years after the Stockholm Convention-Environmental and Analytical Update*, pp. 55–81, InTech, Croatia, 2012.

[7] G. Crini, "Recent developments in polysaccharide-based materials used as adsorbents in wastewater treatment," *Progress in Polymer Science*, vol. 30, no. 1, pp. 38–70, 2005.

[8] M. Vidali, "Bioremediation. An overview," *Pure and Applied Chemistry*, vol. 73, no. 7, pp. 1163–1172, 2001.

[9] K. Y. Foo and B. H. Hameed, "Insights into the modeling of adsorption isotherm systems," *Chemical Engineering Journal*, vol. 156, no. 1, pp. 2–10, 2010.

[10] Y. Salameh, N. Al-Lagtah, M. N. M. Ahmad, S. J. Allen, and G. M. Walker, "Kinetic and thermodynamic investigations on arsenic adsorption onto dolomitic sorbents," *Chemical Engineering Journal*, vol. 160, no. 2, pp. 440–446, 2010.

[11] O. Carmody, R. Frost, Y. Xi, and S. Kokot, "Surface characterisation of selected sorbent materials for common hydrocarbon fuels," *Surface Science*, vol. 601, no. 9, pp. 2066–2076, 2007.

[12] M. Iqbal, A. Saeed, and S. I. Zafar, "FTIR spectrophotometry, kinetics and adsorption isotherms modeling, ion exchange, and EDX analysis for understanding the mechanism of Cd^{2+} and Pb^{2+} removal by mango peel waste," *Journal of Hazardous Materials*, vol. 164, no. 1, pp. 161–171, 2009.

[13] Y. S. Ho, C. T. Huang, and H. W. Huang, "Equilibrium sorption isotherm for metal ions on tree fern," *Process Biochemistry*, vol. 37, pp. 1421–1430, 2002.

[14] L. S. Chan, W. H. Cheung, S. J. Allen, and G. McKay, "Error analysis of adsorption isotherm models for acid dyes onto bamboo derived activated carbon," *Chinese Journal of Chemical Engineering*, vol. 20, no. 3, pp. 535–542, 2012.

[15] J. Febrianto, A. N. Kosasih, J. Sunarso, Y.-H. Ju, N. Indraswati, and S. Ismadji, "Equilibrium and kinetic studies in adsorption of heavy metals using biosorbent: a summary of recent studies," *Journal of Hazardous Materials*, vol. 162, no. 1-2, pp. 616–645, 2009.

[16] S.-J. Park and Y.-S. Jang, "Pore structure and surface properties of chemically modified activated carbons for adsorption mechanism and rate of Cr(VI)," *Journal of Colloid and Interface Science*, vol. 249, no. 2, pp. 458–463, 2002.

[17] M. Salah Azab and P. J. Peterson, "The removal of cadmium from water by the use of biological sorbents," *Water Science and Technology*, vol. 21, no. 12, pp. 1705–1706, 1989.

[18] A. Kapoor and T. Viraraghavan, "Fungal biosorption—an alternative treatment option for heavy metal bearing wastewaters: a review," *Bioresource Technology*, vol. 53, no. 3, pp. 195–206, 1995.

[19] B. Tansel and P. Nagarajan, "SEM study of phenolphthalein adsorption on granular activated carbon," *Advances in Environmental Research*, vol. 8, no. 3-4, pp. 411–415, 2004.

[20] O. B. Belskaya, A. V. Lavrenov, I. G. Danilova, M. O. Kazakov, R. M. Mironenko, and V. A. Likholobov, "FTIR spectroscopy of adsorbed probe molecules for analyzing the surface properties of supported Pt (Pd) catalysts," in *Infrared Spectroscopy-Materials Science, Engineering and Technology*, INTECH Open Access Publisher, London, UK, 2012.

[21] B. Stuart, *Infrared Spectroscopy*, Wiley, Hoboken, NJ, USA, 2005.

[22] S. Yang, J. Li, D. Shao, J. Hu, and X. Wang, "Adsorption of Ni (II) on oxidized multi-walled carbon nanotubes: effect of contact time, pH, foreign ions and PAA," *Journal of Hazardous Materials*, vol. 166, no. 1, pp. 109–116, 2009.

[23] M. Malandrino, O. Abollino, A. Giacomino, M. Aceto, and E. Mentasti, "Adsorption of heavy metals on vermiculite: influence of pH and organic ligands," *Journal of Colloid and Interface Science*, vol. 299, no. 2, pp. 537–546, 2006.

[24] B. B. Johnson, "Effect of pH, temperature, and concentration on the adsorption of cadmium on goethite," *Environmental Science and Technology*, vol. 24, no. 1, pp. 112–118, 1990.

[25] T. Teka and S. Enyew, "Study on effect of different parameters on adsorption efficiency of low cost activated orange peels for the removal of methylene blue dye," *International Journal of Innovation and Scientific Research*, vol. 8, pp. 106–111, 2014.

[26] G. Ks and S. L. Belagali, "Removal of heavy metals and dyes using low cost adsorbents from aqueous medium-, a review," *IOSR Journal of Environmental Science, Toxicology and Food Technology*, vol. 4, no. 3, pp. 56–68, 2013.

[27] I. D. Mall, V. C. Srivastava, and N. K. Agarwal, "Removal of Orange-G and Methyl Violet dyes by adsorption onto bagasse fly ash—kinetic study and equilibrium isotherm analyses," *Dyes and pigments*, vol. 69, no. 3, pp. 210–223, 2006.

[28] B. H. Hameed, I. A. W. Tan, and A. L. Ahmad, "Adsorption isotherm, kinetic modeling and mechanism of 2, 4, 6-trichlorophenol on coconut husk-based activated carbon," *Chemical Engineering Journal*, vol. 144, no. 2, pp. 235–244, 2008.

[29] Y. Liu and Y.-J. Liu, "Biosorption isotherms, kinetics and thermodynamics," *Separation and Purification Technology*, vol. 61, no. 3, pp. 229–242, 2008.

[30] R. M. Allen-King, P. Grathwohl, and W. P. Ball, "New modeling paradigms for the sorption of hydrophobic organic chemicals to heterogeneous carbonaceous matter in soils, sediments, and rocks," *Advances in Water Resources*, vol. 25, no. 8-12, pp. 985–1016, 2002.

[31] C. P. Schulthess and D. L. Sparks, "Equilibrium-based modeling of chemical sorption on soils and soil constituents," in *Advances in Soil Science*, pp. 121–163, Springer, Berlin, Germany, 1991.

[32] V. Patrulea, A. Negrulescu, M. M. Mincea, L. D. Pitulice, O. B. Spiridon, and V. Ostafe, "Optimization of the removal of copper(II) ions from aqueous solution on chitosan and cross-linked chitosan beads," *BioResources*, vol. 8, no. 1, 2013.

[33] R. Rostamian, M. Najafi, and A. A. Rafati, "Synthesis and characterization of thiol-functionalized silica nano hollow sphere as a novel adsorbent for removal of poisonous heavy metal ions from water: Kinetics, isotherms and error analysis," *Chemical Engineering Journal*, vol. 171, no. 3, pp. 1004–1011, 2011.

[34] E. Demirbas, M. Kobya, and A. E. Konukman, "Error analysis of equilibrium studies for the almond shell activated carbon

adsorption of Cr(VI) from aqueous solutions," *Journal of Hazardous Materials*, vol. 154, no. 1–3, pp. 787–794, 2008.

[35] D. Gusain, V. Srivastava, M. Sillanpää, and Y. C. Sharma, "Kinetics and isotherm study on adsorption of chromium on nano crystalline iron oxide/hydroxide: linear and nonlinear analysis of isotherm and kinetic parameters," *Research on Chemical Intermediates*, vol. 42, no. 9, pp. 7133–7151, 2016.

[36] B. Singha and S. K. Das, "Biosorption of Cr(VI) ions from aqueous solutions: Kinetics, equilibrium, thermodynamics and desorption studies," *Colloids and Surfaces B: Biointerfaces*, vol. 84, no. 1, pp. 221–232, 2011.

[37] J. Romero-González, J. R. Peralta-Videa, E. Rodríguez, S. L. Ramirez, and J. L. Gardea-Torresdey, "Determination of thermodynamic parameters of Cr(VI) adsorption from aqueous solution onto *Agave lechuguilla* biomass," *Journal of Chemical Thermodynamics*, vol. 37, no. 4, pp. 343–347, 2005.

[38] E. Mekonnen, M. Yitbarek, and T. R. Soreta, "Kinetic and thermodynamic studies of the adsorption of Cr(VI) onto some selected local adsorbents South African," *Journal of Chemistry*, vol. 68, pp. 45–52, 2015.

[39] Y. Bulut and H. Aydın, "A kinetics and thermodynamics study of methylene blue adsorption on wheat shells," *Desalination*, vol. 194, no. 1–3, pp. 259–267, 2006.

[40] J. Lin and L. Wang, "Comparison between linear and non-linear forms of pseudo-first-order and pseudo-second-order adsorption kinetic models for the removal of methylene blue by activated carbon," *Frontiers of Environmental Science and Engineering in China*, vol. 3, no. 3, pp. 320–324, 2009.

[41] Y.-S. Ho, "Second-order kinetic model for the sorption of cadmium onto tree fern: a comparison of linear and non-linear methods," *Water Research*, vol. 40, no. 1, pp. 119–125, 2006.

[42] D. Mohan and K. P. Singh, "Single-and multi-component adsorption of cadmium and zinc using activated carbon derived from bagasse—an agricultural waste," *Water Research*, vol. 36, no. 9, pp. 2304–2318, 2002.

14

A Comparative Study of Raw and Metal Oxide Impregnated Carbon Nanotubes for the Adsorption of Hexavalent Chromium from Aqueous Solution

Muhammad I. Qureshi,[1] **Faheemuddin Patel,**[2] **Nadhir Al-Baghli,**[1] **Basim Abussaud,**[1] **Bassam S. Tawabini,**[3] **and Tahar Laoui**[2]

[1]*Department of Chemical Engineering, KFUPM, Dhahran 31261, Saudi Arabia*
[2]*Department of Mechanical Engineering, KFUPM, Dhahran 31261, Saudi Arabia*
[3]*Department of Geosciences, KFUPM, Dhahran 31261, Saudi Arabia*

Correspondence should be addressed to Tahar Laoui; tlaoui@kfupm.edu.sa

Academic Editor: Viktor Kochkodan

The present study reports the use of raw, iron oxide, and aluminum oxide impregnated carbon nanotubes (CNTs) for the adsorption of hexavalent chromium (Cr(VI)) ions from aqueous solution. The raw CNTs were impregnated with 1% and 10% loadings (weight %) of iron oxide and aluminum oxide nanoparticles using wet impregnation technique. The synthesized materials were characterized using scanning electron microscopy (SEM) and thermogravimetric analysis (TGA). Batch adsorption experiments were performed to assess the removal efficiency of Cr(VI) ions from water and the effects of pH, contact time, adsorbent dosage, and initial concentration of the Cr(VI) ions were investigated. Results of the study revealed that impregnated CNTs achieved significant increase in the removal efficiency of Cr(VI) ions compared to raw CNTs. In fact, both CNTs impregnated with 10% loading of iron and aluminum oxides were able to remove up to 100% of Cr(VI) ions from aqueous solution. Isotherm studies were carried out using Langmuir and Freundlich isotherm models. Adsorption kinetics of Cr(VI) ions from water was found to be well described by the pseudo-second-order model. The results suggest that metallic oxide impregnated CNTs have very good potential application in the removal of Cr(VI) ions from water resulting in better environmental protection.

1. Introduction

Chromium is mainly found in natural deposits as ores and other compounds such as chrome ochre (Cr_2O_3), crocoite ($PbCrO_4$), and ferric chromite ($FeCr_2O_4$). It is the sixth most abundant transition metal [1, 2]. Chromium is discharged into water bodies from a number of industrial sources such as electroplating and metal cleaning, leather tanning, mining of chrome ore, production of steel and alloys, dyes and pigments, glass industry, wood preservation, and textile industry [2–5].

Chromium is found in different oxidation states such as 2+, 3+, and 6+. In water, it can exist in the form of chromate ion (CrO_4^{2-}), chromic acid (H_2CrO_4), hydrogen chromate ion ($HCrO_4^-$), and dichromate ion ($Cr_2O_7^{2-}$) [6–8]. However, the hexavalent Cr(VI) and trivalent Cr(III) are the two most stable forms present in water in neutral pH range.

The typical concentration of chromium in industrial water ranges from 5.2 to 208,000 mg/L [9, 10]. The maximum allowable limits of chromium in drinking water are 0.05 and 0.1 mg/L, as suggested by the World Health Organization (WHO) and US Environmental Protection Agency (EPA), respectively [11–15].

Due to its carcinogenic and mutagenic nature, Cr(VI) is considered as almost 300 times more toxic than Cr(III) [16]. The toxic effects of Cr(VI) include liver and kidney damage, nausea, dermatitis, diarrhea, vomiting, internal hemorrhage, and repository problems (asthma). Eye and skin contact may cause permanent damage to eye, severe burn, irritation, ulceration, and nasal septum [17, 18].

A number of remediation techniques have been reported to get rid of the Cr(VI) from water including solvent extraction [19], floatation [20], coagulation [21], ion exchange [22–25], membrane technologies [26, 27], adsorption [6, 7, 28] and cyanide treatment [29], and reduction followed by chemical precipitation [30]. However, adsorption is the most versatile, cost effective, and widely used method for removal of different contaminants from water including heavy metals. In the literature, different adsorbents have been reported for the removal of Cr(VI) from water including anaerobic sludge [31], lignocellulosic solid wastes [32], carbon slurry [33], waste slurry [34], agricultural wastes [35], cow dung carbon [36], corncob [37], almond shell carbon [38], zeolite [39], hazelnut shell carbon [40, 41], rice Polish [42], sphagnum moss peat [43], apple residue [44], moss [45], rice husk carbon [46], fly ash [6, 47], pine needles, charcoal, wool, olive stone/cake, cactus [48], used tyre carbon [49], coconut tree sawdust carbon [50], sawdust [51], dust coal, coconut shell and wood activated carbons [52], clay [53], palm pressed fibers and coconut husk [54], activated groundnut husk carbon [55], polyaniline coated on sawdust [56], coniferous leaves [57], leaf mould [58], wheat bran [59], sugar beet pulp [60], seaweeds [61], tannin gel particles [62], seaweed biosorbent [63], chitosan-1,2-cyclohexylenedinitrilotetraacetic acid–graphene oxide (Cs/CDTA/GO) nanocomposite [64], paper mill sludge [65], hydrous concrete particles [66], waste tea [67], activated alumina, rice husk ash, neem bark, saw dust, fuller's earth [6], eucalyptus bark, activated charcoal, and charred rice husk [68], treated waste newspaper [69], and graphene oxide (GO) [70].

Recently, carbon nanotubes (CNTs) have emerged as a novel adsorbent for the removal of various contaminants from water. CNTs offer the advantages of high porous and hallow structure, light mass density, large surface, and strong interaction with the pollutant molecules [28]. Studies have confirmed that surface modification of CNTs significantly enhanced their adsorption capability for the removal of various contaminants from water [71–76].

In the present study, raw CNTs and CNTs impregnated with iron oxide and aluminum oxide nanoparticles were used for the adsorption of Cr(VI) from water. The synthesized materials were characterized using scanning electron microscopy (SEM) and thermogravimetric analysis (TGA). Batch adsorption experiments were performed and the effect of pH, contact time, adsorbent dosage, and initial concentration of the adsorbate on the removal efficiency of Cr(VI) from water was investigated. Isotherm studies were carried out using Langmuir and Freundlich isotherm models.

2. Experimental

2.1. Materials Preparation.
Raw CNTs were acquired from Chengdu Organic Chemicals Co. Ltd. (China), with the following characteristics: 95% purity, outside diameter of 10–20 nm, and length ranging from 1 to 10 μm. These raw

CNTs were impregnated with 1% and 10% loadings (weight %) of iron oxide and aluminum oxide nanoparticles using wet impregnation technique. Specific amount of CNTs was added in ethanol and sonicated to achieve homogenous dispersion of CNTs. Specific amount of metallic salt dissolved separately in ethanol and was sonicated, and then the resultant solution was added dropwise to the CNTs dispersed in ethanol. This dispersion was sonicated for proper mixing with CNTs and subsequently heated at 80–90°C in an oven overnight to evaporate the ethanol. On complete drying, the CNTs were calcined in a furnace at 350°C for 4 hours. This process resulted in the attachment of metal oxide nanoparticles onto the surface of CNTs.

2.2. Characterization of the Adsorbents.
Raw and impregnated CNTs were characterized using various techniques. In order to perform morphological and elemental analysis, samples were coated with about 5 nm thick layer of platinum using Quorum sputter coater (Model: Q150R S). Scanning electron microscope (Model: TESCAN MIRA 3 FEG-SEM) was used to analyze the morphology and structure of raw and metal oxide impregnated CNTs. Thermogravimetric analysis (TGA) of raw and impregnated CNTs was performed using TA Instrument (Model: SDTQ600), in order to evaluate the purity and thermal degradation of materials. Samples were heated to 900°C in air, at heating rate of 10°C/min and air flow rate of 100 mL/min.

2.3. Batch-Mode Adsorption Experiment.
Batch experiments were performed to study the effect of various parameters on the adsorption of Cr(VI) ions by raw and metal oxide impregnated CNTs at room temperature.

The effect of pH, contact time, agitation speed, and adsorbent dosage was investigated on the removal of Cr(VI) ions from aqueous solution. Concentration of Cr(VI) ions was measured using inductively coupled plasma mass spectrometer (Thermo-Fisher, X-Series 2 Q-ICP-MS).

Percentage removal and adsorption capacity were calculated using (1) and (2), respectively:

$$\text{Removal efficiency (\%)} = \frac{C_o - C_t}{C_o} * 100 \tag{1}$$

$$\text{Adsorption capacity } (q) = \frac{(C_o - C_t) V}{m}, \tag{2}$$

where "C_o" is the initial concentration (ppm) at start of the experiment ($t = 0$), while "C_t" is the concentration at time "t". "V" is the volume (L) of the solution and "m" represents the amount (g) of the adsorbent dosage. For the batch adsorption experiments, the stock solution was prepared using the same methodology reported previously [73].

3. Results and Discussion

3.1. Characterization of Raw and Metal Oxide Impregnated CNTs.
Surface morphologies of the raw and metal oxide impregnated CNTs were observed using SEM.

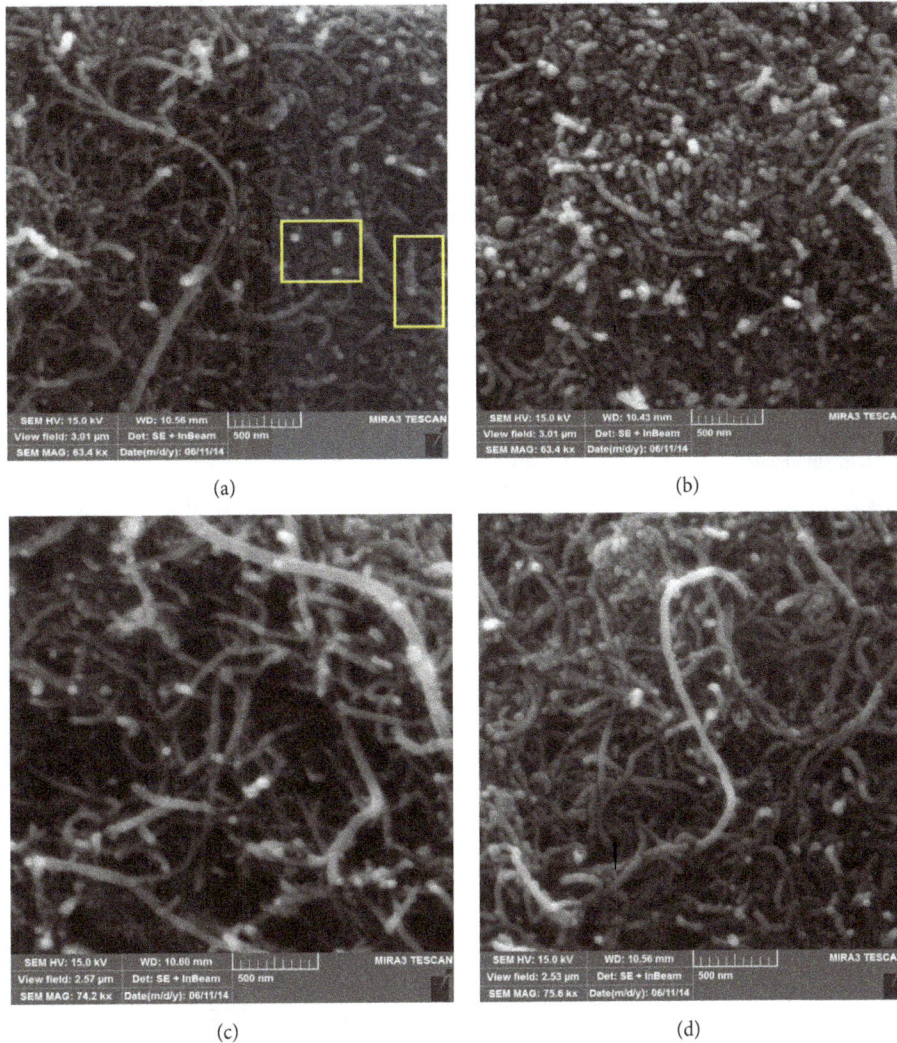

FIGURE 1: SEM images of CNTs with (a) 1% iron oxide (boxes indicate the iron oxide nanoparticles impregnated on CNTs), (b) 10% iron oxide, (c) 1% aluminum oxide, and (d) 10% aluminum oxide.

Figure 1 shows the SEM images for the metal oxides impregnated CNTs. Tubular geometry of the CNTs was observed and no damage was noticed in CNTs structures after impregnation. Metal oxide nanoparticles (highlighted in the box) were clearly observed on the surface of CNTs as displayed in Figures 1(a)–1(d). CNTs were properly dispersed for the low loading of 1% metal oxide (Figures 1(a) and 1(c)); however, at higher loading (10%) a little agglomeration of metal oxide particles could be seen in Figures 1(b) and 1(d). In general, the dispersion of CNTs was improved after impregnation with metal oxide nanoparticles. Metal oxide nanoparticles might help reduce the strong Van der Waals forces between CNTs leading to their improved dispersion.

TGA curves for raw and metal oxide impregnated CNTs are presented in Figure 2. CNTs were heated to 900°C at a rate of 10°C/min under air. All the TGA curves have two main weight loss regions. Initial small weight loss was attributed to the evaporation of physically bound water and some other lighter impurities. The second, steep, and rapid weight loss region represents the combustion of CNTs. Raw CNTs

showed more stability and started degrading around 550°C while degradation of 1% and 10% metal oxide impregnated CNTs started around 450°C and 500°C, respectively. This may be due to the fact that the impregnation of metal oxide nanoparticles on CNTs serves as an impurity hence leading to steep weight loss at lower temperature [77]. The weight of the residue left at the end of the analysis is the indication of metallic oxide nanoparticles. It can be observed that the amount of residue left was higher for the CNTs with 10% metal oxide loading as compared to raw CNTs and CNTs with 1% metal oxide loadings.

3.2. Effect of pH. The removal of Cr(VI) ions by raw and metal oxide impregnated CNTs, as a function of pH, is presented in Figure 3. Solution pH was varied from 3 to 8, while the other variables including adsorbent dosage, contact time, agitation speed, and Cr(VI) initial concentration were kept constant at 200 mg, 2 hours, 50 mg, 200 rpm, and 1 mg/L, respectively.

A maximum removal of Cr(VI) was achieved at pH 3, while the removal was observed to decrease with increase in

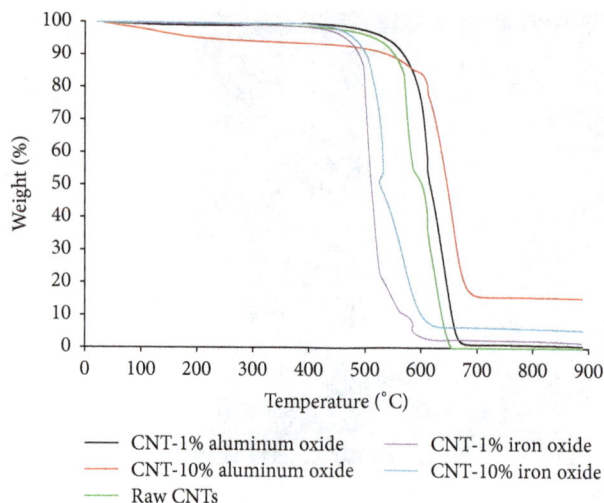

FIGURE 2: TGA curves for raw and metal oxide impregnated CNTs.

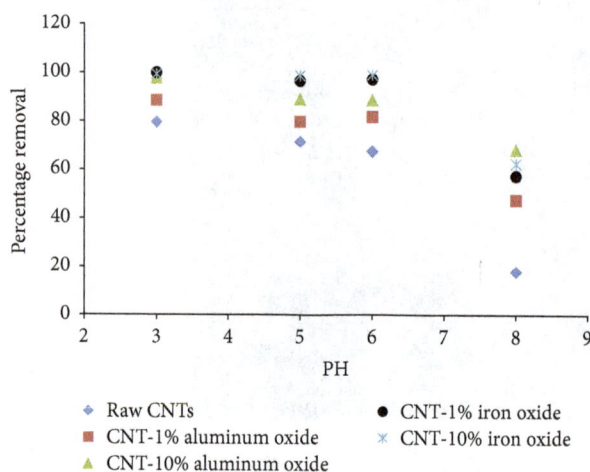

FIGURE 3: Effect of pH on the percentage removal of Cr(VI) (initial concentration = 1 mg/L, agitation speed = 200 rpm, adsorbent dosage = 200 mg, and time = 2 hours).

pH, for all the adsorbents. This phenomenon can be explained on the basis of surface charge of the adsorbents and ionic chemistry of the solution.

Chromium ions may exist in the form of chromate (CrO_4^{2-}), dichromate ($Cr_2O_7^{2-}$), and hydrogen chromate ($HCrO_4^{-}$), depending upon the solution pH and chromate concentration.

The equilibrium between the chromate (CrO_4^{2-}) and dichromate ions ($Cr_2O_7^{2-}$) in aqueous solution is represented by (3) [15, 73].

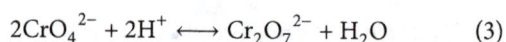

$$2CrO_4^{2-} + 2H^+ \longleftrightarrow Cr_2O_7^{2-} + H_2O \qquad (3)$$

Chromate (CrO_4^{2-}) ions are the dominant species at high pH values, while, at low pH, mainly dichromate ions ($Cr_2O_7^{2-}$) exist in the solution [78, 79].

At low pH, the high removal of Cr(VI) ions is attributed to the electrostatic interaction between the $Cr_2O_7^{2-}$ anions and positively charged CNTs surface. However, at high

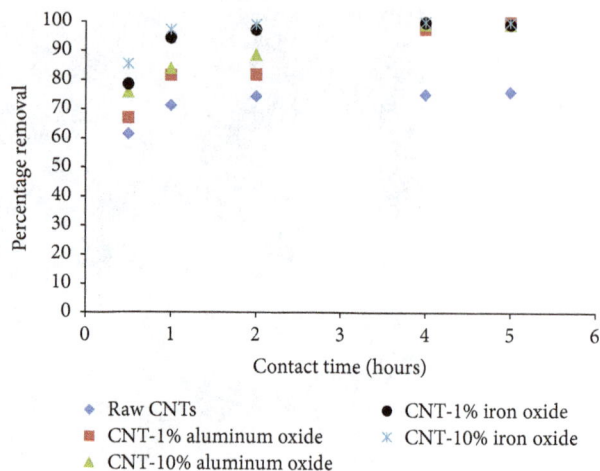

FIGURE 4: Effect of contact time on percentage removal of Cr(VI). (Initial concentration = 1 mg/L, agitation speed = 200 rpm, adsorbent dosage = 200 mg, pH = 6).

pH, surfaces of the CNTs carry more negative charges and repulsion between the CrO_4^{2-} ions and the CNTs surfaces resulted in lower removal of Cr(VI) ions. Furthermore, the low removal might also be due to competition between the OH^- and chromate (CrO_4^{2-}) ions over the limited adsorption sites as well as due to precipitation of $Cr(OH)_3$ that might occur at high pH (here at pH = 8) [73].

Surface impregnation of CNTs with metal oxide was observed to enhance the removal efficiency. The maximum removal of 87.8% was obtained for CNT with 10% aluminum oxide loading at pH 3. Raw CNTs were still able to remove almost 74% Cr(VI) ions at same pH and under similar experimental conditions. Although the maximum removal was obtained at pH 3, however, to evaluate the potential of the adsorbents in real water treatment applications, a pH value of 6 was selected for the remaining experiments.

Because the solution pH has a significant effect on the removal of Cr(VI) ions, we may deduce that the main mechanism is electrostatic interaction. The net surface charge of the adsorbent changes with pH and affects the removal of Cr(VI). In addition to electrostatic interaction, some physical adsorption of Cr(VI) ions is expected on the surfaces of the CNTs due to Van der Walls interactions. Studies also suggest that strong surface complexation and ion exchange are the main mechanisms involved during the adsorption of Cr(VI) ions on CNTs surface [80].

3.3. Effect of Contact Time. The experimental results presenting the effect of time on the removal of Cr(VI) ions by raw and metal oxide impregnated CNTs are shown in Figure 4. Contact time was varied from 0.5 to 5 hours while the solution pH, Cr(VI) initial concentration, adsorbent dosage, and agitation speed were kept constant at 6, 1 mg/L, 200 mg, and 200 rpm, respectively.

It is obvious that Cr(VI) ions removal has improved significantly as the contact time increased from 0.5 to 4 hours. No significant increase in removal was observed after 4 hours of contact time indicating the reach of equilibrium.

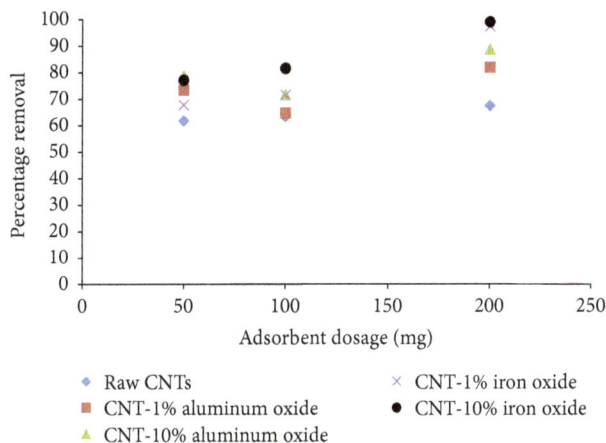

FIGURE 5: Effect of adsorbent dosage on percentage removal of Cr(VI) (initial concentration = 1 mg/L, agitation speed = 200 rpm, contact time = 2 hours, and pH = 6).

It was observed that CNTs impregnated with metal oxide were able to remove more than 97% of Cr(VI) ions after 2 hours of contact time (for CNTs impregnated with iron oxide) and almost 100% after 4 hours of contact time (for CNTs impregnated with both iron and aluminum oxides).

3.4. Effect of Adsorbent Dosage.

The effect of adsorbent dosage on the removal of Cr(VI) ions is depicted in Figure 5. The adsorbent dosage was varied from 50 to 200 mg, while solution pH, contact time, initial concentration of Cr(VI), and agitation speed were kept constant at 6, 2 hours, 1 mg/L, and 200 rpm, respectively.

A direct relationship was observed between the adsorbent dosage and the removal of Cr(VI) ions for all adsorbents. The removal was observed to increase with increase in the adsorbent dosage and the maximum removal was recorded at 200 mg dosage. With increase in the adsorbent dosage, the number of active sites increases; hence more Cr(VI) ions can be adsorbed onto the adsorbent surface. At 200 mg dosage, CNTs with 10% loading of iron oxide yielded a maximum removal of 99% of Cr(VI) ions, as compared to raw CNTs yielding about 67% removal under similar experimental conditions. These results confirmed that metal oxide loading has a significant effect on the removal efficiency of the raw CNTs.

3.5. Effect of Agitation Speed.

Agitation speed is an important parameter that effects and enhances the dispersion of the adsorbent in the solution and reduces the agglomeration. For the two loadings of metal oxides (1% and 10%) used in the present study, the CNTs were found to properly disperse in the solution and no significant agglomeration was observed. Figure 6 displays the effect of agitation speed on the removal of Cr(VI) ions by raw and metal oxides impregnated CNTs. The agitation speed was varied from 50 to 200 rpm, while the solution pH, initial concentration, adsorbent dosage, and contact time were kept constant at 6, 1 mg/L, 200 mg, and 2 hours, respectively. The removal of Cr(VI) ions was observed to increase with increase in agitation speed for all considered

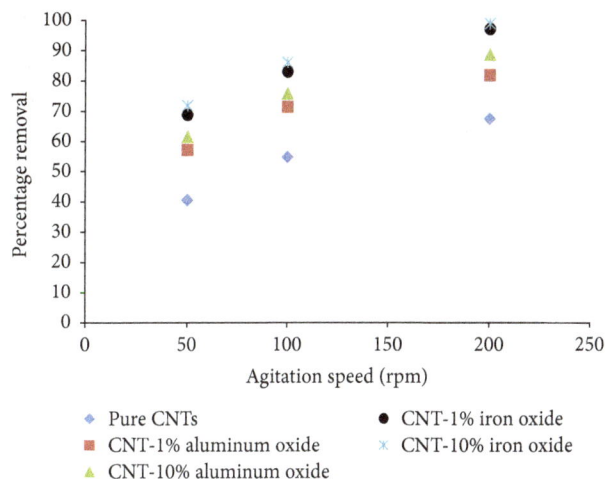

FIGURE 6: Effect of agitation speed on percentage removal of Cr(VI) ions (initial concentration = 1 mg/L, adsorbent dosage = 200 mg, contact time = 2 hours, and pH = 6).

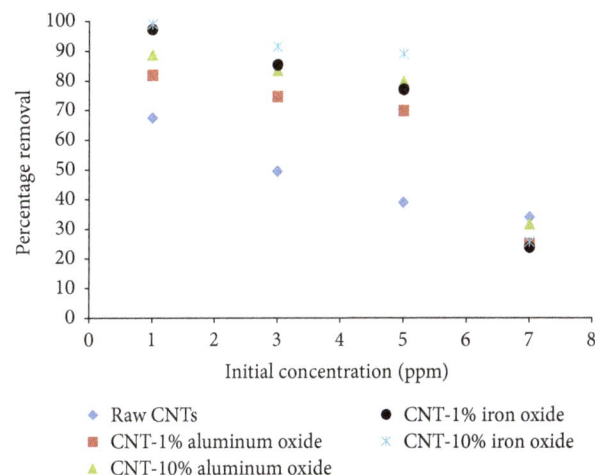

FIGURE 7: Effect of initial concentration on percentage removal of Cr(VI) (adsorbent dosage = 200 mg, contact time = 2 hours, agitation speed = 200 rpm, and pH = 6).

adsorbents. This is due to the fact that agitation facilitates effective diffusion of ions towards the adsorbent surface [73]. At 200 rpm speed, CNTs with 10% loading of iron oxide were able to remove 99% Cr(VI) ions.

3.6. Effect of Initial Concentration.

The removal of Cr(VI) ions was also dependent on the initial concentration of Cr(VI) as shown in Figure 7. The initial concentration was varied from 1 to 7 ppm, while the solution pH, agitation speed, adsorbent dosage, and contact time were kept constant at 6, 200 rpm, 200 mg, and 2 hours, respectively. The maximum removal was achieved at low dosage concentration and the removal was observed to decrease with increase in concentration for all adsorbents. This might be due to the fact that, at high concentration, the adsorption sites are saturated due to availability of surplus Cr(VI) ions. At 1 ppm dosage, a maximum removal 99% of Cr(VI) ions was achieved with CNTs with 10% loading of iron oxide.

TABLE 1: Parameters of Langmuir and Freundlich isotherm models for chromium.

Adsorbent		Freundlich		Langmuir	
	n	K_F (L/mg)	R^2	K_L (L/mg)	R^2
CNT-iron oxide	7.922564	0.628705	0.9980	−7.47535	0.9966
CNT-aluminum oxide	3.907029	0.571687	0.9996	−10.9559	0.9855
Raw CNTs	2.110755	0.291322	0.9975	0.756502	0.9859

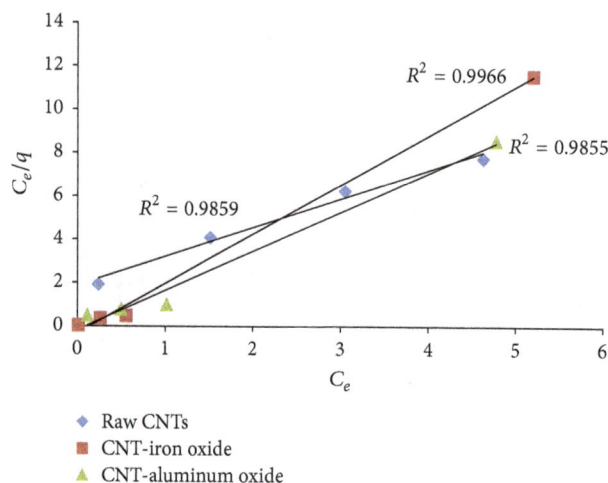

FIGURE 8: Langmuir adsorption model for Cr(VI).

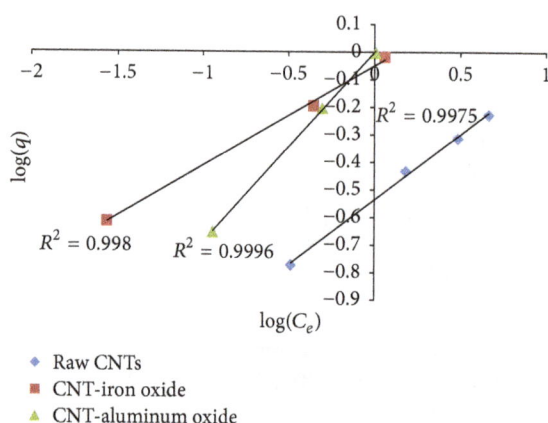

FIGURE 9: Freundlich adsorption model for Cr(VI).

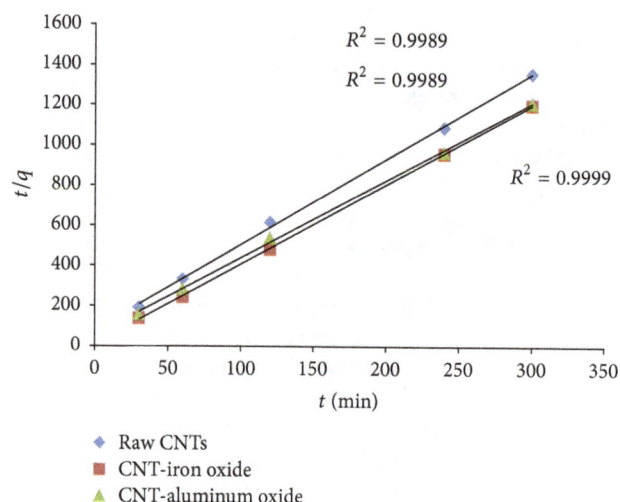

FIGURE 10: Pseudo-second-order kinetics for the adsorption of Cr(VI).

3.7. Freundlich and Langmuir Isotherm Models.

Adsorption equilibrium data was fitted by Langmuir and Freundlich models. Langmuir model best describes the monolayer adsorption while Freundlich model provides information about heterogeneous adsorption on adsorbent surface [81].

Representative equations of the isotherm models are presented below.

Langmuir isotherm model:

$$q_e = \frac{q_m K_L C_e}{1 + K_L C_e};$$ (4)

Freundlich isotherm model:

$$q_e = K_F C_e^{1/n},$$ (5)

where C_e and q_e are the concentrations of contaminants in water and in adsorbent at the adsorption equilibrium, respectively. q_m is the maximum adsorption capacity; K_L is the adsorption equilibrium constant of Langmuir model; K_F and n are Freundlich constants related to the adsorption capacity and surface heterogeneity of the adsorbents, respectively.

Figures 8 and 9 show Langmuir and Freundlich adsorption isotherm models for Cr(VI), respectively, while adsorption parameters and regression data of the models are presented in Table 1. It can be seen that both Langmuir and Freundlich isotherm models show a good fit for both raw and metal oxide impregnated CNTs. However, the value of regression coefficient (R^2) value for Freundlich isotherm model is slightly higher than Langmuir isotherm model.

3.8. Kinetics Modeling.

Adsorption kinetic is one of the most important factors that govern the solute uptake rate and represents the adsorption efficiency of the adsorbent. The pseudo-second-order model was used to model the kinetics of adsorption.

Representative equation of pseudo-second-order model is provided below:

$$\frac{t}{q_t} = \frac{1}{k_2 q_e^2} + \frac{t}{q_e}.$$ (6)

Figure 10 represents the fitting of experimental data with the pseudo-second-order model. Table 2 provides the results of

TABLE 2: Parameters of pseudo-second-order kinetic model for chromium.

Adsorbent	q_e (mg/g)	k_2 (mg g^{-1} min^{-1})	R^2
CNT-iron oxide	0.253062	0.534162	0.9999
CNT-aluminum oxide	0.259575	0.133789	0.9989
Raw CNTs	0.235297	0.109363	0.9989

the kinetics model fittings for the adsorption of Cr(VI) using raw and metal oxide impregnated CNTs.

It can be seen from Figure 10 and Table 2 that the correlation coefficient (R^2) of pseudo-second-order kinetic equation is sufficiently high for all the adsorbents. Therefore, the process of Cr(VI) removal using raw and metal oxide impregnated CNTs can be well described by the pseudo-second-order model.

4. Conclusion

Raw, iron oxide, and aluminum oxide impregnated carbon nanotubes (CNTs) were found to be effective adsorbents for the removal of Cr(VI) ions from aqueous solution. The removal of Cr(VI) ions was strongly dependent on pH, contact time, adsorbent dosage, and initial concentration of the Cr(VI) ions. Solution pH was found to be a critical parameter affecting the adsorption of Cr(VI) ions, in comparison with the other parameters. The removal of Cr(VI) ions was observed to decrease with increase in pH of the solution. It was observed that both CNTs impregnated with 10% of iron and aluminum oxides were able to remove almost 100% of Cr(VI) ions at solution pH 6, Cr(VI) initial concentration of 1 mg/L, adsorbent dosage of 200 mg, agitation speed of 200 rpm, and contact time of 4 hours. The prepared materials were found to exhibit high removal efficiency at pH 6 suggesting their great potential in real water treatment applications.

Acknowledgments

The authors would like to acknowledge the support provided by the Deanship of Scientific Research (DSR) at King Fahd University of Petroleum and Minerals (KFUPM) for funding this work through Project no. IN131065.

References

[1] D. Mohan and C. U. Pittman Jr., "Activated carbons and low cost adsorbents for remediation of tri- and hexavalent chromium from water," *Journal of Hazardous Materials*, vol. 137, no. 2, pp. 762–811, 2006.

[2] B. M. Weckhuysen, I. E. Wachs, and R. A. Schoonheydt, "Surface chemistry and spectroscopy of chromium in inorganic oxides," *Chemical Reviews*, vol. 96, no. 8, pp. 3327–3349, 1996.

[3] S. Kalidhasan, M. Ganesh, S. Sricharan, and N. Rajesh, "Extractive separation and determination of chromium in tannery effluents and electroplating waste water using tribenzylamine as the extractant," *Journal of Hazardous Materials*, vol. 165, no. 1–3, pp. 886–892, 2009.

[4] J. R. Rao, P. Thanikaivelan, K. J. Sreeram, and B. U. Nair, "Green route for the utilization of chrome shavings (chromium-containing solid waste) in tanning industry," *Environmental Science and Technology*, vol. 36, no. 6, pp. 1372–1376, 2002.

[5] S. Vasudevan, G. Sozhan, S. Mohan, R. Balaji, P. Malathy, and S. Pushpavanam, "Electrochemical regeneration of chromium containing solution from metal finishing industry," *Industrial and Engineering Chemistry Research*, vol. 46, no. 9, pp. 2898–2901, 2007.

[6] A. K. Bhattacharya, T. K. Naiya, S. N. Mandal, and S. K. Das, "Adsorption, kinetics and equilibrium studies on removal of Cr(VI) from aqueous solutions using different low-cost adsorbents," *Chemical Engineering Journal*, vol. 137, no. 3, pp. 529–541, 2008.

[7] D. Mohan, K. P. Singh, and V. K. Singh, "Removal of hexavalent chromium from aqueous solution using low-cost activated carbons derived from agricultural waste materials and activated carbon fabric cloth," *Industrial and Engineering Chemistry Research*, vol. 44, no. 4, pp. 1027–1042, 2005.

[8] D. Mohan, K. P. Singh, and V. K. Singh, "Trivalent chromium removal from wastewater using low cost activated carbon derived from agricultural waste material and activated carbon fabric cloth," *Journal of Hazardous Materials*, vol. 135, no. 1–3, pp. 280–295, 2006.

[9] P. Miretzky and A. F. Cirelli, "Cr(VI) and Cr(III) removal from aqueous solution by raw and modified lignocellulosic materials: a review," *Journal of Hazardous Materials*, vol. 180, no. 1–3, pp. 1–19, 2010.

[10] F. C. Richard and A. C. M. Bourg, "Aqueous geochemistry of chromium: a review," *Water Research*, vol. 25, no. 7, pp. 807–816, 1991.

[11] J. Zhu, H. Gu, J. Guo et al., "Mesoporous magnetic carbon nanocomposite fabrics for highly efficient Cr(VI) removal," *Journal of Materials Chemistry A*, vol. 2, no. 7, pp. 2256–2265, 2014.

[12] B. Qiu, C. Xu, D. Sun et al., "Polyaniline coating on carbon fiber fabrics for improved hexavalent chromium removal," *RSC Advances*, vol. 4, no. 56, pp. 29855–29865, 2014.

[13] B. Qiu, C. Xu, D. Sun et al., "Polyaniline coated ethyl cellulose with improved hexavalent chromium removal," *ACS Sustainable Chemistry and Engineering*, vol. 2, no. 8, pp. 2070–2080, 2014.

[14] T. Karthikeyan, S. Rajgopal, and L. R. Miranda, "Chromium(VI) adsorption from aqueous solution by *Hevea Brasilinesis* sawdust activated carbon," *Journal of Hazardous Materials*, vol. 124, no. 1–3, pp. 192–199, 2005.

[15] C. Xu, B. Qiu, H. Gu et al., "Synergistic interactions between activated carbon fabrics and toxic hexavalent chromium," *ECS Journal of Solid State Science and Technology*, vol. 3, no. 3, pp. M1–M9, 2014.

[16] K. K. Krishnani and S. Ayyappan, "Heavy metals remediation of water using plants and lignocellulosic agrowastes," *Reviews of*

Environmental Contamination and Toxicology, vol. 188, pp. 59–84, 2006.

[17] M. Aliabadi, I. Khazaei, H. Fakhraee, and M. T. H. Mousavian, "Hexavalent chromium removal from aqueous solutions by using low-cost biological wastes: equilibrium and kinetic studies," *International Journal of Environmental Science and Technology*, vol. 9, no. 2, pp. 319–326, 2012.

[18] D. E. Kimbrough, Y. Cohen, A. M. Winer, L. Creelman, and C. Mabuni, "A critical assessment of chromium in the environment," *Critical Reviews in Environmental Science and Technology*, vol. 29, no. 1, pp. 1–46, 1999.

[19] E. Salazar, M. I. Ortiz, A. M. Urtiaga, and J. A. Irabien, "Equilibrium and kinetics of chromium(VI) extraction with Aliquat 336," *Industrial and Engineering Chemistry Research*, vol. 31, no. 6, pp. 1516–1522, 1992.

[20] K. A. Matis and P. Mavros, "Recovery of metals by ion flotation from dilute aqueous solutions," *Separation and Purification Method*, vol. 20, no. 1, pp. 1–48, 1991.

[21] F. Akbal and S. Camcı, "Copper, chromium and nickel removal from metal plating wastewater by electrocoagulation," *Desalination*, vol. 269, no. 1–3, pp. 214–222, 2011.

[22] G. Tiravanti, D. Petruzzelli, and R. Passino, "Pretreatment of tannery wastewaters by an ion exchange process for Cr(III) removal and recovery," *Water Science and Technology*, vol. 36, no. 2–3, pp. 197–207, 1997.

[23] S. Rengaraj, K.-H. Yeon, and S.-H. Moon, "Removal of chromium from water and wastewater by ion exchange resins," *Journal of Hazardous Materials*, vol. 87, no. 1–3, pp. 273–287, 2001.

[24] S. Rengaraj, C. K. Joo, Y. Kim, and J. Yi, "Kinetics of removal of chromium from water and electronic process wastewater by ion exchange resins: 1200H, 1500H and IRN97H," *Journal of Hazardous Materials*, vol. 102, no. 2–3, pp. 257–275, 2003.

[25] D. Petruzzelli, R. Passino, and G. Tiravanti, "Ion exchange process for chromium removal and recovery from tannery wastes," *Industrial and Engineering Chemistry Research*, vol. 34, no. 8, pp. 2612–2617, 1995.

[26] C. A. Kozlowski and W. Walkowiak, "Removal of chromium(VI) from aqueous solutions by polymer inclusion membranes," *Water Research*, vol. 36, no. 19, pp. 4870–4876, 2002.

[27] H. F. Shaalan, M. H. Sorour, and S. R. Tewfik, "Simulation and optimization of a membrane system for chromium recovery from tanning wastes," *Desalination*, vol. 141, no. 3, pp. 315–324, 2001.

[28] Ihsanullah, A. Abbas, A. M. Al-Amer et al., "Heavy metal removal from aqueous solution by advanced carbon nanotubes: critical review of adsorption applications," *Separation and Purification Technology*, vol. 157, pp. 141–161, 2016.

[29] L. Monser and N. Adhoum, "Modified activated carbon for the removal of copper, zinc, chromium and cyanide from wastewater," *Separation and Purification Technology*, vol. 26, no. 2–3, pp. 137–146, 2002.

[30] D. Park, Y.-S. Yun, J. Y. Kim, and J. M. Park, "How to study Cr(VI) biosorption: use of fermentation waste for detoxifying Cr(VI) in aqueous solution," *Chemical Engineering Journal*, vol. 136, no. 2–3, pp. 173–179, 2008.

[31] M. Ulmanu, E. Marañón, Y. Fernández, L. Castrillón, I. Anger, and D. Dumitriu, "Removal of copper and cadmium ions from diluted aqueous solutions by low cost and waste material adsorbents," *Water, Air, and Soil Pollution*, vol. 142, no. 1–4, pp. 357–373, 2003.

[32] M. Aliabadi, K. Morshedzadeh, and H. Soheyli, "Removal of hexavalent chromium from aqueous solution by lignocellulosic solid wastes," *International Journal of Environmental Science and Technology*, vol. 3, no. 3, pp. 321–325, 2006.

[33] V. K. Singh and P. N. Tiwari, "Removal and recovery of chromium (VI) from industrial waste water," *Journal of Chemical Technology and Biotechnology*, vol. 69, no. 3, pp. 376–382, 1997.

[34] S. K. Srivastava, R. Tyagi, and N. Pant, "Adsorption of heavy metal ions on carbonaceous material developed from the waste slurry generated in local fertilizer plants," *Water Research*, vol. 23, no. 9, pp. 1161–1165, 1989.

[35] E. Demirbas, M. Kobya, E. Senturk, and T. Ozkan, "Adsorption kinetics for the removal of chromium (VI) from aqueous solutions on the activated carbons prepared from agricultural wastes," *Water SA*, vol. 30, no. 4, pp. 533–539, 2004.

[36] D. D. Das, R. Mahapatra, J. Pradhan, S. N. Das, and R. S. Thakur, "Removal of Cr(VI) from aqueous solution using activated cow dung carbon," *Journal of Colloid and Interface Science*, vol. 232, no. 2, pp. 235–240, 2000.

[37] S. Bosinco, J. Roussy, E. P. Guibal, and P. L. Cloirec, "Interaction mechanisms between hexavalent chromium and corncob," *Environmental Technology*, vol. 17, no. 1, pp. 55–62, 1996.

[38] P. Candela, J. M. M. Martinez, and R. T. Macia, "Chromium(VI) removal with activated carbons," *Water Research*, vol. 29, pp. 2174–2180, 1995.

[39] A. M. El-Kamash, A. A. Zaki, and M. A. El Geleel, "Modeling batch kinetics and thermodynamics of zinc and cadmium ions removal from waste solutions using synthetic zeolite A," *Journal of Hazardous Materials*, vol. 127, no. 1–3, pp. 211–220, 2005.

[40] G. Cimino, A. Passerini, and G. Toscano, "Removal of toxic cations and Cr(VI) from aqueous solution by hazelnut shell," *Water Research*, vol. 34, no. 11, pp. 2955–2962, 2000.

[41] M. Kobya, "Adsorption, kinetic and equilibrium studies of Cr(VI) by hazelnut shell activated carbon," *Adsorption Science and Technology*, vol. 22, no. 1, pp. 51–64, 2004.

[42] K. K. Singh, R. Rastogi, and S. H. Hasan, "Removal of cadmium from wastewater using agricultural waste 'rice polish'," *Journal of Hazardous Materials*, vol. 121, no. 1–3, pp. 51–58, 2005.

[43] D. C. Sharma and C. F. Forster, "Removal of hexavalent chromium using sphagnum moss peat," *Water Research*, vol. 27, no. 7, pp. 1201–1208, 1993.

[44] S. H. Lee, C. H. Jung, H. Chung, M. Y. Lee, and J.-W. Yang, "Removal of heavy metals from aqueous solution by apple residues," *Process Biochemistry*, vol. 33, no. 2, pp. 205–211, 1998.

[45] C. K. Lee, K. S. Low, and K. L. Kek, "Removal of chromium from aqueous solution," *Bioresource Technology*, vol. 54, no. 2, pp. 183–189, 1995.

[46] K. S. Low, C. K. Lee, and A. Y. Ng, "Column study on the sorption of Cr(VI) using quaternized rice hulls," *Bioresource Technology*, vol. 68, no. 2, pp. 205–208, 1999.

[47] W. Jianlong, "Removal of Cr(VI) from aqueous solution by coal flyash adsorption I: characteristics of chromium adsorption on flyash," *Toxicological and Environmental Chemistry*, vol. 68, no. 1–2, pp. 53–62, 1999.

[48] M. Dakiky, M. Khamis, A. Manassra, and M. Mer'eb, "Selective adsorption of chromium(VI) in industrial wastewater using low-cost abundantly available adsorbents," *Advances in Environmental Research*, vol. 6, no. 4, pp. 533–540, 2002.

[49] N. K. Hamadi, X. D. Chen, M. M. Farid, and M. G. Q. Lu, "Adsorption kinetics for the removal of chromium(VI) from aqueous solution by adsorbents derived from used tyres and sawdust," *Chemical Engineering Journal*, vol. 84, no. 2, pp. 95–105, 2001.

[50] K. Selvi, S. Pattabhi, and K. Kadirvelu, "Removal of Cr(VI) from aqueous solution by adsorption onto activated carbon," *Bioresource Technology*, vol. 80, no. 1, pp. 87–89, 2001.

[51] S. S. Shukla, J. Y. Li, K. L. Dorris, and A. Shukla, "Removal of nickel from aqueous solutions by sawdust," *Journal of Hazardous Materials*, vol. 121, no. 1–3, pp. 243–246, 2005.

[52] C. Selomulya, V. Meeyoo, and R. Amal, "Mechanisms of Cr(VI) removal from water by various types of activated carbons," *Journal of Chemical Technology and Biotechnology*, vol. 74, no. 2, pp. 111–122, 1999.

[53] H. Farrah and W. F. Pickering, "The sorption of lead and cadmium species by clay minerals," *Australian Journal of Chemistry*, vol. 30, no. 7, pp. 1417–1422, 1977.

[54] W. T. Tan, S. T. Ooi, and C. K. Lee, "Removal of chromium(VI) from solution by coconut husk and palm pressed fibres," *Environmental Technology*, vol. 14, no. 3, pp. 277–282, 1993.

[55] K. Periasamy, K. Srinivasan, and P. R. Muruganan, "Studies on chromium(VI) removal by activated ground nut husk carbon," *Indian Journal of Environmental Health*, vol. 33, pp. 433–439, 1991.

[56] M. S. Mansour, M. E. Ossman, and H. A. Farag, "Removal of Cd (II) ion from waste water by adsorption onto polyaniline coated on sawdust," *Desalination*, vol. 272, no. 1–3, pp. 301–305, 2011.

[57] M. Ayoma, T. Sugiyama, S. Doi, N. S. Cho, and H. E. Kim, "Removal of Hexavalent chromium from dilute aqueous solution by coniferous leaves," *Holzforschung*, vol. 53, no. 4, pp. 365–368, 1999.

[58] D. C. Sharma and C. F. Forster, "The treatment of chromium wastewaters using the sorptive potential of leaf mould," *Bioresource Technology*, vol. 49, no. 1, pp. 31–40, 1994.

[59] L. Dupont and E. Guillon, "Removal of hexavalent chromium with a lignocellulosic substrate extracted from wheat bran," *Environmental Science and Technology*, vol. 37, no. 18, pp. 4235–4241, 2003.

[60] Z. Reddad, C. Gerente, Y. Andres, and P. Le Cloirec, "Adsorption of several metal ions onto a low-cost biosorbent: kinetic and equilibrium studies," *Environmental Science and Technology*, vol. 36, no. 9, pp. 2067–2073, 2002.

[61] A. C. A. Da Costa and F. P. De França, "Cadmium uptake by biosorbent seaweeds: adsorption isotherms and some process conditions," *Separation Science and Technology*, vol. 31, no. 17, pp. 2373–2393, 1996.

[62] Y. Nakano, M. Tanaka, Y. Nakamura, and M. Konno, "Removal and recovery system of hexavalent chromium from waste water by tannin gel particles," *Journal of Chemical Engineering of Japan*, vol. 33, no. 5, pp. 747–752, 2000.

[63] D. Kratochvil, P. Pimentel, and B. Volesky, "Removal of trivalent and hexavalent chromium by seaweed biosorbent," *Environmental Science and Technology*, vol. 32, no. 18, pp. 2693–2698, 1998.

[64] M. E. A. Ali, "Synthesis and adsorption properties of chitosan-CDTA-GO nanocomposite for removal of hexavalent chromium from aqueous solutions," *Arabian Journal of Chemistry*, In press.

[65] N. Calace, A. Di Muro, E. Nardi, B. M. Petronio, and M. Pietroletti, "Adsorption isotherms for describing heavy-metal retention in paper mill sludges," *Industrial and Engineering Chemistry Research*, vol. 41, no. 22, pp. 5491–5497, 2002.

[66] C.-H. Weng, C. P. Huang, H. E. Allen, and P. F. Sanders, "Cr(VI) adsorption onto hydrous concrete particles from groundwater," *Journal of Environmental Engineering*, vol. 127, no. 12, pp. 1124–1131, 2001.

[67] S. S. Ahluwalia and D. Goyal, "Removal of heavy metals by waste tea leaves from aqueous solution," *Engineering in Life Sciences*, vol. 5, no. 2, pp. 158–162, 2005.

[68] V. Sarin and K. K. Pant, "Removal of chromium from industrial waste by using eucalyptus bark," *Bioresource Technology*, vol. 97, no. 1, pp. 15–20, 2006.

[69] M. H. Dehghani, D. Sanaei, I. Ali, and A. Bhatnagar, "Removal of chromium(VI) from aqueous solution using treated waste newspaper as a low-cost adsorbent: kinetic modeling and isotherm studies," *Journal of Molecular Liquids*, vol. 215, pp. 671–679, 2016.

[70] A. S. K. Kumar, S. S. Kakan, and N. Rajesh, "A novel amine impregnated graphene oxide adsorbent for the removal of hexavalent chromium," *Chemical Engineering Journal*, vol. 230, pp. 328–337, 2013.

[71] Ihsanullah, F. A. Al-Khaldi, B. Abusharkh et al., "Adsorptive removal of cadmium(II) ions from liquid phase using acid modified carbon-based adsorbents," *Journal of Molecular Liquids*, vol. 204, pp. 255–263, 2015.

[72] Ihsanullah, H. A. Asmaly, T. A. Saleh, T. Laoui, V. K. Gupta, and M. A. Atieh, "Enhanced adsorption of phenols from liquids by aluminum oxide/carbon nanotubes: comprehensive study from synthesis to surface properties," *Journal of Molecular Liquids*, vol. 206, pp. 176–182, 2015.

[73] Ihsanullah, F. A. Al-Khaldi, B. Abu-Sharkh et al., "Effect of acid modification on adsorption of hexavalent chromium (Cr(VI)) from aqueous solution by activated carbon and carbon nanotubes," *Desalination and Water Treatment*, vol. 57, no. 16, pp. 7232–7244, 2016.

[74] H. A. Asmaly, B. Abussaud, Ihsanullah, T. A. Saleh, V. K. Gupta, and M. A. Atieh, "Ferric oxide nanoparticles decorated carbon nanotubes and carbon nanofibers: from synthesis to enhanced removal of phenol," *Journal of Saudi Chemical Society*, vol. 19, no. 5, pp. 511–520, 2015.

[75] H. A. Asmaly, B. Abussaud, Ihsanullah et al., "Evaluation of micro- and nano-carbon-based adsorbents for the removal of phenol from aqueous solutions," *Toxicological and Environmental Chemistry*, vol. 97, no. 9, pp. 1164–1179, 2015.

[76] A. Abbas, B. A. Abussaud, Ihsanullah, N. A. H. Al-Baghli, M. Khraisheh, and M. A. Atieh, "Benzene removal by iron oxide nanoparticles decorated carbon nanotubes," *Journal of Nanomaterials*, vol. 2016, Article ID 5654129, 10 pages, 2016.

[77] Ihsanullah, A. M. Al Amer, T. Laoui et al., "Fabrication and antifouling behaviour of a carbon nanotube membrane," *Materials and Design*, vol. 89, pp. 549–558, 2016.

[78] J. Zhu, S. Wei, H. Gu et al., "One-pot synthesis of magnetic graphene nanocomposites decorated with core@double-shell nanoparticles for fast chromium removal," *Environmental Science and Technology*, vol. 46, no. 2, pp. 977–985, 2012.

[79] C. J. Lin, S. L. Wang, P. M. Huang et al., "Chromate reduction by zero-valent Al metal as catalyzed by polyoxometalate," *Water Research*, vol. 43, no. 20, pp. 5015–5022, 2009.

[80] J. Hu, C. Chen, X. Zhu, and X. Wang, "Removal of chromium from aqueous solution by using oxidized multiwalled carbon nanotubes," *Journal of Hazardous Materials*, vol. 162, no. 2-3, pp. 1542–1550, 2009.

[81] H. Ding, X. Li, J. Wang, X. Zhang, and C. Chen, "Adsorption of chlorophenols from aqueous solutions by pristine and surface functionalized single-walled carbon nanotubes," *Journal of Environmental Sciences*, vol. 43, pp. 187–198, 2016.

15

A DFT Study of Structural and Bonding Properties of Complexes Obtained from First-Row Transition Metal Chelation by 3-Alkyl-4-phenylacetylamino-4,5-dihydro-1H-1,2,4-triazol-5-one and its Derivatives

Hubert Jean Nono,[1] **Désiré Bikélé Mama,**[2] **Julius Numbonui Ghogomu,**[1] **and Elie Younang**[3]

[1]*Department of Chemistry, Faculty of Science, University of Dschang, P.O. Box 67, Dschang, Cameroon*
[2]*Department of Chemistry, Faculty of Science, University of Douala, P.O. Box 24157, Douala, Cameroon*
[3]*Department of Inorganic Chemistry, Faculty of Science, University of Yaoundé I, P.O. Box 812, Yaoundé, Cameroon*

Correspondence should be addressed to Desire Bikele Mama; bikelemama@yahoo.fr
and Julius Numbonui Ghogomu; ghogsjuju@hotmail.com

Academic Editor: Konstantinos Tsipis

Density functional calculations were used to explore the complexation of 3-alkyl-4-phenylacetylamino-4,5-dihydro-1h-1,2,4-triazol-5-one (ADPHT) derivatives by first-row transition metal cations. Neutral ADPHT ligand and mono deprotonated ligands have been used. Geometry optimizations have been performed in gas-phase and solution-phase (water, benzene, and N,N-dimethylformamide (DMF)) with B3LYP/Mixed I (LanL2DZ for metal atom and 6-31+G(d,p) for C, N, O, and H atoms) and with B3LYP/Mixed II (6-31G(d) for metal atom and 6-31+G(d,p) for C, N, O, and H atoms) especially in the gas-phase. Single points have also been carried out at CCSD(T) level. The B3LYP/Mixed I method was used to calculate thermodynamic energies (energies, enthalpies, and Gibb energies) of the formation of the complexes analyzed. The B3LYP/Mixed I complexation energies in the gas phase are therefore compared to those obtained using B3LYP/Mixed II and CCSD(T) calculations. Our results pointed out that the deprotonation of the ligand increases the binding affinity independently of the metal cation used. The topological parameters yielded from Quantum Theory of Atom in Molecules (QTAIM) indicate that metal-ligand bonds are partly covalent. The significant reduction of the proton affinity (PA) observed when passing from ligands to complexes in gas-phase confirms the notable enhancement of antioxidant activities of neutral ligands.

1. Introduction

In recent years, the repercussion of free radicals and reactive oxygen species (ROS) in neurodegenerative disorder is more sensitive [1–4]. The contribution of these ROS to the pathophysiology of myocardial reperfusion damage. These ROS can be oxygen-centered radicals [5] or oxygen-centered nonradicals [6]. The removal of electrons from cellular membranes by these ROS and the reaction between these latter ones and proteins [7] provoke the alteration of the structures of these membranes and proteins. Such alterations justify the frailty of these cellular membranes that expose them to be attacked by invaders (viruses and bacteria).

Nevertheless, each cell is naturally equipped by defense systems against any destructive effect of ROS. This statute of protective mechanism against ROS in humans is attributed to antioxidant molecules [8, 9]. In general, antioxidant molecules (tocopherol (vitamin E), ascorbic acid (vitamin C), carotenoids, flavonoids, and polyphenols) prevent the proliferation of free radical reactions in all cell membranes. This explains the emergence of studies on the investigation of antioxidant activities of biologically active compounds. This has led to increased pressure on the need for the newer molecules which may have potentials to curb the spread of this problem. Particularly, 1,2,4-triazole and its derivatives have been reported to possess antioxidant activities [10–13]. The 1,2,4-triazole derivatives have also been known to possess many biological activities (antifungal, analgesic, antiviral, anti-inflammatory, antitumor, anti-HIV properties, etc.) [14, 15]. It is worthwhile to mention that 1,2,4-triazoles have been prepared by different methods. The cyclodehydration of acylthiosemicarbazides with a variety of basic agents is the most common method used. The literature survey revealed that acylthiosemicarbazides are the key intermediates used in the synthesis of 1,2,4-triazol [16, 17].

Our previous research was focused on the theoretical analysis of antioxidant mechanisms of 1,2,4-triazole derivatives [18], more precisely of 3-alkyl-4-phenylacetylamino-4,5-dihydro-1H-1,2,4-triazol-5-one (ADPHT) derivatives. The calculated thermodynamic properties descriptors calculated in gas and solution-phases were Hydrogen Atom Transfer (HAT), Single Electron Transfer-Proton Transfer (SET-PT), and Sequential Proton-Loss Electron Transfer (SPLET) mechanism. Results indicated that, thermodynamically, HAT mechanism is the most predominant process in the gas-phase. But, in solvents (2-propanol, acetonitrile, DMF and water), the SPLET mechanism has shown to be more preferred. It has been shown that the removal of the metals by metal-chelating process causes some problems of toxicity due to the induced charge change [19]. But the examination of the literature demonstrates that the transition metal chelation of the titled ligands has not been done either experimentally or theoretically.

The goal of this work is to do a comprehensive density functional theory (DFT) study on the first-row transition metal (II) chelation by ADPHT derivatives, using both the neutral and mono deprotonated forms of ligands. We have studied the coordination abilities for the first-row transition metal (II) cations used (Fe^{2+}, Ni^{2+}, Cu^{2+}, and Zn^{2+}). The authors evaluated the proton affinity (PA) of each complex and therefore analyzed the comparative impact of the metal chelation on the antioxidant activities. The proton affinity free energy (PAFE) has also been taken into account. The metal interaction is further studied on the basis of NBO charges. The replacement of M^{2+} by the M^+ leads to the evaluation of the influence of the metal charge on analyzed properties. For this last point, the authors have considered only the copper atom as metal. Estimation of the effect of solvents (water, benzene, and N,N-dimethylformamide (DMF)) on the calculated structural parameters has also been done.

2. Computational Details and Theoretical Background

2.1. Computational Details.

All calculations were performed using Gaussian 09 program package [20]. The optimization of gas-phase structure of each molecular system was obtained using DFT method with B3LYP [21] functional. For M^{2+} (M^{2+} = Fe^{2+}, Ni^{2+}, Cu^{2+}, and Zn^{2+}), the authors used the nonrelativistic effective core potential (ECP) LanL2DZ [22]

for metal atom in combination with 6-31+G(d,p) for C, N, O, and H atoms. This generically made basis set is denoted as Mixed I. The importance of using ECPs for transition metal complexes has been emphasized by prior researches [23, 24]. The geometry optimization is followed by vibrational frequency calculations. All of these calculations were carried out in vacuum and in three solvents (water, benzene, and DMF (N,N-dimethylformamide)). The integral equation formalism of the polarized continuum [25, 26] has been taken into account to analyze the solvent effect. To evaluate the influence of the basis set on geometries, an additional geometrical optimization of the gas-phase structures followed by frequency calculations using a new mixed basis set [M^{2+} (Fe^{2+}, Ni^{2+}, Cu^{2+}, and Zn^{2+}) = 6-31G(d) and 6-31+G(d,p) = C, N, O, and H] subsequently denoted as Mixed II has been achieved.

2.2. Theoretical Background

2.2.1. Thermodynamic Energies of Coordination Abilities and Deprotonation of Complexes.

The coordination ability of various cations (D_e) (dissociation energy of the complex noted c) is defined according to (1):

$$D_e = E_c - (E_l + E_m). \tag{1}$$

E_c, E_l, and E_m are, respectively, the energy of the complex, ligand, and metal. Firstly, D_e values are estimated after geometrical optimization calculations at B3LYP/Mixed I in the gas-phase and in various solvents and reevaluated at B3LYP/Mixed II level in the gas-phase Single-point calculations at the CCSD (T) have been carried out to reproduce the gas-phase D_e. The levels of theory have shown to reproduce satisfactorily the first-row transition metal binding energy of glycine and its derivatives [27, 28]. The Metal Ion Affinity (MIA) [29, 30] was assumed to be the negative of the reaction enthalpy (ΔH^0_{298}) defined in (2). We have determined the complexation free energy according to (3):

$$MIA = -\Delta H^0_{298} = -\left[H^0_{298c} - \left(H^0_{298l} + H^0_{298m} \right) \right] \tag{2}$$

$$\Delta G^0_{298} = G^0_{298c} - \left(G^0_{298l} + G^0_{298m} \right). \tag{3}$$

2.2.2. NBO Analysis.

The stabilization energy E_2 associated with $i \rightarrow j$ delocalization for a donor NBO (i) and an acceptor NBO (j) is calculated according to

$$E_2 = q_{ij} \frac{F^2(i, j)}{\varepsilon_j - \varepsilon_i}, \tag{4}$$

where q_{ij} represents the ith donor orbital occupancy and ε_i and ε_j are diagonal elements and $F(i, j)$ off-diagonal elements, respectively, linked to NBO Fock matrix [31].

The thermodynamic energies (proton affinity (PA) and proton affinity fee energy (PAFE)) relative to the deprotonation of each optimized complex are determined as follows:

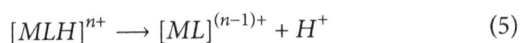

$$[MLH]^{n+} \longrightarrow [ML]^{(n-1)+} + H^+ \tag{5}$$

$$PA = H\left(H^+\right) + H\left(\left[ML^{(n-1)+}\right]\right) \\ - H\left(\left[MLH^{n+}\right]\right) \tag{6}$$

$$PAFE = G\left(H^+\right) + G\left(\left[ML^{(n-1)+}\right]\right) \\ - G\left(\left[MLH^{n+}\right]\right). \tag{7}$$

$[MLH]$, $[ML]^+$, and H^+ are, respectively, complex and deprotonated complexes and the proton. $H(Y)$ is the enthalpy of species Y ($Y = MLH, ML^+$, and H^+). Solvent contribution was determined using an integral equation continuum model (IEF-PCM) method [25, 26].

2.2.3. Atoms in Molecules Analysis Theory.

The Quantum Theory of Atom in Molecules (QTAIM) proposed by Bader [32] performed as implemented in Multiwfn [33] was to analyze the nature of all metal-ligand bonds. A logical approximation of their relative energies eases the specification of the original nature of metal-ligand bonds. The indicators of metal-ligand bond used are the electron densities $\rho(r)$ and their Laplacians $\nabla^2\rho(r)$ calculated at the bond critical points (BCPs). The local kinetic electron energy density and the potential energy density $v(r)$ are then defined, respectively, in (8) and (9) [32, 34]. Consider

$$G(r) = \frac{3}{10}(3\pi)^{2/3}\rho(r)^{5/3} + \frac{1}{6}\nabla^2\rho(r) \tag{8}$$

$$v(r) = \frac{\hbar^2}{4m}\nabla^2\rho(r) - 2G(r). \tag{9}$$

According to the sign of the Laplacian of the electron density ($\nabla^2\rho(r)$), the metal-ligand interactions are covalent and electrostatic, if $\nabla^2\rho(r)$ are, respectively, negative and positive. A further instrument for estimation of the nature of the metal-ligand is the fraction $-G(r)/v(r)$. From this descriptor, the metal-ligand bond is noncovalent or partly covalent in nature, if $-G(r)/v(r) > 1$ and $0.5 < -G(r)/v(r) < 1$, respectively. This ration combined with the electron density ($\nabla^2\rho(r)$) is useful to analyze the intermediate interactions and closed shell interaction: $(-G(r)/v(r) < 1$ and $\nabla^2\rho(r) < 0)$) and $(-G(r)/v(r) > 1$ and $\nabla^2\rho(r) < 0)$), respectively.

This analysis is using B3LYP/Mixed II optimized structures in the gas-phase. QTAIM analysis was then performed as implemented in Multiwfn [32].

3. Results and Discussion

We have optimized the M-ADPHT complexes using all the possible coordination modes. These optimizations yield a unique coordination mode (O_3, O_2) as shown in Figure 1 for neutral or deprotonated ADPHT derivatives.

3.1. Geometrical Details.

All the structures have been optimized without any symmetric constraint. The relevant geometrical parameters of M-ADPHT complexes with neutral ligands, labeled according to convention given in Figure 1, are compiled in Table 1. The optimization generally yields M-O_2 bond distances longer, compared to those of M-O_3 bonds. The bond length difference was in the following range: 0.009–0.077 Å. We attributed such a difference to the fact that the C_3=O_3 carbonyl group is connected to two N_1 and N_2 nitrogen atoms. The induced cumulative electron donating effect of these two neighboring nitrogen atoms then increases the electron density around the O_3 oxygen atom. This atom is consequently more nucleophile than the O_2 homolog which is near only one N_3 nitrogen atom. Our results showed that the M-O_i (i = 2, 3) decreased slightly with the substitution of the hydrogen by donor alkyl group (R = CH_3, C_2H_5). Independently of the substituent, the Cu-O_i (i = 2, 3) distances are the longest in gas-phase. The replacement of Cu^{2+} (d^9) by Cu^+ (d^{10}) leads to the shortening of Cu-O_2 bond lengths but also to the lengthening of Cu-O_3 bond distances. This fact originates predominantly from the interaction between the doubly occupied metal 3d orbital and the lone pairs of donor oxygen atoms, which significantly increases the metal-ligand repulsion in the Cu^+ (d^{10})-complexes. This does not occur for Cu^{2+} (d^9) for which the 3d orbital remains singly occupied. Our results on Cu^+ (d^{10})-complexes are in line with similar theoretical works on monovalent metal cation (Co^+, Ni^+, and Cu^+)-ligands (glycine [35], water [35, 36], ammonia [37, 38], and adenine [39]) complexes. We concluded that this metal-ligand repulsion in Cu^+ (d^{10})-complexes is more pronounced for the more nucleophile oxygen atom O_3. The comparison of the bond distances of Cu^+-complexes with those of Cu^2-complexes (Table 1) reveals the fact that the effects of this substitution are only limited to metal-ligand bond distances. The dissimilarity observed between distances of bonds of ADPHT complexes and those of isolated ligand displays the significant activation of adjacent bonds by the metal cation. For instance, the metal chelation of ligand **a** by Cu^{2+} augments the C_3-O_3 bond distance from 1.216 to 1.250 Å.

The optimized O_2-M^{2+}-O_3 angle in all M^{2+}-complexes in gas-phase varies from 98.3 to 106.4°. The dihedral angles Φ [N_3-C_2-O_2-M^{2+}] and Φ [N_1-N_3-C_2-O_2] were, respectively, in the following range: −57.3–59.1 and −35.6–53.3°. This fact renders chelate rings of M-ADPHT (M = Fe^{2+}, Ni^{2+}, and Cu^{2+}) nonplanar. In addition, the values obtained for the dihedral angles Φ [N_1-N_3-C_2-O_2] revealed that the N_1, N_3, C_2, and O_2 atoms are not located in the same plane (Table 1). Our B3LYP data revealed that the benzene and 1,2,4-triazole ring remained planar (Figure 1S) in the supporting information, available online at https://doi.org/10.1155/2017/5237865). In the same vein, the calculated values of the dihedral angle Φ [N_3-C_2-C_1-C_5] indicated the fact that the C_5 atom is almost in the plane containing N3, C_2, and C1 atoms, which is nearly perpendicular to the benzene ring.

Contrary to the results obtained in gas-phase, the optimization after the solvation of the molecular system yields M-O_3 bond distances longer, compared to those of M-O_2

TABLE 1: Selected B3LYP/Mixed I level: bond lengths (Å), bond angles (degree), and dihedral angles (degree) for neutral ADPHT ligand-metal complexes in various media.

Parameters	Fe^{2+}			Ni^{2+}			Cu^{2+}			Cu^+			Zn^{2+}		
	1a	1b	1c	2a	2b	2c	3a	3b	3c	4a	4b	4c	5a	5b	5c
gas															
M^{2+}-O_2	1.840	1.836	1.835	1.837	1.836	1.835	2.061	2.046	2.040	1.998	1.999	1.999	1.906	1.906	1.905
M^{2+}-O_3	1.818	1.810	1.806	1.827	1.820	1.819	1.969	1.969	1.969	1.995	1.986	1.986	1.899	1.898	1.899
O_2-M^{2+}-O_3	105.4	105.8	105.8	102.6	102.1	102.1	98.5	98.5	98.3	105.6	106.2	105.6	106.1	106.1	106.4
N_1-C_3-O_3-M^{2+}	27.2	24.5	23.4	54.4	−57.3	57.9	57.4	59.1	59.0	51.1	51.6	53.4	42.7	44.9	45.5
N_3-C_2-O_2-M^{2+}	33.7	31.7	31.0	35.0	−35.6	35.5	53.1	53.3	52.1	53.3	53.5	54.0	44.9	42.3	42.4
N_1-N_3-C_2-O_2	12.6	12.3	12.1	11.9	−11.5	11.7	5.7	5.0	6.1	7.7	7.5	7.1	9.8	10.0	9.8
Water															
M^{2+}-O_2	1.920	1.925	1.933	1.854	1.863	1.865	1.954	1.951	1.955	2.029	2.035	2.038	2.048	2.047	2.071
M^{2+}-O_3	1.929	1.916	1.932	1.838	1.829	1.835	1.933	1.927	1.928	2.022	2.016	2.019	2.032	2.055	2.023
O_2-M^{2+}-O_3	99.7	98.2	97.8	97.9	98.3	98.2	101.0	99.7	99.8	103.5	102.9	102.3	90.7	91.1	91.3
N_1-C_3-O_3-M^{2+}	44.4	40.9	48.5	64.3	−66.4	66.7	58.6	60.8	63.0	54.5	56.9	59.8	64.3	57.6	66.5
N_3-C_2-O_2-M^{2+}	47.5	50.9	54.1	22.8	−36.0	37.2	43.8	44.6	47.6	60.6	61.5	61.8	63.2	66.1	62.0
N_1-N_3-C_2-O_2	14.1	14.4	11.4	13.4	52.0	10.1	13.6	14.6	11.4	5.7	4.2	3.2	1.0	5.0	1.0
Benzene															
M^{2+}-O_2	1.865	1.868	1.871	1.833	1.843	1.841	2.042	2.019	2.018	2.012	2.018	2.018	—	1.944	1.949
M^{2+}-O_3	1.850	1.846	1.846	1.829	1.818	1.820	1.971	1.972	1.970	2.007	2.001	2.001	—	1.928	1.932
O_2-M^{2+}-O_3	102.6	102.8	103.3	99.6	100.9	100.9	98.9	99.7	98.5	104.5	103.5	103.2	—	101.9	102.3
N_1-C_3-O_3-M^{2+}	131.8	32.9	35.4	59.9	−58.5	59.8	58.3	59.2	61.3	52.7	55.9	57.5	—	42.5	48.4
N_3-C_2-O_2-M^{2+}	36.1	40.3	38.5	27.9	−31.8	34.7	53.2	50.6	54.7	55.4	58.6	58.6	—	40.8	42.9
N_1-N_3-C_2-O_2	13.9	14.6	13.1	9.3	−14.0	13.6	5.8	7.2	5.2	9.8	4.6	4.7	—	13.6	9.7
DMF															
M^{2+}-O_2	1.918	1.923	1.909	1.852	1.861	1.857	1.958	1.955	1.958	2.028	2.034	2.037	2.043	2.050	2.058
M^{2+}-O_3	1.926	1.912	1.902	1.837	1.828	1.828	1.937	1.930	1.931	2.021	2.015	2.019	2.027	2.042	2.003
O_2-M^{2+}-O_3	99.6	116.8	100.2	98.1	98.4	98.4	101.0	100.9	99.8	103.6	103.0	102.2	90.9	91.0	91.4
N_1-C_3-O_3-M^{2+}	42.4	50.0	38.4	63.9	−66.2	65.7	58.3	60.5	62.9	54.4	56.9	59.8	64.0	57.6	62.8
N_3-C_2-O_2-M^{2+}	48.1	40.4	39.9	23.5	−35.6	34.8	44.4	44.8	47.7	60.5	61.0	61.6	63.1	65.8	64.2
N_1-N_3-C_2-O_2	13.6	14.6	16.6	13.5	−9.5	9.7	13.3	14.5	11.3	5.6	4.5	3.3	1.0	5.0	3.0

bonds especially for complex 1a in benzene and DMF. One can observe that the variations of the geometrical parameters are very versatile in solution-phase.

To investigate the contribution of the metal chelation to the antioxidant activity of ADPHT derivatives, we analyzed the X-H bond distances (X = N_3, N_2, and C_1) of the structures obtained and compared them with those of isolated ligands. From Figure 2S, one can find out that the metal chelation slightly increases the X-H bond distances, then decreases the bond dissociation energies of these bonds, and therefore enhances the antioxidant activities. In M-ADPHT complexes optimized, the longer X-H bond distances are obtained for C_1-H_1 bonds due to the captodative stabilization evoked in our previous work [18]. Figure 3S displays the fact that the solvation of complexes induces an important reduction of the C_1-H_1 bond distances. Such a reduction is more pronounced for Cu^{2+}-complexes in benzene and DMF.

Geometrical parameters of optimized complexes obtained in gas-phase from deprotonated ligands (Tables 1–3S in the supporting information) illustrated the fact that the metal-ligand distances are lower than those of complexes resulting from neutral ligand, with the exception of 8A complex. The main justification of this exception resulted from the optimization of 8A complex that leads to a monodental structure with the O_3 oxygen only effectively binded to the Cu^{2+} cation. Our results also showed that the M-O_3 bond distances are slightly longer than those of M-O_2 bonds. This is basically due to the nearness of the $C_2=O_2$ carbonyl group to proton abstraction site (N_3 atom) that subsequently increases the electron density around the O_2 oxygen atom. Contrary to previous remarks on complexes obtained from neutral ligands, the latter becomes now more nucleophile than the O_3 oxygen atom.

Figure 2 indicates that the M^{2+}-O_i (i = 2, 3) bond distances yielded by B3LYP/Mixed II are shorter than that relative to B3LYP/Mixed I. The bond distance differences obtained are in the ranges 0.023–0.887 and 0.019–0.169 Å, respectively, for M^{2+}-O_2 and M^{2+}-O_3 bond. This fact can be attributed to the fact that the valence orbitals are not properly described by B3LYP/Mixed I. This results from the

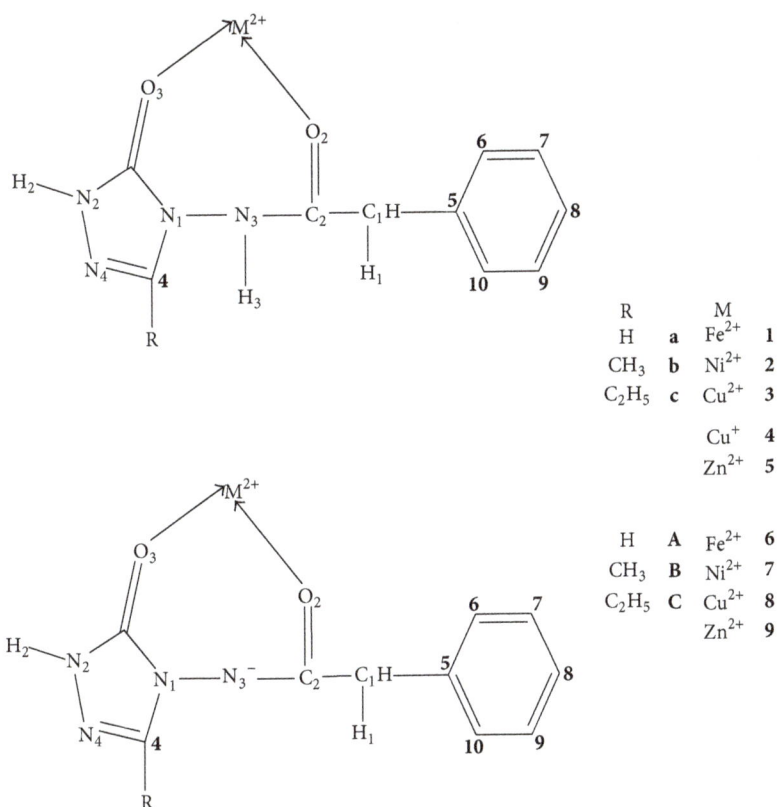

FIGURE 1: Schematic representation of M-ADPHT complexes including the adopted numbering system used.

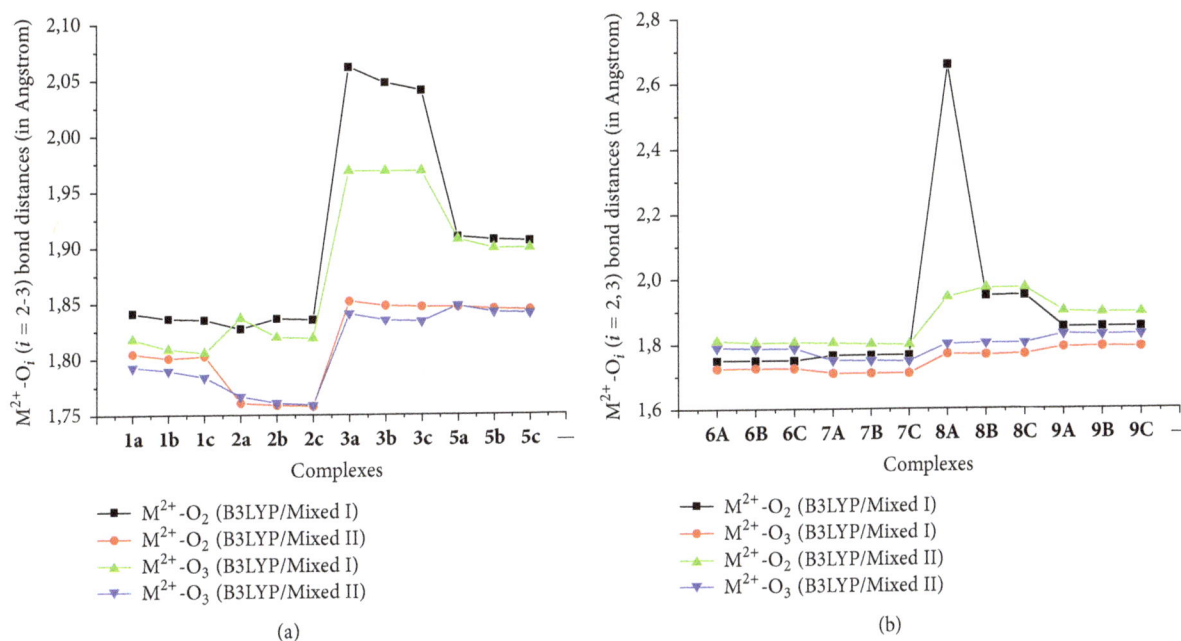

(a)

(b)

FIGURE 2: Superposition of M^{2+}-O_i bond obtained using B3LYP/Mixed I, B3LYP/Mixed II, and CCSD(T) for neutral ADPHT ligands (a) and for deprotonated ADPHT ligands (b).

TABLE 2: Values (in kJ/mol) of metal binding energy (D_e), metal binding enthalpy (ΔH^0_{298}), and metal binding free energy (ΔG^0_{298}) for ADPHT ligand-metal complexes in various media used at B3LYP/Mixed I level.

		Fe^{2+}		Ni^{2+}		Cu^{2+}		Cu^{+}		Zn^{2+}	
		1a	6A	2a	7A	3a	8A	4a	9A	5a	10A
D_e	Gas	−1194	−2081	−1282	−2158	−1215	−2112	−309	−756	−904	−1752
	Water	−277	−435	−267	−443	−421	−642	−121	−211	−57	−177
	Benzene	−642	−1095	−681	−1172	−757	−1284	−201	−451	—	−819
	DMF	−287	−484	−280	−467	−432	−664	−124	−219	−62	−196
ΔH^0_{298}	Gas	−1196	−2083	−1285	−2160	−1218	−2114	−312	−759	−907	−1754
	Water	−279	−438	−270	−446	−424	−645	−124	−214	−59	−179
	benzene	−644	−1097	−683	−1174	−759	−1287	−204	−453	—	−821
	DMF	−289	−486	−282	−469	434	−666	−126	−222	−68	−198
ΔG^0_{298}	Gas	−1152	−2043	−1182	−2121	−274	−2072	−1242	−722	−864	−1714
	Water	−236	−396	−384	−405	−88	−605	−227	−180	−13	−140
	benzene	−598	−1058	−719	−1132	−165	−1245	−634	−416	—	−778
	DMF	−246	−446	−395	−428	−91	−629	−240	−187	−21	−159
		1b	6B	2b	7B	3b	8B	4b	9B	5b	10B
D_e	Gas	−1219	−2093	−1304	−2169	−1231	−2063	−318	−758	−926	−1763
	Water	−274	−383	−269	−448	−422	−606	−119	−211	−56	−825
	Benzene	−653	−1138	−691	−1180	−764	−1241	−204	−452	−382	—
	DMF	−285	−459	−282	−471	−433	−628	−122	−219	−64	—
ΔH^0_{298}	Gas	−1222	−2095	−1307	−2171	−1233	−2066	−320	−760	−929	−1765
	Water	−277	−385	−272	−450	−425	−609	−121	−213	−58	—
	Benzene	−656	−1141	−694	−1183	−767	−1243	−207	−455	−385	−828
	DMF	−288	−461	−284	−474	−435	−630	−124	−222	−67	—
ΔG^0_{298}	Gas	−1177	−2053	−1199	−2130	−283	−2029	−1265	−721	−887	−1726
	Water	−233	−344	−382	−409	−85	−578	−228	−179	−15	—
	Benzene	−608	−1103	−728	−1144	−171	−1211	−647	−421	−334	−789
	DMF	−244	−414	−392	−427	−88	−595	−240	−182	−24	—
		1c	6C	2c	7C	3c	8C	4c	9C	5c	10C
D_e	Gas	−1226	−2093	−1310	−2168	−1234	−2061	−317	−755	−933	−1764
	Water	−274	—	−264	−444	−417	−602	−114	−207	−53	−175
	Benzene	−655	−1140	−692	−1179	−761	−1239	−202	−451	−383	−825
	DMF	−282	−485	−281	−467	−427	−6024	−117	−215	−62	−194
ΔH^0_{298}	Gas	−1229	2096	−1312	−2170	−1237	−2063	−319	−757	−936	−1766
	Water	−277	—	−266	−446	−420	−604	−117	−210	−55	−177
	Benzene	−658	−1142	−695	−1181	−764	−1241	−204	−453	−385	−827
	DMF	−284	−487	−267	−470	−430	−629	−119	−218	−65	−197
ΔG^0_{298}	Gas	−1182	−2055	−1201	−2129	−280	−2027	−1268	−718	−893	−1728
	Water	−231	—	−376	−404	−76	−573	−219	−173	−14	−137
	Benzene	−608	−1103	−722	−1142	−163	−1209	−642	−416	−335	−790
	DMF	−240	−446	−385	−428	−79	−594	−239	−181	−22	−158

poor description of the electron-electron repulsion. In the whole, higher differences are obtained for Cu^{2+}-complexes for both ligands.

3.2. Metal Binding Selectivity.

Table 2 gives the B3LYP/Mixed I complexation energies of the M-ADPHT complexes in various media. All the complexation energies are highly negative showing that the reaction is highly exothermic. One can observe a relevant decrease of the binding energies when passing from the gas-phase to solvent-phases. This drop is directly attributed to solvent effect that hampers the interaction between transition metal cation and ADPHT ligand. This diminution is more pronounced for protic solvents (water and DMF).

The complexation energies for solvent-phase are correlated to M-O$_i$ ($i = 2, 3$) bond distances presented in the previous section. Figure 3 and Table 2 display for neutral ligands (**a, b,** and **c**) an increasing binding selectivity in the following order: Zn^{2+} < Fe^{2+} < Cu^{2+} < Ni^{2+}. This is in line with the previous works which exhibited the highest affinity

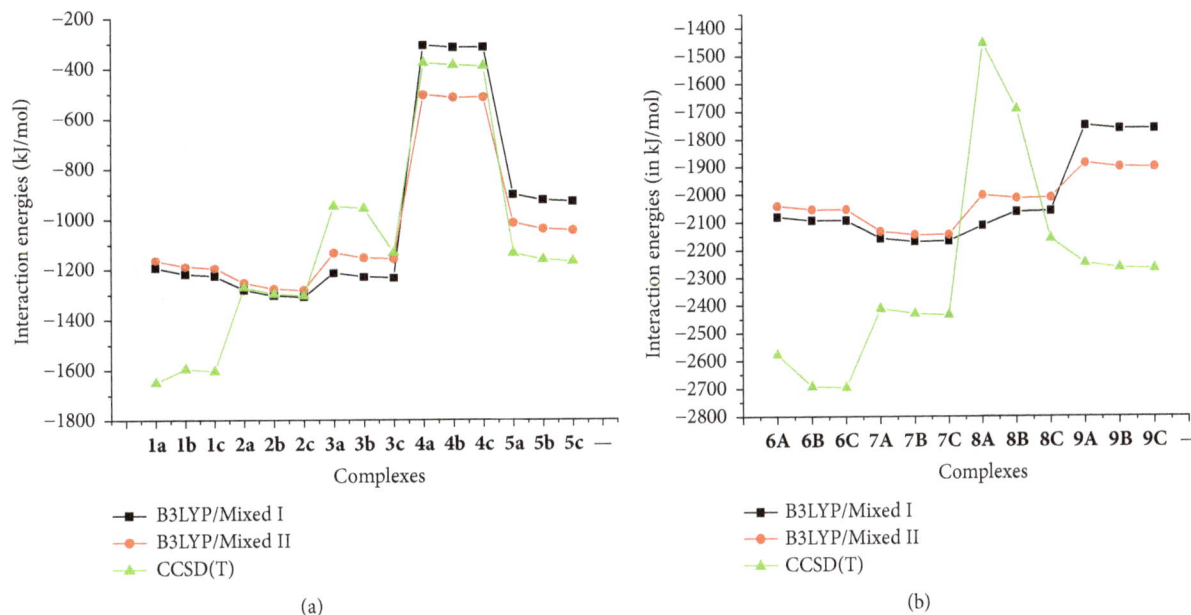

FIGURE 3: Superposition of metal binding energy values (in kJ/mol) obtained using B3LYP/Mixed I, B3LYP/Mixed II, and CCSD(T) for neutral ADPHT ligands (a) and for deprotonated ADPHT ligands (b).

of Ni^{2+} compared to other divalent first-row transition metal cations for glycine, glycine derivatives [40], polyether [41], and polyamine ligand [41]. The M^{2+}-ADPHT complexes have higher metal ion affinities than the Cu^+-ADPHT complexes. For both complexes, the higher binding abilities are obtained for c neutral ligand.

Our data show that the deprotonation of the ligand increases the binding affinity independently of the metal cation used. From Table 2, it is revealed that the binding energy difference between the value obtained from neutral ligand and that from deprotonated ones is in the following ranges: 827–897 kj/mol for M^{2+}-ADPHT and 438–447 kj/mol for Cu^+-ADPHT in gas-phase. The calculated trend for D_e is in agreement with the notable decrease of M-O$_i$ ($i = 2, 3$) bond distances of optimized complexes obtained in gas-phase from deprotonated ligands evoked in previous section (Figure 2 and Tables 1–3S in the supporting information). This increase of binding affinity is also observed in solution-phase. For the deprotonated ligand A, a similar increasing binding selectivity is observed compared to that of neutral ligand, whereas Figure 3 clearly reveals that the B3LYP/Mixed I complexation energies of the divalent ion with deprotonated ligands (B and C) follow the order of $Zn^{2+} < Cu^{2+} < Fe^{2+} < Ni^{2+}$. A similar situation was found for B3LYP/Mixed II values. In the whole, the latter is higher than the former for divalent transition cations with the exception of Zn^{2+}-complexes. The average difference reached 6.2 and 67.0 kJ/mol, respectively, for neutral ADPHT –M^{2+} complexes and deprotonated ADPHT –M^{2+} complexes. For neutral ADPHT –Cu^+ complexes, Figure 3 also displays the drastic drop that is in 197.3–199.7 kJ/mol. In the same vein, difference in complexation energy values in the range 100.5–641.4 kJ is observed for deprotonated ADPHT

–M^{2+} complexes when passing from B3LYP calculations to CCSD(T) ones (Figure 3). This similar larger difference has been found by Constantino et al. in the theoretical study on interactions of Co^+ and Co^{2+} with glycine [35]. We attributed this fact to a bad annulment of self-interaction term by the exchange functional that leads to overstabilization of molecular systems by density functional methods. For neutral ADPHT –M^{2+} complexes, this difference is variable. In the case of Ni^{2+}-complexes, differences are in a very narrow range of 4.4–9.5 kJ/mol for calculations related to B3LYP/Mixed II.

The calculated interaction enthalpies presented in Table 2 do not give any additional relevant information on the metal binding selectivity. In order to get a deep insight into the capacity of ADPHT ligands to be bound to metal cations, we have calculated the interaction free energies. The calculated results obtained in various media inserted in Table 2 are negative showing that the formation of M^{n+}-ADPHT complexes ($n = 1, 2$) is spontaneous. A comparison of the interaction free energies of Cu^{2+} and Cu^+ in various media shows that the preference of both metal cations depends on the nature of the ADPHT ligands.

We note that higher capacity of the neutral ligand to bind to Cu^+ cation is observed. Contrary to monovalent copper cation, higher capacity is attributed to complexation of deprotonated ligand to Cu^{2+}. We find again that our results indicate that all divalent metal cations prefer to be bound to deprotonated ADPHT ligands. In gas-phase, the lower interaction free energies obtained for Ni^{2+}-ADPHT complexes confirms the preference of both ligands to bind to Ni^{2+} previously evoked in this section.

3.3. Electronic Structure and Atoms in Molecules Analysis. In previous studies, it has been found that the HOMO-LUMO

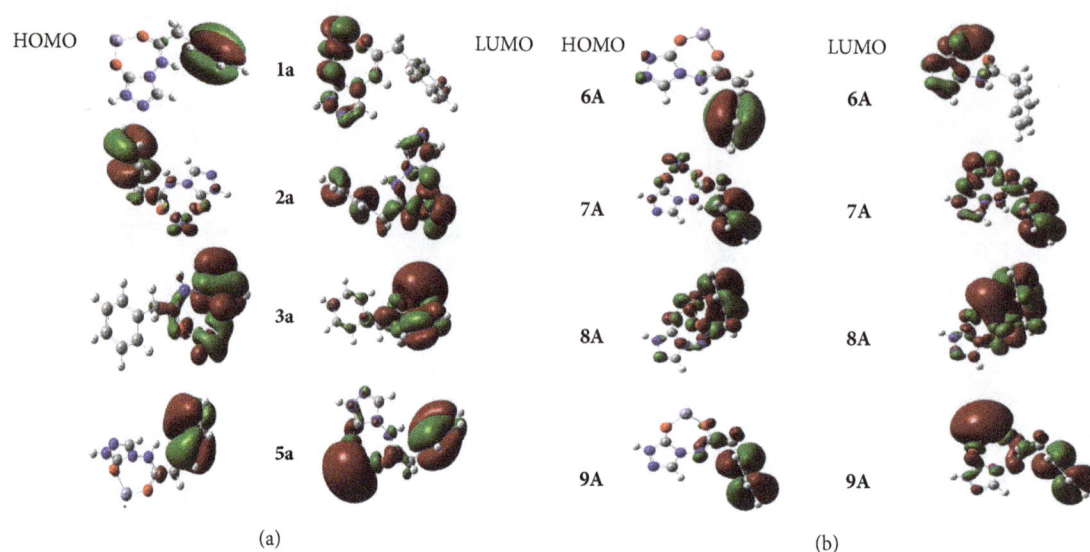

FIGURE 4: Schematic representation of frontier molecular orbitals of M-ADPHT complexes (R = H): neutral ligand-divalent metal cation complexes (a) and deprotonated ligand-divalent metal cation complexes (b).

energy gap is an important stability descriptor [41–44]. The large HOMO-LUMO energy gap is consistent to stable and little reactive systems, but, in the contradictory case, the little stable systems correspond to highly reactive systems. The orbital frontier eigenvalues and HOMO-LUMO gaps of different complexes calculated in various media are inserted in Table 3. Our data reveal that deprotonated ADPHT ligand-M^{2+} complexes are the most stable ones in gas-phase. The highest HOMO-LUMO energy gap matches up with the monodental complex 8A in the same medium. This explains why the electron cloud of HOMO is mainly localized on atoms of benzyl rings and on the metal atom for complex 8A (see Figure 4). On the contrary, for other complexes, this cloud is exclusively located on atoms of benzyl ring. The solvation studied enhances the stability of the complexes studied. This enhancement is more pronounced for protic solvent (water and DMF). The comparison between the HOMO-LUMO gap on isolated ligands (Table 4S in the supporting information) and that of complexes reveals that the metal chelation of isolated ligands reduces their stabilities independently of the medium.

The HOMO eigenvalues are used to characterize the donating ability of the molecule. Higher value of HOMO energy indicates a predisposition of the molecular system to loss electrons [45]. Higher value of HOMO energy is therefore an indication of the higher antioxidant activity [46].

In gas-phase, the values of HOMO energy of the isolated ligands (Table 4S in the supporting information) exhibit $E_c > E_b > E_a$ for neutral ligands and $E_B > E_C > E_A$ for deprotonated ones. One can then conclude that the orders for the antioxidant activity of these ADPHT ligands are c > b > a and B > C > A, respectively, for neutral ligands and deprotonated homologs. For the former, this order remains invariable upon chelation with a metal ion with exception made to Ni^{2+} (c = b > a) in gas-phase. For the latter, the complexation of the deprotonated ligand leads to versatile results. This result

corroborates with previous theoretical research on Hydrogen Atom Transfer in the reaction of metal-associated phenolic acids with $OH^{\bullet}$ radical that showed that the ordering for antioxidant activity of the neutral phenolic acids does not change upon chelation with a divalent metal cation [45]. Our results also demonstrate that the solvation of complexes augments the antioxidant activity of the complexes. This augmentation is more sensitive for protic solvents (water and DMF). These results are consistent with previous studies on neutral ADPHT ligand [18] and other molecular systems [47, 48]. The explanation is that the charge-separation process is quite sensible to polarity of solvent [49]. From Table 3, an increase of dipole of complexes is observed with the increase of the solvent's polarity declines the strength of X-H (X = N, C) bonds and subsequently increases the antioxidant activity of complexes formed. This conclusion is in line with previous works [47–51]. The survey of HOMO eigenvalues and dipole moments of optimized isolated ADPHT ligands (Table 4S in the supporting information) compared to those of its complexes highlights the augmentation of antioxidant activity at the end of the complexation in various media.

Table 3 also shows an authentication of charge donation from ligand to metal atom. This donation is more sensitive for deprotonated ligands. Such a greater electron transfer from these ligands to metal (II) atom is a plausible explanation of their higher stabilities previously evoked. Our data also expose a greater electron transfer Cu(II)-complexes for both ligands. The calculated NBO charges on metal (II) atom are in good agreement with previous theoretical researches [40, 52] that disclosed the fact that greater electron transfer from the electron donor to the acceptor leads to higher stability. On the whole, the value of the metal charge carried by the metal cation in gas-phase was in the following ranges: +(0.81–1.23) for iron, +(0.95–1.04)e for nickel, +(0.75–0.92)e for copper, and +(0.9–1.61)e for zinc. This shows that a real charge transfer leads to an increase of the electronegativity

TABLE 3: NBO charge (q/e) carried by the metal ion, the orbital frontier eigenvalues (eV), LUMO-HOMO gap (ΔE in eV), and dipole moments (Debye) for ADPHT ligand-metal complexes in various media at B3LYP/Mixed I level.

	Q/e	E_{HOMO} (eV)	E_{LUMO} (eV)	ΔE (eV)	μ (Debye)		Q/e	E_{HOMO} (eV)	E_{LUMO} (eV)	ΔE (eV)	μ (Debye)
Gas phase											
1a	1.23	−13.08	−12.42	0.66	6.50	6A	1.06	−9.23	−8.04	1.19	8.09
1b	1.22	−13.02	−12.27	0.75	6.41	6B	0.81	−9.25	−7.89	1.36	7.46
1c	1.22	−12.96	−12.19	0.77	6.95	6C	1.05	−9.23	−7.85	1.38	7.60
2a	1.05	−13.60	−12.70	0.90	1.79	7A	0.95	−9.63	−8.50	1.13	4.63
2b	1.05	−13.49	−12.59	0.99	1.89	7B	0.95	−9.58	−8.41	1.17	4.51
2c	1.04	−13.49	−12.50	0.99	2.52	7C	0.95	−9.57	−8.40	1.17	4.92
3a	0.92	−13.76	−12.85	0.91	7.87	8A	0.75	−10.65	−8.16	2.49	2.27
3b	0.92	−13.63	−12.70	0.93	7.09	8B	0.85	−9.78	−8.41	1.37	2.71
3c	0.92	−13.59	−12.63	0.96	6.93	8C	0.85	−9.75	−8.38	1.37	2.98
5a	1.61	−12.86	−12.17	0.69	7.92	10A	1.50	−9.05	−8.15	0.9	9.57
5b	1.62	−12.80	−12.04	0.76	7.98	10B	1.49	−9.05	−8.03	1.02	9.05
5c	1.61	−12.75	−11.98	0.77	8.58	10C	1.49	−9.03	−7.99	1.04	9.27
Water											
1a	1.58	−7.21	−4.33	2.88	15.03	6A	1.41	−6.68	−3.39	3.29	15.17
1b	1.57	−7.20	−4.28	2.92	14.85	6B	1.37	−6.24	−3.52	2.72	12.45
1c	1.58	−7.21	−4.31	2.90	15.98	6C	0.90	−7.21	−4.31	2.9	15.98
2a	1.36	−7.28	−5.40	1.88	1.36	7A	1.41	−6.74	−4.47	2.27	12.93
2b	1.35	−7.29	−5.37	1.92	1.35	7B	1.37	−6.73	−4.42	2.31	12.14
2c	1.35	−7.30	−5.38	1.92	1.35	7C		−6.74	−4.41	2.33	12.93
3a	1.22	−7.67	−6.49	1.18	4.92	8A	1.25	−7.18	−4.73	2.45	3.64
3b	1.24	−7.48	−6.46	1.02	7.11	8B	1.25	−6.85	−−5.04	1.81	7.60
3c	1.24	−7.48	−6.46	1.02	7.79	8C	1.25	−6.84	−5.03	1.81	7.80
5a	1.61	−7.09	−2.22	4.87	17.97	10A	1.79	−6.50	−1.76	4.74	18.17
5b	1.62	−7.11	−2.30	4.81	17.56	10B					
5c	1.61	−7.11	−2.29	4.82	18.40	10C	1.78	6.48	1.75	4.73	18.3
Benzene											
1a	1.39	−9.58	−8.21	1.37	10.68	6A	1.10	−7.71	−5.63	2.08	10.03
1b	1.49	−9.59	−8.10	1.49	10.24	6B		−7.70	−5.64	2.06	10.7
1c	1.51	−9.60	−8.09	1.51	10.95	6C	1.22	−7.66	−5.62	2.04	11.00
2a	1.19	−9.90	−8.82	1.08	6.90	7A	1.09	7.84	6.45	1.39	8.57
2b	1.18	−9.87	−8.75	1.12	6.70	7B		−7.82	−6.37	2.04	8.29
2c	1.18	−9.83	−8.73	1.10	7.45	7C	1.08	−7.82	−6.36	1.46	8.77
3a	0.97	−10.37	−9.23	1.14	6.43	8A	0.38	−8.73	−6.24	2.49	2.67
3b	0.99	−10.32	−9.18	1.14	5.28	8B	0.42	−7.97	−6.52	1.45	5.51
3c	0.99	−10.29	−9.16	1.13	5.74	8C	0.88	−7.97	−6.51	1.46	5.65
5a						10A	1.65	−7.49	−5.04	2.45	13.39
5b	1.76	−9.44	−6.93	2.51	12.3	10B	1.64	−7.50	−4.97	2.53	12.79
5c	1.75	−9.44	−6.97	2.47	13.4	10C	1.76	−7.50	−5.00	2.50	13.21
DMF											
1a	1.57	−7.28	−4.48	2.80	14.83	6A	—	—	—	—	—
1b	1.56	−7.28	−4.42	2.86	14.65	6B	1.40	−6.68	−3.45	3.23	14.10
1c	1.55	−7.28	−4.39	2.89	16.04	6C	1.47	−6.58	−3.37	3.21	15.57
2a	1.35	−7.36	−5.54	1.82	12.21	7A	1.24	−6.77	−4.55	2.22	12.78
2b	1.34	−7.37	−5.50	1.87	12.02	7B	1.24	−6.76	−4.50	2.26	11.99
2c	1.18	−7.37	−5.48	1.89	12.66	7C	1.24	−6.77	−4.49	2.28	12.75
3a	1.20	−7.81	−6.58	1.23	4.08	8A	0.80	−7.23	−4.77	2.46	3.60
3b	1.22	−7.61	−6.55	1.06	6.28	8B	0.93	−6.88	−5.08	1.8	7.54
3c	1.22	−7.59	−6.55	1.04	6.70	8C	0.92	−6.88	−5.08	1.8	7.73
5a	1.85	−7.16	−2.46	4.70	17.77	10A	1.79	−6.52	−1.91	4.61	18.18
5b	1.85	−7.18	−2.47	4.71	17.37	10B	1.85	−7.18	−2.47	4.71	17.37
5c	1.85	−7.18	−2.45	4.73	18.23	10C	1.78	6.52	1.88	4.64	18.11

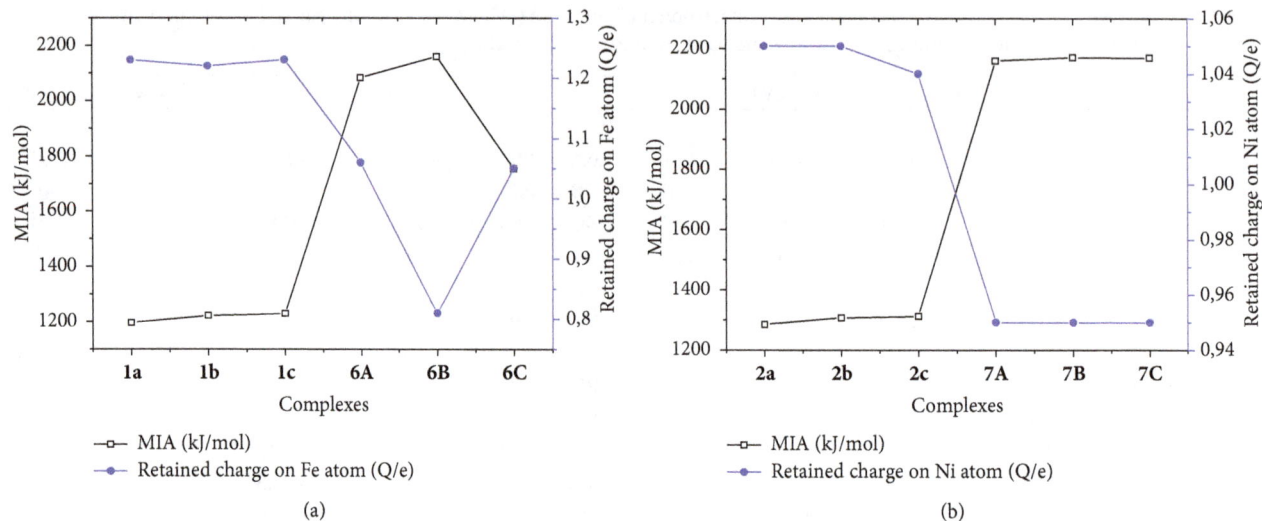

FIGURE 5: Correlation between the MIA (kJ/mol) and retained charge (Q/e) of Fe (a) and correlation between the MIA (kJ/mol) and retained charge (Q/e) of Ni (b).

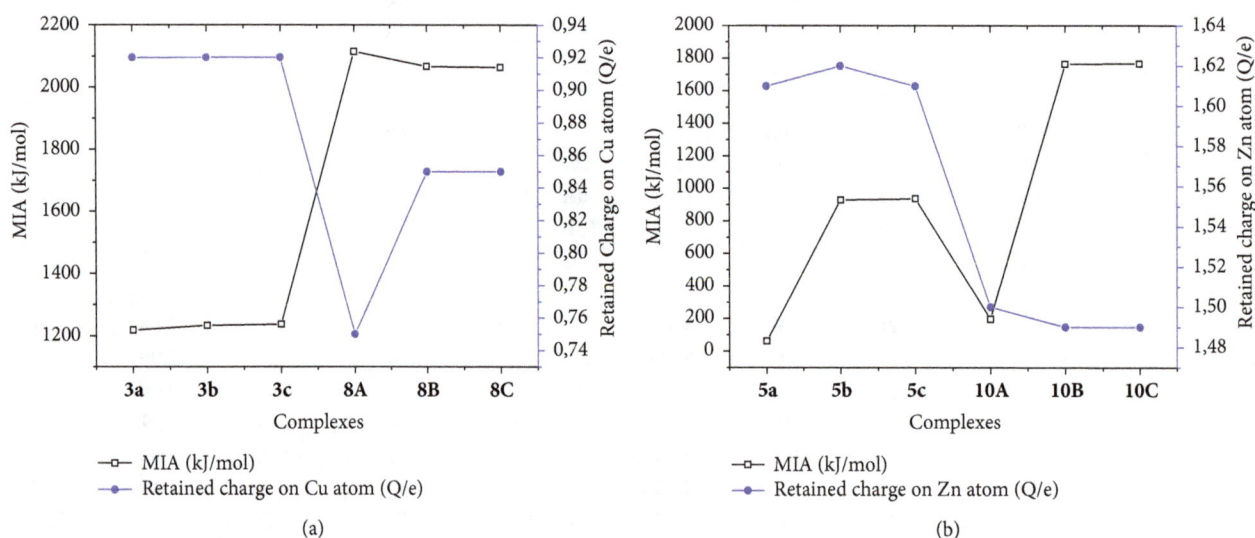

FIGURE 6: Correlation between the MIA (kJ/mol) and retained charge (Q/e) of Cu (a) and correlation between the MIA (kJ/mol) and retained charge (Q/e) of Zn.

of the metal ions that then has undergone a reduction. We then concluded that the metal cation plays an oxidation role towards both ligands. These observations are similar to those made on quercetin [53] and on phenolic acids [52].

To see the possible correlation between the retained NBO charge on the metal atom and metal MIA of each of the metal atoms, the two parameters are plotted (Figures 5 and 6). In the four cases (Fe^{2+}, Ni^{2+}, Cu^{2+}, and Zn^{2+}), MIA values vary proportionally with the retained charge of the metal ion for deprotonated ADPHT ligands. On the contrary, MIA values vary inversely with the retained charge of the metal cation. Our results show that the solvation diminishes the electron donation from the ligand to metal atom, except for $5i$ (i = a, b, and c) complexes in water and for 8C in benzene.

The differences in retained charge on metal atom resulting from the various media were in the following ranges: $(0.15–0.46)e$ for water, $(0.05–0.43)e$ for benzene, and $(0.14–0.59)e$ for DMF (see Table 3). Therefore, one could presume that the diminution of electron transfer is more enunciated in polar solvent.

So as to get detailed information on electron transfer in the coordination sphere complex, we presented in Figure 7 the second-order perturbation energy stabilization (E_2) associated with the electron donation: electron donor (O_2 or O_3) to the acceptor (metal atom). The E_2 values range from 4.4 to 49.68 kJ/mol and from 2.76 to 119.98 kJ/mol, respectively, for metal-neutral ligand complexes and metal-deprotonated ligand complexes. The higher gap between the contribution

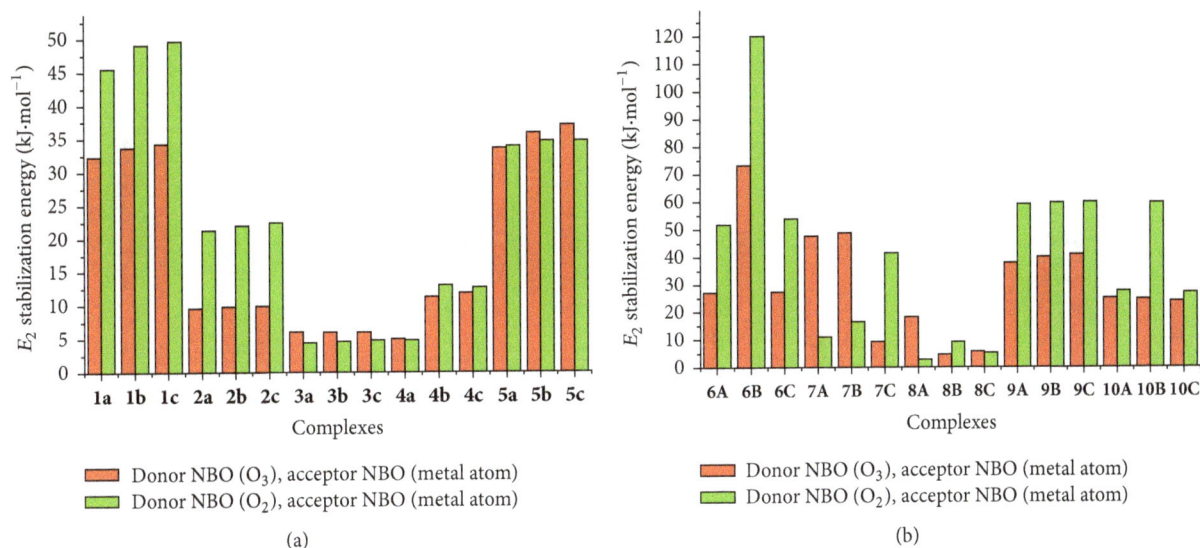

FIGURE 7: Second perturbative interaction energies E_2 (in kJ·mol^{-1}) between the first-row transition metals and lone pair electron ligand atoms (oxygen atom O$_3$ and oxygen atom O$_2$) for each complex: E_2 energies for neutral ligands (a) and E_2 energies for deprotonated ligands (b).

of O$_2$ → metal donation and that of O$_3$ → metal ones is obtained for Fe-complexes in both cases (with exception to 6C) in favor of the former one. This higher gap varies from 12.24 to 46.78 kJ/mol. This higher contribution of O$_2$ → metal donation is also observed when the Cu$^+$ and Zn^{2+} were chelated to the deprotonated ligands. Figure 6 also exhibits a drastic drop of both contributions for Cu^{2+}-complexes. The negligible contribution of O$_2$ → metal donation (2,76 kJ/mol) highlighted the monodental nature of 8A structure evoked in the prior geometrical analysis. To better appreciate the impact of the metal chelation beyond the coordination inner-sphere, we made a comparison between calculated energies of hyperconjugative interaction for neutral ADPHT ligands and those of its complexes (Table 5S in supporting information). In the whole, this comparison exhibits the noteworthy influence of the metal chelation on the interaction within both ligands used in this work. This fact is in agreement with the significant activation of adjacent bonds of metal cations previously underlined.

The topological parameters obtained at B3LYP/Mixed II level are inserted in Tables 4 and 5. The positive $\nabla^2(\rho(r))$ values and negative $H(r)$ values obtained illustrate that the M^{2+}-O$_i$ ($i = 2, 3$) bonds are partly covalent. This result is confirmed by the fact $0.5 < -G(r)/\nu(r) < 1$. Close inspection of Table 4 reveals that the $\rho(r)$ values of M^{2+}-O$_3$ bond are larger than that for M^{2+}-O$_2$ bond for M^{2+}-neutral ADPHT complexes with exception of 2a. This observation points out that M^{2+}-O$_3$ bonds are stronger than M^{2+}-O$_2$ bond of these complexes. This corroborates the fact that M^{2+}-O$_2$ bonds are stronger than M^{2+}-O$_2$ bonds. Nevertheless, Table 4S displays contrary information for M^{2+}-deprotonated ADPHT complexes. The $\rho(r)$ values of M^{2+}-O$_i$ ($i = 2, 3$) in Cu^{2+}-deprotonated ADPHT complexes are higher than that for Cu$^+$ complexes. This confirms the preference of Cu^{2+} to

bind deprotonated ligand. The $\rho(r)$ values of M^{2+}-O$_i$ ($i = 2, 3$) in M^{2+}-deprotonated ADPHT complexes are sensitively higher than its homolog in M^{2+}-neutral ADPHT complexes with exception to Fe^{2+}-complexes. The $\rho(r)$ value difference ranges from 0.006 to 0.038. Consequently, the M^{2+}-O$_i$ ($i = 2, 3$) in M^{2+}-deprotonated ADPHT complexes are stronger than that in M^{2+}-neutral ADPHT complexes. This result is in line with the fact that metal ligand distances in the former are lower.

3.4. Proton Affinity. In order to investigate the possibility of deprotonation of metal-neutral ligand complexes in various media, the authors analyzed the PA and PAFE values of the complexes. We absolutely need the enthalpy and free energy of H$^+$ to determine the thermodynamic energies. The calculated values of these thermodynamic energies used in this work are compiled in Table 6S. In the interest of comparison, we presented in Table 7S the PA and PAFE values of isolated neutral ligands and its complexes in various media. A major drop of the PA values is observed when passing from ligands to complexes in gas-phase. This drop is lower for Cu$^+$-complexes.

Such a drop discloses the notable enhancement of antioxidant activities of neutral ligands in gas-phase. This fact is in good agreement with the increase of the X-H bond distances induced by the metal chelation (shown in Figure 1S). This augmentation of the antioxidant activities has been pointed out in our previous researches [54]. The difference in PA values from ligands to complexes declines in solution-phase.

We therefore concluded that the solvation reduces the enhancement of the antioxidant activities induced by the metal chelation. The positive PA values obtained for all the complexes revealed that the deprotonation of metal-neutral ligand complexes is endothermic in gas-phase, water,

TABLE 4: Topological parameters of the metal-ligand for optimized structures at B3LYP/Mixed II level for neutral ADPHT-M^{2+}-complexes.

	M^{2+}-O_2					M^{2+}-O_3				
	1a	**2a**	**3a**	**4a**	**5a**	**1a**	**2a**	**3a**	**4a**	**5a**
$\rho(r)$	0.121	0.133	0.108	0.107	0.112	0.134	0.133	0.113	0.106	0.113
$\nabla^2(\rho(r))$	0.852	0.902	0.564	0.550	0.510	0.852	0.878	0.589	0.526	0.510
$H(r)$	−0.009	−0.022	−0.036	−0.038	−0.047	−0.018	−0.022	−0.037	−0.039	−0.048
$G(r)$	0.222	0.247	0.176	0.175	0.175	0.230	0.247	0.176	0.170	0.176
$v(r)$	−0.231	−0.269	−0.212	−0.213	−0.222	−0.248	−0.269	−0.212	−0.209	−0.224
$-G(r)/v(r)$	0.962	0.920	0.833	0.823	0.787	0.929	0.920	0.833	0.815	0.785
	1b	**2b**	**3b**	**4b**	**5b**	**1b**	**2b**	**3b**	**4b**	**5b**
$\rho(r)$	0.124	0.133	0.109	0.108	0.113	0.133	0.135	0.115	0.106	0.115
$\nabla^2(\rho(r))$	0.846	0.908	0.572	0.552	0.516	0.856	0.888	0.600	0.527	0.521
$H(r)$	−0.011	−0.022	−0.036	−0.038	−0.048	−0.017	−0.023	−0.037	−0.039	−0.049
$G(r)$	0.223	0.249	0.179	0.176	0.177	0.231	0.245	0.187	0.170	0.229
$v(r)$	−0.234	−0.270	−0.214	−0.213	−0.225	−0.248	−0.269	−0.225	−0.209	−0.229
$-G(r)/v(r)$	0.953	0.920	0.834	0.823	0.787	0.931	0.913	0.834	0.815	0.785
	1c	**2c**	**3c**	**4c**	**5c**	**1c**	**2c**	**3c**	**4c**	**5c**
$\rho(r)$	0.122	1.34	0.110	0.108	0.113	0.137	0.136	0.116	0.107	0.115
$\nabla^2(\rho(r))$	0.856	0.911	0.576	0.552	0.517	0.877	0.890	0.602	0.528	0.522
$H(r)$	−0.009	−0.022	−0.036	−0.038	−0.048	−0.019	−0.024	−0.038	−0.039	−0.049
$G(r)$	0.223	0.250	0.180	0.176	0.177	0.238	0.246	0.188	0.171	0.180
$v(r)$	−0.233	−0.271	−0.215	−0.213	−0.225	−0.257	−0.270	−0.226	−0.209	−0.230
$-G(r)/v(r)$	0.959	0.920	0.834	0.823	0.787	0.926	0.912	0.834	0.815	0.785

TABLE 5: Topological parameters of the metal-ligand for optimized structures at B3LYP/Mixed II level for deprotonated ADPHT-M^{2+}-complexes.

	M^{2+}-O_2				M^{2+}-O_3			
	6A	**7A**	**8A**	**9A**	**6A**	**7A**	**8A**	**9A**
$\rho(r)$	0.162	0.158	0.138	0.132	0.131	0.139	0.125	0.120
$\nabla^2(\rho(r))$	1.019	1.004	0.759	0.634	0.865	0.935	0.679	0.552
$H(r)$	−0.035	−0.036	−0.045	−0.058	−0.015	−0.024	−0.039	−0.052
$G(r)$	0.290	0.287	0.235	0.216	0.232	0.258	0.210	0.190
$v(r)$	−0.325	−0.322	−0.281	−0.274	−0.247	−0.282	−0.248	−0.242
$-G(r)/v(r)$	0.893	0.889	0.838	0.790	0.937	0.914	0.842	0.784
	6B	**7B**	**8B**	**9B**	**6B**	**7B**	**8B**	**9B**
$\rho(r)$	0.162	0.158	0.138	0.132	0.133	0.140	0.126	0.121
$\nabla^2(\rho(r))$	1.017	1.003	0.762	0.633	0.869	0.932	0.684	0.556
$H(r)$	−0.035	−0.036	−0.046	−0.058	−0.016	−0.025	−0.040	−0.053
$G(r)$	0.289	0.286	0.236	0.216	0.234	0.258	0.211	0.191
$v(r)$	−0.324	−0.322	−0.282	−0.274	−0.250	−0.283	−0.251	−0.244
$-G(r)/v(r)$	0.892	0.889	0.838	0.789	0.934	0.911	0.841	0.784
	6C	**7C**	**8C**	**9C**	**6C**	**7C**	**8C**	**9C**
$\rho(r)$	0.162	0.158	0.138	0.132	0.133	0.141	0.126	0.121
$\nabla^2(\rho(r))$	1.017	1.005	0.758	0.634	0.870	0.933	0682	0.554
$H(r)$	−0.035	−0.036	−0.046	−0.058	−0.017	−0.025	−0.040	−0.053
$G(r)$	0.289	0.287	0.235	0.216	0.234	0.259	0.210	0.191
$v(r)$	−0.324	−0.323	−0.281	−0.274	−0.251	−0.284	−0.250	−0.244
$-G(r)/v(r)$	0.892	0.889	0.837	0.789	0.933	0.911	0.840	0.784

and DMF (3a complex excepted). It is fair to note that this deprotonation remains endothermic in benzene only for Cu^+ and Zn^{2+}-complexes. The positive PAFE values obtained show that the deprotonation of metal-neutral ligand complexes is not spontaneous with exception made to minority of endothermic deprotonation previously mentioned.

4. Conclusions

In this study, we have presented the B3LYP/Mixed I and B3LYP/Mixed II calculations which allowed us to treat the complexation of 3-alkyl-4-phenylacettylamino-4,5-dihydro-1h-1,2,4-triazol-5-one derivatives by metal cation (Fe^{2+}, Ni^{2+}, Cu^{2+}, Cu^+, and Zn^{2+}). The optimizations yield a unique coordination mode (O_3, O_2) independent of the nature of the ligand. In the whole, the optimization leads to longer M-O_2 bond distances, compared to those of M-O_3 bonds in gas-phase. Our results indicate that the M^{2+}-O_i ($i = 2, 3$) bond distances yielded by B3LYP/ Mixed II are shorter than those relative to B3LYP/Mixed I due to the poor description of the electron-electron repulsion by this latter. The variations of the B3LYP/Mixed I geometrical parameters are very variable in solution-phase. The metal chelation slightly induces an increase of the X-H bond distances that leads to the enhancement of the antioxidant activities of ligands. This shortening of the X-H bond is in agreement with the major drop of the PA values observed when passing from ligands to complexes in gas-phase. The highly negative values of the complexation energies of the M-ADPHT complexes in various media showed that the metal chelation is exothermic. The authors' data revealed that the solvation of complexes drops these complexation energies. This diminution is more pronounced for protic solvents. The highest affinity is obtained for Ni^{2+}. From our calculations, we conclude that the formation of M^{n+}-ADPHT complexes ($n = 1, 2$) is spontaneous. The HOMO-LUMO gap values reveal that deprotonated ADPHT ligand-M^{2+}-complexes are the most stable ones in gas-phase. The highest value of the HOMO-LUMO gap is obtained for the monodental Cu^{2+} complexes obtained. The variation of the values of this gap revealed the fact that the solvation enhances the stability of the complexes. This enhancement is more pronounced in protic solvents. For the neutral ligands, the metal chelation does not affect the increasing order of the HOMO values (with exception made to Ni^{2+}) in gas-phase. Our results also show that the solvation of complexes augments the HOMO values and therefore enhances the antioxidant activity of the complexes. The calculated NBO charge on metal (II) atom illustrates a clear relationship between greater electron transfer from the electron donor to the acceptor and higher stability. The metal cation then plays an oxidation role towards both ligands. The MIA value varies proportionately with the retained charge of the metal ion for deprotonated ADPHT ligands. The comparison of calculated energies of hyperconjugative interaction for neutral ADPHT ligands and those of its complexes exhibits the noteworthy influence of the metal chelation on the interaction within both ligands used. The topological parameters yielded from Quantum Theory of Atom in Molecules (QTAIM) indicate that metal-ligand bonds are partly covalent.

Acknowledgments

The authors are grateful to Bénoît Champagne (Université Catholique de Louvain (UCL), Belgium) for his support. This work has been supported by the Ministry of Higher Education of Cameroon.

References

[1] I. Dalle-Donne, R. Rossi, D. Giustarini, A. Milzani, and R. Colombo, "Protein carbonyl groups as biomarkers of oxidative stress," *Clinica Chimica Acta*, vol. 329, no. 1-2, pp. 23–38, 2003.

[2] J. K. Andersen, "Oxidative stress in neurodegeneration: cause or consequence?" *Nature Medicine*, vol. 10, pp. S18–S25, 2004.

[3] E. Esposito, D. Ratilio, V. Di Glulia, M. Cacchio, and S. Algeri, "A review of specific dietary antioxidants and the ... for novel therapeutic approach," *Progress in Neurobiology*, vol. 23, p. 719, 2002.

[4] C. Behl, "Alzheimer's disease and oxidative stress: Implications for novel therapeutic approaches," *Progress in Neurobiology*, vol. 57, no. 3, pp. 301–323, 1999.

[5] M. Senevirathne, S.-H. Kim, N. Siriwardhana, J.-H. Ha, K.-W. Lee, and Y.-J. Jeon, "Antioxidant potential ofecklonia cavaon reactive oxygen species scavenging, metal chelating, reducing power and lipid peroxidation inhibition," *Food Science and Technology International*, vol. 12, no. 1, pp. 27–38, 2006.

[6] A. Michalak, "Phenolic compounds and their antioxidant activity in plants growing under heavy metal stress," *Polish Journal of Environmental Studies*, vol. 15, no. 4, pp. 523–530, 2006.

[7] T. P. A. Devasagayam and J. P. Kamat, "Biological significance of singlet oxygen," *Indian Journal of Experimental Biology*, vol. 40, no. 11, pp. 680–692, 2002.

[8] S. Cuzzocrea, D. P. Riley, A. P. Caputi, and D. Salvemini, "Antioxidant therapy: a new pharmacological approach in shock, inflammation, and ischemia/reperfusion injury," *Pharmacological Reviews*, vol. 53, no. 1, pp. 135–159, 2001.

[9] Y. Gilgun-Sherki, Z. Rosenbaum, E. Melamed, and D. Offen, "Antioxidant therapy in acute central nervous system injury: Current state," *Pharmacological Reviews*, vol. 54, no. 2, pp. 271–284, 2002.

[10] W. W. Hope, E. M. Billaud, J. Lestner, and D. W. Denning, "Therapeutic drug monitoring for triazoles," *Current Opinion in Infectious Diseases*, vol. 21, no. 6, pp. 580–586, 2008.

[11] G. R. Kokil, P. V. Rewatkar, S. Gosain et al., "Synthesis and in vitro evaluation of novel 1, 2, 4-triazole derivatives as antifungal agents," *Letters in Drug Design and Discovery*, vol. 7, no. 1, pp. 46–49, 2010.

[12] N. Upmanyu, S. Kumar, M. D. Kharya, K. Shah, and P. Mishra, "Synthesis and anti-microbial evaluation of some novel 1,2,4-triazole derivatives," *Acta Poloniae Pharmaceutica - Drug Research*, vol. 68, no. 2, pp. 213–221, 2011.

[13] O. D. Cretu, S. F. Barbuceanu, G. Saramet, and C. Draghici, "Synthesis and characterization of some 1,2,4-triazole-3-thiones obtained from intramolecular cyclization of new 1-(4-(4-X-phenylsulfonyl)benzoyl)-4-(4- iodophenyl)-3-thiosemicarba-zides," *Journal of the Serbian Chemical Society*, vol. 75, no. 11, pp. 1463–1471, 2010.

[14] S. Hussain, J. Sharma, and M. Amir, "Synthesis and antimicrobial activities of 1,2,4-triazole and 1,3,4-thiadiazole derivatives of 5-amino-2-hydroxybenzoic acid," *E-Journal of Chemistry*, vol. 5, no. 4, pp. 963–968, 2008.

[15] L. P. Guan, Q. H. Jin, G. R. Tian, K. Y. Chai, and Z. S. Quan, "Synthesis of some quinoline-2(1H)-one and 1, 2, 4 - triazolo [4, 3 -a] quinoline derivatives as potent anticonvulsants," *Journal of Pharmacy and Pharmaceutical Sciences*, vol. 10, no. 3, pp. 254–262, 2007.

[16] M. Oyaizu, "Studies on products of the browning reaction," *Japanese Journal of Nutrition*, vol. 44, no. 6, pp. 307–315, 1986.

[17] M. S. Blois, "Antioxidant determinations by the use of a stable free radical," *Nature*, vol. 181, no. 4617, pp. 1199-1200, 1958.

[18] J. H. Nono, D. M. Bikélé, J. N. Ghogomu et al., "DFT-study on antioxydant of 3-alkyl-4-phenylacetylamino-1h-1,2,4-triazol-5-ones and its derivatives," *Journal of Chemistry and Chemical Engineering*, vol. 8, pp. 1109–1124, 2014.

[19] M. Leopoldini, N. Russo, S. Chiodo, and M. Toscano, "Iron chelation by the powerful antioxidant flavonoid quercetin," *Journal of Agricultural and Food Chemistry*, vol. 54, no. 17, pp. 6343–6351, 2006.

[20] M. J. Frisch, G. W. Trucks, H. B. Schlegel et al., *Gaussian 03, Revision A.1*, Gaussian, Inc., Pittsburgh, Pa, USA, 2003.

[21] C. Lee, W. Yang, and R. G. Parr, "Development of the Colle-Salvetti correlation-energy formula into a functional of the electron density," *Physical Review B*, vol. 37, no. 2, pp. 785–789, 1988.

[22] P. J. Hay and W. R. Wadt, "*Ab initio* effective core potentials for molecular calculations. Potentials for the transition metal atoms Sc to Hg," *The Journal of Chemical Physics*, vol. 82, no. 1, pp. 270–283, 1985.

[23] J. Sabolović, C. S. Tautermann, T. Loerting, and K. R. Liedl, "Modeling anhydrous and aqua copper(II) amino acid complexes: a new molecular mechanics force field parametrization based on quantum chemical studies and experimental crystal data," *Inorganic Chemistry*, vol. 42, no. 7, pp. 2268–2279, 2003.

[24] T. J. M. De Bruin, A. T. M. Marcelis, H. Zuilhof, and E. J. R. Sudhölter, "Geometry and electronic structure of bis-(glycinato)-Cu(II) ·2H_2O complexes as studied by density functional B3LYP computations," *Physical Chemistry Chemical Physics*, vol. 1, no. 18, pp. 4157–4163, 1999.

[25] E. Cances and B. Mennucci, "New applications of integral equations methods for solvation continuum models: ionic solutions and liquid crystals," *Journal of Mathematical Chemistry*, vol. 23, no. 3-4, pp. 309–326, 1998.

[26] E. Cances, B. Mennucci, and J. Tomasi, "A new integral equation formalism for the polarizable continuum model: theoretical background and applications to Isotropic and anisotropic dielectrics," *The Journal of Chemical Physics*, vol. 107, pp. 3032–3041, 1997.

[27] L. Rodrìguez-Santiago, M. Sodupe, and J. Tortajada, "The different nature of bonding in Cu^+-glycine and Cu^{2+}-glycine," *The Journal of Physical Chemistry B*, vol. 103, pp. 2310–2317, 1999.

[28] T. Shoeib, C. F. Rodriquez, K. W. M. Siu, and A. C. Hopkinson, "A comparison of copper(I) and silver(I) complexes of glycine,

[29] M. S. F. Jahromi and A. Fattahi, "DFT study of the interaction of thymine with Cu^+ and Zn^{2+}," *Transactions C: Chemistry and Chemical Engineering*, vol. 16, no. 2, pp. 75–80, 2009.

[30] R. Parajuli, "DFT study of Cu^+-thymine and Zn^{2+}-thymine complexes in the gas phase: HOMO-LUMO approach," *Acta Chimica and Pharmaceutica Indica*, vol. 2, no. 2, pp. 85–94, 2012.

[31] R. P. Gangadharana and S. S. Krishnanb, "Natural Bond Orbital (NBO) population analysis of 1-azanapthalene-8-ol," *Acta Physica Polonica A*, vol. 125, no. 1, pp. 18–22, 2014.

[32] R. F. W. Bader, *Atoms in Molecules-A Quantum Theory*, Oxford University Press, Oxford, UK, 1990.

[33] L. Tian and C. Fei-Wu, "Calculation of molecular orbital composition," *Acta Chimica Sinica*, vol. 69, no. 20, pp. 2393–2406, 2011.

[34] G. V. Baryshnikov, B. F. Minaev, V. A. Minaeva, A. T. Podgornaya, and H. Ågren, "Application of Bader's atoms in molecules theory to the description of coordination bonds in the complex compounds of Ca^{2+} and Mg^{2+} with methylidene rhodanine and its anion," *Russian Journal of General Chemistry*, vol. 82, no. 7, pp. 1254–1262, 2012.

[35] E. Constantino, L. Rodríguez-Santiago, M. Sodupe, and J. Tortajada, "Interaction of Co+ and Co2+ with glycine. A theoretical study," *Journal of Physical Chemistry A*, vol. 109, no. 1, pp. 224–230, 2005.

[36] M. Rosi and C. W. Bauschlicher Jr., "The binding energies of one and two water molecules to the first transition-row metal positive ions," *The Journal of Chemical Physics*, vol. 90, no. 12, pp. 7264–7272, 1989.

[37] M. Rosi and C. W. Bauschlicher Jr., "The binding energies of one and two water molecules to the first transitionrow metal positive ions. II," *The Journal of Chemical Physics*, vol. 92, no. 3, pp. 1876–1878, 1990.

[38] E. Magnusson and N. W. Moriarty, "Binding Patterns in Single-Ligand Complexes of NH3, H2, OH-, and F- with First Series Transition Metals," *Inorganic Chemistry*, vol. 35, no. 19, pp. 5711–5719, 1996.

[39] D. Walter and P. B. Armentrout, "Sequential bond dissociation energies of M+(NH3)(x) (x = 1-4) for M=Ti-Cu," *Journal of the American Chemical Society*, vol. 120, no. 13, pp. 3176–3187, 1998.

[40] M. Remko and B. M. Rode, "Catalyzed peptide bond formation in the gas phase," *Physical Chemistry Chemical Physics*, vol. 3, no. 21, pp. 4667–4673, 2001.

[41] F. C. A. Lima, R. B. Viana, J. W. M. Carneiro, M. Comar Jr., and A. B. F. Da Silva, "Coordination ability of polyether and polyamine ligands: A density functional theory study of first- and second-row transition metals," *Journal of Computational and Theoretical Nanoscience*, vol. 10, no. 9, pp. 2034–2040, 2013.

[42] F. Boukli-Hacene, S. Ghalem, M. Merad, and W. Soufi, "Theoretical study of interaction of some heavy metals with carbohydrates in aqueous solution," *Asian Journal of Chemistry*, vol. 28, no. 6, pp. 1293–1296, 2016.

[43] F. Lamchouri, H. Toufi, S. M. Bouzzine, M. Hamidi, and M. Bouachrine, "Experimental and computational study of biological activities of alkaloids isolated from *Peganum harmala* seeds," *Journal of Materials and Environmental Science*, vol. 1, pp. 343–352, 2010.

[44] S. Singh, J. Singh, S. Gulia, and R. Kakkar, "Metal ion selectivity of Kojate complexes: theoretical study," *Journal of Theoretical Chemistry*, pp. 1–9, 2013.

diglycine and triglycine," *Physical Chemistry Chemical Physics*, vol. 3, no. 5, pp. 853–861, 2001.

[45] M. Nsangou, J. J. Fifen, Z. Dhaouadi, and S. Lahmar, "Hydrogen atom transfer in the reaction of hydroxycinnamic acids with radical ˙OH and radical ˙HO_2 radicals: DFT study," *Journal of Molecular Structure*, vol. 862, no. 1–3, pp. 53–59, 2008.

[46] E. G. Bakalbassis, A. Chatzopoulou, V. S. Melissas, M. Tsimidou, M. Tsolaki, and A. Vafiadis, "*Ab initio* and density functional theory studies for the explanation of the antioxidant activity of certain phenolic acids," *Lipids*, vol. 36, no. 2, pp. 181–190, 2001.

[47] M. Najafi, "On the antioxidant activity of the ortho and meta substituted daidzein derivatives in the gas phase and solvent environment," *Journal of the Mexican Chemical Society*, vol. 58, no. 1, pp. 36–45, 2014.

[48] O. Holtomo, M. Nsangou, J. J. Fifen, and O. Motapon, "Antioxidative potency and UV-Vis spectra features of the compounds resulting from the chelation of Fe^{2+} by caffeic acid phenethyl ester and two of its derivatives," *Computational and Theoretical Chemistry*, vol. 1067, pp. 135–147, 2015.

[49] J. Rimarčík, V. Lukeš, E. Klein, and M. Ilčin, "Study of the solvent effect on the enthalpies of homolytic and heterolytic nh bond cleavage in p-phenylenediamine and tetracyano-p-phenylenediamine," *Journal of Molecular Structure*, vol. 952, no. 1–3, pp. 25–30, 2010.

[50] O. Holtomo, M. Nsangou, J. J. Fifen, and O. Motapon, "DFT study of the effect of solvent on the H-atom transfer involved in the scavenging of the free radicals ˙HO_2 and ˙$O_2{}^-$ by caffeic acid phenethyl ester and some of its derivatives," *Journal of Molecular Modeling*, vol. 20, no. 11, pp. 1–13, 2014.

[51] J. J. Fifen, M. Nsangou, Z. Dhaouadi, O. Motapon, and N. Jaidane, "Solvent effects on the antioxidant activity of 3,4-dihydroxyphenylpyruvic acid: DFT and TD-DFT studies," *Computational and Theoretical Chemistry*, vol. 966, no. 1–3, pp. 232–243, 2011.

[52] J. J. Fifen, M. Nsangou, Z. Dhaouadi, O. Motapon, and S. Lahmar, "Single or double hydrogen atom transfer in the reaction of metal—associated phenolic acids with ˙OH radical: DFT Study," *Journal of Molecular Structure: THEOCHEM*, vol. 901, no. 1–3, pp. 49–55, 2009.

[53] S. Fioruci, *Acquisition et analyse des données pour l'expérience de recherche de matière noire (EDELWEISS) [M.S. thesis]*, Université de Nice-Sophia Antipolis, 2006.

[54] D. M. Bikélé, J. M. Zobo, D. Lissouck, E. Younang, J. H. Nono, and L. M. Mbaze, "DFT investigation of chelation of divalent cations by some 4-benzylidenamino-4, 5-dihydro-1h-1, 2, 4-triazol-5-one derivatives," *Journal of Chemistry and Chemical Engeneering*, vol. 9, pp. 1–14, 2015.

16

Synthesis, Characterization, and Biological Evaluation of Unimetallic and Heterobimetallic Complexes of Bivalent Copper

Anshul Singh and Ashu Chaudhary (iD)

Department of Chemistry, Kurukshetra University, Kurukshetra 136 119, India

Correspondence should be addressed to Ashu Chaudhary; ashuchaudhary21@gmail.com

Academic Editor: Enrico Rizzarelli

We present an inclusive characterization of the unimetallic and heterobimetallic complexes of copper synthesized using $CuCl_2$ and diamine (4-fluoro 1,2-phenylenediamine) resulting in monometallic complex which further undergoes treatment with organometallic dichlorides of group 4 and 14 in 1:2 molar ratio resulting in heterobimetallic complexes. These complexes thoroughly characterized using various physical, analytical, and spectroscopic techniques indicate square planar and distorted octahedral geometry for the synthesized unimetallic and heterobimetallic complexes, respectively. These complexes were evaluated for their antimicrobial efficacy against various bacterial and fungal strains while hepatoprotective activity was also examined in male albino rats.

1. Introduction

The advances in the field of bioinorganic chemistry have augmented the interest in heterobimetallic complexes since it has been recognized that many of these complexes may serve as biologically significant species [1]. The biomedical inorganic chemistry has been a captivating research area due to extensive applications of inorganic pharmaceuticals in innumerable clinical therapy and diagnosis [2]. Transition metal complexes have received much attention because of their biological activities, including antiviral [3], anticarcinogenic [4], antifertility [5], antibacterial [6], and antifungal activities [7].

Among the transition metals, copper has been the most studied metal ion. It is one of the trace elements essential for human life, vital for the function of enzymes, protein and DNA synthesis, and also in the regulation of intracellular redox potential. Ceruloplasmin, the major copper-carrying protein in the blood is bound to almost all the copper prevailing in the serum of living organisms. There is a reduction of copper from oxidation state +2 to +1 by metalloreductases at the cell surface, before uptake into the cell [8]. The chemistry of copper has been ruled essentially by copper(II) compounds,

chiefly due to the difficulty arising in stabilizing copper(I) species. Due to the ineffable biological significance, an enormous amount of copper(II) complexes have been synthesized and explored for their biological activities [9–14].

The emphasis has been laid down on the design of heterobimetallic complexes based on the inference that combining two different metal centers in the same molecule results into the substantial modification of reactivity over that of their monometallic species. The provocative properties of heterobimetallic complexes ascend from the possible "synergistic" effect of two different metal ions held together in close proximity [15]. These annotations have emboldened chemists to synthesize new heterobimetallic complexes comprising two different metal centers with an impression being that such a system will be more efficacious than the unimetallic species involving the individual metal centers for numerous fundamental reactions [16]. Considering the significant role of Cu(II) played in human beings, in addition to its proficient biological activity, we have synthesized and characterized the mononuclear and heterobimetallic Cu(II) complexes with organometallic dichlorides of group 4 and 14 in the presence of 4-fluoro 1,2-phenylenediamine in the current study. Besides the characterization of complexes by various physicochemical techniques, their examination against bacterial strains, namely, *Escherichia coli* (MTCC1687) and *Staphylococcus aureus* (MTCC96) and some fungal strains, namely, *Fusarium oxysporum* and *Aspergillus niger* was carried out with antibiotic streptomycin and Bavistin used as standards.

Apart from antimicrobial activity, hepatoprotectivity of the complexes was also evaluated using male albino rats. The imperative functions in regulating and maintaining homeostasis of the body are performed by the liver, which is also known for its vital role in fat metabolism, bile secretion, storage of vitamins, and detoxification of endogenous and exogenous substances. Therefore, to remain healthy is crucial to maintain a healthy liver [17]. Hepatotoxicity entails chemical-driven liver damage. The considerable cause for the occurrence of iatrogenic diseases is drug-induced hepatotoxicity, and hence, there is an acute necessity for developing reliable hepatoprotective drugs contributing towards the current medical practice. Hence, the synthesized heterobimetallic complexes of copper were inspected for their hepatoprotective activity in male albino rats, and the results obtained were quite encouraging.

2. Experimental

2.1. Materials.
All the chemicals used were of AnalaR grade. $CuCl_2$, 4-fluoro 1,2-phenylenediamine, and organometallic dichlorides were purchased from Sigma-Aldrich and used as obtained. The solvents used were purchased from HiMedia and were distilled and dried before use.

2.2. Physical Measurements and Analytical Methods.
The IR spectra (4000–200 cm^{-1}) were recorded on a Nicolet Magna FTIR-550 spectrophotometer in the form of KBr pellets while the far infrared spectra of the complexes were recorded on the same spectrophotometer in Nujol mulls using CsI cell. Conductivity measurements of the compounds were measured in 10^{-3} M DMF solution on a Century Digital conductivity meter model CC601. Electronic spectra were recorded on a UV-visible spectrophotometer 752/752N. X-ray powder diffraction spectra of the compound were obtained on PANalytical X'pert Pro 3040 Almelo, while EPR spectra of the complexes were recorded on a Varian E-4 EPR spectrometer. Molecular weights were determined by the Rast Camphor method. Metal contents were estimated gravimetrically. Nitrogen and chlorine were estimated by the Kjeldahl's method and the Volhard's method, respectively.

2.3. Synthesis of Monometallic Complex $[Cu(C_6H_7N_2F)_2]Cl_2$.
The monometallic complex was prepared by dissolving anhydrous $CuCl_2$ (1.34 g, 0.01 mol) in hot methanol in a 100 mL round-bottomed flask to which the required amount of diamine, that is, 4-fluoro 1,2-phenylenediamine (2.52 g, 0.02 mol) was added drop wise in 1 : 2 stoichiometric ratio under constant stirring. The product thus obtained was washed with methanol and recrystallized from a 1 : 1 solution of benzene and methanol and then dried in vacuo.

2.4. Synthesis of Heterobimetallic Complex $[Cu(C_6H_5N_2F)_2Sn_2(Ph)_4Cl_2]$.
This complex was obtained by mixing a methanolic solution of monometallic complex $[Cu(C_6H_7N_2F)_2]Cl_2$ (3.86 g, 0.01 mol) to a methanolic solution of Ph_2SnCl_2 (6.88 g, 0.02 mol) in 1 : 2 stoichiometric proportions under constant stirring. The mixture resulted into a solid product, obtained after keeping the reaction mixture overnight at room temperature. The colored product thus attained was washed with methanol and then dried in vacuo. All the other complexes were synthesized using the same procedure by replacing Ph_2SnCl_2 with $(CH_3)_2SnCl_2$, Ph_2SiCl_2, $(Cp)_2TiCl_2$, and $(Cp)_2ZrCl_2$, respectively.

The monometallic and heterobimetallic complexes reported in the current work were synthesized as outlined in Scheme 1.

2.5. Biological Assay

2.5.1. Test Microorganisms.
Unimetallic complex of Cu(II) and its corresponding heterobimetallic complexes have been examined for their fungicidal and bactericidal efficacy. The bacterial and fungal species used for evaluating antimicrobial activity of the synthesized complexes were *Escherichia coli*, *Staphylococcus aureus*, *Fusarium oxysporum*, and *Aspergillus niger*.

(1) Antibacterial Activity. The synthesized complexes were appraised for their antibacterial activity against the bacterial strains *Escherichia coli* (MTCC1687) and *Staphylococcus aureus* (MTCC96) by employing the inhibition zone technique [18]. In this technique, a nutrient agar medium comprising 5 g peptone, 3 g beef extract, 20 g agar-agar, and 5 g NaCl were suspended in 1000 mL distilled water, which was further boiled, allowing all the ingredients to dissolve completely. The prepared agar medium was then poured into the petriplates thereby allowing it to solidify. Solutions of the test compounds in methanol with 500 and 1000 ppm concentrations were prepared. The Whatman No. 1 paper discs with 5 mm diameter, soaked in different solutions of the compounds were dried and then placed in petriplates previously seeded with the test organisms. The petri dishes were then incubated at a temperature of $28 \pm 2°C$ for duration of 24 hrs. DMSO was used as a negative control while a disc of streptomycin was used as a positive control for antibacterial activity. The zone of inhibition thus formed around each disc encompassing the test compounds was measured accurately and used as indices of antimicrobial actions.

(2) Antifungal Activity. The efficacy of the synthesized complexes was evaluated against two of the fungal strains *Fusarium oxysporum* and *Aspergillus niger* using the agar plate technique [19]. In the current technique, potato dextrose agar media is prepared consuming 20 g starch, 20 g dextrose, and 20 g agar-agar dissolved in 1000 mL of distilled water. The compounds were directly mixed with the medium in 50, 100, and 200 ppm concentrations dissolved in methanol. The medium was transferred into the petriplates

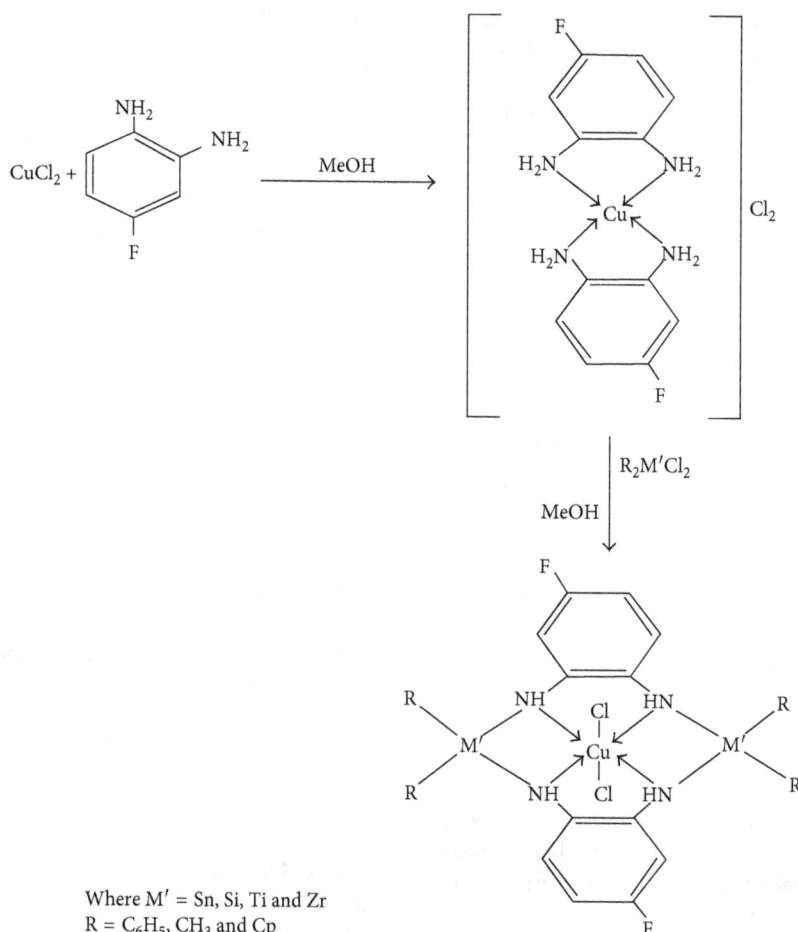

Where M′ = Sn, Si, Ti and Zr
R = C$_6$H$_5$, CH$_3$ and Cp

SCHEME 1

onto which spores of fungi were placed with the help of an inoculum needle. These petriplates were wrapped in the polythene bags containing a few drops of alcohol and were placed in an incubator at $25 \pm 2°C$. The standard drug Bavistin was used as a positive control for the antifungal activity while DMSO was used as a negative control. The linear growth of the fungus was obtained by measuring the diameter of the fungal colony after four days (96 hrs).

The growth of fungi was analyzed, and the percentage of inhibition was calculated by the following equation:

$$\% \text{ inhibition} = \frac{C - T}{C} \times 100, \qquad (1)$$

where C and T are the diameters of the fungal colony in the control and the test plates, respectively.

(3) Hepatoprotective Activity. The hepatoprotective activity was evaluated by carrying out an experiment with the male albino rats weighing about 180–200 g distributed into four groups of 10 rats each. The group I was treated as control, while hepatotoxicity was introduced in the rats belonging to groups II, III, and IV via oral administration of CCl$_4$ of about 0.25 mL/100 g body weight. Animals were fed twice in a week for about 4 weeks. Commencing the fifth day, the animals of groups II and III were provided with an oral

dose (50 mg/100 g bw for 30 days) of the compounds [Cu(C$_6$H$_5$N$_2$F)$_2$Sn$_2$(Ph)$_4$Cl$_2$] and [Cu(C$_6$H$_5$N$_2$F)$_2$Sn$_2$(CH$_3$)$_4$Cl$_2$], respectively. All the animals were fed on commercial standard pellet diet (Hindustan Lever Ltd., Mumbai) with water ad libitum and were maintained in the animal house at a temperature of $25 \pm 2°C$ under 12 hr light/dark cycle with a relative humidity of about $60 \pm 5\%$.

Further experiment required blood of the animals for which three rats were scarified every week. These blood samples were collected by direct heart puncture into a sterilized dried out centrifuge tube while serum was collected for the evaluation of total bilirubin, protein, and albumin/globulin ratio [20]. It is observed that the toxic effect of CCl$_4$ causes harm to the liver resulting in its disfuctioning in the experimental animals, and analogous results are anticipated in the human viral hepatitis model. In CCl$_4$-induced toxic hepatitis, a toxic reactive metabolite, trichloromethyl radical was produced by the microsomal oxidase system that binds covalently to the macromolecules of the lipid membranes of the endoplasmic reticulum resulting in peroxidative degradation of lipids. As an outcome, adipose tissue from the fats gets translocated and further accumulates down in the liver. In majority of cases, this toxic chemical is used as a tool to instigate hepatotoxicity in experimental animals.

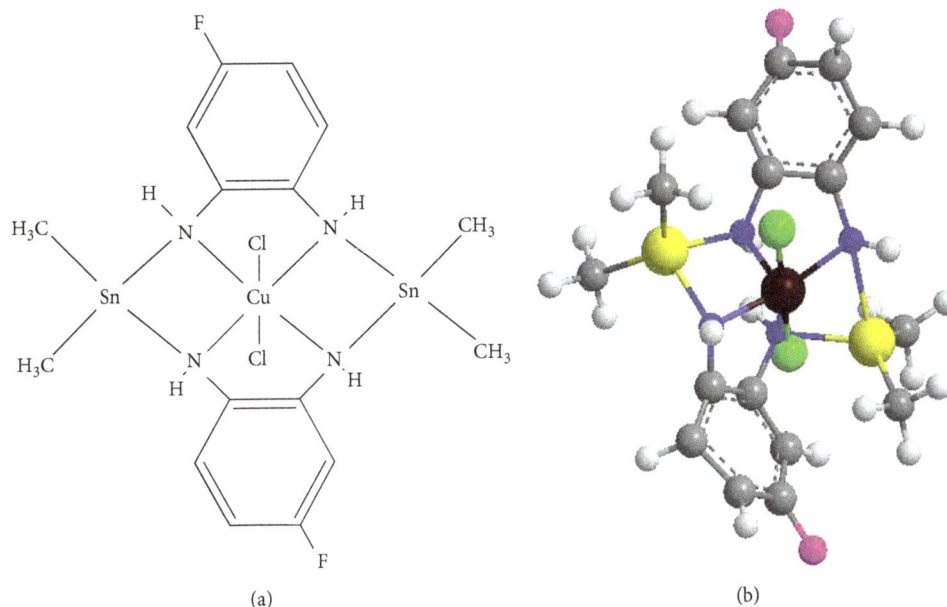

(a)   (b)

FIGURE 1: Proposed structure for heterobimetallic complex $[Cu(C_6H_5N_2F)_2Sn_2(CH_3)_4Cl_2]$.

TABLE 1: Bacterial screening data of the synthesized Cu(II) complexes (percent growth inhibition after 24 hrs at $28 \pm 2°C$).

Complex	Escherichia coli		Staphylococcus aureus	
	500 ppm	1000 ppm	500 ppm	1000 ppm
$[Cu(C_6H_7N_2F)_2]Cl_2$	6	8	6	7
$[Cu(C_6H_5N_2F)_2Sn_2(Ph)_4Cl_2]$	12	15	10	12
$[Cu(C_6H_5N_2F)_2Sn_2(CH_3)_4Cl_2]$	10	11	8	10
$[Cu(C_6H_5N_2F)_2Si_2(Ph)_4Cl_2]$	9	11	8	9
$[Cu(C_6H_5N_2F)_2Ti_2(Cp)_4Cl_2]$	8	10	7	8
$[Cu(C_6H_5N_2F)_2Zr_2(Cp)_4Cl_2]$	8	9	6	7
Streptomycin	17	18	15	17

3. Results and Discussion

3.1. Chemistry. The resulting unimetallic and hetero-bimetallic complexes are color solids with sharp melting point. The synthesized complexes are stable at room temperature and sparingly soluble in cold organic solvents like methanol, ethanol, and benzene but completely soluble in hot solvents. The molar conductance measurements of 10^{-3} M solution in DMF indicate that the synthesized complexes are 1:2 electrolytes. The conductivity measured for the unimetallic complex have conductance of $232\,ohm^{-1}\cdot mol^{-1}\cdot cm^2$ holding electrolytic character while heterobimetallic complexes exhibit conductance in the range of $20-35\,ohm^{-1}\cdot mol^{-1}\cdot cm^2$ suggesting the non-electrolytic character of these complexes [21]. The analytical and physical data of the synthesized complexes have been precised in Supplementary Table 1.

3.1.1. Infrared Spectra. The IR spectra of all the newly synthesized complexes were recorded and summarized in Supplementary Table 2. However, a comparative study was made between the spectra of unimetallic and hetero-bimetallic complexes of Cu(II). The primary amine exhibits a band at higher frequency than that of the corresponding secondary amine. In case of unimetallic complex, a broad and strong band appears for ν(N–H) in the range of $3180-3256\,cm^{-1}$, which swipes over to a lower frequency region in case of heterobimetallic complexes approving the formation of bond between metal and nitrogen, while bands due to δ(N–H) appear in the region of $1535-1546\,cm^{-1}$ with no apparent change after chelation [22]. Bands at 1643, 1528, and $1453\,cm^{-1}$ appear due to (C=C) stretching in case of aromatic ring, while bands due to (C–H) and (C–N) stretching appear at 3057 and $844\,cm^{-1}$, respectively. Bands of medium intensity appearing in the range of $420-585\,cm^{-1}$ are attributed to (M–N) vibrations [23]. In case of hetero-bimetallic complexes of copper with titanium and zirconium, the presence of cyclopentadienyl ring is confirmed by the IR bands appearing at 3000 for ν(C–H), 1433 for ν(C–C), 1030 for δ(C–H) in plane, and 812 for δ(C–H) out of plane vibrations. In addition, bands due to (Ti–C_5H_5) and (Zr–C_5H_5) appear at 445 and $442\,cm^{-1}$, respectively. A medium intensity band observed in the far IR region of the metal complexes $(460-470)\,cm^{-1}$ was assigned to (Cu–N). The far IR spectra show a distinct band at $317\,cm^{-1}$ attributed to (Cu–Cl) band [24], which clearly indicates the presence of

TABLE 2: Fungicidal screening data of the synthesized Cu(II) complexes (percent growth inhibition after 96 hrs at $25 \pm 2°C$, conc. in ppm).

Complex	Fusarium oxysporum			Aspergillus niger		
	50 ppm	100 ppm	200 ppm	50 ppm	100 ppm	200 ppm
$[Cu(C_6H_7N_2F)_2]Cl_2$	32	38	46	28	35	41
$[Cu(C_6H_5N_2F)_2Sn_2(Ph)_4Cl_2]$	52	56	64	46	51	58
$[Cu(C_6H_5N_2F)_2Sn_2(CH_3)_4Cl_2]$	43	50	52	40	45	49
$[Cu(C_6H_5N_2F)_2Si_2(Ph)_4Cl_2]$	40	46	49	37	41	45
$[Cu(C_6H_5N_2F)_2Ti_2(Cp)_4Cl_2]$	38	42	44	35	38	40
$[Cu(C_6H_5N_2F)_2Zr_2(Cp)_4Cl_2]$	36	40	43	33	35	37
Bavistin	91	100	100	89	98	100

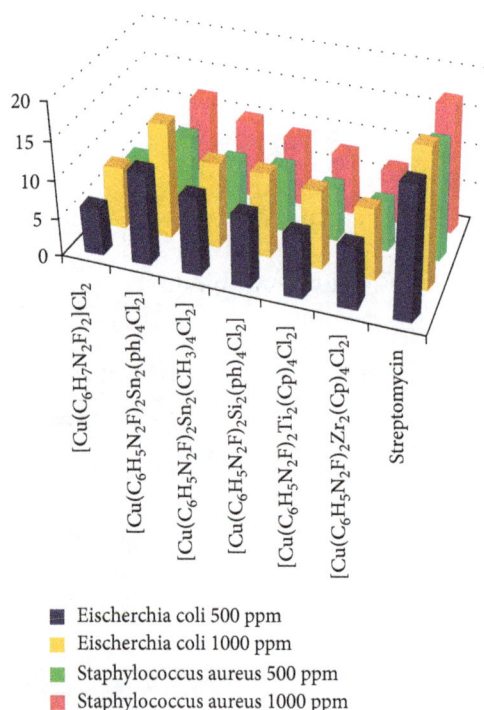

FIGURE 2: Bactericidal activity of the synthesized Cu(II) complexes.

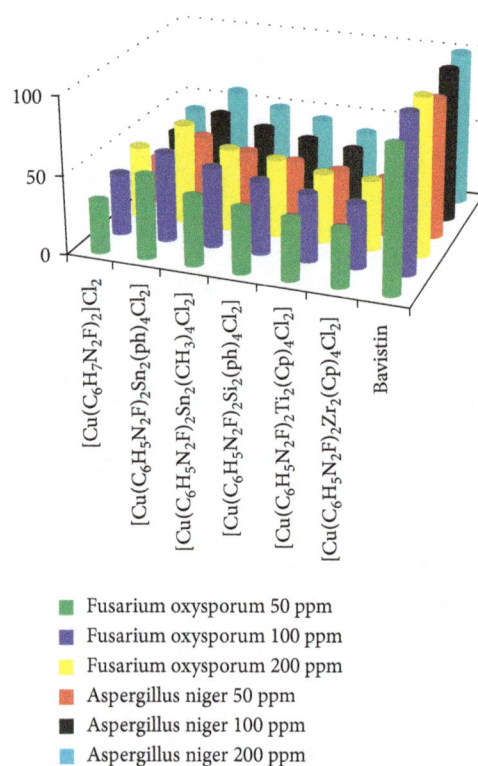

FIGURE 3: Fungicidal activity of the synthesized Cu(II) complexes.

chlorine bonded with central metal (i.e., Cu) ion. Hence, the IR spectra support an octahedral geometry in case of heterobimetallic complexes further verified by the ESR spectra.

3.1.2. Electron Spin Resonance Spectra. EPR studies of synthesized complexes were carried out on the X-band at 9.1 GHz under the magnetic field strength 3000 G. It is well known that in the case of tetragonal and square planar complexes, the unpaired electron lies in the $d_{x^2-y^2}$ orbital giving $^2B_{1g}$ as the ground state with $g_\| > g\perp$ [1]. The data obtained in case of unimetallic complex shows $g_\| = 2.09$ and $g\perp = 2.05$, respectively. Thus, for the copper complexes with $g_\| > g\perp$ supports the fact that the ground state of Cu(II) is $^2B_{1g}$ with the unpaired electron in the $d_{x^2-y^2}$ orbital. The observed g values for the synthesized monometallic copper complex lie in the range reported for square planar complexes; thus, the EPR spectral studies strongly support the square planar structure for unimetallic complex. Their square planar geometry has also been verified from electronic spectra. However, in case of heterobimetallic

complexes, $g\perp > g\|$ is observed thereby proposing a distorted octahedral geometry with d_{z^2} as the ground state of the system [25]. On this basis, it is decided that chloride ions coordinate with Cu(II) on complexation with organometallic dichlorides in case of heterobimetallic complexes thus achieving an octahedral environment for copper in these complexes.

3.1.3. Electronic Spectra. In the electronic spectrum of unimetallic complex, a strong band is observed at 589 nm, which is due to d-d transitions, and is a characteristic feature of the square planar complexes while a strong band at 297 nm and shoulders arise due to the intraligand transitions [26]. In case of heterobimetallic complexes of copper, two bands are witnessed at 920 nm and 875 nm, respectively. This displacement of d-d bands to lower energies is basically due to the distortion in the octahedral geometry of copper(II) complexes.

TABLE 3: Hepatoprotective activity of heterobimetallic complexes of Cu(II).

Group of animals	Time in weeks				F value	C.D at 5%
	I	II	III	IV		
Total bilirubin						
Control	0.85 ± 0.10	0.86 ± 0.79	0.73 ± 0.06	0.86 ± 0.69	2.74	0.018
$[Cu(C_6H_5N_2F)_2Sn_2(Ph)_4Cl_2]$	1.32 ± 0.75	1.48 ± 0.38	1.56 ± 0.45	1.61 ± 0.18	15.1	0.071
$[Cu(C_6H_5N_2F)_2Sn_2(CH_3)_4Cl_2]$	1.46 ± 0.50	1.52 ± 0.08	1.75 ± 0.10	1.80 ± 0.08	46.8	0.084
CCl_4	1.80 ± 0.05	2.30 ± 0.60	2.75 ± 0.15	3.70 ± 0.41	55.2	0.315
Total protein						
Control	7.24 ± 0.23	6.87 ± 0.22	6.90 ± 0.25	7.08 ± 0.10	0.451	0.442
$[Cu(C_6H_5N_2F)_2Sn_2(Ph)_4Cl_2]$	6.67 ± 0.25	6.44 ± 0.20	6.40 ± 0.30	6.30 ± 0.24	3.10	0.348
$[Cu(C_6H_5N_2F)_2Sn_2(CH_3)_4Cl_2]$	6.50 ± 0.16	6.39 ± 0.27	6.36 ± 0.25	6.15 ± 0.22	4.69	0.342
CCl_4	5.98 ± 0.24	5.20 ± 0.17	4.82 ± 0.29	4.19 ± 0.17	39.6	0.350
Albumin						
Control	4.68 ± 0.25	4.50 ± 0.20	4.50 ± 0.20	4.78 ± 0.30	0.45	0.376
$[Cu(C_6H_5N_2F)_2Sn_2(Ph)_4Cl_2]$	4.23 ± 0.37	3.97 ± 0.17	3.75 ± 0.25	3.55 ± 0.20	4.49	0.366
$[Cu(C_6H_5N_2F)_2Sn_2(CH_3)_4Cl_2]$	3.72 ± 0.24	3.62 ± 0.40	3.60 ± 0.25	3.42 ± 0.17	1.61	0.416
CCl_4	3.50 ± 0.20	2.50 ± 0.25	1.97 ± 0.40	1.35 ± 0.12	414.9	0.405
Globulin						
Control	2.40 ± 0.81	2.35 ± 0.10	2.34 ± 0.73	2.30 ± 0.17	0.18	0.207
$[Cu(C_6H_5N_2F)_2Sn_2(Ph)_4Cl_2]$	2.53 ± 0.14	2.53 ± 0.24	2.67 ± 0.19	2.57 ± 0.14	1.44	0.150
$[Cu(C_6H_5N_2F)_2Sn_2(CH_3)_4Cl_2]$	2.67 ± 0.19	2.73 ± 0.18	2.74 ± 0.31	2.79 ± 0.12	1.23	0.295
CCl_4	2.58 ± 0.10	2.62 ± 0.10	2.80 ± 0.16	2.80 ± 0.15	5.80	0.183
Aspartate aminotransaminase						
Control	0.45 ± 0.10	7.50 ± 0.24	7.78 ± 0.50	9.20 ± 0.20	12.9	0.598
$[Cu(C_6H_5N_2F)_2Sn_2(Ph)_4Cl_2]$	11.55 ± 0.40	14.25 ± 0.35	13.90 ± 0.55	15.10 ± 0.35	56.10	0.580
$[Cu(C_6H_5N_2F)_2Sn_2(CH_3)_4Cl_2]$	13.0 ± 0.40	14.15 ± 0.24	14.32 ± 0.25	16.40 ± 0.65	52.61	0.602
CCl_4	16.80 ± 0.40	22.20 ± 0.80	25.76 ± 0.70	32.20 ± 1.0	245.7	1.160
Alanine aminotransaminase						
Control	9.65 ± 0.65	9.85 ± 0.51	8.65 ± 0.20	8.40 ± 0.65	7.90	6.729
$[Cu(C_6H_5N_2F)_2Sn_2(Ph)_4Cl_2]$	14.32 ± 0.27	14.82 ± 0.63	14.82 ± 0.79	15.93 ± 0.45	8.95	0.896
$[Cu(C_6H_5N_2F)_2Sn_2(CH_3)_4Cl_2]$	12.73 ± 0.62	13.24 ± 0.66	14.04 ± 0.52	14.42 ± 0.27	5.18	0.990
CCl_4	16.10 ± 0.30	20.90 ± 0.24	26.00 ± 0.43	33.40 ± 1.5	265.0	1.376
Alkaline phosphatase						
Control	9.67 ± 0.50	9.85 ± 0.50	8.65 ± 0.20	8.45 ± 0.64	7.75	0.728
$[Cu(C_6H_5N_2F)_2Sn_2(Ph)_4Cl_2]$	12.85 ± 0.50	14.83 ± 0.52	14.06 ± 0.52	14.48 ± 0.30	5.18	0.985
$[Cu(C_6H_5N_2F)_2Sn_2(CH_3)_4Cl_2]$	13.40 ± 0.60	14.77 ± 0.43	14.87 ± 0.43	15.61 ± 0.42	8.95	0.886
CCl_4	16.10 ± 0.34	20.24 ± 0.24	25.00 ± 0.44	33.60 ± 1.46	263.5	1.278

3.1.4. Mass Spectroscopy. The FAB mass spectra of the heterobimetallic complex $[Cu(C_6H_5N_2F)_2Sn_2(Ph)_4Cl_2]$ exhibited molecular ion peaks at m/z 928, 890, 813, 736, 617, and 344 assigned to $[Cu(C_6H_5N_2F)_2Sn_2(Ph)_4Cl_2]^+$, $[Cu(C_6H_5N_2)_2Sn_2(Ph)_4Cl_2]^+$, $[Cu(C_6H_5N_2)_2Sn_2(Ph)_3Cl_2]^+$, $[Cu(C_6H_5N_2)_2Sn_2(Ph)_2Cl_2]^+$, $[Cu(C_6H_5N_2)_2Sn(Ph)_2Cl_2]^+$, and $[Cu(C_6H_5N_2)_2Cl_2]^+$. The two coordinated chlorides are removed with a mass loss of $m/z = 71$ with peak obtained at $m/z = 273$ (Supplementary Figure 1).

3.1.5. X-Ray Powder Diffraction Studies. The possible geometry of the finely powdered mononuclear complex $[Cu(C_6H_7N_2F)_2]Cl_2$ has been inferred on the basis of X-ray powder diffraction studies (Supplementary Figure 2). The interplanar spacing values ("d" in Å) of this complex have been measured from the diffractogram and the Miller indices h, k, and l have been assigned to each d value as reported in Supplementary Table 3. The results display that compounds belong to "orthorhombic" crystal system with

the unit cell parameters $a = 9.480$, $b = 13.416$, $c = 21.193$, and $\alpha = \beta = \gamma = 90°$.

From the data obtained using various physicochemical techniques such as IR, ESR, mass, and X-ray powder diffraction studies, the structure proposed for heterobimetallic complex is depicted in Figure 1.

3.2. Biological Assay

3.2.1. Antimicrobial Activity. The data obtained after evaluating the synthesized complexes against two bacterial (*E. coli* and *S. aureus*) strains and two fungal (*F. oxysporum* and *A. niger*) strains evidently divulges the amplified activity of heterobimetallic complexes in comparison to unimetallic complexes of copper as summarized in Tables 1 and 2. This can be explained on the basis of Tweedy's chelation theory [27] according to which, the polarity of metal ion reduces due to chelation owing to the partial sharing of its positive charge with the donor groups and

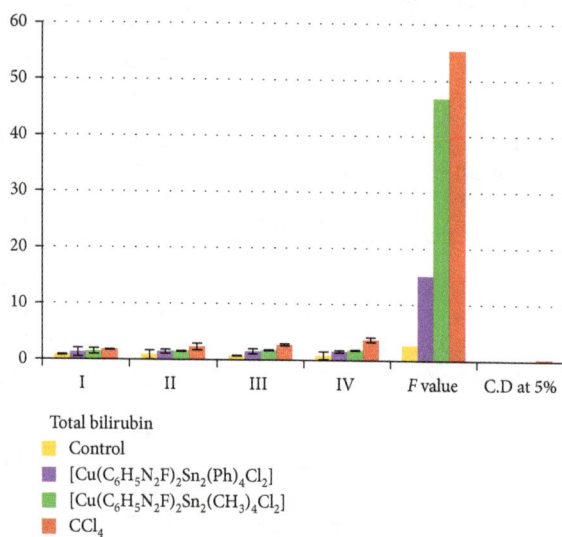

Total bilirubin
- Control
- $[Cu(C_6H_5N_2F)_2Sn_2(Ph)_4Cl_2]$
- $[Cu(C_6H_5N_2F)_2Sn_2(CH_3)_4Cl_2]$
- CCl_4

(a)

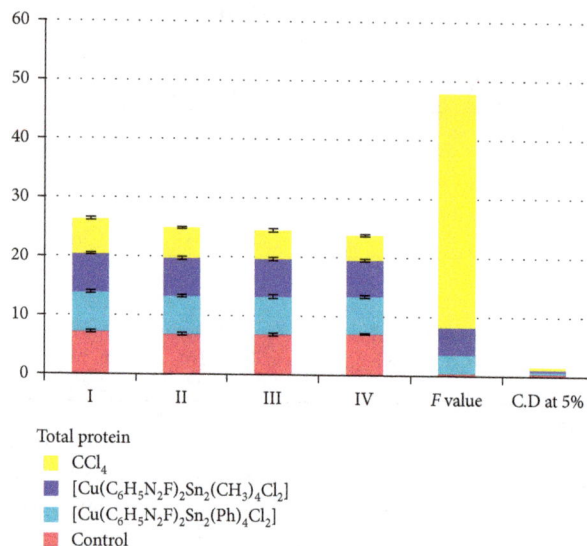

Total protein
- CCl_4
- $[Cu(C_6H_5N_2F)_2Sn_2(CH_3)_4Cl_2]$
- $[Cu(C_6H_5N_2F)_2Sn_2(Ph)_4Cl_2]$
- Control

(b)

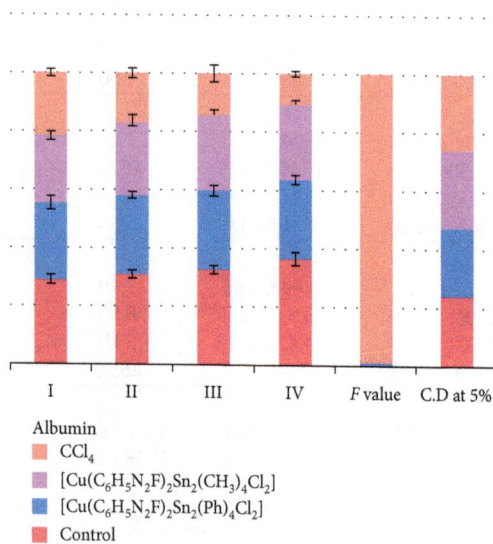

Albumin
- CCl_4
- $[Cu(C_6H_5N_2F)_2Sn_2(CH_3)_4Cl_2]$
- $[Cu(C_6H_5N_2F)_2Sn_2(Ph)_4Cl_2]$
- Control

(c)

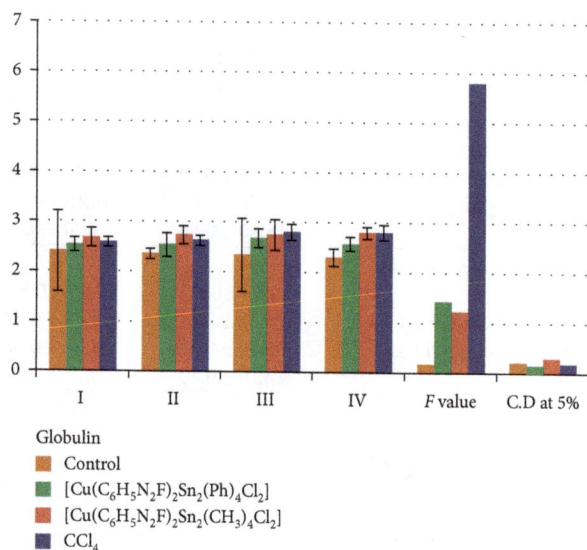

Globulin
- Control
- $[Cu(C_6H_5N_2F)_2Sn_2(Ph)_4Cl_2]$
- $[Cu(C_6H_5N_2F)_2Sn_2(CH_3)_4Cl_2]$
- CCl_4

(d)

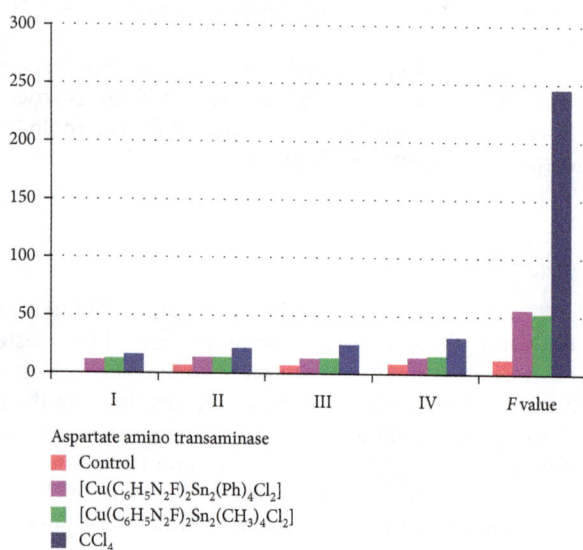

Aspartate amino transaminase
- Control
- $[Cu(C_6H_5N_2F)_2Sn_2(Ph)_4Cl_2]$
- $[Cu(C_6H_5N_2F)_2Sn_2(CH_3)_4Cl_2]$
- CCl_4

(e)

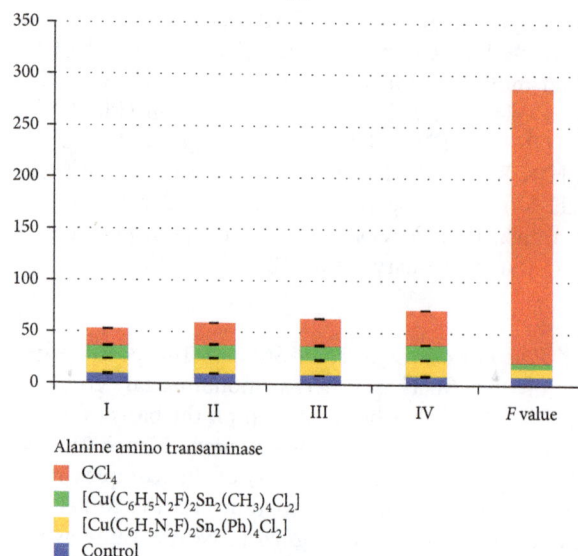

Alanine amino transaminase
- CCl_4
- $[Cu(C_6H_5N_2F)_2Sn_2(CH_3)_4Cl_2]$
- $[Cu(C_6H_5N_2F)_2Sn_2(Ph)_4Cl_2]$
- Control

(f)

FIGURE 4: Continued.

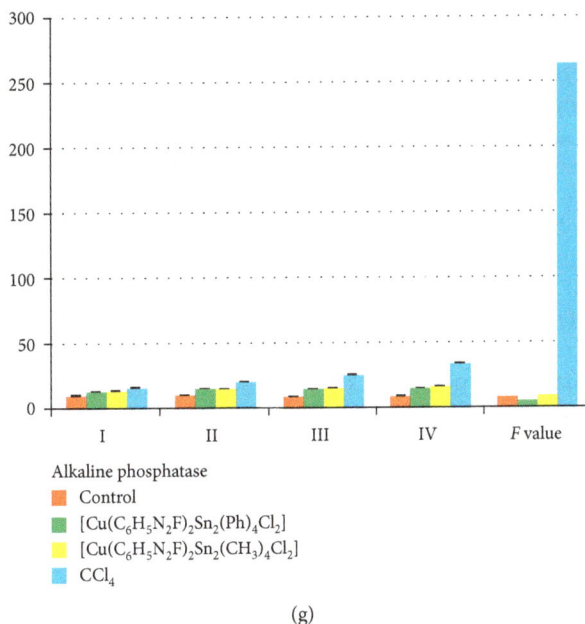

(g)

FIGURE 4: Graphical representation of various biochemical parameters: (a) total bilirubin, (b) total protein, (c) albumin, (d) globulin, (e) aspartate aminotransaminase, (f) alanine aminotransaminase, and (g) alkaline phosphatase as compared to the control.

possible π-electron-delocalization over the whole chelation ring, which intensifies the lipophilic character of the metal complex, consecutively approving its penetration through the lipid layers of the organism cell membrane, resulting in reduction of the normal cell process along with de-activation of numerous cellular enzymes that play dicta-torial roles in the various metabolic pathways of these microorganisms. Hence, it is clear that amid all the syn-thesized complexes, unimetallic complexes of Cu(II) are lesser active, that is, having less significant antimicrobial efficacy as compared to heterobimetallic complexes where the synergistic effect of two metals displays remarkable biological activity.

In the present series, heterobimetallic complex $[Cu(C_6H_5N_2F)_2Sn_2(Ph)_4Cl_2]$ exhibits the unsurpassed antifungal and antibacterial activity thus presenting enriched antimicrobial efficacy as equated to the unimetallic complex of Cu(II) (Figures 2 and 3).

3.2.2. Appraisal of Hepatoprotective Activity. There are number of biochemical parameters used for the evaluation of hepatoprotective activity in carbon tetrachloride-induced toxicity which comprises bilirubin, protein, albumin, glob-ulin, aspartate aminotransaminase, alkaline phosphatase, and alanine aminotransaminase (Table 3). The prominent in-ferences drawn from the experiment are as follows:

(1) The present study revealed a substantial escalation in the levels of bilirubin, globulin, aspartate amino-transaminase, alanine aminotransaminase, and al-kaline phosphatase in the blood samples of the group bearing animals treated with CCl₄. The rise in the levels of these biochemical parameters is a clear

indication of cellular leakage with a loss of functional integrity of the cell membrane.

(2) The intensity of jaundice is confirmed from the total amount of bilirubin present with a normal range of 0.2 to 1 mg/100 mL of serum. However, it is ob-served that hyperbilirubinemia occurs more in case of hepatitis patients with the excretion of bilirubin present in the liver into the canaliculi and then re-gurgitated into the blood stream. The group of animals treated with compounds $[Cu(C_6H_5N_2F)_2Sn_2(Ph)_4Cl_2]$ and $[Cu(C_6H_5N_2F)_2Sn_2(CH_3)_4Cl_2]$ exhibits reduced levels of total bilirubin in the blood samples.

(3) A number of serum proteins are being synthesized by the liver which plays an imperative role in the diagnosis of jaundice. Further, there is a correlation between the degree of serum hypoalbuminemia and hyper-globulinemia [28]. Normally, the ratio of albumin and globulin lies in the range of 2 : 1. However, in case of CCl₄-treated animals, there is a significant reduction in the levels of total protein and albumin with the increase in globulin level. Even with the toxic effect of CCl₄, the complexes $[Cu(C_6H_5N_2F)_2Sn_2(Ph)_4Cl_2]$ and $[Cu(C_6H_5N_2F)_2Sn_2(CH_3)_4Cl_2]$ were highly proficient in refurbishing the reduced and increased levels of serum total protein and globulin, respectively.

The percentage of serum protein level restored was found to be more in case of heterobimetallic complex $[Cu(C_6H_5N_2F)_2Sn_2(Ph)_4Cl_2]$.

(4) The normal values of aspartate aminotransaminase (ALT), alanine aminotransaminase (AST), and alkaline phosphatase (ALP) ranges from 5 to 20, 5 to 15, and 7 to 9 IU/mg of protein, respectively.

These enzymes present in the serum are supportive in the diagnosis of hepatitis disease while an upsurge in the concentration of these enzymes is detected with the damage in the liver tissue, which is ostensibly due to the release of these enzymes from the damaged cells. In acute hepatic necrosis, the level of AST and ALT are estimated to rise by 2 to 20 folds over that of controls while on the other hand, in case of obstructive and posthepatic jaundice, escalation of ALP was more. In the current research, the group of animals treated with CCl_4 exhibits remarkable increase in the activities of these enzymes parallelly from 1st to 4th week of treatment. While complexes $[Cu(C_6H_5N_2F)_2Sn_2(Ph)_4Cl_2]$ and $[Cu(C_6H_5N_2F)_2Sn_2(CH_3)_4Cl_2]$ prohibited the increase in the levels of these enzymes, presenting the pattern of recovery from the toxic effect (Figure 4).

4. Conclusion

The present study unveiled the biological potency of heterobimetallic complexes of copper synthesized using group 4 and 14 organometallic dichlorides. All these synthesized unimetallic and heterobimetallic complexes have been structurally characterized using varied techniques such as IR, ESR, electronic spectra, mass spectra, and X-ray diffraction studies. The spectroscopic data revealed square planar and distorted octahedral geometry for mononuclear and heterobimetallic complexes, respectively. These heterobimetallic complexes possess high biological activity emerging from the synergistic effect arising due to the two different metal centers seized together in close proximity. The antimicrobial-screening data of these complexes indicate that heterobimetallic complexes are more active against these microbes as compared to the unimetallic complexes with heterobimetallic complex of copper with tin proving out to be the best. The heterobimetallic complexes with tin were also evaluated for the hepatoprotective activity in the male albino rats taking into account different biochemical parameters.

Acknowledgments

The authors are grateful to the Council of Scientific and Industrial Research (CSIR), New Delhi, India, for financial assistance in the form of Senior Research Fellowship (SRF), vide letter no. 09/105(0221)/2015-EMR-I.

References

[1] H. Zafar, A. Kareem, A. Sherwani et al., "Synthesis and characterization of Schiff base octaazamacrocyclic complexes and their biological studies," *Journal of Photochemistry and Photobiology B: Biology*, vol. 142, pp. 8–19, 2015.

[2] L. Ronconi and P. J. Sadler, "Using coordination chemistry to design new medicines," *Coordination Chemistry Reviews*, vol. 251, no. 13-14, pp. 1633–1648, 2007.

[3] T. H. Sucipto, S. Churrotin, H. Setyawati, T. Kotaki, F. Martak, and S. Soegijanto, "Antiviral activity of copper(II) chloride dihydrate against dengue virus type-2 in vero cell," *Indonesian Journal of Tropical and Infectious Disease*, vol. 6, no. 4, pp. 84–87, 2017.

[4] M. Jopp, J. Becker, S. Becker et al., "Anticancer activity of a series of copper(II) complexes with tripodal ligands," *European Journal of Medicinal Chemistry*, vol. 132, pp. 274–281, 2017.

[5] A. Chaudhary and A. Singh, "Synthesis, characterization, and evaluation of antimicrobial and antifertility efficacy of heterobimetallic complexes of copper(II)," *Journal of Chemistry*, vol. 2017, Article ID 5936465, 9 pages, 2017.

[6] A. K. Patel, J. C. Patel, H. R. Dholariya, K. S. Patel, V. K. Patel, and K. D. Patel, "Synthesis, characterization and biological evaluation: copper(II) complexes of hydroxy coumarins with ciprofloxacin," *Research Journal of Pharmaceutical, Biological and Chemical Sciences*, vol. 4, no. 3, pp. 564–579, 2013.

[7] A. Kostelidou, S. Kalogiannis, O. A. Begou, F. Perdih, I. Turel, and G. Psomas, "Synthesis, structure and biological activity of copper(II) complexes with gatifloxacin," *Polyhedron*, vol. 119, pp. 359–370, 2016.

[8] J. Lopes, D. Alves, T. S. Morais et al., "New copper(I) and heteronuclear copper(I)–ruthenium(II) complexes: synthesis, structural characterization and cytotoxicity," *Journal of Inorganic Biochemistry*, vol. 169, pp. 68–78, 2017.

[9] C. Santini, M. Pellei, V. Gandin, M. Porchia, F. Tisato, and C. Marzano, "Advances in copper complexes as anticancer agents," *Chemical Reviews*, vol. 114, no. 1, pp. 815–862, 2014.

[10] S. K. Meena and M. Jain, "Synthesis, spectral studies and antimicrobial screening of some transition metal complexes with substituted acetophenone thiosemicarbazone," *International Journal of Pharmaceutical Sciences and Research*, vol. 5, no. 11, pp. 4880–4886, 2014.

[11] T. Aiyelabola, E. Akinkunmi, E. Obuotor, I. Olawuni, D. Isabirye, and J. Jordaan, "Synthesis characterization and biological activities of coordination compounds of 4-hydroxy-3-nitro-2H-chromen-2-one and its aminoethanoic acid and pyrrolidine-2-carboxylic acid mixed ligand complexes," *Bioinorganic Chemistry and Applications*, vol. 2017, Article ID 6426747, 9 pages, 2017.

[12] Y. Xiao, Q. Wang, Y. Huang, X. Ma, X. Xiong, and H. Li, "Synthesis, structure, and biological evaluation of a copper(II) complex with fleroxacin and 1,10-phenanthroline," *Dalton Transactions*, vol. 45, pp. 10928–10935, 2016.

[13] S. Alfonso, S. Gonzalez, A. R. Higuera-Padilla et al., "A new complex of copper-phosphole. Synthesis, characterization and evaluation of biological activity," *Inorganica Chimica Acta*, vol. 453, pp. 538–546, 2016.

[14] A. A. Faheim and A. N. M. A. Alaghaz, "Synthesis, characterization, biological activity and molecular modeling studies of Cu(II) complex with N,O-chelating Schiff's base ligand incorporating Azo and sulfonamide moieties," *Current Synthetic and Systems Biology*, vol. 2, no. 113, pp. 1–7, 2014.

[15] H. Molaee, S. M. Nabavizadeh, M. Jamshidi, M. Vilsmeier, A. Pfitzner, and M. S. Sangaria, "Phosphorescent heterobimetallic complexes involving platinum(IV) and rhenium (VII) centers connected by an unsupported μ-oxido bridge," *Dalton Transactions*, vol. 46, pp. 16077–16088, 2017.

[16] T. Huang, X. Wu, X. Song et al., "Ferromagnetic coupling in d^1-d^3 linear oxido-bridged heterometallic complexes: ground-state models of metal-to-metal charge transfer excited states," *Dalton Transactions*, vol. 44, pp. 18937–18944, 2015.

[17] K. Lahon and S. Das, "Hepatoprotective activity of *Ocimum sanctum* alcoholic leaf extract against paracetamol-induced liver damage in Albino rats," *Pharmacognosy Research*, vol. 3, no. 1, pp. 13–18, 2011.

[18] K. M. Chauhan and S. C. Joshi, "Synthesis, spectral, antimicrobial and antiandrogenic studies of main group metal complexes with biologically potent benzothiazoline," *Inorganic Letters*, vol. 1, no. 1, pp. 23–35, 2014.

[19] I. Masih, N. Fahmi, and Rajkumar, "Template synthesis, spectroscopic studies, antimicrobial, nematicidal and pesticidal activities of chromium(III) macrocyclic complexes," *Journal of Enzyme Inhibition and Medicinal Chemistry*, vol. 28, no. 1, pp. 33–40, 2013.

[20] A. Chaudhary and R. V. Singh, "In Vitro hepatoprotective evaluation of organolead(II) complexes on carbon tetrachloride induced hepatoxicity: synthesis, characterization and biopotency," *Main group Metal Chemistry*, vol. 28, no. 4, pp. 223–233, 2005.

[21] S. Chandra, M. Tyagi, and S. Aggarwal, "Spectral and antimicrobial studies on tetraaza macrocyclic complexes of Pd(II), Pt(II), Rh(III) and Ir(III) metal ions," *Journal of Saudi Chemical Society*, vol. 15, no. 1, pp. 49–54, 2011.

[22] S. K. Tripathy, A. Panda, P. K. Das, N. K. Behera, A. Mahapatra, and A. K. Panda, "Homo and bimetallic dinuclear macrocyclic complexes of 1,4,5,8,11,14,15,18,19,21, decaaza 2,3,6,7, 12,13,16,17,-octaphenyl-20-thia-undodecane-1,3,5,7,11,13,15,17-octaene (DOTUO): its synthesis, characterization and biological properties," *International Journal of Science, Environment and Technology*, vol. 3, no. 1, pp. 208–223, 2014.

[23] R. Odhiambo, G. K. Muthakia1, and S. M. Kagwanja, "Synthesis, characterization and electrochemistry of heterobimetallic complexes containing molybdenum(II) nitrosyl and manganese(II)-Schiff base centers," *Bulletin of the Chemical Society of Ethiopia*, vol. 24, no. 1, pp. 47–58, 2010.

[24] S. Chandra, M. Tyagi, and K. Sharma, "Mn(II), Co(II), Ni(II) and Cu(II) complexes of a tetraaza macrocyclic ligand: synthesis, characterization and biological screening," *Journal of the Iranian Chemical Society*, vol. 6, no. 2, pp. 310–316, 2009.

[25] E. E. Aranda, T. A. Matias, K. Araki et al., "Design, syntheses, characterization, and cytotoxicity studies of novel heterobinuclear oxindolimine copper(II)-platinum(II) complexes," *Journal of Inorganic Biochemistry*, vol. 165, pp. 108–118, 2016.

[26] C. J. Dhanaraj and J. Johnson, "Synthesis, characterization, electrochemical and biological studies on some metal(II) Schiff base complexes containing quinoxaline moiety," *Spectrochimica Acta Part A: Molecular and Biomolecular Spectroscopy*, vol. 118, no. 24, pp. 624–631, 2014.

[27] T. R. Arun, R. Subramanian, S. Packianathan, and N. Raman, "Fluorescence titrations of bio-relevant complexes with DNA: synthesis, structural investigation, DNA binding/cleavage, antimicrobial and molecular docking studies," *Journal of Fluorescence*, vol. 25, no. 4, pp. 1127–1140, 2015.

[28] A. Chaudhary, E. Rawat, and R. C. Kamboj, "Microwave assisted synthesis, antimicrobial and hepatotoxic activity of tetraamide functionalized macrocyclic complexes of tin(II)," *International Journal of Current Research in Chemistry and Pharmaceutical Sciences*, vol. 1, no. 8, pp. 25–33, 2014.

1717

Hydrothermal Synthesis, Structural Characterization, and Interaction Mechanism with DNA of Copper(II) Complex Containing 2,2′-Bipyridine

Ting Liu⬤, **Yi-An Wang, Qing Zang, and Guo-Qing Zhong**⬤

School of Material Science and Engineering, Southwest University of Science and Technology, Mianyang 621010, China

Correspondence should be addressed to Guo-Qing Zhong; zgq316@163.com

Academic Editor: Concepción López

A Cu(II) complex [Cu(bipy)(H$_2$O)$_2$(SO$_4$)]$_n$ (bipy = 2,2′-bipyridine) was synthesized by hydrothermal method and characterized structurally by elemental analyses, single crystal X-ray diffraction, infrared spectra, and thermogravimetry and differential scanning calorimetry. The Cu(II) was hexacoordinated by two N atoms from bipy, two O atoms from different sulfate radical anions, and two O atoms from two water molecules, forming a slightly distorted octahedral geometry, and bridged by sulfato groups into polymeric chains. Under the condition of physiological pH, the interaction mechanism between the complex and hsDNA was explored with acridine orange as a fluorescence probe by spectroscopic methods. The binding modes between the complex and hsDNA were the electrostatic and embedded modes.

1. Introduction

Design and synthesis of organometallic complexes have become an active research area because of their novel topologies, fascinating functionalities, special properties, and potential applications, such as biomedical utilization, multifunctional materials, molecular adsorption, gas storage, catalysis, magnetism, and so on [1–6]. Chemists have synthesized many organic-inorganic hybrid materials with nitrogen heterocyclic compounds as organic building blocks [7–10]. Trace element copper plays an important role in endogenous oxidative DNA damage associated with aging and cancer [11, 12]. Cu(II) complexes have many bioactivities such as antitumor [13, 14], antimicrobial [15–17], and oxidation of ascorbic acid in the presence of oxygen [18]. In addition, Cu(II) complexes can bind to DNA through noncovalent and covalent interactions [19]. Many researchers found that Cu(II) complexes had potential in the treatment of cancers and other diseases [20]. 2,2′-Bipyridine is a potential antitumor agent and often acts as ancillary ligand to strengthen the binding ability of a complex through enhancing the molecule planarity [21]. The complexes of

2,2′-bipyridine and its derivatives have been reported by a number of authors [22, 23]. The method of hydrothermal synthesis has been used to produce various solids, such as oxide ceramics, microporous crystals, metal complexes, nanomaterials, and so forth [24–29]. In particular, the molecular structures obtained by this method are unexpected compared with those obtained by the common solution method [30].

The modes of noncovalent interaction for metal complexes with DNA include intercalation, electrostatic effect, groove binding, and so on, and the effectiveness mainly depends on the binding modes and affinities between complexes and DNA [31–34]. There is continuing interest in some metal complexes that interact with DNA [35]. Furthermore, the studies of interaction of metal coordination polymers with DNA have been of great interest [36, 37]. However, examples of such metal coordination polymers are still few. Therefore, it is of great significance to explore the binding modes of DNA with metal coordination polymers containing rigid ligands.

We herein report the X-ray single crystal structure, Fourier transform infrared spectra (FTIR), and thermogravimetry and differential scanning calorimetry (TG-DSC) of the Cu(II) complex containing 2,2′-bipyridine, which is synthesized by hydrothermal method, and explore its bioactivities with herring sperm DNA (hsDNA) using acridine orange (AO) as a fluorescence probe by spectral methods.

2. Materials and Methods

2.1. Materials. All chemicals were of analytical reagents and used as received without further purification. 2,2′-Bipyridine and AO were purchased from Alfa Aesar, and copper sulfate pentahydrate and other reagents were purchased from Merck. The hsDNA was purchased from Sigma Biological Co., its purity was monitored by the ratio of absorbance at 260–280 nm, and the ratio of 1.8–1.9 indicated that the hsDNA was free from protein. The hsDNA was dissolved in double-distilled water with 50 mmol·L^{-1} sodium chloride and dialyzed at 4°C for 48 h [38]. The hsDNA concentration was measured by UV-Vis at 260 nm. Tris-HCl buffer solution (pH 7.40) was prepared by using triple-distilled water.

2.2. Physical Measurements. The C, H, and N in the complex were analyzed with a Vario EL CUBE elemental analyzer, and the copper was determined by EDTA titration. FTIR spectra were obtained with KBr pellets on a Perkin-Elmer Spectrum One-Spectrometer in the range 4000–400 cm^{-1}. The thermal analysis was performed by a SDT Q600 thermogravimetric analyzer from 30 to 800°C at a heating rate of 10°C·min^{-1} under air flow of 100 mL·min^{-1}. UV-Vis spectra in Tris-HCl buffer solution (pH 7.40) were measured with a Unico spectrophotometer (UV-2102) in the range 200–600 nm. Fluorescence spectra were recorded on a PE LS-55 spectrofluorophotometer. Viscosities were measured with an Ubbelohde capillary viscometer having diameters of 0.40–0.50 nm and 0.50–0.60 nm, respectively. The viscometers were selected on the basis of the flow time of the complex, and the flow time was at least 120 s.

2.3. Synthesis of the Title Complex. 2,2′-Bipyridine (0.2 mmol, 31.7 mg) and NaOH (0.2 mmol, 8.0 mg) were dissolved in a minimum amount of distilled water. Copper sulfate (0.2 mmol, 51.3 mg) was added to the above solution, and the volume of the mixed solution was increased to 18 mL with distilled water. The mixture was transferred to a 30 mL Teflon-lined stainless steel reactor and heated to 140°C for 72 h, and then, it was cooled at a rate of 10°C·h^{-1} to room temperature. Blue stick single crystals suitable for X-ray diffraction analysis were obtained. The crystals were washed by a small amount of distilled water and anhydrous ethanol and dried naturally (yield 84%). Anal. Calc. (%) for CuC$_{10}$H$_{12}$N$_2$O$_6$S: C, 34.14; H, 3.44; N, 7.96; Cu, 18.06. Found (%): C, 34.12; H, 3.30; N, 7.77; Cu 18.14.

2.4. X-Ray Crystallography. A blue crystal with dimensions 0.374 mm × 0.199 mm × 0.117 mm was installed on a Bruker Apex II CCD diffractometer with graphite monochromated Mo K$_\alpha$ radiation ($\lambda = 0.71073$ Å). Diffraction data were collected at 296(2) K in the θ range 3.273–27.639°. The programs of the SHELXL-97 and SHELXTL-97 were used for the structure determination and refinement [39, 40]. The structure was solved by direct methods, and all nonhydrogen atoms were obtained from the difference Fourier map and subjected to anisotropic refinement by full-matrix least squares on F^2. Crystallographic data have been deposited with the Cambridge Crystallographic Data Centre, CCDC, UK. Copies of the data can be obtained free of charge on quoting the depository CCDC-1028718 for the title complex (deposit@ccdc.cam.ac.uk, http://www.ccdc.cam.ac.uk).

2.5. Procedures of Biological Activity. The experimental methods of biological activity were mainly referred to [38, 41]. The specimens for absorption and fluorescence were obtained through diluting the stock solutions of the title complex (abbreviated as Cu-bipy) and hsDNA with Tris-HCl buffer solution to the required concentrations. Under the condition of the fixed Cu-bipy concentration and changing only the hsDNA concentration, the spectra of UV-Vis and fluorescence were tested with the quartz cuvettes of 1 cm. The excitation wavelength of the fluorescence measurement was 411.7 nm.

The samples of viscosity measurement were filled into the cleaned and dried viscometers. A thermostat was used to keep the temperature constant with the deviations within ±0.01°C. Double-distilled water was used in the calibration experiments, and the viscosity of pure water was derived from Lange's Handbook of Chemistry [42]. The time was recorded on a digital stopwatch with the deviations within ±0.01 s, and the average deviation of the three experimental results was within ±0.2 s. The hsDNA of 1.0×10^{-5} mol·L^{-1} was mixed with different concentrations of Cu-bipy. The flow time was recorded at 20 ± 0.1°C after the reaction mixture was placed in the darkness for 0.5 h. The relative viscosities of hsDNA were measured with molar ratio ($r = c$(Cu-bipy)/c(hsDNA)) from 0.0 to 4.0 at atmospheric pressure and 15°C ambient temperature.

3. Results and Discussion

3.1. Crystal Structure Analysis. The molecular structure diagram of the complex is shown in Figure 1. The crystallographic data and structure refinement parameters are given in Table 1, and the selected bond distances and angles are shown in Table 2.

The unit of the complex is composed of one Cu(II), one 2,2′-bipyridine, two water molecules, and one sulfate radical anion. The Cu(II) is hexacoordinated by two O atoms from the coordinated water molecules and two pyridyl N atoms from bipy which are located at equatorial sites and two O atoms from different bidentate bridging sulfato groups which are located at axial positions. The coordination configuration is a slightly distorted octahedron, and the Cu(II) is bridged by sulfato groups into polymeric chains. The distances of Cu–O with the aqua ligands and the sulfato group are 1.9728 and 2.455 Å, respectively, and the distances of Cu–N are 1.9947 Å. Because of Jahn–Teller effect of Cu(II)

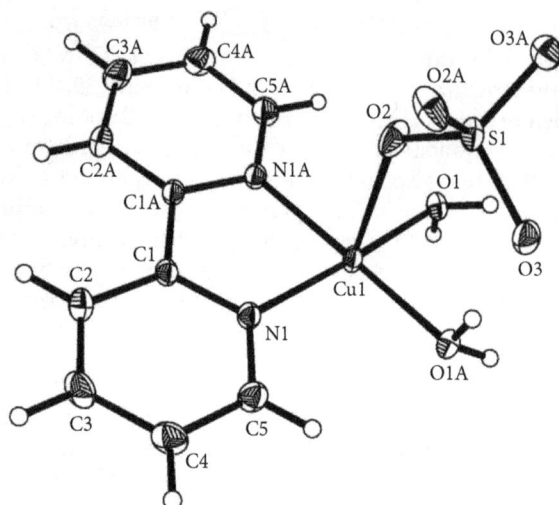

FIGURE 1: Molecular structure of the title complex.

TABLE 1: Crystal data and structure refinement parameters for the title complex.

Empirical formula	$CuC_{10}H_{12}O_6N_2S$
Formula weight (g·mol^{-1})	351.82
Temperature (K)	296(2)
Wavelength (Å)	0.71073
Crystal system	Monoclinic
Space group	$C2/c$
a (Å)	15.1279(7)
b (Å)	12.4488(6)
c (Å)	6.9987(3)
β (°)	105.9576(13)
V (Å^3)	1267.23(10)
Z	4
Calculated density (g·cm^{-3})	1.844
Absorption coefficient (mm^{-1})	1.916
$F(000)$	716
Crystal size (mm^3)	0.374 × 0.199 × 0.117
θ range for data collection (°)	3.273–27.639
Index ranges	$-19 \le h \le 19$, $-16 \le k \le 16$, $-9 \le l \le 8$
Reflections collected/unique	9535
Data/restraints/parameters	1473/3/97
Goodness of fit on F^2	0.974
Final R indices ($I > 2\sigma (I)$)	$R_1 = 0.0226$, $wR_2 = 0.0579$
R indices (all data)	$R_1 = 0.0238$, $wR_2 = 0.0589$
Largest differential peak and hole (e·Å^{-3})	0.356 and −0.442

TABLE 2: Selected bond lengths (Å) and angles (°) for the title complex.

Cu(1)–O(1)	1.9728(11)
Cu(1)–O(1)$^{\#1}$	1.9728(11)
Cu(1)–N(1)	1.9947(12)
Cu(1)–N(1)$^{\#1}$	1.9948(13)
Cu(1)–O(2)	2.455
Cu(1)–O(2)$^{\#1}$	2.455
O(1)$^{\#1}$–Cu(1)–O(1)	93.07(7)
O(1)$^{\#1}$–Cu(1)–N(1)	93.11(5)
O(1)–Cu(1)–N(1)	172.31(5)
O(1)$^{\#1}$–Cu(1)–N(1)$^{\#1}$	172.31(5)
O(1)–Cu(1)–N(1)$^{\#1}$	93.11(5)
N(1)–Cu(1)–N(1)$^{\#1}$	81.12(7)
O(2)–Cu(1)–O(1)	85.52
O(2)–Cu(1)–O(1)$^{\#1}$	92.25
O(2)–Cu(1)–N(1)	89.66
O(2)–Cu(1)–N(1)$^{\#1}$	92.79
O(2)–Cu(1)–O(2)$^{\#1}$	176.77
O(2)$^{\#1}$–Cu(1)–O(1)	92.25
O(2)$^{\#1}$–Cu(1)–N(1)	92.97
O(2)$^{\#1}$–Cu(1)–N(1)$^{\#1}$	89.66
O(2)$^{\#1}$–Cu(1)–O(1)$^{\#1}$	85.52

Symmetry transformations used to generate equivalent atoms: $^{\#1}-x+2$, y, $-z+1/2$; $^{\#2}-x+2$, y, $-z+3/2$.

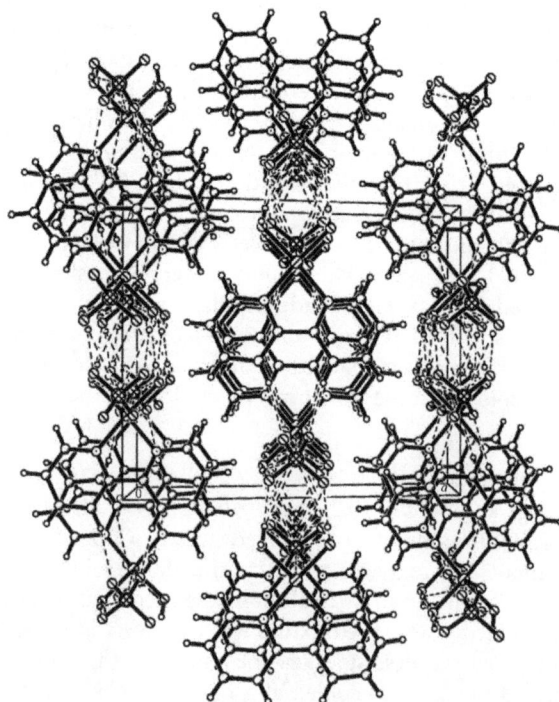

FIGURE 2: Crystal packing diagram of the title complex.

with d^9 electron configuration, the axial distances of Cu (1)–O(2) and Cu(1)–O(2)$^{\#1}$ are stretched. As shown in Figure 2, the generated chains extend along (001) plane direction, and the crossing of the chelate bipy forms the polymeric chains formulated as [Cu(bipy)(H$_2$O)$_2$(SO$_4$)]$_n$. In Figure 3, the molecular structure shows the existence of face-to-face π–π stacking weak interaction. The interplanar distances of 3.495–3.627 Å between two adjacent mirror planes of bipy are normal for weak π–π interaction. The hydrogen bond lengths and bond angles are given in Table 3, and the molecules of the complex are linked together by intermolecular hydrogen bonds. It is obvious that the

formation of the interchain hydrogen bonds is related to the layers parallel to the (100) plane. A weak hydrogen bond with $d(C \cdots O) = 3.282$ Å is formed between the outer C–H bonds of one chain and the coordinated sulfato O atoms of the adjacent chain, and the chains are stabilized by interchain π–π interaction and interchain C–H$\cdots$O hydrogen bonds [23]. Each water molecule in the complex nearly forms a linear intrachain and interchain hydrogen bonds with the uncoordinated sulfato O atoms.

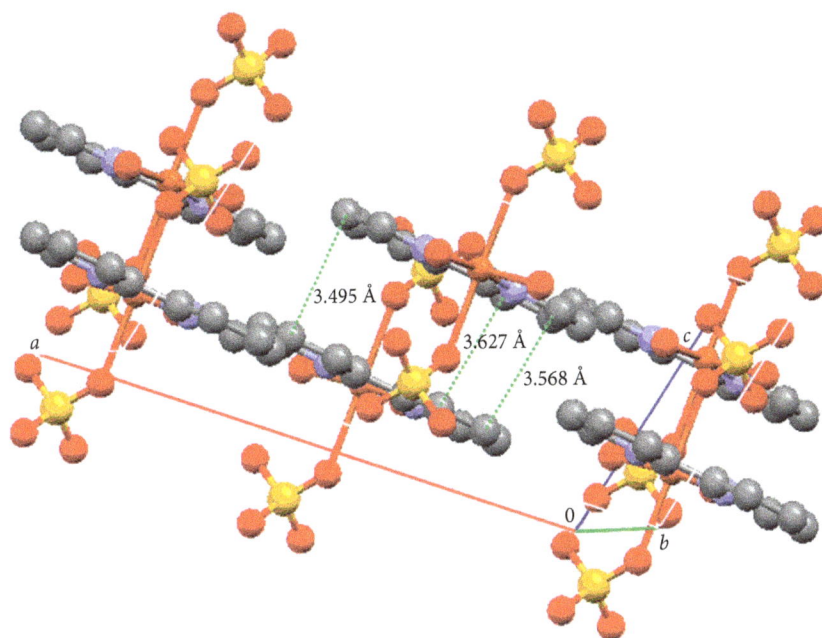

FIGURE 3: Weak π–π stacking interactions of the title complex.

TABLE 3: Hydrogen bond lengths (Å) and bond angles (°) for the title complex.

D–H$\cdots$A	d(D–H)	d(H$\cdots$A)	d(D$\cdots$A)	$\angle$DHA
O(1)–H(1W)$\cdots$O(3)[#3]	0.852(9)	1.843(11)	2.6763(16)	165.4(19)
O(1)–H(1W)$\cdots$S(1)[#3]	0.852(9)	2.926(16)	3.6599(12)	145.4(18)
C(5)–H(5)$\cdots$O(1)[#1]	0.93	2.50	3.026(2)	115.9
C(2)–H(2)$\cdots$O(2)[#4]	0.93	2.43	3.282(2)	152.1
O(1)–H(1)$\cdots$O(3)[#1]	0.82	1.81	2.6198(16)	169.5
O(1)–H(1)$\cdots$S(1)[#5]	0.82	2.74	3.4143(11)	140.8

Symmetry transformations used to generate equivalent atoms: [#1]$-x+2$, y, $-z+1/2$; [#2]$-x+2$, y, $-z+3/2$; [#3]$-x+2$, $-y+1$, $-z+1$; [#4]$-x+2$, $-y+2$, $-z+1$; [#5]x, y, $z-1$.

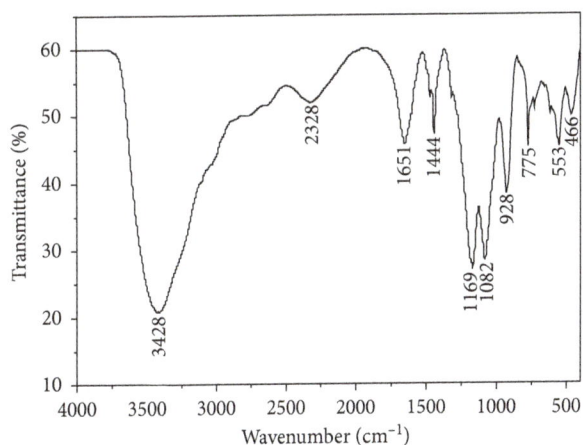

FIGURE 4: FTIR spectrum of the title complex.

3.2. FTIR Spectroscopy.

The FTIR spectrum of the complex is shown in Figure 4. Few number of the absorption bands in the FTIR spectrum means that the symmetry of the complex is very good. A wide intense absorption band around $3428 \, cm^{-1}$ is due to stretching vibration of hydroxyl [43, 44]. This indicates that there are the coordinated water or lattice water molecules in the complex. The band corresponding to the stretching vibration of the cumulative double bond (C=C–C=C) of pyridine ring is situated at $2328 \, cm^{-1}$. The absorption peaks at 1651 and $1444 \, cm^{-1}$ are assigned to the stretching vibrations of the C=N and C=C bonds, respectively [45].

As a free anion, sulfate has tetrahedral symmetry, whereas, if sulfate forms a bidentate binuclear (bridging) complex, the symmetry is lowered and the band splits into two bands [46]. As shown in Figure 4, the FTIR spectrum of the complex makes out peaks at 1169 and $1082 \, cm^{-1}$. The absorption peaks around 928 and $775 \, cm^{-1}$ are assigned to the rocking and wagging vibrations of the hydroxyl, which indicate the existence of the coordinated water molecules in the complex [47]. As a result of the formation of the Cu–N

bond, the corresponding C–N bond becomes so weak that disappeared in the FTIR spectrum [48]. The absorption peaks at 553 and $466 \, cm^{-1}$ are assigned to the Cu–N bond and Cu–O bond, respectively [49], which agrees with the X-ray crystal structure of the complex.

3.3. Thermal Analysis.

The TG-DSC curves of the title complex are shown in Figure 5, and there are one endothermic peak and two exothermic peaks in the DSC curve. The endothermic peak at 169°C is accompanied by obvious mass loss, and the sample loses two H_2O molecules. The experimental mass loss (10.31%) is close to the calculated one (10.24%). Due to the high temperature of water loss, the molecules should be the coordinated water. After the water molecules are lost, the complex becomes [Cu($C_{10}H_8N_2$)(SO_4)]. The sequential exothermic peaks at 393 and 423°C in

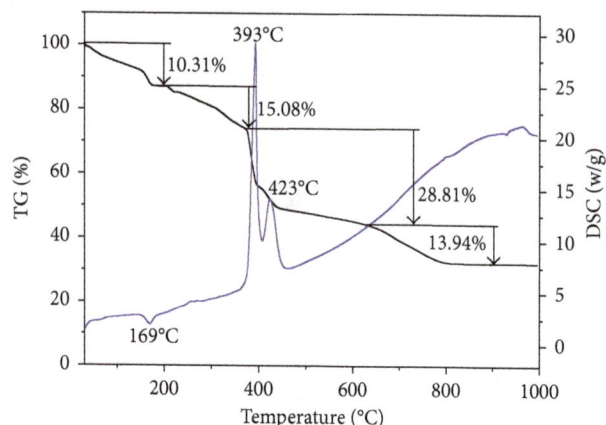

FIGURE 5: TG-DSC curves of the title complex.

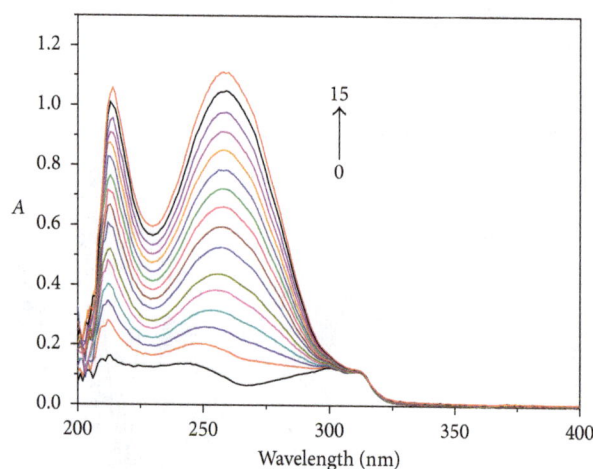

FIGURE 6: Influence of hsDNA on the UV-Vis spectra of Cu-bipy (pH 7.40). From curves 0–15, c(Cu-bipy) $= 4.0 \times 10^{-6}$ mol·L^{-1}; c(hsDNA) $= 0.00, 0.17, 0.33, 0.50, 0.67, 0.83, 1.00, 1.17, 1.33, 1.50, 1.67, 1.83, 2.00, 2.17, 2.33,$ and 2.50×10^{-6} mol·L^{-1}, respectively.

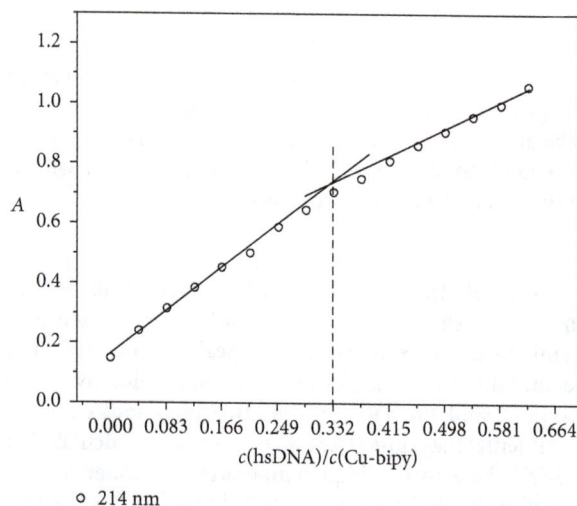

FIGURE 7: Mole ratio plots of Cu-bipy with hsDNA (pH $= 7.40$, $\lambda = 214$ nm), c(Cu-bipy) $= 4.0 \times 10^{-6}$ mol·L^{-1}.

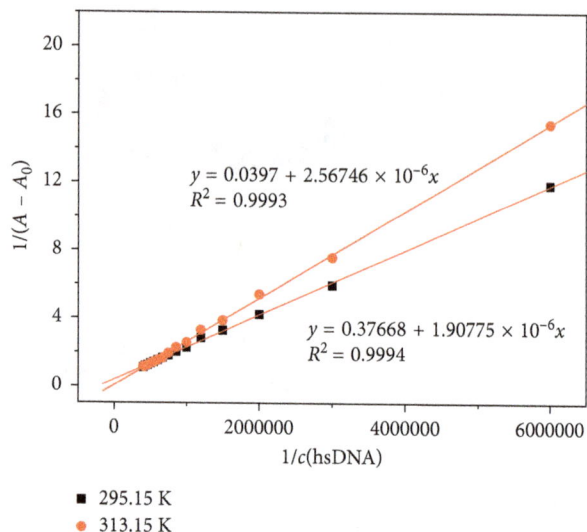

FIGURE 8: Double reciprocal plots of hsDNA-Cu-bipy at 295.15 K and 313.15 K (pH 7.40). c(Cu-bipy) $= 4.0 \times 10^{-6}$ mol·L^{-1}; c(hsDNA) $= 0.00, 0.60, 1.20, 1.80, 2.40, 3.00, 3.60, 4.20, 4.80, 5.40, 6.00, 6.60, 7.20, 7.80,$ and 8.40×10^{-6} mol·L^{-1}, respectively.

TABLE 4: Thermodynamic parameters at two different temperatures.

T (K)	$K^{\ominus}$ (L·mol^{-1})	$\Delta_r G_m^{\ominus}$ (J·mol^{-1})	$\Delta_r S_m^{\ominus}$ (J·mol^{-1}·K^{-1})	$\Delta_r H_m^{\ominus}$ (J·mol^{-1})
295.15	1.97×10^5	-3.53×10^4	115.26	-1280
313.15	1.55×10^4	-2.96×10^4	90.44	-1280

the DSC curve correspond to the oxidation and decomposition of bipy, the decomposition product of this step is CuSO$_4$, and the mass loss of 43.89% is in agreement with the calculated result of 44.39%. The mass loss remains constant until ca. 800°C, the final remnant mass is 31.86%, and the residue is Cu$_2$SO$_4$ (calculated as 31.72%).

3.4. Biological Activity

3.4.1. Binding Ratio. As shown in Figure 6, the UV-Vis spectra were obtained by determination of the Cu-bipy solution with an independent variable of hsDNA concentration. The wavelength which is obtained from Figure 6 and used in the mole ratio method is 214 nm, and the binding ratio [n(Cu-bipy) : n(hsDNA) $= 3 : 1$] is shown in Figure 7.

3.4.2. Double Reciprocal Method. The double reciprocal equation [50] is listed as follows to express the relationship between Cu-bipy and hsDNA:

$$(A - A_0)^{-1} = A_0^{-1} + [K^{\ominus} \cdot A_0 \cdot c(\text{hsDNA})]^{-1}. \quad (1)$$

In (1), c(hsDNA) is the hsDNA concentration, A and A_0 are the absorbance of Cu-bipy in the presence and lack of hsDNA, respectively, and $K^{\ominus}$ is the binding constant of hsDNA-Cu-bipy. In Figure 8, $1/c$(hsDNA) is used as an abscissa and $1/(A - A_0)$ is used as an ordinate. The binding constants are, respectively, calculated: $K^{\ominus}$

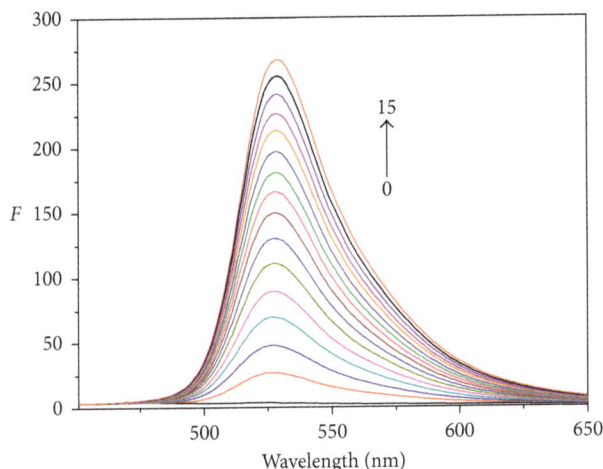

FIGURE 9: Emission spectra of hsDNA-Cu-bipy mixture in different concentrations of AO (pH = 7.40, λ_{ex} = 411.7 nm). From curves 0–15, c(hsDNA-Cu-bipy) = 1.0×10^{-5} mol·L^{-1}; c(Cu-bipy) = 0.00, 0.33, 0.67, 1.00, 1.33, 1.67, 2.00, 2.33, 2.67, 3.00, 3.33, 3.67, 4.00, 4.33, 4.67, and 5.00×10^{-6} mol·L^{-1}, respectively.

FIGURE 10: Emission spectra of hsDNA-AO mixture in different concentrations of Cu-bipy (pH = 7.40, λ_{ex} = 411.7 nm). From curves 0–15, c(hsDNA-AO) = 1.0×10^{-5} mol·L^{-1}; c(Cu-bipy) = 0.00, 0.33, 0.67, 1.00, 1.33, 1.67, 2.00, 2.33, 2.67, 3.00, 3.33, 3.67, 4.00, 4.33, 4.67, and 5.00×10^{-6} mol·L^{-1}, respectively.

(295.15 K) = 1.97×10^5 L·mol^{-1} and $K^\ominus$ (313.15 K) = 1.55×10^4 L·mol^{-1}. As is known to all, there are some ways in which macromolecule bind to small molecule, including hydrogen bond, van der Waals force, hydrophobic force, electrostatic interaction, and so on. According to the following equations, we can calculate a series of thermodynamic parameters ($\Delta_r H_m^\ominus$, $\Delta_r S_m^\ominus$, and $\Delta_r G_m^\ominus$) to confirm the interaction forces.

$$\ln \frac{K_2^\ominus}{K_1^\ominus} = \frac{\Delta_r H_m^\ominus \left((1/T_1) - (1/T_2) \right)}{R},$$

$$\Delta_r G_m^\ominus = -RT \ln K^\ominus,$$

(2)

$$\Delta_r G_m^\ominus = \Delta_r H_m^\ominus - T \Delta_r S_m^\ominus,$$

where T_1 is 295.15 K, T_2 is 313.15 K, and $\Delta_r H_m^\ominus$ and $\Delta_r G_m^\ominus$ are the standard molar reaction enthalpy and the standard molar reaction Gibbs free energy, respectively. The calculated results in Table 4 indicate that the interaction between Cu-bipy and hsDNA is driven by entropy [51]. The values of $\Delta_r H_m^\ominus$ and $\Delta_r G_m^\ominus$ indicate that this is an exothermic reaction, and there is a spontaneous interaction between Cu-bipy and hsDNA.

3.4.3. Competitive Binding Experiments. AO as a fluorescence probe was widely used to study the binding way between small molecule and DNA [38], and it can embed between two adjacent base pairs of DNA helix and enhance the fluorescence intensity. As the concentration of AO (Cu-bipy) increases, the fluorescence intensity of Cu-bipy-hsDNA (hsDNA-AO) reduces gradually at the maximum wavelength of 528 (531) nm in Figures 9 and 10. The experimental result indicates that the reaction competition between Cu-bipy and AO with hsDNA is conspicuous, and the bonding mode between Cu-bipy and hsDNA mainly includes insertion binding.

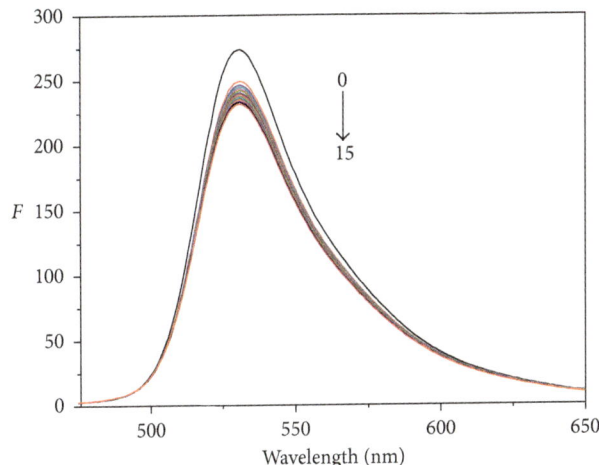

3.4.4. Scatchard Method. The Scatchard equation (3) can be used to study the binding mode between hsDNA and AO with Cu-bipy, whose concentration is gradually changing.

$$\frac{r}{c} = K(n - r),$$

(3)

where r is the mole number of AO bound per mole of DNA, c is the AO concentration, K is the binding constant, and n is the maximum value of a binding site with AO. Generally, if the n value in the absence of Cu-bipy is the same with the presence of Cu-bipy, the binding mode is an insertion mode. If the K value in the absence of Cu-bipy is the same with the presence of Cu-bipy, there is noninsertion in the binding mode. If the K value is different from the n value, the binding mode between Cu-bipy and hsDNA is a mixed mode of noninsertion and insertion binding. The Scatchard plots in the absence and the presence of sodium chloride are shown in Figures 11 and 12, and the data of n and K are listed in Table 5. It can be seen from Table 5 that both values of n and K vary with the concentrations of Cu-bipy. The results show the presence of the mixed interaction. The n values in the presence of sodium chloride are lower than that of no sodium chloride, and this indicates that there is an electrostatic interaction between Cu-bipy and hsDNA.

3.4.5. Influence of Phosphate Group. The above conclusion is further demonstrated by the phosphate experiment. If Cu-bipy binds to phosphate radical, then there is an electrostatic interaction between Cu-bipy and hsDNA by changing the Na$_2$HPO$_4$ concentration while keeping the Cu-bipy concentration fixed. As shown in Figure 13, when the amounts of Na$_2$HPO$_4$ are increased, UV-Vis spectra of Cu-bipy are slightly changed. The result hints that the electrostatic interaction exists between Cu-bipy and hsDNA.

TABLE 5: Data from the Scatchard equation of the interaction between Cu-bipy and hsDNA.

Curve	c(Cu-bipy)/c(hsDNA)	NaCl (mol·L^{-1})	Scatchard	K (L·mol^{-1})	n
a	0.00	0	$1173.6 - 2.44 \times 10^5 r$	2.44×10^5	4.81×10^{-3}
		0.50	$813.4 - 2.09 \times 10^5 r$	2.09×10^5	3.89×10^{-3}
b	0.40	0	$1176.5 - 1.81 \times 10^5 r$	1.81×10^5	6.50×10^{-3}
		0.50	$1241.5 - 2.20 \times 10^5 r$	2.20×10^5	5.64×10^{-3}
c	0.80	0	$1441.2 - 2.25 \times 10^5 r$	2.25×10^5	6.41×10^{-3}
		0.50	$1318.6 - 2.21 \times 10^5 r$	2.21×10^5	5.97×10^{-3}
d	1.20	0	$1641.4 - 2.27 \times 10^5 r$	2.27×10^5	7.23×10^{-3}
		0.50	$1113.9 - 2.56 \times 10^5 r$	2.56×10^5	4.35×10^{-3}

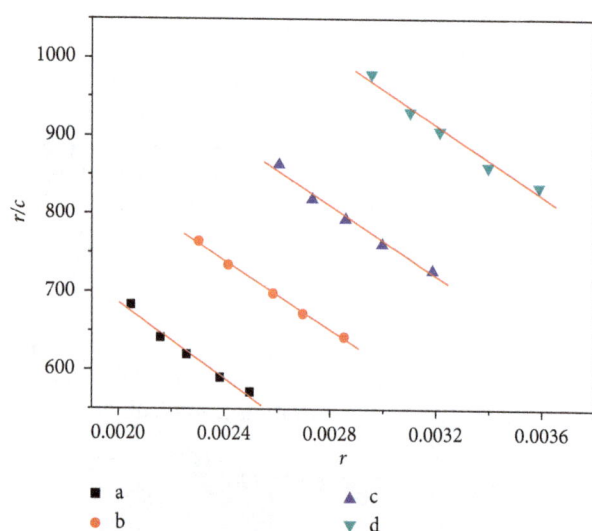

FIGURE 11: Scatchard plots of the interaction between Cu-bipy and hsDNA-AO (pH = 7.40, without NaCl). c(hsDNA) = 1.0×10^{-5} mol·L^{-1}; Rt = c(Cu-bipy)/c(hsDNA); Rt = a, 0.00; b, 0.40; c, 0.80; d, 1.20.

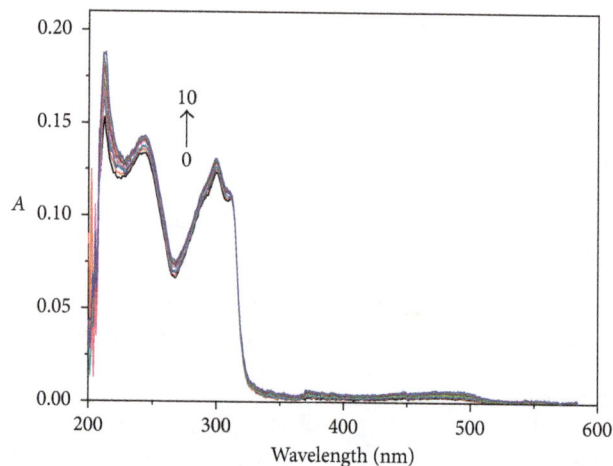

FIGURE 13: Influence of phosphate on the UV-Vis spectra of Cu-bipy (pH 7.40). c(Cu-bipy) = 1.0×10^{-5} mol·L^{-1}; c(Na$_2$HPO$_4$) = 0.00, 0.33, 0.67, 1.00, 1.33, 1.67, 2.00, 2.33, 2.67, 3.00, 3.33, 3.67, 4.00, 4.33, 4.67, and 5.00×10^{-5} mol·L^{-1}, respectively.

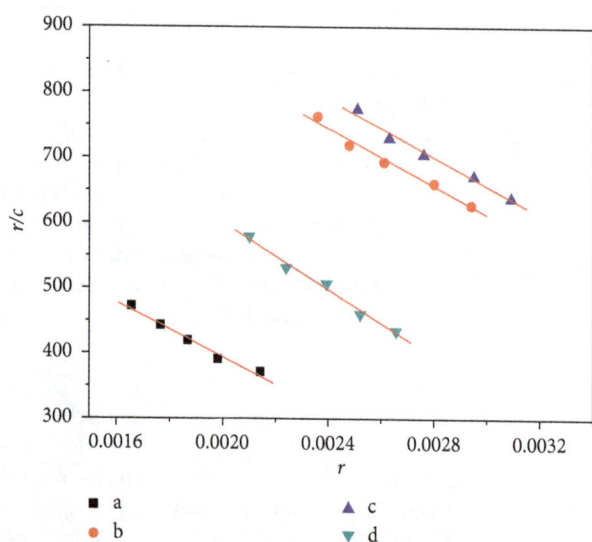

FIGURE 12: Scatchard plots of the interaction between Cu-bipy and hsDNA-AO (pH = 7.40, with NaCl). c(hsDNA) = 1.0×10^{-5} mol·L^{-1}; Rt = c(Cu-bipy)/c(hsDNA); Rt = a, 0.00; b, 0.40; c, 0.80; d, 1.20.

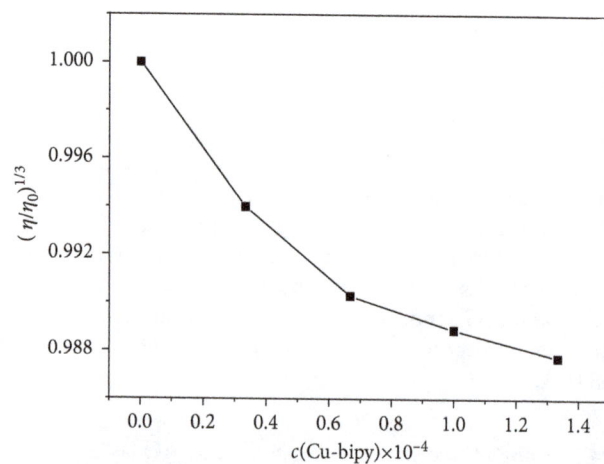

FIGURE 14: Effect of increasing amounts of Cu-bipy on the relative viscosity of hsDNA, c(hsDNA) = 1.00×10^{-4} mol·L^{-1}.

3.4.6. *Viscosity Measurements.* The viscosity measurements of complexes at different concentrations can obtain useful data for identifying binding mode [41, 52]. If a micromolecule is inserted in the interspace of base pairs, the DNA helix will be extended because the separated base pairs can accommodate the bound ligand. Conversely, the viscosity will not increase if the binding with DNA is in other ways;

the groove binding does not obviously change viscosity, whereas a partial intercalation of the complex causes a bend in the DNA helix, reducing its viscosity [53, 54]. The viscosity was determined by the fixed hsDNA concentration and changing the Cu-bipy concentration in the experiment. In Figure 14, the relative viscosity of hsDNA reveals a consistent decrease during the addition of Cu-bipy, which may be due to partial inserting of the complex. According to the result of viscosity measurement, the interaction between Cu-bipy and hsDNA is in insertion mode.

4. Conclusions

The complex [Cu(bipy)(H$_2$O)$_2$(SO$_4$)]$_n$ was synthesized by hydrothermal method and characterized by EA, single crystal X-ray diffraction, FTIR, and TG-DSC. The complex crystallizes in the monoclinic system with C2/c space group. The Cu(II) was hexacoordinated by two N atoms and four O atoms, forming a slightly distorted octahedron, and bridged by sulfato groups into polymeric chains. Under the physiological pH, the interaction between the complex and hsDNA was studied with AO as a fluorescent probe by spectral method. The interaction mechanism of the complex with hsDNA is electrostatic and intercalative binding. The calculated thermodynamic parameters indicate that the interaction of the complex and hsDNA is driven by entropy. The influence of phosphate radical and Scatchard method reveals that the complex is combined with hsDNA in the electrostatic and intromittent modes.

Acknowledgments

This work was supported by the Scientific Research Funds of Education Department of Sichuan Province (10ZA016) and the Longshan Academic Talent Research Supporting Program of SWUST (17LZX414).

References

[1] K. A. Kounavi, A. A. Kitos, E. E. Moushi et al., "Supramolecular features in the engineering of 3d metal complexes with phenyl-substituted imidazoles as ligands: the case of copper (II)," *CrystEngComm*, vol. 17, no. 39, pp. 7510–7521, 2015.

[2] K. Soulis, C. Gourlaouen, C. Daniel et al., "New luminescent copper(I) complexes with extended π-conjugation," *Polyhedron*, vol. 140, pp. 42–50, 2018.

[3] P. A. Tanner and C. K. Duan, "Luminescent lanthanide complexes: selection rules and design," *Coordination Chemistry Reviews*, vol. 254, no. 23-24, pp. 3026–3029, 2010.

[4] S. Sanram, J. Boonmak, and S. Youngme, "Structural diversity, single-crystal to single-crystal transformation and photocatalytic properties of Cu(II)-metal-organic frameworks based on 1,4-phenylenedipropionic acid," *Inorganica Chimica Acta*, vol. 469, pp. 11–19, 2018.

[5] J. Cui, Y. Li, Z. Guo, and H. Zheng, "A porous metal-organic framework based on Zn$_6$O$_2$ clusters: chemical stability, gas adsorption properties and solvatochromic behavior," *Chemical Communications*, vol. 49, no. 6, pp. 555–557, 2013.

[6] R. Yang, H. H. Li, K. V. Hecke, and G. H. Cui, "Cobalt(II) and copper(II) complexes constructed from bis(benzimidazole) and 2,6-pyridinedicarboxylate Co-ligands: synthesis, crystal structures, and catalytic properties," *Zeitschrift für Anorganische und Allgemeine Chemie*, vol. 641, no. 3-4, pp. 642–649, 2015.

[7] H. Paul, T. Mukherjee, M. G. B. Drew, and P. Chattopadhyay, "Synthesis, characterization, crystal structure, and DNA-binding of ruthenium(II) complexes of heterocyclic nitrogen ligands resulting from a benzimidazole-based quinazoline derivative," *Journal of Coordination Chemistry*, vol. 65, no. 8, pp. 1289–1302, 2012.

[8] J. Luo, L. Qiu, B. Liu, X. Zhang, F. Yang, and L. Cui, "Synthesis, structure and magnetic properties of two cobalt(II) dicyanamide (dca) complexes with heterocyclic nitrogen donors tetra(2-pyridyl)pyrazine (tppz) and 2,4,6-tri(2-pyridyl)-1,3,5-triazine (tptz): [Co$_2$(tppz)(dca)$_4$]·CH$_3$CN and [Co(tptz)(dca)(H$_2$O)](dca)," *Chinese Journal of Chemistry*, vol. 30, no. 3, pp. 522–528, 2012.

[9] W. Levason, G. Reid, and W. Zhang, "Synthesis, properties, and structures of chromium(VI) and chromium(V) complexes with heterocyclic nitrogen Ligands," *Zeitschrift Für Anorganische und Allgemeine Chemie*, vol. 640, no. 1, pp. 35–39, 2014.

[10] J. Devi and N. Batra, "Synthesis, characterization and antimicrobial activities of mixed ligand transition metal complexes with isatin monohydrazone Schiff base ligands and heterocyclic nitrogen base," *Spectrochimica Acta Part A: Molecular and Biomolecular Spectroscopy*, vol. 135, pp. 710–719, 2014.

[11] H. Wu, F. Kou, F. Jia, B. Liu, J. Yuan, and Y. Bai, "Synthesis, characterization, DNA-binding properties, and antioxidant activity of a copper(II) complex with 1,3-bis-(1-allaylbenzimidazol-2-yl)-2-oxopropane," *Zeitschrift für Anorganische und Allgemeine Chemie*, vol. 638, no. 2, pp. 443–450, 2012.

[12] X. Qiao, Z.-Y. Ma, C.-Z. Xie et al., "Study on potential antitumor mechanism of a novel Schiff base copper(II) complex: synthesis, crystal structure, DNA binding, cytotoxicity and apoptosis induction activity," *Journal of Inorganic Biochemistry*, vol. 105, no. 5, pp. 728–737, 2011.

[13] Y.-M. Ji, Y. Fang, P.-P. Han, M.-X. Li, Q.-Q. Chen, and Q.-X. Han, "Copper(II) and cadmium(II) complexes derived from Strandberg-type polyoxometalate clusters: synthesis, crystal structures, spectroscopy and biological activities," *Inorganic Chemistry Communications*, vol. 86, pp. 22–25, 2017.

[14] M. Jiang, Y.-T. Li, Z.-Y. Wu, Z.-Q. Liu, and C.-W. Yan, "Synthesis, crystal structure, cytotoxic activities and DNA-binding properties of new binuclear copper(II) complexes bridged by N,N'-bis(N-hydroxyethylaminoethyl)oxamide," *Journal of Inorganic Biochemistry*, vol. 103, no. 5, pp. 833–844, 2009.

[15] Y. Huang, E. Kong, J. Zhan et al., "Synthesis and cytotoxic evaluation of steroidal copper (Cu(II)) complexes," *Bioinorganic Chemistry and Applications*, vol. 2017, Article ID 4276919, 6 pages, 2017.

[16] J. D. C. Almeida, D. A. Paixão, I. M. Marzano et al., "Copper (II) complexes with β-diketones and N-donor heterocyclic ligands: crystal structure, spectral properties, and cytotoxic activity," *Polyhedron*, vol. 89, pp. 1–8, 2015.

[17] M. Nasrollahzadeh, A. Zahraei, A. Ehsani, and M. Khalajc, "Synthesis, characterization, antibacterial and catalytic activity of a nanopolymer supported copper(II) complex as a highly active and recyclable catalyst for the formamidation of arylboronic acids under aerobic conditions," *RSC Advances*, vol. 4, no. 39, pp. 20351–20357, 2014.

[18] A. D. Naik, P. A. N. Reddy, M. Nethaji, and A. R. Chakravarty, "Ternary copper(II) complexes of thiosemicarbazones and heterocyclic bases showing N_3OS coordination as models for the type-2 centers of copper monooxygenases," *Inorganica Chimica Acta*, vol. 349, no. 3, pp. 149–158, 2003.

[19] K. Ghosh, P. Kumar, N. Tyagi et al., "DNA interaction, superoxide scavenging and cytotoxicity studies on new copper (II) complexes derived from a tridentate ligand," *Polyhedron*, vol. 30, no. 16, pp. 2667–2677, 2011.

[20] Z. Chen, J. Zhang, P. Zeng, S. Zhang, and C. Jin, "Evaluation of DNA binding, protein interaction, and cytotoxic activity of a mononuclear copper(II) complex," *Zeitschrift für Anorganische und Allgemeine Chemie*, vol. 640, no. 7, pp. 1506–1513, 2014.

[21] D. Dey, A. De, S. Pal et al., "Synthesis, crystal structure, catecholase activity, DNA cleavage and anticancer activity of a dinuclear manganese(III)-bipyridine complex," *Indian Journal of Chemistry*, vol. 54A, no. 2, pp. 170–178, 2015.

[22] T. S. Lobana, S. Indoria, A. K. Jassal, H. Kaur, D. S. Arora, and J. P. Jasinski, "Synthesis, structures, spectroscopy and antimicrobial properties of complexes of copper(II) with salicylaldehyde N-substituted thiosemicarbazones and 2,2′-bipyridine or 1,10-phenanthroline," *European Journal of Medicinal Chemistry*, vol. 76, pp. 145–154, 2014.

[23] Y.-Q. Zheng and J.-L. Lin, "Crystal structures of $[Cu_2(bpy)_2(H_2O)(OH)_2(SO_4)]\cdot4H_2O$ and $[Cu(bpy)(H_2O)_2]SO_4$ with bpy = 2,2′-bipyridine," *Zeitschrift für Anorganische und Allgemeine Chemie*, vol. 629, no. 9, pp. 1622–1626, 2003.

[24] S. Sanotra, R. Gupta, H. N. Sheikh, B. L. Kalsotra, V. K. Gupta, and Rajnikant, "Hydrothermal synthesis and crystal structure of a supramolecular dinuclear cobalt(II) complex containing the dianion of pyridine-2,6-dicarboxylic acid," *Monatshefte fur Chemie*, vol. 144, no. 12, pp. 1807–1813, 2013.

[25] Y. V. Kolen'ko, V. D. Maximov, A. V. Garshev, P. E. Meskin, N. N. Oleynikov, and B. R. Churagulov, "Hydrothermal synthesis of nanocrystalline and mesoporous titania from aqueous complex titanyl oxalate acid solutions," *Chemical Physics Letters*, vol. 388, no. 4–6, pp. 411–415, 2004.

[26] M.-S. Chen, Y.-F. Deng, X. Nie, and C.-H. Zhang, "Hydrothermal synthesis, crystal structure, and luminescence property of an Mn-Tb complex based on 5-(isonicotinamido)-isophthalate," *Synthesis and Reactivity in Inorganic, Metal-Organic, and Nano-Metal Chemistry*, vol. 44, no. 6, pp. 868–872, 2014.

[27] Y.-F. Wang, Z. Li, Y.-C. Sun, and J.-S. Zhao, "Hydrothermal synthesis, crystal structures, and characterization of two copper complexes with 3-amino-1,2,4-triazole," *Synthesis and Reactivity in Inorganic, Metal-Organic, and Nano-Metal Chemistry*, vol. 44, no. 2, pp. 277–281, 2014.

[28] P. Zhao, W. Jing, L. Jing, F. Jian, and Y. Li, "Hydrothermal synthesis, crystal structures and properties of zinc(II) dinuclear complex and copper(I) coordination polymer based on building block 2-phenyl-4,6-di(pyridine-2-yl)pyrimidine," *Bulletin of the Korean Chemical Society*, vol. 34, no. 12, pp. 3743–3748, 2013.

[29] H. A. J. L. Mourão, O. F. Lopes, A. R. Malagutti, E. C. Paris, and C. Ribeiro, "Hydrothermal synthesis and photocatalytic properties of anatase TiO_2 nanocrystals obtained from peroxytitanium complex precursor," *Materials Science in Semiconductor Processing*, vol. 25, pp. 320–329, 2014.

[30] Y.-H. Wen, J.-K. Cheng, Y.-L. Feng, J. Zhang, Z.-J. Li, and Y.-G. Yao, "Hydrothermal syntheses, crystal structures and characterizations of three new copper coordination polymers," *Inorganica Chimica Acta*, vol. 358, no. 12, pp. 3347–3354, 2005.

[31] X. Li, C.-F. Bi, Y.-H. Fan, X. Zhang, N. Zhang, and X.-C. Yan, "Synthesis, crystal structure and DNA interaction of a novel three-nuclear cobalt(II) complex with Schiff base derived from 4-chloroanthranilic acid and 2,4-dihydroxybenzaldehyde," *Journal of Inorganic and Organometallic Polymers and Materials*, vol. 24, no. 3, pp. 582–590, 2014.

[32] M. Hazra, T. Dolai, A. Pandey, S. K. Dey, and A. Patra, "Synthesis and characterisation of copper(II) complexes with tridentate NNO functionalized ligand: density function theory study, DNA binding mechanism, optical properties, and biological application," *Bioinorganic Chemistry and Applications*, vol. 2014, Article ID 104046, 13 pages, 2014.

[33] P. Haribabu, R. K. Hussain, Y. P. Patil, and M. Nethaji, "Synthesis, crystal structure, DNA binding and cleavage activities of oximato bridged cationic dinuclear copper(II) complex having labile ligands," *Indian Journal of Chemistry*, vol. 52A, no. 3, pp. 327–333, 2013.

[34] M. Pragathi and K. H. Reddy, "Synthesis, spectral characterization and DNA interactions of copper(II) and nickel(II) complexes with unsymmetrical Schiff base ligands," *Indian Journal of Chemistry*, vol. 52A, no. 7, pp. 845–853, 2013.

[35] X.-G. Shu, C.-L. Wu, C.-J. Li, M. Zhang, K. Wan, and X. Wu, "Synthesis, crystal structure, spectroscopic properties, and interaction with Ct-DNA of Zn(II) with 2-aminoethanethiol hydrochloride ligand," *Bioinorganic Chemistry and Applications*, vol. 2016, Article ID 2691253, 7 pages, 2016.

[36] Z.-Q. Liu, Y.-T. Li, Z.-Y. Wu, and S.-F. Zhang, "$[Cu_4(H_2O)_4(dmapox)_2(btc)]_n\cdot10nH_2O$: the first two-dimensional polymeric copper(II) complex with bridging μ-trans-oxamidate and μ_4-1,2,4,5-benzentetracarboxylato ligands: synthesis, crystal structure and DNA binding studies," *Inorganica Chimica Acta*, vol. 362, no. 1, pp. 71–77, 2009.

[37] R. S. Kumar and S. Arunachalam, "DNA binding and antimicrobial studies of polymer–copper(II) complexes containing 1,10-phenanthroline and l-phenylalanine ligands," *European Journal of Medicinal Chemistry*, vol. 44, no. 5, pp. 1878–1883, 2009.

[38] J. Long, X.-M. Wang, D.-L. Xu, and L.-S. Ding, "Spectroscopic studies on the interaction mechanisms of safranin T with herring sperm DNA using acridine orange as a fluorescence probe," *Journal of Molecular Recognition*, vol. 27, no. 3, pp. 131–137, 2014.

[39] G. M. Sheldrick, *SHELXS-97, Program for the Solution of Crystal Structure*, University of Göttingen, Göttingen, Germany, 1997.

[40] G. M. Sheldrick, *SHELXL-97, Program for the Refinement of Crystal Structure*, University of Göttingen, Göttingen, Germany, 1997.

[41] Q. Zang, G.-Q. Zhong, and M.-L. Wang, "A copper(II) complex with pyridine-2,6-dicarboxylic acid: synthesis, characterization, thermal decomposition, bioactivity and interactions with herring sperm DNA," *Polyhedron*, vol. 100, pp. 223–230, 2015.

[42] W. Kirk-Othmer, *Encyclopedia of Chemical Technology*, Vol. 9, Wiley, New York, NY, USA, 1998.

[43] D. Li and G.-Q. Zhong, "Synthesis and crystal structure of the bioinorganic complex [Sb(Hedta)]·2H$_2$O," *Bioinorganic Chemistry and Applications*, vol. 2014, Article ID 461605, 7 pages, 2014.

[44] E. H. Lizarazo-Jaimes, P. G. Reis, F. M. Bezerra et al., "Complexes of different nitrogen donor heterocyclic ligands with SbCl$_3$ and PhSbCl$_2$ as potential antileishmanial agents against Sb[III]-sensitive and -resistant parasites," *Journal of Inorganic Biochemistry*, vol. 132, pp. 30–36, 2014.

[45] K. Nakamoto, *Infrared and Raman Spectra of Inorganic and Coordination Compounds*, John Wiley & Sons Inc., New York, NY, USA, 6th edition, 2009.

[46] J. J. Murcia, M. C. Hidalgo, J. A. Navío, J. Araña, and J. M. Doña-Rodríguez, "In situ FT–IR study of the adsorption and photocatalytic oxidation of ethanol over sulfated and metallized TiO$_2$," *Applied Catalysis B: Environmental*, vol. 142-143, pp. 205–213, 2014.

[47] G.-Q. Zhong and Q. Zhong, "Solid–solid synthesis, characterization, thermal decomposition and antibacterial activities of zinc(II) and nickel(II) complexes of glycine–vanillin Schiff base ligand," *Green Chemistry Letters and Reviews*, vol. 7, no. 3, pp. 236–242, 2014.

[48] I. Das, S. A. Ansari, and N. R. Agrawal, "Growth of nanosized cupric sulphate–urea complex in aqueous medium and its characterization," *Indian Journal of Chemistry*, vol. 50A, no. 6, pp. 798–801, 2011.

[49] W. Zhong, G.-Q. Zhong, and J.-H. He, "Preparation of tribasic copper chloride nanopowders by solid state reaction at room temperature," *Journal of Synthetic Crystals*, vol. 39, no. 6, pp. 1581–1585, 2010.

[50] A. Rescifina, M. G. Varrica, C. Carnovale, G. Romeo, and U. Chiacchio, "Novel isoxazole polycyclic aromatic hydrocarbons as DNA-intercalating agents," *European Journal of Medicinal Chemistry*, vol. 51, pp. 163–173, 2012.

[51] P. D. Ross and S. Subramanian, "Thermodynamics of protein association reactions: forces contributing to stability," *Biochemistry*, vol. 20, no. 11, pp. 3096–3102, 1981.

[52] S. J. Kharat and P. S. Nikam, "Density and viscosity studies of binary mixtures of aniline + benzene and ternary mixtures of (aniline + benzene + N,N-dimethylformamide) at 298.15, 303.15, 308.15, and 313.15 K," *Journal of Molecular Liquids*, vol. 131-132, pp. 81–86, 2007.

[53] S. Satyanarayana, J. C. Dabrowiak, and J. B. Chaires, "Neither Δ- nor Λ-tris(phenanthroline)ruthenium(II) binds to DNA by classical intercalation," *Biochemistry*, vol. 31, no. 39, pp. 9319–9324, 1992.

[54] Z.-G. Zhang and X.-D. Dong, "Interaction of DNA with a novel photoactive platinum diimine complex," *BioMetals*, vol. 22, no. 2, pp. 283–288, 2009.

18

Structural basis for pH-Dependent Oligomerization of Dihydropyrimidinase from *Pseudomonas aeruginosa* PAO1

Jen-Hao Cheng,[1] **Chien-Chih Huang,**[1] **Yen-Hua Huang,**[1] **and Cheng-Yang Huang**[1,2]

[1]*School of Biomedical Sciences, Chung Shan Medical University, No. 110, Sec. 1, Chien-Kuo N. Rd., Taichung, Taiwan*
[2]*Department of Medical Research, Chung Shan Medical University Hospital, No. 110, Sec. 1, Chien-Kuo N. Rd., Taichung, Taiwan*

Correspondence should be addressed to Cheng-Yang Huang; cyhuang@csmu.edu.tw

Academic Editor: Luigi Casella

Dihydropyrimidinase, a dimetalloenzyme containing a carboxylated lysine within the active site, is a member of the cyclic amidohydrolase family, which also includes allantoinase, dihydroorotase, hydantoinase, and imidase. Unlike all known dihydropyrimidinases, which are tetrameric, pseudomonal dihydropyrimidinase forms a dimer at neutral pH. In this paper, we report the crystal structure of *P. aeruginosa* dihydropyrimidinase at pH 5.9 (PDB entry 5YKD). The crystals of *P. aeruginosa* dihydropyrimidinase belonged to space group $C222_1$ with cell dimensions of $a = 108.9$, $b = 155.7$, and $c = 235.6$ Å. The structure of *P. aeruginosa* dihydropyrimidinase was solved at 2.17 Å resolution. An asymmetric unit of the crystal contained four crystallographically independent *P. aeruginosa* dihydropyrimidinase monomers. Gel filtration chromatographic analysis of purified *P. aeruginosa* dihydropyrimidinase revealed a mixture of dimers and tetramers at pH 5.9. Thus, *P. aeruginosa* dihydropyrimidinase can form a stable tetramer both in the crystalline state and in the solution. Based on sequence analysis and structural comparison of the dimer-dimer interface between *P. aeruginosa* dihydropyrimidinase and *Thermus* sp. dihydropyrimidinase, different oligomerization mechanisms are proposed.

1. Introduction

Dihydropyrimidinase is a key enzyme for pyrimidine catabolism [1, 2]. Dihydropyrimidinase catalyzes the reversible cyclization of dihydrouracil to *N*-carbamoyl-β-alanine in the second step of the pyrimidine degradation pathway (Figure 1). Dihydropyrimidinase can also detoxify xenobiotics with an imide functional group, ranging from linear imides to heterocyclic imides [3–9]. Homologous enzymes from microorganisms are known as hydantoinase, used as biocatalyst for hydrolysis of 5-monosubstituted hydantoins in the synthesis of D- and L-amino acids [10, 11]. Optically pure amino acids have been widely used as intermediates for semisynthesis of antibiotics, active peptides, hormones, antifungal agents, pesticides, and sweeteners. Dihydropyrimidinase and hydantoinase generally possess a similar active site, but their overall sequence identity and substrate specificity may differ [3, 12]. For example, hydantoinase purified from *Agrobacterium* species has no 5,6-dihydropyrimidine amidohydrolase activity [13]. Dihydropyrimidinases from the yeast *Saccharomyces kluyveri*

and the slime mold *Dictyostelium discoideum* do not hydrolyze hydantoin [14]. Thus, several bacterial hydantoinases are still named and identified as dihydropyrimidinase because of their catalytic activity toward natural substrates, namely, dihydrouracil and dihydrothymine. These bacterial enzymes include *Pseudomonas aeruginosa* and *Thermus* sp. dihydropyrimidinases [15, 16].

Dihydropyrimidinase, hydantoinase, imidase, allantoinase, and dihydroorotase belong to the cyclic amidohydrolase family because of their functional and structural similarities [17]. Members of this enzyme family catalyze the ring-opening hydrolysis of the cyclic amide bond of each substrate in either five- or six-membered rings. Even if these enzymes have similar functions, they have relatively low amino acid sequence identity. In addition, the substrate selectivity and specificity of these enzymes highly differ [18, 19]. Most of the active sites of dihydropyrimidinases, hydantoinases, allantoinases, and dihydroorotases contain four histidines, one aspartate, and one carboxylated lysine residue, which are required for metal binding and catalytic

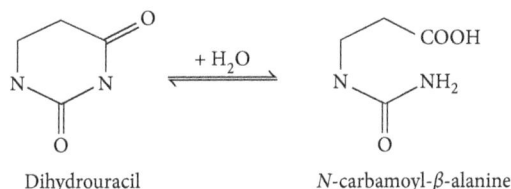

FIGURE 1: The physiological reaction of dihydropyrimidinase. Dihydropyrimidinase catalyzes the reversible cyclization of dihydrouracil to N-carbamoyl-β-alanine in the second step of the pyrimidine degradation pathway.

activity [8, 15, 18, 20, 21]. The presence of a carboxylated lysine in hydantoinase is also required for the self-assembly of the binuclear metal center [12, 20, 22] and increases the nucleophilicity of the hydroxide for catalysis [23]. The global architecture of the dihydropyrimidinase monomer consists of two domains, namely, a large domain with a classic $(\beta/\alpha)_8$-barrel structure core embedding the catalytic dimetal center and a small β-sandwich domain [16, 22, 24, 25].

All known dihydropyrimidinases are tetramers except pseudomonal enzymes. Hydantoinase from *P. putida* YZ-26 functions as a dimer [26, 27]. Recently, we identified that dihydropyrimidinase from *P. aeruginosa* PAO1 also forms a dimer [28]. In addition, the crystal structure of *P. aeruginosa* PAO1 dihydropyrimidinase indicated that several residues crucial for tetramerization are not found in *P. aeruginosa* dihydropyrimidinase [28]. In this study, we found that the oligomerization of *P. aeruginosa* PAO1 dihydropyrimidinase is a pH-dependent process. At pH 5.9, *P. aeruginosa* PAO1 dihydropyrimidinase mainly formed a tetramer. To confirm this result and determine how this enzyme can also form a tetramer, we also determined the crystal structure of *P. aeruginosa* PAO1 dihydropyrimidinase at 2.17 Å resolution at acidic environment. Structural comparison indicated that although *P. aeruginosa* PAO1 dihydropyrimidinase can also form a tetramer, the residues being crucial for tetramerization are different from those in *Thermus* sp. dihydropyrimidinases.

2. Materials and Methods

2.1. Cloning, Protein Expression, and Purification. Construction of the *P. aeruginosa* dihydropyrimidinase expression plasmid has been reported [15]. Recombinant *P. aeruginosa* dihydropyrimidinase was expressed and purified using the protocol described previously [15]. The protein purified from the soluble supernatant by Ni^{2+}-affinity chromatography (HiTrap HP; GE Healthcare Bio-Sciences, Piscataway, NJ, USA) was eluted with Buffer A (20 mM Tris-HCl, 250 mM imidazole, and 0.5 M NaCl, pH 7.9) and dialyzed against a dialysis buffer (20 mM HEPES and 100 mM NaCl, pH 7.0; Buffer B). Protein purity remained > 97% as determined by SDS-PAGE (Mini-PROTEAN Tetra System; Bio-Rad, CA, USA).

2.2. Gel Filtration Chromatography. Gel filtration chromatography was carried out by the AKTA-FPLC system (GE Healthcare Bio-Sciences, Piscataway, NJ, USA). In brief, purified protein (5 mg/mL) in Buffer C (20 mM MES and

TABLE 1: Data collection and refinement statistics.

Data collection	
Crystal	*P. aeruginosa* dihydropyrimidinase
Wavelength (Å)	0.975
Resolution (Å)	30–2.17
Space group	$C222_1$
Cell dimension (Å)	$a = 108.9,\ \alpha = 90$ $b = 155.7,\ \beta = 90$ $c = 235.6,\ \gamma = 120$
Completeness (%)	99.8 (100)*
$<I/\sigma I>$	15.13 (3.7)
R_{sym} or R_{merge} (%)	0.122 (0.599)
Redundancy	7.1 (7.3)
Refinement	
Resolution (Å)	30–2.17
Number of reflections	100197
R_{work}/R_{free}	0.1759/0.2312
Number of atoms	
Protein	1912
Water	312
RMS deviation	
Bond lengths (Å)	0.0151
Bond angles (°)	1.6495
Ramachandran plot	
In preferred regions	1345 (94.19%)
In allowed regions	68 (4.76%)
Outliers	15 (1.05%)
PDB entry	5YKD

*Values in parentheses are for the highest resolution shell.

100 mM NaCl, pH 5.9) was applied to a Superdex 200 prep grade column (GE Healthcare Bio-Sciences, Piscataway, NJ, USA) equilibrated with the same buffer [29]. The column was operated at a flow rate of 0.5 mL/min, and the proteins were detected at 280 nm. The column was calibrated with proteins of known molecular weight: thyroglobulin (670 kDa), γ-globulin (158 kDa), ovalbumin (44 kDa), myoglobin (17 kDa), and vitamin B_{12} (1.35 kDa).

2.3. Crystallography. Before crystallization, *P. aeruginosa* dihydropyrimidinase was concentrated to 20 mg/mL in Buffer C. Crystals were grown at room temperature by hanging drop vapor diffusion in 10% PEG 8000, 100 mM HEPES, 200 mM calcium acetate, pH 5.9. Data collection and refinement statistics for the crystal of *P. aeruginosa* dihydropyrimidinase are shown in Table 1. Data were collected using an ADSC Quantum-315r CCD area detector at SPXF beamline BL13C1 at NSRRC (Taiwan, ROC). All data integration and scaling were carried out using HKL-2000 [30]. There were four *P. aeruginosa* dihydropyrimidinase monomers per asymmetric unit. The crystal structure of *P. aeruginosa* dihydropyrimidinase was solved at 2.17 Å resolution with the molecular replacement software AMoRe [31] using the dihydropyrimidinase (PDB entry 5E5C) [28] as

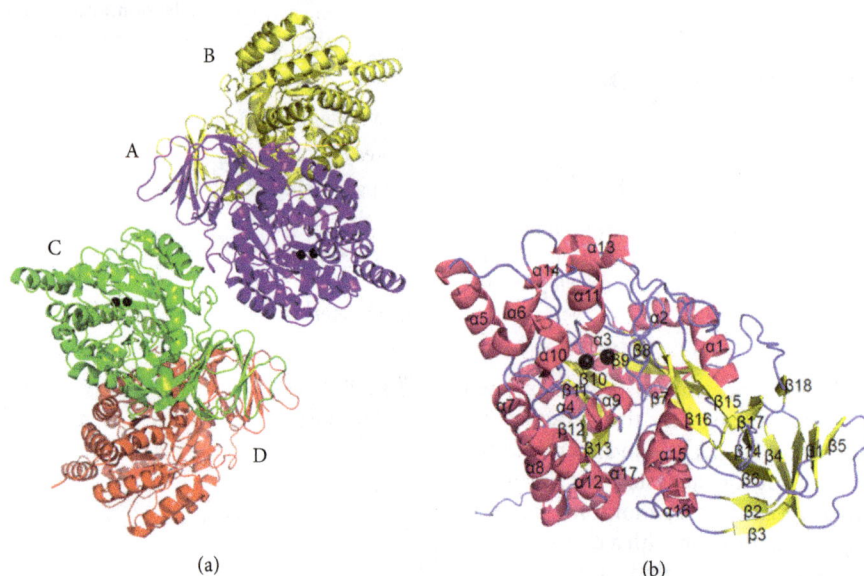

FIGURE 2: Crystal structure of *P. aeruginosa* dihydropyrimidinase. (a) Ribbon diagram of a *P. aeruginosa* dihydropyrimidinase tetramer. Each *P. aeruginosa* dihydropyrimidinase monomer is color-coded. Two zinc ions in the active site are presented as black spheres. (b) Ribbon diagram of a *P. aeruginosa* dihydropyrimidinase monomer with the secondary structures labeled.

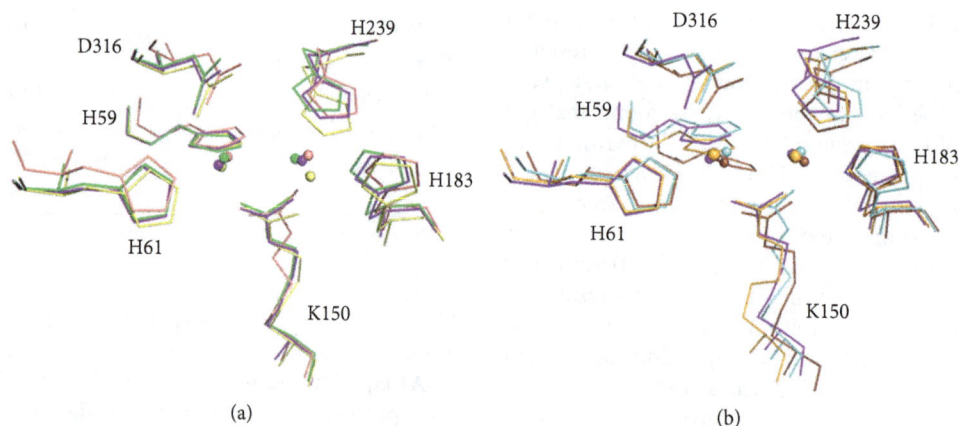

FIGURE 3: Structural comparison. (a) Superposition of the active site of dihydropyrimidinases. Their active sites contain four histidines, one aspartate, and one carboxylated lysine residue, which are required for metal binding and catalytic activity. Dihydropyrimidinases from *P. aeruginosa* (PDB entry 5E5C; green), *Thermus* sp. (PDB entry 1GKQ; salmon), *Tetraodon nigroviridis* (PDB entry 4H01; pale yellow), and the structure (PDB entry 5YKD; purple blue) in this study are shown. The architecture of these active sites is similar. (b) Superposition of the active site of members of the amidohydrolase family. Their active sites contain four histidines, one aspartate, and one carboxylated lysine residue, which are required for metal binding and catalytic activity. *P. aeruginosa* dihydropyrimidinase (PDB entry 5YKD; purple blue), *Escherichia coli* allantoinase (PDB entry 3E74; bright orange), *Burkholderia pickettii* hydantoinase (PDB entry 1NFG; aquamarine), and *E. coli* dihydroorotase (PDB entry 1J79; brown) are shown. The architecture of these active sites is similar.

model. After molecular replacement, model building was carried out using XtalView [32]. CNS was used for molecular dynamics refinement [33]. The final structure was refined to an *R*-factor of 0.1759 and an R_{free} of 0.2312. Atomic coordinates and related structural factors have been deposited in the PDB with accession code 5YKD.

3. Results and Discussion

3.1. Structure of the *P. aeruginosa* Dihydropyrimidinase Monomer. Crystals of *P. aeruginosa* dihydropyrimidinase

were grown at room temperature by hanging drop vapor diffusion in 10% PEG 8000, 100 mM HEPES, 200 mM calcium acetate, pH 5.9. The crystals of *P. aeruginosa* dihydropyrimidinase grown under this condition belonged to space group $C222_1$ with cell dimensions of $a = 108.9$, $b = 155.7$, and $c = 235.6$ Å. The crystal structure of *P. aeruginosa* dihydropyrimidinase was solved at 2.17 Å resolution (Table 1). The unit cell contained eight molecules. An asymmetric unit of the crystal contained four crystallographically independent *P. aeruginosa* dihydropyrimidinase monomers, in which two zinc ions were found in the active site per

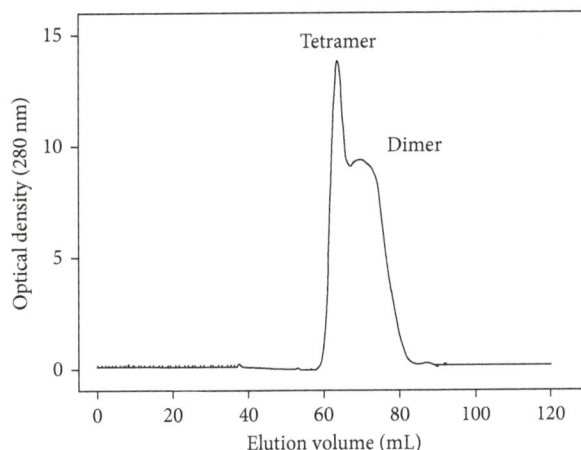

FIGURE 4: Gel filtration chromatographic analysis. Gel filtration chromatography was carried out by the AKTA-FPLC system in Buffer C (20 mM MES and 100 mM NaCl, pH 5.9). The corresponding peaks show the eluting *P. aeruginosa* dihydropyrimidinase. The column was calibrated with proteins of known molecular weight: thyroglobulin (670 kDa), γ-globulin (158 kDa), ovalbumin (44 kDa), myoglobin (17 kDa), and vitamin B_{12} (1.35 kDa).

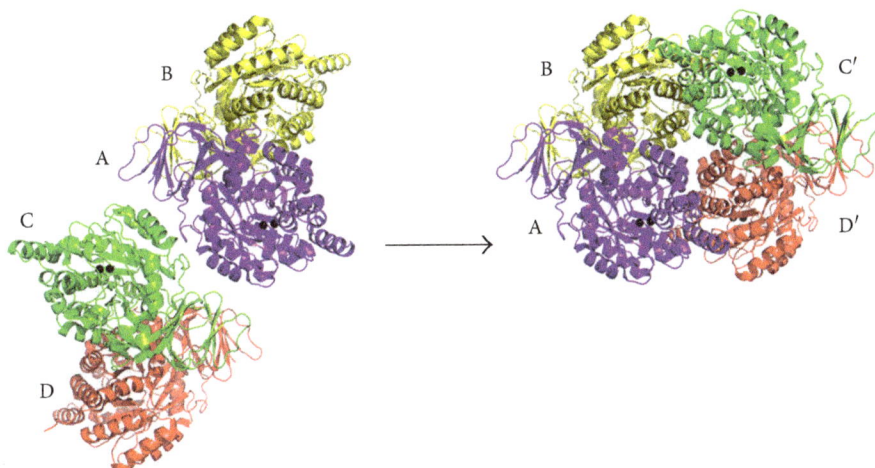

FIGURE 5: The structure of *P. aeruginosa* dihydropyrimidinase tetramer. An asymmetric unit contains four crystallographically independent *P. aeruginosa* dihydropyrimidinase monomers B-A-C-D. Crystallographically related tetramer B-A-C′-D′ was formed and further stabilized via many hydrogen bonds and salt bridges. This tetramerization structure was similar to that of *Thermus* sp. dihydropyrimidinase (PDB entry 1GKQ).

monomer (Figure 2(a)). The majority of the electron density for *P. aeruginosa* dihydropyrimidinase exhibited good quality, and no discontinuity was observed. Briefly, the overall structure of each *P. aeruginosa* dihydropyrimidinase unit consists of 17 α-helices, 19 β-sheets, and two zinc ions (Figure 2(b)). At pH 5.9, the architecture of the *P. aeruginosa* dihydropyrimidinase monomer consists of two domains, namely, a large domain with a classic $(\beta/\alpha)_8$-barrel structure core embedding the catalytic dimetal center and a small β-sandwich domain.

3.2. Structural Comparison.
The overall structure and architecture of the active site of *P. aeruginosa* dihydropyrimidinase are similar to those of other dihydropyrimidinases (Figure 3(a)) and other members of the amidohydrolase family of enzymes, such as hydantoinases, dihydroorotases, and allantoinases (Figure 3(b)). The active sites of these enzymes contain four histidines, one aspartate, and one carboxylated lysine residue, which are required for metal binding and catalytic activity [12, 14, 15, 19, 20, 34, 35].

3.3. pH-Dependent Oligomerization of P. aeruginosa Dihydropyrimidinase.
It was noted that the crystals of the dimeric *P. aeruginosa* dihydropyrimidinase belonged to space group $P3_121$ grown at the condition of 28% PEG 6000, 100 mM HEPES, 200 mM lithium acetate, pH 7.5 [28]. Due to the different crystallization condition, we attempted to test whether the oligomerization of *P. aeruginosa* dihydropyrimidinase is pH-dependent. All known dihydropyrimidinases are tetramers. However, pseudomonal dihydropyrimidinase/hydantoinase forms a dimer at neutral pH [26–28]. Given that the structure implies that

TABLE 2: The formation of hydrogen bonds at the dimer-dimer interface of *P. aeruginosa* dihydropyrimidinase.

Subunit 1	Distance [Å]	Subunit 2
A: K374 [NZ]	3.00	B: E14 [OE1]
A: H13 [NE2]	2.88	B: E14 [OE1]
A: R386 [NH2]	3.86	B: E14 [OE2]
A: R386 [NH1]	2.81	B: E15 [OE2]
A: R386 [NH2]	2.83	B: E15 [OE2]
A: R468 [NH2]	3.61	B: Q306 [OE1]
A: R253 [NH1]	3.27	B: S307 [O]
A: R253 [NH2]	3.13	B: S307 [O]
A: R467 [NH1]	2.92	B: V354 [O]
A: R468 [NE]	2.95	B: G357 [O]
A: R468 [NH2]	3.09	B: G357 [O]
A: R468 [NH2]	3.40	B: R358 [O]
A: R467 [NH1]	3.24	B: L359 [O]
A: E14 [OE1]	3.09	B: K374 [NZ]
A: E14 [OE1]	2.47	B: H13 [NE2]
A: E15 [OE2]	2.70	B: R386 [NH1]
A: S307 [O]	3.30	B: R253 [NH1]
A: S307 [O]	3.55	B: R253 [NH2]
A: V354 [O]	2.91	B: R467 [NH1]
A: G357 [O]	2.94	B: R468 [NH2]
A: G357 [O]	2.94	B: R468 [NE]
A: R358 [O]	3.56	B: R468 [NH2]
A: L359 [O]	3.16	B: R467 [NH1]
C′: H13 [NE2]	2.79	D′: E14 [OE1]
C′: K374 [NZ]	3.25	D′: E14 [OE1]
C′: R386 [NH1]	2.85	D′: E15 [OE1]
C′: R386 [NH2]	2.59	D′: E15 [OE2]
C′: R468 [NH2]	3.26	D′: Q306 [OE1]
C′: R253 [NH1]	3.13	D′: S307 [O]
C′: R253 [NH2]	3.16	D′: S307 [O]
C′: R468 [NE]	2.71	D′: G357 [O]
C′: R468 [NH2]	3.11	D′: R358 [O]
C′: E14 [OE1]	2.88	D′: H13 [NE2]
C′: E14 [OE1]	2.89	D′: K374 [NZ]
C′: E15 [OE2]	2.88	D′: R386 [NH1]
C′: E15 [OE2]	2.73	D′: R386 [NH2]
C′: Q306 [OE1]	3.53	D′: R468 [NH2]
C′: S307 [O]	3.21	D′: R253 [NH1]
C′: S307 [O]	3.59	D′: R253 [NH2]
C′: G357 [O]	2.65	D′: R468 [NE]
C′: R358 [O]	3.33	D′: R468 [NH2]

TABLE 3: The formation of salt bridges at the dimer-dimer interface of *P. aeruginosa* dihydropyrimidinase.

Subunit 1	Distance [Å]	Subunit 2
A: K374 [NZ]	3.00	B: E14 [OE1]
A: H13 [NE2]	2.88	B: E14 [OE1]
A: R386 [NH2]	3.86	B: E14 [OE2]
A: H13 [NE2]	3.75	B: E14 [OE2]
A: R386 [NH1]	3.55	B: E15 [OE1]
A: R386 [NH1]	2.81	B: E15 [OE2]
A: R386 [NH2]	2.83	B: E15 [OE2]
A: E14 [OE1]	3.09	B: K374 [NZ]
A: E14 [OE1]	2.47	B: H13 [NE2]
A: E14 [OE2]	3.93	B: H13 [NE2]
A: E15 [OE1]	3.69	B: R386 [NH1]
A: E15 [OE2]	3.00	B: R386 [NH2]
A: E15 [OE2]	2.70	B: R386 [NH1]
C′: H13 [NE2]	2.79	D′: E14 [OE1]
C′: K374 [NZ]	3.25	D′: E14 [OE1]
C′: H13 [NE2]	3.86	D′: E14 [OE2]
C′: R386 [NH1]	2.85	D′: E15 [OE1]
C′: R386 [NH2]	3.84	D′: E15 [OE1]
C′: R386 [NH1]	2.96	D′: E15 [OE2]
C′: R386 [NH2]	2.59	D′: E15 [OE2]
C′: E14 [OE1]	2.88	D′: H13 [NE2]
C′: E14 [OE1]	2.89	D′: K374 [NZ]
C′: E14 [OE2]	3.78	D′: H13 [NE2]
C′: E15 [OE1]	3.34	D′: R386 [NH1]
C′: E15 [OE2]	2.88	D′: R386 [NH1]
C′: E15 [OE2]	2.73	D′: R386 [NH2]

elution volume of 63.25 and 69. 26 mL did coexist. The molecular mass of a *P. aeruginosa* dihydropyrimidinase monomer, as calculated from the amino acid sequence, is 53 kDa. Assuming that these two forms of *P. aeruginosa* dihydropyrimidinase have a shape and partial specific volume similar to the standard proteins, the native molecular masses of *P. aeruginosa* dihydropyrimidinase were estimated to be 105 and 180 kDa, approximately 1.9 and 3.5 times the molecular mass of a *P. aeruginosa* dihydropyrimidinase monomer, respectively. In comparison at pH 7.5, gel filtration chromatographic analysis of *P. aeruginosa* dihydropyrimidinase revealed a single peak; the native molecular mass was estimated to be 117 kDa [28]. The two forms of this enzyme obtained from the gel filtration chromatography at pH 5.9 had similar specific activity (data not shown). Thus, *P. aeruginosa* dihydropyrimidinase did exist as a mixture of dimers and tetramers at pH 5.9.

3.4. Structural Insights into Dimer of Dimer (Tetramer) Formation of Dihydropyrimidinase.

In this study, we have identified that *P. aeruginosa* dihydropyrimidinase did exist as a mixture of dimers and tetramers at pH 5.9. To assess how *P. aeruginosa* dihydropyrimidinase can form a stable tetramer, the dimer-dimer interface was analyzed. In the

P. aeruginosa dihydropyrimidinase may also form a tetramer in the crystalline state at pH 5.9 (Figure 2(a)), we performed biochemical verification to confirm the oligomerization state. To confirm whether or not the oligomerization of *P. aeruginosa* dihydropyrimidinase is pH-dependent, we conducted gel filtration chromatography at pH 5.9. As shown in Figure 4, the results revealed that two species with

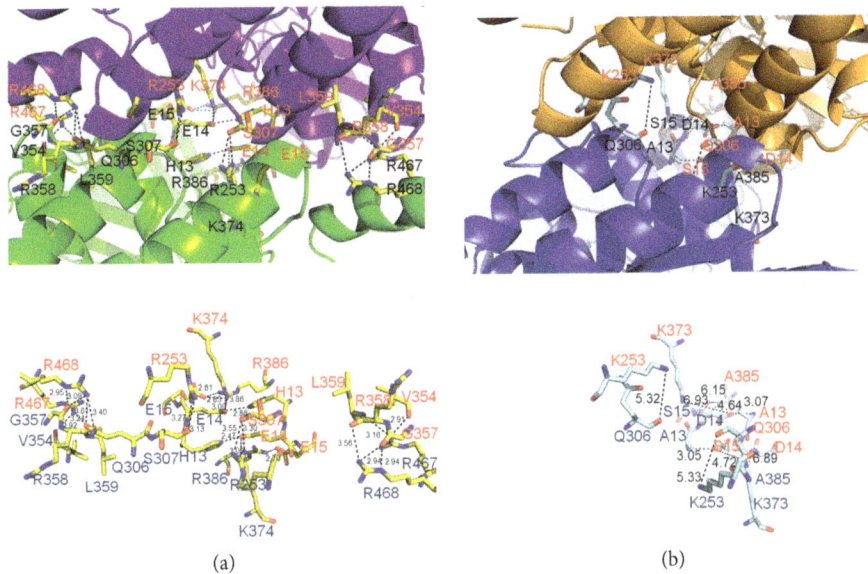

FIGURE 6: Comparison of the tetrameric structures of *Thermus* sp. dihydropyrimidinase and *P. aeruginosa* dihydropyrimidinase. (a) Structural analysis of the dimer-dimer interface of *P. aeruginosa* dihydropyrimidinase. The distance (Å) of the residues is shown. (b) Many residues crucial for forming hydrogen bonds at the dimer-dimer interface of *P. aeruginosa* dihydropyrimidinase were not found in the dimer-dimer interface of *Thermus* sp. dihydropyrimidinase.

R   *Pseudomonas aeruginosa* PAO1
-    *Thermus* sp.

FIGURE 7: Sequence alignment of dihydropyrimidinases from *P. aeruginosa* and *Thermus* sp. The amino acids that are involved in dimer-dimer interface of *P. aeruginosa* and *Thermus* sp. dihydropyrimidinase are boxed, respectively.

crystal of *P. aeruginosa* dihydropyrimidinase, the four molecules formed two pairs of dimers, B-A and C-D, respectively (Figure 5). Since the two dimers of *P. aeruginosa* dihydropyrimidinase associate via few contacts to create the

tetramer, it was thought that the tetrameric state may be possibly due to crystal packing forces. We noted that in the crystal, another crystallographically related tetramer B-A-C′-D′ (Figure 5) was formed and further stabilized via many hydrogen bonds and salt bridges (Tables 2 and 3). This tetramerization structure was similar to that of *Thermus* sp. dihydropyrimidinase (PDB entry 1GKQ).

We also compared the residues important for tetramerization located at the B-A-C'-D' dimer-dimer interface with those of *Thermus* sp. dihydropyrimidinase (Figure 6). Although their overall structures are similar, the important residues for tetramer (dimer B-C' with dimer A-D') formation are quite different. For the tetramer formation of *P. aeruginosa* dihydropyrimidinase, many hydrogen bonds with close distance were found: these bonds (<3 Å) include K374(A)–E14(B), H13(A)–E14(B), R386(A)–E14(B), R386(A)–E15(B), R467(A)–V354(B), R468(A)–G357(B), E14(A)–H13(B), E15(A)–R386(B), V354(A)–R467(B), G357(A)–R468(B), H13(C')–E14(D'), R386(C')–E15(D'), R468(C')–G357(D'), E14(C')–H13(D'), E14(C')–K374(D'), E15(C')–R386(D'), and G357(C')–R468(D'); however, these residues were not found for the tetramer formation of *Thermus* sp. dihydropyrimidinase (Figure 6). Only A13–D14 hydrogen bond was found in *Thermus* sp. dihydropyrimidinase (i.e., H13–E14 in *P. aeruginosa* dihydropyrimidinase). Thus, the dimer-dimer interface between *P. aeruginosa* dihydropyrimidinase and *Thermus* sp. dihydropyrimidinase was significantly different (Figure 7). Comparison by superimposition indicated that many Arg residues (R253, R358, R386, R467, and R468) found in *P. aeruginosa* dihydropyrimidinase, but not in *Thermus* sp. dihydropyrimidinase, may play a crucial role for the pH-dependent oligomerization. If consider the pK_a, a much better candidate is His13, which is involved in intermolecular interactions and, dependent on the environment of its side chain, which may easily change protonation state between pH 5.9 and pH 7.5. However, this speculation needs to be confirmed by further biochemical experiments.

3.5. Different Mechanisms for Tetramer Formation of Dihydropyrimidinases. In this study, we identified *P. aeruginosa* dihydropyrimidinase can be a tetramer both in the crystalline state and in solution (Figure 4). The structure of the tetrameric *Thermus* sp. dihydropyrimidinase and *P. aeruginosa* dihydropyrimidinase was compared (Figure 6). Many important residues for *Thermus* sp. dihydropyrimidinase tetramer formation are different from those for *P. aeruginosa* dihydropyrimidinase (Figure 7). On the basis of these results, we concluded that *P. aeruginosa* dihydropyrimidinase could form a tetramer, but its oligomerization mechanism differed from those of other dihydropyrimidinases such as *Thermus* sp. dihydropyrimidinase.

Acknowledgments

The authors thank the technical services provided by the Synchrotron Radiation Protein Crystallography Facility of the National Core Facility Program for Biotechnology, Ministry of Science and Technology, and the National Synchrotron Radiation Research Center, a national user facility supported by the Ministry of Science and Technology, Taiwan, ROC. This research was supported by a grant from the Ministry of Science and Technology, Taiwan (MOST 106-2320-B-040-004) to Cheng-Yang Huang.

References

[1] K. D. Schnackerz and D. Dobritzsch, "Amidohydrolases of the reductive pyrimidine catabolic pathway purification, characterization, structure, reaction mechanisms and enzyme deficiency," *Biochimica et Biophysica Acta*, vol. 1784, no. 3, pp. 431–444, 2008.

[2] D. P. Wallach and S. Grisolia, "The purification and properties of hydropyrimidine hydrase," *Journal of Biological Chemistry*, vol. 226, no. 1, pp. 277–288, 1957.

[3] C. C. Hsu, L. Y. Lu, and Y. S. Yang, "From sequence and structure of sulfotransferases and dihydropyrimidinases to an understanding of their mechanisms of action and function," *Expert Opinion on Drug Metabolism & Toxicology*, vol. 6, no. 5, pp. 591–601, 2010.

[4] C. Y. Huang and Y. S. Yang, "Discovery of a novel N-iminylamidase activity: substrate specificity, chemicoselectivity and catalytic mechanism," *Protein Expression and Purification*, vol. 40, no. 1, pp. 203–211, 2005.

[5] C. Y. Huang and Y. S. Yang, "A novel cold-adapted imidase from fish *Oreochromis niloticus* that catalyzes hydrolysis of maleimide," *Biochemical and Biophysical Research Communications*, vol. 312, no. 12, pp. 467–472, 2003.

[6] C. Y. Huang, S. K. Chiang, Y. S. Yang, and Y. J. Sun, "Crystallization and preliminary X-ray diffraction analysis of thermophilic imidase from pig liver," *Acta Crystallographica Section D Biological Crystallography*, vol. 59, no. 5, pp. 943–945, 2003.

[7] C. Y. Huang, Y. P. Chao, and Y. S. Yang, "Purification of industrial hydantoinase in one chromatographic step without affinity tag," *Protein Expression and Purification*, vol. 30, no. 1, pp. 134–139, 2003.

[8] C. Y. Huang and Y. S. Yang, "The role of metal on imide hydrolysis: metal content and pH profiles of metal ion-replaced mammalian imidase," *Biochemical and Biophysical Research Communications*, vol. 297, no. 4, pp. 1027–1032, 2002.

[9] Y. S. Yang, S. Ramaswamy, and W. B. Jakoby, "Rat liver imidase," *Journal of Biological Chemistry*, vol. 268, no. 15, pp. 10870–10875, 1993.

[10] H. E. Schoemaker, D. Mink, and M. G. Wubbolts, "Dispelling the myths–biocatalysis in industrial synthesis," *Science*, vol. 299, no. 5613, pp. 1694–1697, 2003.

[11] J. Altenbuchner, M. Siemann-Herzberg, and C. Syldatk, "Hydantoinases and related enzymes as biocatalysts for the synthesis of unnatural chiral amino acids," *Current Opinion in Biotechnology*, vol. 12, no. 6, pp. 559–563, 2001.

[12] C. Y. Huang, C. C. Hsu, M. C. Chen, and Y. S. Yang, "Effect of metal binding and posttranslational lysine carboxylation on the activity of recombinant hydantoinase," *Journal of Biological Inorganic Chemistry*, vol. 14, no. 1, pp. 111–121, 2009.

[13] S. M. Runser and P. C. Meyer, "Purification and biochemical characterization of the hydantoin hydrolyzing enzyme from Agrobacterium species. A hydantoinase with no 5,6-dihydropyrimidine amidohydrolase activity," *European Journal of Biochemistry*, vol. 213, no. 3, pp. 1315–1324, 1993.

[14] Z. Gojkovic, L. Rislund, B. Andersen et al., "Dihydropyrimidine amidohydrolases and dihydroorotases share the same origin and several enzymatic properties," *Nucleic Acids Research*, vol. 31, no. 6, pp. 1683–1692, 2003.

[15] C. Y. Huang, "Inhibition of a putative dihydropyrimidinase from *Pseudomonas aeruginosa* PAO1 by flavonoids and substrates of cyclic amidohydrolases," *PLoS One*, vol. 10, no. 5, Article ID e0127634, 2015.

[16] J. Abendroth, K. Niefind, and D. Schomburg, "X-ray structure of a dihydropyrimidinase from *Thermus* sp. at 1.3 A resolution," *Journal of Molecular Biology*, vol. 320, no. 1, pp. 143–156, 2002.

[17] L. Holm and C. Sander, "An evolutionary treasure: unification of a broad set of amidohydrolases related to urease," *Proteins: Structure, Function, and Genetics*, vol. 28, no. 1, pp. 72–82, 1997.

[18] W. F. Peng and C. Y. Huang, "Allantoinase and dihydroorotase binding and inhibition by flavonols and the substrates of cyclic amidohydrolases," *Biochimie*, vol. 101, pp. 113–122, 2014.

[19] C. C. Wang, H. W. Tsau, W. T. Chen, and C. Y. Huang, "Identification and characterization of a putative dihydroorotase KPN01074, from *Klebsiella pneumoniae*," *Protein Journal*, vol. 29, no. 6, pp. 445–452, 2010.

[20] Y. Y. Ho, Y. H. Huang, and C. Y. Huang, "Chemical rescue of the post-translationally carboxylated lysine mutant of allantoinase and dihydroorotase by metal ions and short-chain carboxylic acids," *Amino Acids*, vol. 44, no. 4, pp. 1181–1191, 2013.

[21] D. R. Evans and H. I. Guy, "Mammalian pyrimidine biosynthesis: fresh insights into an ancient pathway," *Journal of Biological Chemistry*, vol. 279, no. 32, pp. 33035–33038, 2004.

[22] Y. C. Hsieh, M. C. Chen, C. C. Hsu, S. I. Chan, Y. S. Yang, and C. J. Chen, "Crystal structures of vertebrate dihydropyrimidinase and complexes from *Tetraodon nigroviridis* with lysine carbamylation: metal and structural requirements for post-translational modification and function," *Journal of Biological Chemistry*, vol. 288, no. 42, pp. 30645–30658, 2013.

[23] V. Kumar, N. Saxena, M. Sarma, and K. V. Radha Kishan, "Carboxylated lysine is required for higher activities in hydantoinases," *Protein & Peptide Letters*, vol. 18, no. 7, pp. 663–669, 2011.

[24] S. Martinez-Rodriguez, A. I. Martinez-Gomez, J. M. Clemente-Jimenez et al., "Structure of dihydropyrimidinase from *Sinorhizobium meliloti* CECT4114: new features in an amidohydrolase family member," *Journal of Structural Biology*, vol. 169, pp. 200–208, 2010.

[25] B. Lohkamp, B. Andersen, J. Piskur, and D. Dobritzsch, "The crystal structures of dihydropyrimidinases reaffirm the close relationship between cyclic amidohydrolases and explain their substrate specificity," *Journal of Biological Chemistry*, vol. 281, no. 19, pp. 13762–13776, 2006.

[26] X. Y. Zhang, L. X. Niu, Y. W. Shi, and J. M. Yuan, "The flexibility of the non-conservative region at the C terminus of D-hydantoinase from Pseudomonas putida YZ-26 is extremely limited," *Applied Biochemistry and Biotechnology*, vol. 144, no. 3, pp. 237–247, 2008.

[27] L. Niu, X. Zhang, Y. Shi, and J. Yuan, "Subunit dissociation and stability alteration of D hydantoinase deleted at the terminal amino acid residue," *Biotechnology Letters*, vol. 29, no. 2, pp. 303–308, 2007.

[28] C. T. Tzeng, Y. H. Huang, and C. Y. Huang, "Crystal structure of dihydropyrimidinase from *Pseudomonas aeruginosa* PAO1: insights into the molecular basis of formation of a dimer," *Biochemical and Biophysical Research Communications*, vol. 478, no. 3, pp. 1449–1455, 2016.

[29] Y. H. Huang, Y. Lien, C. C. Huang, and C. Y. Huang, "Characterization of *Staphylococcus aureus* primosomal DnaD protein: highly conserved C-terminal region is crucial for ssDNA and PriA helicase binding but not for DnaA protein-binding and self-tetramerization," *PLoS One*, vol. 11, no. 6, Article ID e0157593, 2016.

[30] Z. Otwinowski and W. Minor, "Processing of X-ray diffraction data collected in oscillation mode," *Methods in Enzymology*, vol. 276, pp. 307–326, 1997.

[31] J. Navaza, "AMoRe: an automated package for molecular replacement," *Acta Crystallographica Section A Foundations of Crystallography*, vol. 50, no. 2, pp. 157–163, 1994.

[32] D. E. McRee, "XtalView/Xfit–A versatile program for manipulating atomic coordinates and electron density," *Journal of Structural Biology*, vol. 125, no. 2-3, pp. 156–165, 1999.

[33] A. T. Brunger, P. D. Adams, G. M. Clore et al., "Crystallography & NMR system: a new software suite for macromolecular structure determination," *Acta Crystallographica Section D Biological Crystallography*, vol. 54, no. 5, pp. 905–921, 1998.

[34] Y. H. Huang and C. Y. Huang, "Creation of a putative third metal binding site in type II dihydroorotases significantly enhances enzyme activity," *Protein & Peptide Letters*, vol. 22, no. 12, pp. 1117–1122, 2015.

[35] Y. Y. Ho, H. C. Hsieh, and C. Y. Huang, "Biochemical characterization of allantoinase from *Escherichia coli* BL21," *Protein Journal*, vol. 30, no. 6, pp. 384–394, 2011.

19

An Overview of the Potential Therapeutic Applications of CO-Releasing Molecules

Aiten Ismailova, David Kuter, D. Scott Bohle, and Ian S. Butler ⓘ

Department of Chemistry, McGill University, 801 Sherbrooke Street West, Montreal, QC, Canada H3A 3K6

Correspondence should be addressed to Ian S. Butler; ian.butler@mcgill.ca

Academic Editor: Francesco Paolo Fanizzi

Carbon monoxide (CO) has long been known as the "silent killer" owing to its ability to form carboxyhemoglobin—the main cause of CO poisoning in humans. Its role as an endogenous neurotransmitter, however, was suggested in the early 1990s. Since then, the biological activity of CO has been widely examined via both the direct administration of CO and in the form of so-called "carbon monoxide releasing molecules (CORMs)." This overview will explore the general physiological effects and potential therapeutic applications of CO when delivered in the form of CORMs.

1. Introduction

Carbon monoxide (CO) is a colorless, odorless gas that is endogenously produced in the human body through heme oxygenase, which is present in two forms: the constitutive (HO-2 and HO-3) and inducible (HO-1) isoforms [1]. It has a higher affinity for hemoglobin than does O_2, when it forms carboxyhemoglobin, thus preventing O_2 transport throughout the body. A few years ago, it was discovered that CO gas in small doses exhibits some anti-inflammatory and antimicrobial properties making it useful as a potential therapeutic agent for disease control [2]. Unfortunately, using CO gas under clinical conditions is not practical. The gas has a low solubility in water, and so it has only a limited solubility in body fluids, which means that a person would need to inhale a high concentration of the gas to attain a meaningful concentration in the body [3]. Moreover, the delivery of gaseous CO cannot be precisely controlled and overexposure of body tissue to the gas could be harmful [4].

It is the well-known synergistic bonding (σ-donor and π-acceptor) ability of CO to transition metals that accounts for both its stability and reactivity [5]. Countless transition metal carbonyl complexes are now known, and some of these can release CO within the human body without affecting the level of carboxyhemoglobin produced. These

complexes are now referred to as "carbon monoxide releasing molecules (CORMs)." A CORM is made up of two parts: a *CORM sphere* and a *drug sphere*. The CORM sphere determines the number of CO molecules that can be released, the kinetics of the CO release, and the trigger mechanism necessary to cause the CO release. The drug sphere, defined by the periphery of the coligands surrounding the transition metal center, affords the most critical advantage of a CORM over CO gas. This advantage is because the CORM can modulate the partition ratio between the various body fluids and tissues, thus allowing it to be targeted to specific tissues. Most of the recent research on CORMs has been focused on the CO-release properties, and the drug sphere has been essentially ignored. Since the drug sphere is essential for targeting the CORM into a desired area of the body, more research is needed in this area [6].

The Ph.D. thesis of Joao Daniel de Silva Seixas from the Institute of Chemistry and Biology at the New University of Lisbon in 2011 entitled *Development of CO-Releasing Molecules for the Treatment of Inflammatory Diseases* is particularly informative about the utility of CORMs in some medical situations [7]. Interestingly, one of the first CORM molecules ever identified was $Et_4N[Mo(CO)_5Br]$ (ALF062), and this was prepared during the Ph.D. thesis work of one of the authors of this overview during the early 1960s [8].

Several important review articles on CORMs have been published over the years [9–14]. The present overview is focused on the pharmacological uses of CORMs and their impact on various human pathologies with the aim of determining whether CORMs will indeed be useful as therapeutic agents as was originally suggested in the pioneering paper by Motterlini in 2002 [1]. Reviews somewhat related to ours have recently been published by Ward [15] and Ling et al. [16]. The structures of the CORMs mentioned here are shown in Scheme 1, and a summary of the potential therapeutic applications is listed in Table 1.

2. Effect of CORMs on Bacteria

CORMs possess the ability to accumulate inside bacterial cells before they release CO, and this fact has led them to become useful CO donors to bacterial targets, such as *Escherichia coli, Staphylococcus aureus, Helicobacter pylori,* and *Pseudomonas aeruginosa*. For instance, Bang et al. [17] have demonstrated that dimeric CORM-2, $Ru_2Cl_4(CO)_6$, decreases the bacterial viability of multidrug resistant uropathogenic isolates of *E. coli* (UPEC) under biofilm conditions as well as in the colonization of human bladder epithelial cells. Biofilm is the term given to bacteria enclosed in a matrix that adheres to an abiotic or a biotic surface. There is increased resistance to antimicrobial agents by many bacteria when grown in biofilms. The formation of biofilm on medical devices, such as urinary catheters, increases the likelihood of urinary tract infections (UTIs). In patients with UTI, UPEC, and biofilm-producing UPEC are frequently seen. An example of a bacterium that causes biofilm formation is extended-spectrum beta-lactamase-(ESBL-) producing *Enterobacteriaceae*. The objective of this 2016 study of Bang et al. [17] was to investigate the antibacterial effects of CORM-2 on ESBL-producing UPEC in biofilm and in the colonization of human bladder epithelial cells. The results revealed that CORM-2 exhibited antibacterial properties against drug-resistant UPEC both under biofilm and host cell colonization conditions. CORM-2 reduced planktonic bacterial counts in the exponential growth phase by more than 3 log units within 4 hours. Although CORM-2 showed a delayed inhibitory response in the stationary growth phase, an antibacterial effect was observed after 24 hours. To confirm that CO is necessary for the antibacterial effect of CORM-2, the CO-free molecule, $Ru(DMSO)_4Cl_2$, was exposed to ESBL biofilms and epithelial cells, and unlike CORM-2, it demonstrated no reduction.

Another study on the efficacy of CORMs against *E. coli* has been reported by Tinajero-Trejo et al. [18]. Instead of using CORM-2, however, they employed the novel water-soluble, Mn-based photoactivated CORM, $Mn(CO)_3(tpa-k^3N)$, known as PhotoCORM. Upon irradiation at 325 nm, the complex releases CO. Tested specifically against the EC958 strain of *E. coli*, it was shown that the illuminated PhotoCORM suppressed respiration of EC958 membranes when compared to the untreated samples at a 50% O_2 tension and even more so at a much lower O_2 tension (15%). The PhotoCORM also inhibited EC958 growth on glucose

upon illumination, but not in the dark. Although the antimicrobial effects of PhotoCORM depend only on light activation, it is independent of O_2, suggesting that PhotoCORM is toxic in anoxic cultures as well. A related photoactivated CORM, Trypto-CORM, also displays marked cytotoxicity toward *E. coli* and, more recently, has been shown to be effective against *Neisseria gonorrhoeae* [19, 20].

In addition to *E. coli,* the effect of CORMs against the leading pathogen in skin infections, *Staphylococcus aureus,* has been explored. This bacterium can form biofilms that are almost impossible to eradicate using regular antibiotics. The goal of a study by Klinger-Strobel et al. [89] was to test the ability of CO to eliminate this highly resistant bacterium. The CO-treated biofilms showed elevated counts of dead bacteria compared to other biofilms that were not exposed to CO gas. Bacterial cells within methicillin-resistant *S. aureus* biofilms, the most common pathogen that causes skin infections, were killed after 135 min exposure to 405 nm light in the presence of CO gas. Furthermore, based on the work of Tavares et al. [21], CORMs prevented the survival of *Helicobacter pylori,* a pathogen responsible for chronic gastric ulcers. Treatment with CORM-2, which is soluble in DMSO, for 15 hours resulted in a 4-log loss of cell viability; however, when treated with water-soluble CORM-3, $Ru(CO)_3Cl(glycinate)$, the viability was attenuated by 2-log only. Thus, a more potent effect against *H. pylori* is achieved by CORM-2. This difference is presumably due to the different reactivity, hydrophobicity, polarity, and/or hydrogen bonding ability. Specifically, the DMSO-soluble CORM-2 is more hydrophobic than is the water-soluble CORM-3, which may favour the interaction of CORM-2 with the medium and the *H. pylori* cells. Additionally, exposure to CORM-2 for 5 min caused a decrease of greater than 50% in O_2; this suggests that CORM-2 inhibits *H. pylori* cell respiration, thereby inducing bacterial death. The mechanism for bacterial extermination ability of CORM-2 can be traced to the ability of *H. pylori* to express a nickel-containing urease, an enzyme responsible for the capability of the bacteria to cause disease. Since CO can bind transition metals, the effect of CORM-2 on the urease activity was analyzed. The results indicated that *H. pylori* grown in the presence of CORM-2 demonstrated a 65% decrease in urease activity. Furthermore, when the bacterial cells were incubated with increasing concentrations of CORM-2 for 15 min, the urease activity was completely abolished. Thus, the antibacterial effect induced by CORM-2 is most probably due to the ability of CO to bind to nickel.

The enteropathogen, *S. typhimurium,* which is associated with gastroenteritis, survives and proliferates within innate immune cells. Research conducted by Rana et al.co [22] has assessed the effects of CORMs on the bacterium, *Salmonella enterica serova typhimurium*. With an increasing concentration of CORM-3, both the growth and viability of *S. typhimurium* were reduced. The CO released from CORM-3 binds to the terminal oxidases of *S. typhimurium* and is rapidly taken up by the *Salmonella* cells. The ruthenium complex, CORM-3, speedily fluxes into the *Salmonella* cells. After 80 min, the level of ruthenium inside the cells exceeded the final concentration added to the extracellular medium,

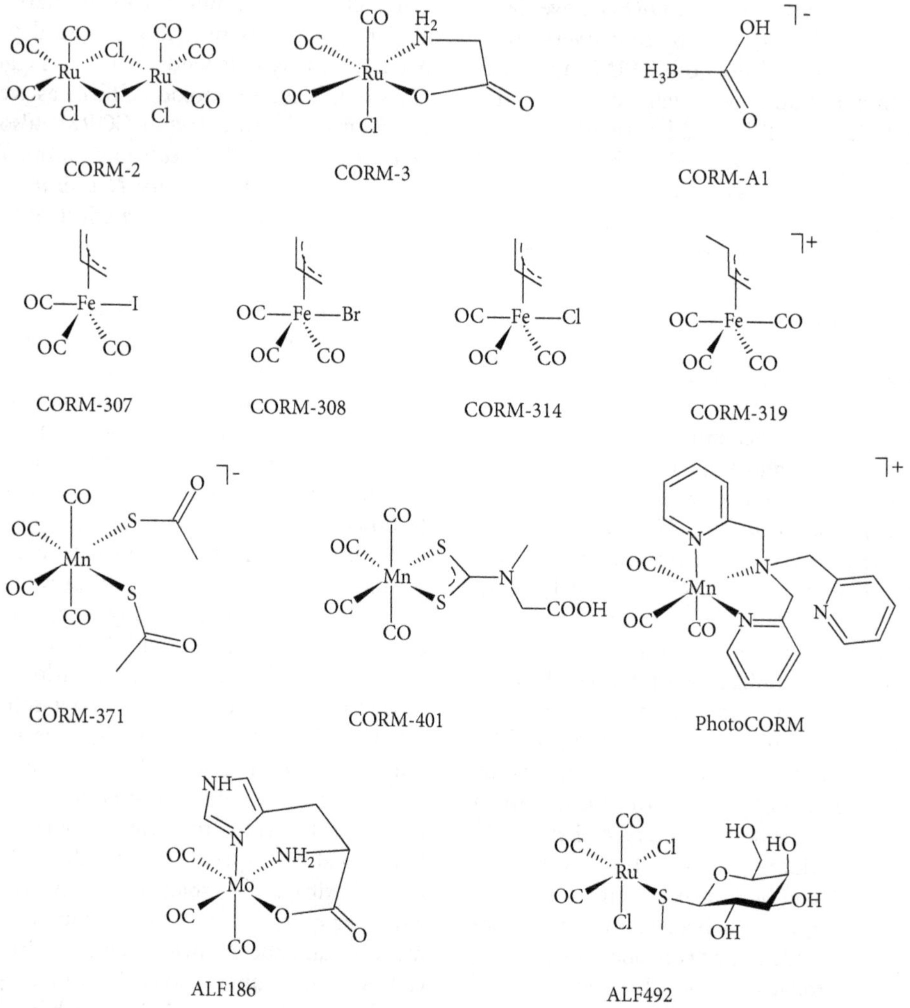

SCHEME 1

indicating that CORM-3 is actively concentrated into bacterial cells.

Desmard et al. [23] have demonstrated that different antibacterial activities occur against *Pseudomonas aeruginosa* depending on the structural nature of the CORM used. Both CORM-2 and CORM-3 contain ruthenium and are rapid CO liberators—CO release occurs within 1 min after their addition to biological systems. These two compounds only differ in their solubility in DMSO for CORM-2 and in water for CORM-3. However, another CORM, CORM-A1, which does not contain any transition metal, but boron instead, slowly releases CO. In this work, a new compound was also tested, CORM-371. This CORM contains manganese and liberates CO slowly. While all three CORMs inhibit growth and metabolism of *P. aeruginosa*, as determined by a decreased oxygen consumption, CORM-A1 reduced O_2 consumption less than did CORM-2 and CORM-3 and only slowed the initiation of bacterial growth. Moreover, CORM-A1 had no bactericidal properties but acted more as a bacteriostatic agent (i.e., it only stopped bacteria from reproducing and it did not kill any bacteria). Although both CORM-A1 and CORM-371 are both slow CO releasers,

CORM-371 was found to be intermediate between CORM-2/CORM-3 and CORM-A1 in this respect. It elicited a stronger antibacterial effect and inhibition of O_2 consumption than did CORM-A1, but the ruthenium-based compounds still showed the largest sustained effect. Owing to the similar rates of CO release from CORM-371 and CORM-A1, but the ability of CORM-A1 to only exert a bacteriostatic effect, the kinetics of CO release from CORMs is not necessary for their antibacterial effects. Instead, as indicated by the results obtained with the ruthenium-based CORM-2 and CORM-3, which rapidly decrease bacterial growth, the presence of a metal center that favours CO targeting bacteria and the nature of transition metal are more likely to be the causes of the antibacterial activity exhibited by these CORMs. A recent investigation by Flanagan et al. has also shown that $[Et_4N][MnBr_2(CO)_4]$ is active against a variety of *P. aeruginosa* strains [24].

Although the best understood targets of CORMs are hemes, Wilson et al. [90] have reported that a CORM can have enhanced toxicity against nonclassical targets, that is, nonheme bacteria. When exposed to *E. coli* hemA mutants and *Lactococcus lactis*, two bacteria that lack heme, the

TABLE 1: A summary of the potential therapeutic applications of CORMs.

CORM	Effect	Ref.
Bacteria:		
E. coli CORM-2	Decreased viability of uropathogenic isolates and reduced colonization of human bladder epithelial cells	[17]
PhotoCORM	Suppressed cell membrane respiration in the EC598 strain	[18]
TryptoCORM	Reduces cell viability by > 99.9%	[19]
N. gonorrhoeae TryptoCORM	Reduces cell viability by > 99%	[20]
H. pylori CORM-2	Reduced cell viability via inhibition of Ni-containing urease	[21]
S. typhimurium CORM-3	Reduced growth and viability	[22]
CORM-2		
P. aeruginosa CORM-3	CORM-2, -3, and -371 reduced bacterial O_2 consumption and displayed bactericidal properties	[23]
CORM-371		
CORM-A1	CORM-A1 slowed bacterial growth (bacteriostatic)	
$[MnBr_2(CO)_4]$	Varied reduction of cellular growth for a variety of strains	[24]
Neurodifferentiation and neuroprotection:		
Neurodifferentiation CORM-A1	Improved neuronal differentiation and yield in NT2 cell line by promoting oxidative metabolism	[25]
CORM-2	Increased viability of neural stem cells and reduced number of apoptotic cells	[26]
Neuroprotection	Lessened mitochondrial damage and improved neurological function of mice after induced cardiac arrest	[27]
ALF186	Prevented apoptosis in nerve cells simulating ischemic respiratory arrest by increasing cellular cGMP levels	[28]
ALF492	Protected mice against cerebral malaria	[29]
Cochlear inflammation CORM-2	Inhibited MCP-1/CCL2 upregulation, reducing oxidative stress and protecting against cochlear inflammation	[30]
Neuroinflammation CORM-3	Reduced inflammatory response in BV-2 microglial cells by reducing NO production	[31]
	Suppresses interleukin-1β-induced inflammatory responses	[32]
Nociception and diabetes:		
Neuropathic pain CORM-2	Attenuated mechanical allodynia, thermal hyperalgesia, and thermal allodynia when used in combination with the antinociceptive JWH-015	[33]
	Reduced sciatic nerve injury-induced mechanical and thermal hypersensitivity by attenuating spinal microglial activation and expression of NOS1/2 and CD11b/c proteins	[34]
Diabetes CORM-A1	Facilitated beta cell regeneration by reducing T-helper cell counts and TGF-β and Ki-67 expression	[35]
Inflammatory disease:		
Colitis CORM-2	Reduced cell survival of colitis-inducing cells	[36]
CO-HbV	Reduced tissue damage and prolonged survival of mice with induced colitis	[37]
Bacterial LPS-induced inflammation CORM-2	Prevented LPS-mediated inflammation by reducing TLR4/MD2 expression on dendritic cell surfaces and protected mice against increased neutrophil counts associated with septic inflammation	[38]
Tumour necrosis factor α-induced inflammation CORM-2	Induced p65 glutathionylation which protects cysteinyl residues from irreversible oxidation	[39]
Inflammatory disease (cont.):		
Inflammation-induced blood clotting CORM-2	Decreased blood clotting in human umbilical vein endothelial cells by suppressing MAPK and NF-κB signaling pathways	[40]
Uveitis CORM-A1	Improved retina morphology and expression of IFNgamma and IL-17A was lowered and IL-10 raised in uveitis-induced mice	[41]
Chronic inflammatory pain CORM-2	Reduced mechanical allodynia and thermal hyperalgesia in mice and diminished upregulation of NOS1 expression	[42]
Intestinal barrier function CORM-2	Improved barrier function of intestinal epithelial cells by suppressing phosphorylation of the myosin light chain	[43]
Periodontal disease CORM-3	Inhibited nuclear translocation of NF-κB and reduced DNA binding of p65/p50 subunits	[44]

TABLE 1: Continued.

	CORM	Effect	Ref.
Vascular inflammation	CORM-3	Inhibited neutrophilic myeloperoxidase activity	[45]
Sepsis and associated conditions:			
Oxidative stress	CORM-2	Reduced oxidative stress during sepsis by increasing HO-1 expression	[46]
NO-induced lipid peroxidation	CORM-2	Attenuated inducible NO synthase and NO production	[46]
CLP-induced sepsis	CORM-2	Improved morphology of intestinal mucosa during sepsis, protecting against LPS-induced intestinal damage	[47]
		Reduced mortality of mice with sepsis-induced acute kidney injury by reducing biomarkers	[48]
Septic lung injury	CORM-3	Restored downregulated annexin A2 levels to normal in LPS-induced lung sepsis	[49]
Myocardial dysfunction	CORM-3	Improved myocardial function in cardiac fibroblasts of septic mice by inhibiting activation of the NLRP3 inflammasome	[50]
Abnormal platelet coagulation	CORM-2	Abnormal platelet activation was reduced by inhibition of glycoprotein-mediated HS1 phosphorylation	[51]
Hyperglycemia	CORM-2	Suppression of hepatic glucose metabolism in mice	[52]
Recruitment of PMN leukocytes	CORM-3	Reduced leukocyte infiltration and attenuated several (but not all) proteins expressed during sepsis	[53]
Obesity:			
Dietary-induced	CORM-A1	Reduced weight gain, aided weight loss, and increased lean body mass in mice receiving a high-fat diet	[54]
	CORM-2	Reduced leptin resistance and led to lower body weight of animals fed high-fat diet	[55]
Hyperglycemia	CORM-A1	Decreased hyperglycemia and reduced plasma insulin levels	[54]
Angiogenesis, aggregation, and cancer:			
Angiogenesis	CORM-2	Prevented endothelial cell migration and proliferation induced by vascular endothelial growth factor and suppressed phosphorylation of retinoblastoma protein, halting extreme cell replication	[56]
Cancer	CORM-401	Promoted vasorelaxation of precontracted aortic rings	[57]
	CORM-2	Increased survival of mice with A20 lymphoma tumours when encapsulated by folic acid-tagged protein nanoemulsions	[58]
		Prevented global protein synthesis in pancreatic stellate cells	[59]
	PhotoCORM	Reduced cell biomass upon irradiation at 365 nm	[60]
Cell aggregation	CORM-2	Decreased binding affinity of a integrin-specific ligand and lead to reduced cellular aggregation	[61]
Hemorrhagic shock and postresuscitation injuries:			
Hemorrhagic shock	CORM-A1	Maintained levels of fenestrations, cells, and adherent leukocytes by reducing expression of cytokines	[62]
	CORM-3	Increased frequency of live human umbilical vein endothelial cells; reduced apoptosis and decreased mitochondrial transmembrane potential and reduced tissue necrosis	[63]
Postresuscitation myocardial injury	CORM-2	Reduced myocytolysis and damage from myocardial fibers and decreased cardiac mitochondrial ROS	[64]
Hypoxia reoxygenation	CORM-3	Conserved cell viability	[65]
Cardiac transplantation	CORM-3	Prolonged survival of rats after heart transplantation	[65]
		Improved coronary flow in mice following heart transplantation	[66]
	CORM-2	Pretreating donor rats improved renal histology and function in recipients and long term treatment, however, produced excess lymphocyte accumulation and glomerulus atrophy	[67]
Kidney transplantation	CORM-3		
Gastric, intestinal, kidney, and liver disorders:			
Gastric disorder	CORM-2	Reduced formation of mucosal lesions caused by alendronate (osteoclast inhibitor) in rats stressed by water immersion	[68]
Liver injury	CORM-A1	Reduced hepatocyte cell death by decreasing CK18 cleavage products and lowering RIP3 expression	[69]
Hepatitis	CORM-A1	Significantly reduced deaths in a murine model of autoimmune hepatitis	[70]

TABLE 1: Continued.

	CORM	Effect	Ref.
Nephrotoxicity	CORM-3	Reduced cell damage induced by cisplatin in renal epithelial cells by suppressing caspase-3 activity and prevented apoptosis and kidney mass loss	[71]
Renoprotection	CORM-3	Increased viability of normal and cancerous human renal cells that were subjected to cisplatin-induced toxicity and ischemia-reperfusion injury	[72]
Intestinal disorder	CORM-3	Partially restored intestinal contractility in mice presenting postoperative ileus and reduced oxidative stress levels	[73]
Lungs: Pulmonary hypertension	CORM-3	Prevented ventricular hypertrophy and distal pulmonary artery muscularization in hypoxia-induced mice	[74]
		Resulted in irreversible pulmonary vasoconstriction in an in vitro hypoxic pulmonary vasoconstriction model	[75]
Ocular system: Intraocular pressure	CORM-3	Lowered intraocular pressure in rabbits	[76]
Cardiovascular effects: Cardioprotection/ toxicity	CORM-2	Decreased oxidative stress and apoptosis induced by DXR (antitumor agent) in a narrow therapeutic window	[77]
		Attenuated angiotensin II-induced aortic smooth muscle cell migration by inhibiting matrix metalloproteinase-9 expression and ROS/interleukin-6 generation	[78]
	CORM-3	Improved recovery of cardiac structure and function following myocardial infarction in rats	[79]
Pro- and anticoagulant effects: Procoagulation	CORM-2	Increased strength and velocity of clot formation	[80–82]
		Attenuates snake venom with fibrinogenolytic and thrombin-like activity	[83–86]
Anticoagulation	CORM-2	Reduced platelet aggregation in aortic allograft recipient mice	[87]
	CORM-3 CORM-A1	Decreased arterial thrombus formation	[88]

CORMs reduced their growth considerably. CORMs were also found to disturb the cell membranes of these two bacteria and deplete iron levels in wild type and heme-deficient *E. coli*.

3. Effect of CORMs in Neuroprotection and Neuronal Differentiation

An investigation by Verma et al. [91] has indicated that CO gas may function as a neurotransmitter. These researchers found high concentrations of heme oxygenase in the brain. Moreover, since heme oxygenase degrades heme to biliverdin and releases CO in the process, there was a substantial amount of CO in the brain as well. It was discovered that CO is a weak activator of guanylyl cyclase (GC) thus allowing CO to effect smooth muscle relaxation and potentially block platelet aggregation. What therefore is the role of CO in neural activity? Two independent research groups have found that CO is linked to the functions of nonadrenergic, noncholinergic neurons of the enteric nervous system and, therefore, has an impact on spontaneous slow waves of the smooth muscle [92, 93]. Furthermore, it has been reported that CO may influence long-term potentiation (LTP) in a tonic manner [94]; through the strengthening of synapses, LTP has been shown to be involved in learning and memory. In addition, endogenous CO has been implicated in attenuating endotoxin-stimulated arginine vasopressin release. Excessive vasopressin release leads to greatly increased water retention by the kidneys and contributes to the dilation of blood; this dilation consequently leads to a low salt concentration in the blood, a condition known as hyponatremia [95]. However, CO may also be associated with pathological states, and a few studies have concluded that CO may be linked to onset of Parkinson's disease [96–98].

Two research teams have examined the effect of exogenous CO on the brain via CO inhalation. In the study by Piantadosi et al. [99], awake rats were exposed to high concentrations (2500 ppm) of CO gas for 60 min. This exposure led to delayed histological damage in the brain, the cerebral cortex being the most affected followed by the basal ganglia and cerebellum. However, the performance of the CO-exposed rats improved gradually and approached that of control rats one month after the exposure. Exposure to high concentrations of CO led to interstitial glutamate accumulation which was attributed to increased hydroxyl radical formation. In addition, CO interfered with mitochondrial redox state and energy production. Glial damage was also seen upon CORM administration. These observed effects led to the conclusion that CO exhibited toxic effects in the brain [99]. Another study confirmed the same finding by investigating the effects of CO poisoning on the brain in the acute and chronic periods using magnetic resonance imaging. After examining the brains of 16 previously CO-exposed patients, the imaging showed necrosis of the globus pallidus, white matter lesions with demyelination or necrosis, lesions of the cerebral cortex, and necrotic lesions of the hippocampus [100]. Again, this study validated the notion of CO poisoning leading to brain dysfunction. However, CO under lower concentrations produced an opposite effect on the brain. The research by Quieroga et al. [101] was aimed at evaluating the effect of CO on astrocyteneuronal communication. The communication between neurons and astrocytes is crucial for neurogenesis, normal dendritic maturation, spine formation, and integration of adult born neurons. Astrocytes play a role in neuroinflammation and neuroprotection—both processes are compromised in diseases, such as Alzheimer's in which there is a substantial amount of neuronal cell death. Upon the addition of 40 μm of t-BHP (a neuronal cell death inducer) on neurons and cocultures (primary cultures of astrocytes and neurons), a considerable amount of neuronal death occurred. Though, once CO was added in the coculture, neuronal survival levels increased. Moreover, CO influenced ATP extracellular content; extracellular ATP is a signaling molecule involved in the communication between astrocytes and neurons. Following CO treatment, intracellular ATP concentration increased. Not only did CO boost ATP content, but it also raised rates of consumption of the amino acids, serine and cysteine, considerably. This result indicated that amino acids were being converted into pyruvate at a much higher rate thus setting in motion the citric acid cycle, which results in oxidative phosphorylation and ATP production. To determine the role of ATP in astrocyteneuron communication, the ATP extracellular content was measured at different times. The results indicated that CO stimulated ATP metabolism, which in turn triggered increased communication between astrocytes and neurons, therefore causing the astrocytes to protect the neurons from apoptosis.

Because of the increasing occurrence of neurodegenerative diseases and ischemic stroke, balancing the amount of neurogenesis is a hopeful method to improve the detection of the symptoms and ultimately cure those disorders. Almeida et al. [25] have demonstrated that CO promotes neuronal differentiation and increased NT2 neuronal yield by boosting cell metabolism. When the cultures were treated with CORM-A1, the final number of postmitotic neurons was duplicated. Kinetic parameters for cell growth displayed an increase on growth rate, doubling time, and fold increase when CORM-A1 was added into the medium. Moreover, in the presence of CORM-A1, the total number of a mixed cell population (composed of progenitor cells and postmitotic neurons) was higher after 24 days of differentiation than without CORM-A1.

Higher levels of mRNA expression of the specific neuronal differentiation markers nestin, tuj1, and microtubule-associated protein 2 (MAP2) were observed when treated with CORM-A1 by comparison with treatment with retinoic acid alone [25]. This CORM was also shown to modulate cell metabolism by increasing neuronal yield. The expression of mRNA (Glut1 and MCT2), pyruvate dehydrogenase, and lactate dehydrogenase membrane transporters all increased upon administration of CORM-A1. In addition, when supplemented with CORM-A1, the ratio of lactate concentration/glucose concentration was lowered, which validated the conclusion that the differentiation process was

elevated owing to a switch from oxidative to glycolytic metabolism that occurred early on.

A promising approach to repair the impaired brain following a hemorrhagic stroke is neural stem cell (NSC) transplantation, but there is reduced cell survival because of iron overload following the transplantation. According to the data of Xie et al. [26], addition of CORM-2 increased the viability of NSC in a concentration-dependent manner, suggesting a neuroprotective effect of the CORM against iron overload. CORM-2 also reduced the number of apoptotic cells, whereas the ferrous iron-treated and iCORM-2 (inactive CORM) groups increased it. Iron overload generated nuclear translocation of NF-κB p65 in neural stem cells but when CORM-2 was administered, the nuclear translocation was diminished. Additionally, the levels of the molecules, which induced apoptosis (Bax and cleaved caspase-3), were elevated when exposed to excess iron. Treatment with CORM-2 attenuated the levels of these molecules to half that of the FeCl$_2$ group. As well as investigating the viability of transplanted neural stem cells, Wang et al. [27] examined the effect of CO treatment on brain mitochondrial impairment and brain injury in rats. Following the return of spontaneous circulation after cardiac arrest or return of spontaneous circulation (ROSC), the introduction of CO resulted in an increase in the 3-day survival time of the rats. After 24 hours in ROSC-challenged rats, CORM-2 treatment led to an elevated neurologic deficit score when measuring arousal, reflex, motor, sensory, and balance responses. Furthermore, CO was observed to lower S-100B levels; S-100B refers to a marker of neurological outcomes after cardiac arrest. CO supplementation also elevated ATP levels, the mitochondrial respiratory ratio, $\Delta\Psi m$, and mitochondrial DNA; these factors are all indicators of mitochondrial function, suggesting that CO lessens mitochondrial damage associated with ROSC [27].

A different type of CORM, known as CORM ALF492, has been shown to fully protect mice against experimental cerebral malaria and acute lung injury via controlling CO delivery *in vivo* without affecting O$_2$ transport by hemoglobin. Through altering the ligand in the Ru-core scaffold and metal-ligand linkage, Pena et al. [29] have managed to synthesize CORM ALF492. This CORM has a galactose-derived ligand coordinated to the Ru center via a thioether linkage, which helped to improve water solubility and compatibility with biological molecules. The presence of the galactose ligand may also allow an elevated degree of specificity by targeting appropriate receptors in the liver. Owing to its increased solubility in water and high level of specificity, CORM ALF492 was able to protect mice from death due to cerebral malaria, which is mainly caused by the parasite, *Plasmodium falciparum* [29]. CORM ALF186, was also able to prevent nerve cell apoptosis in a simulation of ischemic insult to neuronal cells. When SH-SY5Y-cells were exposed to rotenone to simulate ischemic respiratory arrest, only CORM ALF186 and not the CO-free inactivated ALF186 (iALF 186) prevented nerve cell apoptosis induced by rotenone by increasing cellular cGMP levels and sGC expression [28]. Furthermore, based on the study by Woo et al. [30], CO was found to increase the expression of oxidative stress-protecting enzymes within the developing inner ear (cochlea) in a mouse model and thus demonstrated that mild CO exposure may be beneficial for diseases caused by oxidative stress.

Excessive inflammatory responses can be harmful and the modulation of inflammation in microglia by CO has been shown to be necessary for the control of neuro-inflammation. Several studies concerning the anti-inflammatory effect of CO have been performed in vitro using BV-2 microglial cells. Bani-Hani, for example, has found that CORM-3 decreases NO production and TNF-α release in response to LPS, thrombin, and IFN-γ stimuli, thereby limiting the inflammatory response of neurons [31]. Lin et al. have also recently reported that CORM-3 sequentially activates a c-Src/Pyk2/PKCα/Erk1/2 pathway in rat brain astrocytes causing induction of HO-1 and subsequent suppression of interleukin-1β-mediated neuro-inflammation [32].

4. Contributions of CORMs in Nociception and Diabetes

Painful neuropathy is the most common complication of diabetes mellitus affecting 40% of people with type-1 diabetes. Administration of cannabinoid receptor 1 (CB1R) and 2 (CB2R) agonists does reduce pain in many animal pain models and may be a target for treatment of neuropathy. In a mice model of streptozotocin- (STZ-) induced diabetic neuropathy, a study was undertaken to evaluate the anti-nociceptive effects of the CB2R agonist, JWH-015, when combined with CORM-2 and when with CORM-2 alone. A STZ injection induced mechanical allodynia (pain due to stimuli that do not normally provoke a painful response), thermal hyperalgesia (increased sensitivity to pain), and thermal allodynia (pain from normally mild skin temperatures in the affected area). Upon administration of CORM-2, mechanical allodynia, thermal hyperalgesia, and thermal allodynia were significantly attenuated compared to the control group. Furthermore, low levels of HO-1, CB2R, and overexpression of nitric oxide synthase (NOS1) are characteristic of diabetes. Supplementation of CORM-2, combined with JWH-015, increased protein levels of HO-1 and CB2R and decreased the NOS1 protein levels. Taken together, these results suggest that CORM-2 could act as a potential treatment for painful neuropathy [33].

Additional evidence of CORMs being able to treat neuropathic pain has come from a study in which the antinociceptive effect of CORMs was evaluated in a murine model of sciatic nerve injury [34]. Sciatic nerve injury elevates NOS1, NOS2, and CD11b/c protein levels and increases spinal microglial activation; however, expression of these proteins and activation of microglial cells were attenuated upon CORM-2 administration. It was also observed that CORM treatment reduced the mechanical and thermal hypersensitivity caused by sciatic nerve injury. Next, the effects of CORM-A1 on preventing diabetes via beta cell regeneration were examined by Nikolic et al. [35]. In diabetes, the immune system of the body attacks and destroys the beta cells; this situation is characterized by a high interleukin and T-helper cell count as well as a low TGF-ß count and Ki-67 protein expression. When mice were treated with CORM-A1, the frequency of T-helper cells was noted to be considerably lowered in the spleen and pancreatic-draining lymph nodes of the mice when compared to the values for the non-CORM-A1-treated control group. CORM-A1 also resulted in an elevation in TGF-ß production and Ki-67 protein expression; hence, it counteracts the effects of diabetes.

Another study was undertaken to investigate the ability of CO in preventing diabetes via injecting dendritic cells treated with CO into diabetic rats [102]. This approach resulted in an inhibition of diabetes independent of IL-10 secretion and CD4+ T cell expression in dendritic cells; thus, this denoted that the protection was not mediated by dendritic cells (DCs), but instead was mediated by CO.

5. Role of CORMs in Inflammatory Disease

The CO gas liberated by CORMs also has an important role in modulating inflammation, a symptom observed in multiple pathological conditions. An example of such a pathological condition is colitis, which is described as an inflammatory reaction in the colon. One study exposed NIH 3T3 fibroblasts, cells that contribute to colitis, to varying concentrations of CORM-2 for 24 hours [36]. A significant decrease in cell survival was noted suggesting an anti-inflammatory effect of CORM-2 on colitis-inducing cells.

Another study has focused on the potential uses of a new type of nanosized CO donor, CO-HbV (CO bound hemoglobin encapsulated liposomes), with respect to providing therapeutic benefits for the symptoms of colitis [37]. CO-HbV administration was seen to suppress the progression of symptoms associated with colitis as well as prolonging the survival duration of dextran sulphate sodium- (DSS-) induced mice. Tissue damage, such as necrosis and ulcers, was found in DSS-induced colitis mice, whereas DSS mice that were treated with CO-HbV displayed lower amounts of tissue damage and a histological appearance that was like that of the control mice. Similarly, the administration of DSS generated anemia due to bleeding in the colon; reduced levels of red blood cells (RBC) and hemoglobin (Hb) were observed in the DSS-induced colitis mice. In contrast, the CO-HbV treatment overturned this reduction in RBC and Hb. Therefore, the overall results from the two studies described here [36, 37] support the notion of CORMs being able to negate cell death and the harmful effects induced by colitis. Certain inflammatory diseases work by targeting dendritic cells thus resulting in an inflammatory response. The roles of dendritic cells are to recognize, destroy, and present harmful antigens to T-cells; hence, they protect the host from various infections. When exposed to lipopolysaccharide (LPS) derived from bacteria, however, the dendritic cells undergo an inflammatory response via binding to toll-like receptor 4 (TLR4)/myeloid differentiation factor-2 (MD2) complex present on the surface of the cell.

Research undertaken by Riquelme et al. [38] showed that CO was able to prevent LPS-mediated inflammation.

The expression of TLR4/MD2 on the surface of dendritic cells was reduced upon CO treatment, whereas in the control cells, no changes were observed. Moreover, after CO exposure, mice that received LPS preserved their ability to move with no changes in their velocity. Conversely, the groups of mice that received LPS and were subjected to iCORM-2 displayed a decrease in their mobility, with values falling more than 70% compared with the values noted before LPS exposure. LPS mice supplemented with either iCORM-2 or CORM-2 maintained their body weight and had an increased survival rate. However, mice treated with iCORM-2 and then challenged with LPS experienced a decrease in body weight by almost 18% and had low survival rates. Moreover, CO treatment was found to protect mice from the toxic increase of neutrophils in blood associated with septic inflammation. Mice treated with iCORM-2 and then exposed to LPS demonstrated a 25-fold significant increase in neutrophil count. This increase in neutrophils was not seen in LPS mice exposed with CORM-2.

On a molecular level, how do CORMs mediate inflammation? Research conducted by Jung et al. [103] investigated these mechanisms, and the contribution of CO to NLRP3 (nucleotide-binding domain, leucine-rich-containing family, pyrin domain-containing-3) inflammasome activation in macrophages was explored. Inflammasomes are protein complexes in the cytosol that cause cleavage of caspase-1, which results in the maturation and secretion of proinflammatory cytokines, such as interleukin-1 (IL-1) and IL-18. This cascade of events consequently leads to inflammation in the body. The study found that CO treatment inhibited the secretion of IL-1 and IL-18 in response to LPS and ATP in macrophages by preserving the mitochondrial membrane potential. Moreover, mitochondrial ROS generation induced by LPS and ATP was inhibited by CO treatment. CO also inhibited mtDNA translocation into the cytosol, which is associated with the obstruction of cytokine secretion. Therefore, these findings demonstrate the ability of CO to reduce inflammation via negatively regulating LRP3 inflammasome activation in macrophages.

Yeh et al. [39] have also evaluated the molecular mechanisms of anti-inflammatory effects of CORMs. It was hypothesized that CORMs could promote NF-κB-p65 glutathionylation. When the study was conducted, however, CORMs were found to repress TNF-α-induced monocyte adhesion to endothelial cells and reduce the expression of ICAM-1. Furthermore, CORMs were noted to block NF-κB-p65 nuclear translocation. CORMs also induced Nrf2 activation and HO-1 expression, which elevated p65 glutathionylation. Therefore, these findings indicate that the glutathionylation of p65 is responsible for the NF-κB inactivation mediated by CORMs.

A common phenomenon associated with inflammatory diseases is excess blood clotting. The effect of CORM-2 on blood coagulation was investigated by testing it on the expression of TF and PAI-1 and on signaling pathways (MAPK and NF-κB), which all stimulate thrombosis [40]. It was demonstrated that CORM-2 subdued TNF-α-induced TF and PAI-1 upregulation in HUVECs (human umbilical vein endothelial cells). Moreover, CORM-2 treatment suppressed MAPK and NF-κB signaling pathways activation by TNF-α. Based on these results, CORM-2 supplementation was seen to decrease blood clotting induced by inflammation.

According to Fagone et al. [41], CORMs may also pose an effective treatment for yet another inflammatory disorder—uveitis, the leading cause of blindness due to an inflammation of the eye. CORM-A1 was noted to improve the morphology of the retina in the eye. Furthermore, rats treated with CORM-A1 expressed lower levels of IFN-gamma and IL-17A and increased amounts of IL-10 when compared to the non-CORM-treated uveitis-induced mice.

An additional complication due to inflammatory diseases is the onset of chronic inflammatory pain. Negrete et al. [42] found that CORM administration reduced this pain. When exposed to CORM-2, the inflammatory pain symptoms induced by complete Freund's adjuvant in mice, which consist of mechanical allodynia and thermal hyperalgesia, were considerably reduced. Moreover, inflammatory pain induced enlarged NOS1 protein expression in mice. This upregulation was prohibited by CORM-2 treatment in only wild-type cells; it did not decrease NOS1 levels in the NOS2-knockout mice.

Intestinal epithelial cells (IECs) play a key role in protecting against pathogens in the intestinal tract as well as maintaining gut barrier function. However, when inflamed, the function of the cells is greatly reduced, making them more vulnerable to microbes. Mu et al. have observed that CORM-2 treatment results in an improvement of the barrier function of IEC-6 cells [43]. In addition, the CO released from CORM-2 inhibited the secretion of proinflammatory cytokines (TNF-a and IL-1b) produced by LPS. One of the underlying mechanisms of inflammation includes the induction of myosin light-chain phosphorylation in the IECs. CORM-2 suppressed this phosphorylation in the cells; taken together, these data suggest that CORM-2 carries the ability to restore function of the IECs in the instance of inflammation.

Periodontal disease, yet another example of an inflammatory disease, has been linked to other potential disorders, such as cardiovascular disease, diabetes, and stroke, and a study by Choi et al. [44] has concluded that CORM-3 possesses anti-inflammatory activity against the periodontal-disease inducing the pathogen *Prevotella intermedia*. Upon the addition of CORM-3, HO-1 expression in LPS-exposed cells was enhanced. Although LPS-induced phosphorylation of p38 and JNK and the degradation of IκB-α were not affected by CORM-3, nuclear translocation of NF-κB p65 and p50 subunits was blocked by CORM-3 in LPS-treated cells. This result is significant due to the activation of NF-κB being associated in inflammation and disease. In addition, CORM-3 was noted to diminish LPS-induced p65 and p50 binding to DNA and phosphorylation of STAT1.

What may be the main underlying cause of several inflammatory disorders? According to Patterson et al. [45], polymorphonuclear leukocyte-derived myeloperoxidase (MPO) is known for its direct peroxidation as well as generation of potent oxidizing compounds, namely hypochlorous acid, which contributes to increased oxidant damage, which in turn leads to tissue injury and inflammation. These researchers found that CORM-3 suppresses MPO oxidative activity. The inactivated CORM-3 (iCORM-3) was also tested, and it showed no attenuation in the MPO activity, suggesting that binding the released CO to the MPO heme is crucial for its effects. When the activity of MPO in peroxidation and halogenation was tested, CORM-3 showed an inhibitory effect in both cases. Surprisingly, iCORM-3 also was able to decrease the chlorination activity of MPO, but to a lesser extent when compared to active CORM-3. These may be due to a CO-independent downstream reactivity of the iCORM metal centers with hypochlorite.

6. Role of CORMs in Sepsis: A Systemic Inflammation

As mentioned by Woo et al. [30], oxidative stress is believed to contribute to tissue damage, and nitrosative stress is often suggested to be initiated by an inflammatory cascade that consists of acute phase protein synthesis, upregulation of inflammatory adhesion molecules, and proinflammatory cytokine liberation. Severe injury is concerned with lipid peroxidation mediated by reactive oxygen species (ROS) and NO; lipid peroxidation is known to be an essential cause of oxidative damage to cellular membranes. In another study, Sun et al. [46] found that upon *in vivo* administration of CORM-2, the protein expression of inducible NOS and the overabundance of NO were considerably attenuated in thermal injury-induced septic mice. Furthermore, both the production of ROS and NO were decreased in LPS-stimulated HUVECs when incubated with CORM-2. In addition, recent findings suggest that the stress-inducible gene HO-1 is an important factor in protecting against oxidative stress. Sun et al. have also shown that the induction of HO-1 effectively provides protection both *in vivo* and *in vitro* against oxidative stress. They found that HO-1 is upregulated in HUVECs by LPS stimulation, and they went further by testing it with CORM-2. Their results indicated that not only was LPS able to induce the expression of HO-1, but also that an increase in HO-1 expression can be enhanced by CORM-2 treatment. Hence, the study concluded that CORM-2 ultimately leads to cytoprotection and inhibition of oxidative stress during sepsis.

When mice were impaired with cecal ligation and puncture (CLP), which led to systemic inflammation known as sepsis, CO was found to increase survival via strengthening the processes of autophagy (degradation of harmful or damaged intracellular contents) and phagocytosis (engulfment of bacteria by cells) [48]. Treatment with CO induced the expression of the autophagy proteins, beclin 1, Atg7, and LC3B; CO was also observed to increase the number of autophagosomes, the structure in which autophagy occurs, in the lungs of mice. Further, CLP-induced Becn1+/+ mice (mice with the eclin protein) that were then treated with CO demonstrated increased survival compared with the room air-exposed control mice. Next, the ability of CO to mediate

phagocytosis was assessed. It was noted that CO elevated the phagocytic activity of bone marrow-derived macrophages induced with LPS. The process by which CO can increase phagocytosis was revealed by its observed ability to induce SLAM (a signaling molecule responsible for killing bacteria by macrophages) mRNA expression in bone marrow derived-macrophages.

An additional study that focused on the role of CO in sepsis was undertaken by Zhang et al. [47], who discovered that CORM-2 could protect intestinal epithelial tight junctions when damaged with LPS in a rat CLP model. CORM-2 improved the morphology of the intestinal mucosa during sepsis when compared to CLP alone or treatment with iCORM, which showed ulceration and hemorrhage on the mucosa. Additionally, CORM-2 was seen to improve CLP-induced, tight-junction disruption and increase the proximity of the intercellular connection. Not only did CORM-2 ameliorate the physical appearance of the tight junctions, but it also raised the expressions of occludin, claudin-1, and ZO-1 proteins; in contrast, these proteins were observed to be reduced during sepsis. By improving the morphology of tight junctions and increasing tight junction protein levels, the study established that CORM-2 could protect the LPS-induced damaged intestinal tight junctions.

A common complication induced by sepsis is known as AKI (acute kidney injury), and, based on the results of Nagao et al. [37], CO was found to reverse the harmful effects of AKI. Blood urea nitrogen and serum creatinine, frequently used markers for AKI onset, were increased following CLP-induced sepsis. In contrast, when treated with CORM-2, levels of these markers were decreased. Sepsis with AKI resulted in a greater mortality, killing about 50% of the rats within three days. However, upon CORM-2 administration, the rats were seen to possess a higher longevity. Through its capability of decreasing markers for AKI onset and increasing survival rates, CORM-2 can be noted to counteract AKI-induced toxicity.

In yet another examination of the role of CO in sepsis, Liu et al. [51] have noted that CORM-2 reversed the LPS-induced activation of platelets. The LPS groups had abnormal platelet function, namely, spreading and aggregation. In the CORM-2 group, however, the platelets were only mildly activated; CORM-2 treatment reduced spreading as well as aggregation. To resolve whether LPS induced platelet secretion, LPS-induced ATP release was inspected in human platelets. LPS stimulation increased ATP release; in contrast, CORM-2 treatment decreased ATP release in response to LPS stimulation. Furthermore, while the platelet membrane glycoproteins, GPIbα and GPVI, were upregulated in sepsis, CORM-2 decreased glycoprotein upregulation. CORM-2 treatment also inhibited LPS-induced HS1 phosphorylation, which is also linked to platelet activation. Furthermore, the effect of CORM on hepatic mitochondria and abnormal glucose metabolism in septic mice was evaluated [52]. With respect to the histopathological changes of the liver, CLP-induced mice suffered hepatocellular damage and swelling; in contrast, CLP mice that later received CORM-2 treatment demonstrated less severe liver damage. Hyperglycemia, a condition associated with sepsis, is concerned with high activities of glucose metabolism.

The CLP group of mice showed an increase in the level of hepatic glucose metabolism and GK, an enzyme involved in the metabolism; these functions were suppressed by CORM-2 treatment. Plasma alanine transaminase (ALT) and aspartate transaminase (AST) levels were utilized to indicate the degree of hepatocellular damage. Although the CLP group of mice exhibited higher activities of ALT and AST, the CLP + CORM-2 group reduced those levels. Another indicator of sepsis, lactic acid, was considered: the CLP and CLP + iCORM groups had increased levels of lactic acid, whereas the CLP + CORM group had lower levels of the acid. The results from the study also indicated that CORM-2-exposed mice had higher survival rates than the septic CLP-induced mice.

CORM-3 has also recently been reported to alleviate myocardial dysfunction due to sepsis as well as to provide therapeutic benefit for sepsis in lung injury [50]. In the case of the former, CO released by CORM-3 is proposed to inhibit the activation of the NLRP3 inflammasome in cardiac fibroblasts preventing myocardial apoptosis. In the latter, CORM-3 restored annexin A2 (a membrane-associated protein involved in fibrinolysis homeostasis which is downregulated in LPS-induced sepsis) to normal levels when introduced intraperitoneally. While this observation was ascribed to the presence of CO, it is not entirely conclusive since the influence of the inactivated CORM was not investigated [49].

A substantial recruitment of polymorphonuclear (PMN) leukocytes to the affected organs is another important characteristic associated with severe sepsis [53]. Exposure of the hCMEC/D3 cell line with LPS resulted in an increase in PMN rolling and adhesion upon fluid shear stress. Treatment of hCMEC/D3 with CORM-3 was seen to suppress both LPS-induced PMN rolling and adhesion to hCMEC/D3. Stimulation of hCMEC/D3 with LPS resulted in upregulation of cell adhesion molecules, including eselectin, ICAM-1, and VCAM-1 expression. Upon the addition of hCMEC/D3 with CORM-3 but not with iCORM-3, VCAM-1 expression was effectively reduced. However, CORM-3 expression failed to reduce LPS-induced expression of E-selectin and ICAM-1. Although CORM-3 was able to decrease leukocyte infiltration and attenuate expression of some of the proteins expressed in sepsis, it was not able to modulate them all; this indicates that whether CORM-3 may be an effective treatment for sepsis is questionable based on these results.

7. Obesity Affected by CORMs

Chronic obesity, identified with increased inflammation, increases the risk of health problems such as diabetes, heart disease, and arthriti, and the main objective of the study by Hosick et al.co [54] was to investigate whether chronic treatment with CORM-A1 could reverse dietary-induced obesity, hyperglycemia, and insulin resistance. The study used mice that were divided into different treatment groups: (i) a control high-fat diet group; (ii) a group receiving a fat diet and saline intraperitoneal injection; (iii) a group exposed to a fat diet and CORM-A1; and (iv) a last group with a fat diet exposed to iCORM. CORM-A1 treatment caused a lack of weight gain in the high-fat-induced mice over the

first 18 weeks of the study compared to other groups; over the last 12 weeks of the study, the mice started to lose weight, up to 33% of their initial body weight. Not only did CORM-A1 induce fat reduction and weight loss, but it also provided a 45% increase in lean body mass as a percent of total body weight at 30 weeks. In addition, fasting blood glucose and plasma insulin levels were elevated in all groups before treatment. However, CORM-A1 treatment resulted in a considerable reduction of hyperglycemia after 6 weeks of treatment, which continued until the end of the study. After 30 weeks of treatment, the fasting blood glucose levels in the CORM-A1-treated mice were 55% of those observed in the control mice. Moreover, CORM-A1 treatment resulted in a significant decrease in plasma insulin levels as compared to all other groups after both 24 and 30 weeks of treatment. HMGB1 is a protein responsible for high levels of inflammation that is linked to obesity. Following CORM-A1 treatment, HMGB1 levels were reduced in dietary-induced obese mice in comparison to those found for the other groups suggesting an antiinflammatory effect induced by CORM-A1 on obesity.

Zheng et al. [55] have also studied the effect of CORM's on obesity. They discovered that obesity may be due to leptin resistance caused by endoplasmic reticulum (ER) stress. Leptin is a hormone that circulates in the bloodstream and serves the function of regulating food intake and body weight. They hypothesized that CO could prevent leptin resistance during ER stress. Thapsigargin or tunicamycin was used to cause ER stress in human cells, and they were found to hinder leptin-induced STAT3 phosphorylation, establishing that ER stress induces leptin resistance. CORM-2 treatment of these cells blocked the STAT3 phosphorylation and induced the phosphorylation of protein kinase R-like ER kinase and eukaryotic translation initiation factor-2 during ER stress. When tested *in vivo*, CO exposure reduced the body weight of animals being fed diets high in fat. Based on the results of the two studies, it is evident that CO does have a role in attenuating obesity.

8. Effect of CORMs on Angiogenesis, Aggregation, and Cancer

According to Ahmad et al. [56], CORM-2 halts angiogenesis caused by the vascular endothelial grown factor (VEGF). When exposed to VEGF, endothelial cells experience an increase in actin stress fiber production indicated by specific spindle-like processes. This production was significantly inhibited by CORM-2, but not as much by iCORM-2. Moreover, CORM-2 prevented VEGF-dependent endothelial cell migration and proliferation. The retinoblastoma protein (Rb) is a tumour suppressor protein whose purpose is to restrict the ability of the cell to progress through the cell cycle. In this way, Rb prevents over replication in cells and hence cancer. However, when Rb is continuously phosphorylated on serine residues, as is the case in VEGF, it can no longer suppress the cell cycle and, as a result, the cell replicates in excess. CORM-2 was noted to attenuate VEGF-induced phosphorylation of Rb and, in this way, it was able to stop extreme cell replication and therefore angiogenesis caused by VEGF. Another study by Fayad-Kobeissi et al. [57] focused on examining the biochemical and potential anticancer properties of CORM-401, a novel CO-releasing molecule containing manganese in its metal center. CORM-401 was observed to promote vasorelaxation of precontracted aortic rings (3 times higher than that evoked by CORM-A1), and this was enhanced in combination with H_2O_2. In addition, CORM-401 was found to exhibit proangiogenic properties, as determined by augmented gene expression of VEGF and IL-8. Furthermore, as demonstrated by Loureiro et al. [58], folic acid-tagged protein nanoemulsions prefer to be internalized on the B-cell lymphoma cell line (A20 cell line). The folic acid-tagged protein nanoemulsions that contained CORM-2 resulted in an increased antitumor activity and increased the survival of mice with A20 lymphoma tumours.

Pancreatic fibrosis is associated with pancreatitis and pancreatic cancer; pancreatic stellate cells (PSCs) are mainly responsible for the development of fibrosis [59] and, owing to its ability to induce cell cycle arrest, CORM-2 can prevent global protein synthesis in PSCs. PSCs were found to induce a decrease in the amounts of both phosphorylated eEF2 and eIF2 (components of the translational machinery) as well as an increase in the quantities of phosphorylated eIF4E and 4E-BP1. Exposure of CORM-2 to these cells resulted in no effect on the phosphorylation of eIF2 or eIF4E, but rather it prevented the decrease of phosphorylated eEF2 caused by the cells and reduced the level of phosphorylated 4EBP1. Next, CORM-2 was tested as to whether it affects translation in PSCs. Effectively, CORM-2 treatment resulted in a 51% reduction in protein synthesis. This repression led to further testing of CORM-2 on its effect with regards to the proliferation of PSCs; the levels of cyclin and phosphorylated Rb, both of which cause proliferation, were compared. The results indicated that CORM-2 was able to abrogate the increase in cyclin D1 and cyclin E and Rb phosphorylation induced by PSCs. An important property of cancer cells is their ability to aggregate. Integrins are cell adhesion receptors capable of modulating rapid adhesion and deadhesion events, without altering the number of molecules expressed. Ligand interactions with integrins suggest integrin-dependent cell adhesion, which is regulated by multiple signaling pathways initiated by other cellular receptors. This signaling allows for fast leukocyte arrest on endothelium, cell migration, and mobilization. $\alpha4\beta1$-integrin (CD49d/CD29, Very Late Antigen-4, VLA-4) is expressed on leukocytes, dendritic cells as well as cancer cells. When a CORM was tested on VLA-4, the results demonstrated that the CORM induced a rapid decrease in the binding of the VLA-4-specific ligand [61]. Moreover, exposure to the CORM prevented VLA-4/VCAM-1-dependent aggregation. These findings suggest that CORMs seem to possess the capability to avert aggregation, a striking feature of cancer cells.

Photochemotherapy is a promising strategy for the treatment of several types of cancers. PhotoCORMs, which can release CO once activated by light, as mentioned by Tinajero-Trejo et al. [18], can be efficiently internalized in HT29 human colon cancer cells, as demonstrated by Niesel et al. [60]. Based on the study, no change in the number of cells was observed with the control group nor with the cells incubated with PhotoCORM in the dark for up to 48 hours. However, when irradiated at 365 nm for 10 min in the middle of the incubation

period, PhotoCORM treatment led to a considerable reduction in cell biomass, whereas direct cell damage by irradiation in the control group displayed a relatively smaller reduction.

9. Contribution of CORMs in Hemorrhagic Shock and Postresuscitation Injuries due to Transplantation and Surgical Operation

Hemorrhagic shock is a condition that results when an insufficient amount of oxygen and nutrients is delivered to the tissues and cells due to certain injuries such as to the liver and kidney. Resuscitation, the process of correcting injuries, can potentially result in even more trauma to the patient (e.g., CPR can induce broken ribs) [104]. Nassour et al. [62] have developed a surgical rat model for hemorrhagic shock which is induced by withdrawing blood and returning it back to the animal to maintain a relatively low blood pressure. After the shock, mice were resuscitated with Ringer's lactate solution using two times the volume of maximum shed blood. The control group of mice had the operation performed but did not undergo the resuscitation. The noncontrol group had CORM-421 intravenously administered with the Ringer solution during the resuscitation. Hemorrhagic shock and resuscitation (HS/R) contributed to a reduction in normal endothelium fenestrations, rounding of cells, and increased adherent leukocyte count in mice not exposed to the CORM. When the CORM was delivered to the rat, it was observed that the number of fenestrations, cells, and adherent leukocytes was maintained owing to the ability of the CORM to reduce the expression of HS/R-induced cytokines. Hence, CO managed to protect against HS/R in a rat model. In addition, when CORM-3 was administered to HUVECs, which were subjected to 20 hours of cold storage, an increased frequency of live cells, reduced apoptosis, and decreased mitochondrial transmembrane potential in HUVECs was noted [63]. This result was in direct contrast to the control group, which displayed an opposite effect. Regarding the morphology of the renal tissue, a reduction in glomerular atrophy and necrosis was observed in the CORM-3 group versus the control group. Furthermore, on a molecular level, CORM-3 exposure led to a 2.1-fold upregulation in Bcl-2 gene expression, a gene involved with regulating apoptosis. One interpretation of this study is that CO increases the viability of cells in cold storage for future transplantation.

In another study, CORMs were examined by Yao et al. [64] in a rat model to determine whether they can offer protection of the heart from postresuscitation myocardial injury and cardiac mitochondrial dysfunction. The rats were induced with ventricular fibrillation and then resuscitated; it was found that one hour following resuscitation, the mean aortic pressure, ventricular contractility assessment, and isovolumetric contractility values were increased in the group exposed to low concentrations of CORM-2 compared to the control group. In terms of the myocardium morphology, 3 hours after resuscitation, the control group displayed myocytolysis and transverse contraction bands. Rats exposed to CORM-2 depicted less severe myocytolysis and relief from damaged myocardial fibers. Additionally, the CORM-2-exposed rats showed decreased production of cardiac mitochondrial ROS indicating that CORM-2 reduced oxidative stress in the resuscitated rats by uncoupling mitochondrial respiration. However, high concentrations of CORM-2 caused the reverse effect; here, there is increased mitochondrial ROS generation, most likely because of its excessive uncoupling action. Taken together, these findings imply that CORM-2 lowers oxidative stress in the heart and ameliorates cardiac function after hemorrhagic shock and resuscitation.

The bioactive properties of CORM-3 have been investigated using two models: an in vitro model of hypoxia-reoxygenation and oxidative stress and a cardiac transplant rejection model [65]. In the first model, it was found that CORM-3 conserved cell viability against reoxygenation-induced damage in a concentration-dependent manner. This cytoprotective effect was observed when CORM-3 was applied to the cells either during the hypoxic event or at reoxygenation. The inactive compound (iCORM-3) did not demonstrate any protection against hypoxia-reoxygenation; this suggests that CO released from the ruthenium carbonyl is necessary to exert the observed effect. For the second model, CORM-3 was seen to prevent cardiac allograft rejection. This was evidenced by the prolonged survival time of hearts transplanted into rats exposed to CORM-3; all hearts were still beating 18 days after the transplantation, whereas the non-CORM-treated mice were subjected to rejection within 9 days of transplantation.

Similar findings by Clark [65] and Musameh et al. [66] have demonstrated that CORM-3 protected isolated hearts against the combined effects of cold ischemic storage/preservation and warm reperfusion injury when transplanted into mice. A higher dP/dt recovery, which represents increased diastolic function, was seen in hearts treated with CORM-3 when compared with iCORM-3, suggesting that CO release is necessary for enhanced diastolic function of the heart. After a cold storage lasting both 4 and 6 hours, the coronary flow rates were similar at both times with CORM-3-treated hearts, whereas there was a decrease in coronary flow when the hearts were exposed to iCORM-3 after 6 hours when compared to the flow after 4 hours. Owing to the CORM-3-induced persistent blood flow supplying the hearts, they survived for prolonged times, thus demonstrating a successful transplantation in mice.

Caumartin et al. [67] have investigated the ability of pretreating the kidney donor with CORM-2 to prevent ischemia—reperfusion injury in a transplant rat model. The histological analysis of the kidney grafts demonstrated that CORM-2-treated kidneys had decreased damage when compared to the control group (no CORM-2 treatment) postoperation. Ten days after the transplantation, infiltrated lymphocytes and tubular necrosis were observed in the control group, whereas the CORM-2-treated rats exhibited entirely normal histology. On day 70, CORM-2-administered animals showed optimal renal function but an excess accumulation of lymphocytes (indicating an infection or inflammation) and atrophy of the glomerulus. This observation reveals that although CORM-2 administration was noted to lead to a successful kidney transplantation for the first several days, it can have negative effects over a much longer term. Because of this result, the water-soluble CORM-3 was tested as well. The CORM-3-treated mice were only followed for 3 days; at the third day of posttransplantation, improved renal histology and function were noted. Because CORM-3 was not tested for as long a time as CORM-2, unfortunately, no definitive conclusion about the benefit of CORM-3 on posttransplantation survival could be reached.

10. Effect of CORMs in Gastric, Intestinal, Kidney, and Liver Disorders

CORMs have been shown to improve lesions and ulcerations found in various parts of the body, and Magierowski et al. [68] have found that the administration of alendronate, despite serving as an inhibitor of osteoclast-mediated bone resorption, can result in gastrointestinal complications including bleeding erosions. The gastric mucosa of rats, pretreated with alendronate, induced damage to the mucosa. In rats preexposed to mild WRS (mild stress by water immersion and restraint in cold water), administration of alendronate produced various gastric hemorrhagic dot-like erosions and band-like lesions. However, CORM-2 treatment diminished mucosal lesions in rats preexposed to mild WRS with subsequent alendronate administration. Moreover, administration of acetic acid to mice was seen to result in the development of deep gastric ulceration in the abdomen. In contrast, mice treated with an aqueous CO gas-saturated solution demonstrated a smaller gastric ulcer, and their wounds were reepithelialized [105].

Excess consumption of ethanol invokes liver injuries in biological organisms. In the study by Bakhautdin et al. [69], the hypothesis that CORM treatment can protect ethanol-induced liver injury was investigated. Exposure to CORM-A1 was observed to reduce hepatocyte cell death, indicated by decreased build-up of CK18 cleavage products and lowered RIP3 expression in hepatocytes. CORM-A1 administration also decreased the effects of ethanol on plasma ALT and AST. Notably, CORM-A1 had positive effects on mRNA regulation for inflammatory mediators, as assessed by its ability to decrease TNFa, IL6, and CXCL10 expression. Moreover, although ethanol exposure increased Ly6C+ cell count, a marker of penetrating immune cells in liver, CORM-A1 reduced this effect. Additionally, CORM-A1 decreased the number of inflammatory foci observed on hematoxylin and eosin sections. The growth of c-ketoaldehydes adducts, such as isolevuglandins (iso[4]LGE2) adducts, in the liver suggests oxidative stress. Ethanol supplementation increased the accumulation of iso[4]LGE2 protein adducts; however, CORM-A1 decreased the build-up. Owing to these results, it was deduced that CORM-A1 could prevent hepatocyte death when the liver is compromised by an increased ethanol intake. However, more recent work by Mangano et al. suggests that CORM-A1 may also possibly be applied for the treatment of autoimmune hepatitis owing to its ability to significantly reduce mortality in a murine model of this disease [70].

Not only does CO reduce liver damage, but it also demonstrates protective effects against nephrotoxicity, an adverse effect caused by the chemotherapy agent cisplatin (CP). When confluent renal tubule epithelial cells (LLC-PK1) were exposed to CP for a period of 16 h, an increase in caspase-3 activity was observed [71]. This effect was followed by cell damage—there was an increased number of cells floating in the culture media. Upon the addition of CORM-3 at concentrations of $1-50 \mu M$, the increase in caspase-3 activity was abolished in a concentration-dependent fashion, suggesting its protective action on CP-induced cells. When CORM-3 was added to cells in the presence of an inhibitor of the GC pathway, the caspase-3 activity was not suppressed. This observation indicated that the protective effect of CO appears to be mediated by cGMP. However, biliverdin, which is also generated along with CO during the degradation of heme by HO, did not have any effect on CP-induced caspase-3 activation, demonstrating that CO specifically is necessary to prevent the nephrotoxicity. When tested in vivo, CP caused an increase in plasma urea and creatinine levels. This effect was completely prevented by CORM-3; the plasma urea creatinine levels were reduced to similar ones to those for the control group. Unlike CORM-3, iCORM-3 was unable to improve the renal impairment caused by CP. When analyzed histologically, the CP-treated kidneys displayed severe tissue damage that primarily affected the tubules, whereas the glomeruli structure was maintained. On the other hand, CORM-3 ameliorated the tubular tissue, and its appearance resembled to that of the control kidneys. Furthermore, CP exposure resulted in the apoptosis of kidney cells, whereas addition of CORM-3, but not iCORM-3, prevented the CP-induced apoptosis. Also, CP-treatment caused a 1.2% weight loss in the kidney; this was counteracted by CORM-3, but not by iCORM-3. Furthermore, recent findings have confirmed the in vitro renoprotective effect of CORM-3 in normal and cancerous human renal cell lines when subjected to CP-induced toxicity and ischemia-reperfusion injury [72].

Another severe condition is postoperative ileus, known as a temporary impairment of bowel motility that can commonly occur following a major abdominal surgery. The severity of postoperative ileus was noted to be reduced in mice exposed to CORMs; this was evidenced by a partial restoration of intestinal contractility and, hence, a less distended bowel [73]. Three groups of mice were examined: one group had an intestinal manipulation (IM) procedure performed to induce the ileus, the second was a control group (no manipulation or any CORM treatment), and the last group was treated with CORM-3 at 3 hours and 1 hour before the surgical procedure. An increase in leukocytes was seen in the intestinal muscularis beginning between 3 and 6 hours in the IM mice. The excess leukocytes were reduced with mice that were exposed to CORM-3 prior to the operation. Moreover, intestinal contractility was observed to be restored in the CORM-treated mice. When tested for expression levels of IL-6, monocyte chemoattractant protein-1, and intercellular adhesion molecule-1, as well as inducible NOS activity, were all seen to be

decreased following CORM-3 treatment. In contrast, all these parameters were increased with the non-CORM-treated mice. The IM-treated mice demonstrated higher levels of oxidative stress when compared to the mice treated with CORM-3, which displayed significantly reduced oxidative stress levels. From these results, it was concluded that CORM-3 reversed the damaging effects due to postoperative ileus.

11. Effect of CORMs in the Lungs

The beneficial effect of CO on lung diseases is currently under much debate. Although in vivo models of lung injury induced by acid aspiration [106], aeroallergens [107], hyperoxia [108], and mechanical stretch [109] indicated efficacy of inhaled CO, and other studies [110, 111] did not prove any protective impact of CO. Abid et al.[74] have performed an experiment that proposed beneficial effects of CO on reverse pulmonary hypertension (PH), a complication that occurs as a result of certain diseases. CORM-3 exposure was seen to prevent PH ventricular hypertrophy and distal pulmonary artery muscularization in hypoxia-induced mice. CORM-3 treatment also reversed PH in smooth muscle promoter 22 serotonin transporter mice by decreasing Ki67. Furthermore, CORM-3 increased p21 mRNA and protein levels in lungs. Taken together, the data suggested that through its modulation of p21, CORM-3 may serve as an effective treatment for PH. However, when CORMs were tested under the influence of hypoxic pulmonary vasoconstriction (HPV) using an in vitro model, it was observed that the CORMs did not significantly diminish HPV [75]. Furthermore, CORM administration did not result in the inhibition of CYP, a cytochrome that regulates HPV. When higher concentrations of CORMs were used, irreversible pulmonary vasoconstriction resulted. However, inhaled CO led to a decrease in HPV and CYP. Thus, using CORMs for the purpose of treating hypoxic pulmonary vasoconstriction is not particularly effective. Nevertheless, it is possible that the application of CORMs in vivo may lead to different results, and they may even have inhibitory effects on HPV.

12. Influence of CORMs on the Ocular System

That CO is a molecule crucial to the ocular system has only recently been noted. Low concentrations of CO can exert helpful effects in various ocular conditions, for example, in glaucoma [112]. Furthermore, Stagni et al. [76] have reported that CORM-3 administration lowered intraocular pressure in rabbits. A considerable decrease of intraocular pressure was noted 30 min after CORM-3 exposure for up to 24 hours. A maximum effect was observed 6 hours after addition of a 1% dose of CORM-3; this produced a maximum drop in pressure of 12 mm Hg. Also, treatment with 0.01% and 0.1% CORM-3 solutions generated IOP drops of 3 and 7 mm Hg, respectively. Treatment with the inactive form, iCORM-3, had no effect on the intraocular pressure, suggesting that the CO released by CORM-3 is responsible for the observed this decrease.

13. Cardiovascular Effects of CORMs

Anderson [113] have utilized an electrocardiogram to evaluate myocardial damage due to CO poisoning. Five of six CO-poisoned patients were observed to have considerable electrocardiographic abnormalities. ST-segment and T-wave abnormalities were detected, although atrial fibrillation and intraventricular block were also sometimes seen. The persistence of electrocardiographic abnormalities was variable; in one patient, minor changes were observed and then only for a few hours, whereas, in another, the changes were more severe and persisted for more than 24 days. Some patients had durations of electrocardiographic abnormalities lasting up to 4 months. Myocardial injury and necrosis, focal areas of leukocyte infiltration, and punctate hemorrhages were seen histologically as well. Furthermore, CO exposure of 250 ppm induced endothelium-dependent and -independent vascular relaxation abnormalities in rats; these abnormalities were caused by a decrease in cardiac cGMP/cAMP ratio [114]. The anomalies in coronary vascular relaxation were observed in the presence of an increase in heart contractility and cardiac mitochondrial respiration inhibition. It was concluded that CO promoted deformities in coronary vascular relaxation, myocardial contractility, and mitochondrial respiration and that these may induce heart hypoxia. In another study by Andre et al. [115], it was observed that chronic CO exposure reduced the contraction of single rat ventricular myocyte due to the decrease of both systolic Ca^{2+} release and myofilament Ca^{2+} sensitivity. Furthermore, the velocity of both the onset and the relaxation of contraction was significantly reduced, and a slowing of the Ca^{2+} transient decay was seen. Another finding was that CO in high concentrations could induce cardiac arrhythmia. Since CO had no effect on the QT interval of the PQRST deflection seen in the electrocardiogram, its effect on prolonging the duration of the action potential and increasing diastolic Ca^{2+} concentration all together support the hypothesis that the ventricular arrhythmia observed upon CO exposure is caused by an overload of Ca^{2+}.

From these three reports, exposure to CO results in severe cardiac arrhythmia, decreased heart contractility, and heart hypoxia. But when CORMs were utilized, different effects were observed due to their low concentration and controlled targeting in the body. For instance, a recent investigation by Segersvärd et al. has shown CORM-3 (but not iCORM-3) improves cardiac structure and function of rats following experimentally induced myocardial infarction [79]. CORM-2 has also displayed cardioprotective behaviour. Tsai et al. [78] have demonstrated that aortic smooth muscle cell migration induced by angiotensin II is attenuated by CORM-2 through inhibition of matrix metalloproteinases-9 expression and ROS/interleukin-6 generation. In addition, Soni et al. [77] have suggested that CORM-2 treatment may improve the heart's function when it is jeopardized by doxorubicin (DXR), a commonly used antitumor agent for heart cancer. DXR can cause cardiotoxic effects, which can lead to cardiomyopathy and heart failure. The mechanism of DXR-induced cardiotoxicity

has been associated with excess ROS formation, mitochondrial and DNA damage, and apoptosis. Chronic treatment with CORM-2 can protect the myocardium from DXR exposure. It was also observed that CORM-2 exerted beneficial cardioprotective effects against DXR-induced cardiotoxicity in vivo by means of decreasing oxidative stress and apoptosis. Although CORM-2 may possess this effect, it can only exert it under certain concentrations; there is a thin line between the therapeutic benefits and toxic effects of CORM-2, which suggests that it has a narrow therapeutic index. However, there was no increase in the amount of CO-Hb at higher concentrations of CORM-2, indicating that the toxicity at higher doses of CORM-2 may be due to some mechanism other than the increased CO-Hb. Moreover, decreased amounts of serum creatine kinase and lactate dehydrogenase were found for the CORM-2 + DXR-treated groups when compared to DXR alone. Furthermore, increased antioxidant and decreased malondialdehyde contents were seen for the CORM-2 + DXR-treated hearts when compared to those for DXR alone.

Finally, Abramochkin et al. [116] have evaluated the electrophysiological effects of CORMs on murine myocardium. Two effects were noted: action potential shortening in the working myocardium and an acceleration of the sinus rhythm. Thus, an increase in heart rate was observed after CO exposure; this property may suggest that CORMs could have a potential use as a positive chronotropic drug.

14. Procoagulant and Anticoagulant Effects of CORMs

Several studies have suggested that CORMs are procoagulant agents and Nielsen et al. [80] have shown that exposure of human plasma to CORM-2 caused an increased speed of growth and strength of thrombi. After tissue factor activation, both CORM-2 and iCORM-2 treatment resulted in a decrease in the time to onset of coagulation, an increase in the velocity of clot growth and an increase in the clot strength in all three types of plasma. Plasma exposed to CORM-2 had a velocity of clot growth that was more than double that of the unexposed samples; however, the samples treated with iCORM only displayed a 50% increase. After celite activation, CORM-2 exposure increased the velocity of clot growth by 56% and clot strength by 57%, when compared with the data for the untreated plasma samples. With respect to the mechanism of CORM-2 enhancement of coagulation, it appears to be due to the increased thrombin-fibrinogen interactions and to be independent of FXIII activation.

Whether CORM-2 could enhance plasmatic coagulation and decrease bleeding times were assessed further in the work conducted by Nielsen et al. [81] and Machovec et al. [82]. Contrary to the findings in Nielsen et al.'s 2009 article [80], there was no significant effect seen on time to maximum rate of thrombus generation induced by CORM-2 in either study. However, the velocity of thrombus formation and the clot strength both increased with CORM-2 administration, in agreement with the study by Nielsen et al.

[80]. In addition, CORM-2 treatment caused a considerable decrease in the bleeding time values. The results from these studies suggest that CORM-2 has the potential to improve hemostasis due to its ability to increase the strength of clots and the velocity of clot formation [80–82]. More recently, Nielsen et al. have demonstrated that CORM-2 (but not iCORM-2) is able to attenuate snake venoms with fibrinogen- or thrombin-like activity [83–86].

In addition to displaying procoagulant effects, CORMs also possess anticoagulant properties. Thrombosis, the local blood clotting condition that leads to the obstruction of blood flow through the circulatory system, is a major complication in many vascular pathological conditions. Chen et al. [87] have shown that CO, a product of the HO-1 reaction, can result in antithrombotic effects when platelet mediated thrombus formation was inhibited in the graft. HO-1-deficient mice were seen to result in a 0% survival rate within 4 days due to the development of arterial thrombosis after aortic transplantation. However, recipients that normally expressed HO-1 demonstrated 100% graft acceptance and survival for more than 56 days. When analyzed histologically, treatment with CORM-2 revealed a reduced platelet aggregation within the graft, whereas the non-CORM-2-treated recipients exhibited an elevated aggregation. Therefore, it appears that CO can indeed protect against vascular arterial thrombosis in murine aortic allotransplantation.

Another study that has illustrated the anticoagulant impact of CORMs is that performed by Kramkowski et al. in which they demonstrated that CORM-3 and CORM-A1 inhibit platelet aggregation in vivo [88]. When both CORMs were administered intravenously at micromolar concentrations, there was an inhibition of arterial thrombus formation. The effect of CORM-A1 is dependent on the concentration, and this compound is more potent than CORM-3. In comparison to CORM-A1, CORM-3 only inhibited thrombus formation the best at the highest concentration ($30\,\mu$mol/kg). A considerable decrease in thrombus weight was correlated with the inhibition of platelet aggregation only for the CORM-A1-treated group, perhaps due to its constant and gradual mechanism of CO release. Neither CORM-A1 nor CORM-3 had any effect on the plasma concentration of active tissue plasminogen activator. CORM-3, but not CORM-A1, was noted to lead to a decrease in the concentration of fibrinogen, fibrin generation, and prolonged prothrombin time. Although both CORMs decreased platelet accumulation in thrombus, it was only CORM-A1 that could inhibit platelet activation to phosphatidylserine on their surface.

15. CORM Toxicity

The toxicity of CO is thought to primarily arise owing to its preferential binding to Hb. This is proposed to reduce oxygen delivery to tissues. In general, though, any five-coordinate ferrous heme can bind CO, and thus its toxicity will be a result of the combination of these interactions. Indeed, almost all ferrous metalloproteins are potential substrates for CO, and thus the number of metabolic pathways to be considered in these questions of toxicity is enormous. However, CO in low physiological doses is considered safe, with endogenous CO production (arising from heme catabolism) producing CO-Hb levels of about 1% [117, 118]. In fact, low-dose exposure via inhalation has been reported as therapeutic for several conditions [119], although the effects of chronic low-dose exposure are still unclear [118]. High-dose CO exposure, on the other hand, is well established as being toxic. Any CO-Hb levels in the range 15–20% are associated with mild symptoms, including headache and nausea, but more severe effects can be induced at levels >20% and death can occur when levels are greater than 60% [118, 120]. As is common with other gaseous metabolic products such as NO or H_2O, the excretion pathway is via gas exchange in the lung. From this perspective, CO-Hb is an excretion transporter.

Owing to the potential toxicity of CO, CORMs have been proposed to provide increased CO concentrations for therapeutic purposes without generating toxic CO-Hb levels, which would otherwise be produced through inhalation. This approach has largely been achieved, for example, with CORM-2 [121] and CORM-3 [122], which produce CO-Hb levels lower than the acceptable limits set by the FDA (12–14%) [123]. There is a concern, however, regarding the cytotoxicity of CORMs themselves, particularly of the CO-carrying scaffolds that persist after CO release. Indeed, Seixas et al. have shown that despite the anti-inflammatory activity of CORM-2 and CORM-3 in reducing NO production, these CORMs induced hemagglutination and were hemolytic [124]. CORM-2 has also been associated with disturbing vitamin D3 metabolism in mice [125]. It can produce excessive eryptosis in erythrocytes, which may lead to pathological conditions such as anemia [126] and it has shown significant cellular toxicity by reducing cell viability, decreasing mitochondrial enzyme activity, and causing necrosis in cardiomyocytes and kidney cells [127]. Furthermore, there appears to be rather narrow therapeutic windows for CORM-2 with regard to cardioprotection [77] and for CORM-2 and -3 in treating oxidative stress and ROS production (as described above) [64]. Of perhaps more concern are the effects of consecutive CORM dosing. CORM-3 and other ruthenium-based CORMs as well as a variety of chromium-, molybdenum-, and tungsten-containing CORMs have been shown to cause severe liver and kidney damage in rats after consecutive exposure [128] and, as mentioned above, long-term exposure of CORM-2 in mice following kidney transplant caused lymphocyte accumulation and atrophy of the glomerulus [67].

A frequent test for CORM toxicity and for their CO release physiology is the use of their expected decarbonylation products (iCORMs). It is important to remember though that these derived complexes will have metals in lower oxidation states and thus potentially are able to bind and perhaps activate O_2. In addition, many of these centers will have accessible redox chemistry and the potential for their interfering with electron transfer pathways in cell membranes or even the cytosol is thus an important consideration.

It is thus clear that CORM cytotoxicity is an essential property to consider when developing these types of compounds for therapeutic purposes and it is an important growing area of study. Motterlini et al. have shown that increasing the water solubility and controlling the rate of CO release of several iron-containing CORMs (CORM-307, -308, -314, and -319) can assist in limiting potential harmful side-effects [129]. Development of compounds that release CO in specific tissue is another approach being considered to reduce unwanted toxicity. Such compounds include PhotoCORMs which release CO only when irradiated with light [13]; enzyme triggered-CORMs which are activated intracellularly, for example, by esterase activity [130, 131]; and CORMs covalently attached to carriers such as nanoparticles [132], micelles [133], or dendrimers [134]. Others have opted to circumvent possible metal-based toxicity by developing transition metal-free CORMs such as CORM-A1 [135]. The continued work in this area and rational design of CORMs, particularly with careful consideration of the "drug sphere" of these molecules, are vital to the development of safe and therapeutically useful CORMs.

16. Summary and Future Outlook

This overview of the recent literature on CORMs illustrates their potential benefits for treating a wide array of diseases and conditions (Table 1), but also highlights some important disadvantages and limitations that need be addressed in their continued development as possible therapeutics. Further research into the *drug sphere* proposed by Romão et al. [6] will certainly aid this goal. An additional consideration for the future development of CORMs as therapeutic agents is obtaining better mechanistic understanding of the modes of action for these compounds under physiological and pathological conditions. While *in vitro* measurements of CO release have been routinely investigated spectrophotometrically, using deoxymyoglobin, they can be somewhat unreliable [22] and are not necessarily representative of the biochemistry that occurs within a cellular environment. Information obtained via direct in situ measurement of the rates and amounts of CO released within cells are of greater value. Recently developed CO sensors [27] are proving promising in this regard, as are fluorescent/luminescent CORM derivatives, which provide a means to track CORMs and monitor their CO release [136–138]. Continued studies on the metabolism and retention/clearance of CORMs, as well as the cytotoxicity of these molecules and their by-products, are also urgently needed.

One question that needs to be addressed is whether CO or its carrier scaffolds are responsible for the observed effects, be they beneficial or detrimental. Several investigations have made use of inactivated CORMs as controls; however, it remains to be seen if these iCORMs are equivalent to the by-products generated in situ following CO release or if they are metabolized in the same manner. It has not yet been established whether the effect due to a CORM is due to released CO gas, the CORM itself or any of the intermediates, or final breakdown products [139]. An important proviso is that even the most direct release of CO will generate a 16-electron coordinatively unsaturated intermediate that can engage in a wide range of organometallic chemistry in situ. There is no necessary connection between these reactive intermediates and their ultimate decomposition products made ex vivo. A credible attempt to address the problem of the unknown structures of the iCORM intermediates has been recently been reported by the Schatzschneider group [140]. As mentioned earlier, the photoactivable complex, $[Mn(CO)_3(tpa-\kappa^3N)]Br$, is a CORM prodrug that is stable in solution in the dark. However, photoactivation at 365 nm leads to CO release transfer to heme proteins, as demonstrated by the standard myoglobin assay. Interestingly, several different iCORM intermediates could be detected by solution IR spectroscopy and assigned using DFT vibrational calculations. Further information regarding the association of CORMs (and iCORMs) with proteins would provide important insights into the mechanism of action or toxicity of these molecules. Such investigations on CORM-3 have already provided evidence in this regard [141]. Limiting the nonspecific interactions of CORMs with proteins and tissues, however, is a strong driving force in the development of new compounds and this will continue to be an important avenue of research.

There is a need for more concrete information pertaining to the identity of specific CO targets, their binding mechanism, and the subsequent effects induced by the next generation of CORMs that lead to the observed therapeutic benefits, such as those described for the variety of conditions above. Many studies have implicated heme-containing proteins (other than Hb and Mb) as targets, despite their generally low affinity for CO [142]. This does not preclude their involvement, however, as molecules that enhance CO binding affinity have been reported for mitochondrial cytochrome c (naturally occurring cardiolipin) [143] and sGC (synthetic compound YC-1) [144]. These findings lead to the question if CORMs (or iCORMs) themselves could be acting as CO-binding enhancers or if there are other, as yet, unidentified biomolecules that augment the CO-binding affinity of proteins. In any event, efforts to further our understanding of the biological behaviour of CORMs will certainly be worthwhile and important in the advancement towards using these compounds in therapeutic applications.

Another area to be examined is the reversibility of CORM interactions. There does not seem to have been much research yet on the effect of hyperoxia on CORM treatment, that is, the exposure of tissue and organs to excess amounts of O_2. Does the introduction of O_2 reverse the effect of CO production from CORMs? There has been one study on the exposure of the tricarbonyldichlororuthenium(II) dimer (CORM-2; $\geq 5\,\mu M$) to erythrocytes for 24 h that significantly increased the formation of carboxyhemoglobin, which could be partially reversed by the introduction of O_2 [126].

Finally, the latest development in CO delivery to biological systems is the new field of organic CO prodrugs. A summary of this new area has just been published in *Accounts of Chemical Research* in an article entitled "Strategies Toward Organic Carbon Monoxide Prodrugs" [145]. A specific example of this topic from the same research group at Georgia State University is their study of organic CO prodrugs activated by endogenous ROS [146]. The absence of a metal center from the delivery system may well prove to be an important factor in future clinical studies in terms of avoiding the potential toxicity of the organometallic CORMs. Two types of CO prodrug have been developed, which are precursors to norbornadien-7, ones that release CO under mild conditions [145].

In conclusion, despite all the research activity focussing on the possibility of controlling CO delivery in clinical situations, one must be be conscious of the fact that the use of CORMs and now organic CO prodrugs may not always be beneficial as even such limited CO exposure may result in damage of internal organs and possibly have serious effects on the fetuses of pregnant women [147].

Abbreviations

AKI:	Acute kidney injury
IECs:	Intestinal epithelial cells
ALT:	Alanine transaminase
IL:	Interleukin
AST:	Aspartate transaminase
LPS:	Lipopolysaccharide
CB1R/CB2R:	Cannabinoid receptor 1/2
LTP:	Long-term potentiation
CLP:	Cecal ligation and puncture
Mb:	Myoglobin
CO:	Carbon monoxide
MD2:	Myeloid differentiation factor-2
CO-Hb:	Carboxyhemoglobin
MPO:	Myeloperoxidase
CO-HbV:	CO-bound hemoglobin encapsulated liposomes
NO:	Nitric oxide
NOS1/NOS2:	Nitric oxide synthase 1/2
CORM:	CO-releasing molecule
NSC:	Neural stem cell
CP:	Cisplatin
PH:	Pulmonary hypertension
DCs:	Dendritic cells
PMN:	Polymorphonuclear leukocytes
DSS:	Dextran sulphate sodium
PSCs:	Pancreatic stellate cells
DXR:	Doxorubicin
Rb:	Retinoblastoma protein
ER:	Endoplasmic reticulum
RBC:	Red blood cells
ESBL:	Extended-spectrum beta-lactamase
ROS:	Reactive oxygen species
ROSC:	Return of spontaneous circulation
Hb:	Hemoglobin
sGC:	soluble guanylate cyclase
HO:	Heme oxygenase
STZ:	Streptozotocin
HPV:	Hypoxic pulmonary vasoconstriction
TLR4:	Toll-like receptor 4
UPEC:	Uropathogenic isolates of *E. coli*
HS/R:	Hemorrhagic shock and resuscitation
UTI:	Urinary tract infections
VEGF:	Vascular endothelial growth factor
HUVECs:	Human umbilical vein endothelial cells
WRS:	Stress by water immersion and restraint in cold water
iCORM:	inactivated CORM.

Acknowledgments

This research was supported by a discovery grant from the NSERC (Canada).

References

[1] R. Motterlini, "Carbon monoxide-releasing molecules: characterization of biochemical and vascular activities," *Circulation Research*, vol. 90, no. 2, pp. 17e–24e, 2002.

[2] R. Motterlini, B. Haas, and R. Foresti, "Emerging concepts on the anti-inflammatory actions of carbon monoxide-releasing molecules (CO-RMs)," *Medical Gas Research*, vol. 2, no. 1, p. 28, 2012.

[3] S. H. Heinemann, T. Hoshi, M. Westerhausen, and A. Schiller, "Carbon monoxide–physiology, detection and controlled release," *Chemical Communications*, vol. 50, no. 28, pp. 3644–3660, 2014.

[4] T. S. Murray, C. Okegbe, Y. Gao et al., "The carbon monoxide releasing molecule CORM-2 attenuates *Pseudomonas aeruginosa* biofilm formation," *PLoS One*, vol. 7, no. 4, Article ID e35499, 2012.

[5] A. W. Adamson, *Advanced Inorganic Chemistry*, Vol 2, Wiley, New York, NY, USA, 6th edition, 1963.

[6] C. C. Romão, W. A. Blättler, J. D. Seixas, and G. J. L. Bernardes, "Developing drug molecules for therapy with carbon monoxide," *Chemical Society Reviews*, vol. 41, no. 9, p. 3571, 2012.

[7] J. Daniel and S. Seixas, *Development of CO-releasing molecules for the treatment of inflammatory diseases alfama development of CO-releasing molecules for the treatment of inflammatory diseases*, Ph.D. dissertation, New University of Lisbon, Lisbon, Portugal, 2011.

[8] E. W. Abel, I. S. Butler, and J. G. Reid, "382. The anionic halogenopentacarbonyls of chromium, molybdenum, and tungsten," *Journal of the Chemical Society*, p. 2068, 1963.

[9] A. C. Kautz, P. C. Kunz, and C. Janiak, "CO-releasing molecule (CORM) conjugate systems," *Dalton Transactions*, vol. 45, no. 45, pp. 18045–18063, 2016.

[10] M. A. Wright and J. A. Wright, "PhotoCORMs: CO release moves into the visible," *Dalton Transactions*, vol. 45, no. 16, pp. 6801–6811, 2016.

[11] U. Schatzschneider, "Novel lead structures and activation mechanisms for CO-releasing molecules (CORMs)," *British Journal of Pharmacology*, vol. 172, no. 6, pp. 1638–1650, 2015.

[12] S. García-Gallego and G. J. L. Bernardes, "Carbon-monoxide-releasing molecules for the delivery of therapeutic co in vivo," *Angewandte Chemie International Edition*, vol. 53, no. 37, pp. 9712–9721, 2014.

[13] M. A. Gonzales and P. K. Mascharak, "Photoactive metal carbonyl complexes as potential agents for targeted CO delivery," *Journal of Inorganic Biochemistry*, vol. 133, pp. 127–135, 2014.

[14] C. Steiger, C. Hermann, and L. Meinel, "Localized delivery of carbon monoxide," *European Journal of Pharmaceutics and Biopharmaceutics*, vol. 118, pp. 3–12, 2017.

[15] J. S. Ward, "Carbon monoxide-releasing molecules: therapeutic molecules with a wide variety of medical applications," in *Organometallic Chemistry*, Vol. 40, pp. 140–176, The Royal Society of Chemistry, London, UK, 2016.

[16] K. Ling, F. Men, W. C. Wang, Y. Q. Zhou, H. W. Zhang, and D. W. Ye, "Carbon monoxide and its controlled release—therapeutic application, detection and development of carbon monoxide-releasing molecules (CO-RMs)," *Journal of Medicinal Chemistry*, vol. 61, no. 7, pp. 2611–2635, 2017.

[17] C. S. Bang, R. Kruse, K. Johansson, and K. Persson, "Carbon monoxide releasing molecule-2 (CORM-2) inhibits growth of multidrug-resistant uropathogenic *Escherichia coli* in biofilm and following host cell colonization," *BMC Microbiology*, vol. 16, no. 1, p. 64, 2016.

[18] M. Tinajero-Trejo, N. Rana, C. Nagel et al., "Antimicrobial activity of the manganese photoactivated carbon monoxide-releasing molecule [Mn(CO)$_3$(tpa-κ^3N)]$^+$ against a pathogenic *Escherichia coli* that causes urinary infections," *Antioxidants and Redox Signaling*, vol. 24, no. 14, pp. 765–780, 2016.

[19] J. S. Ward, J. M. Lynam, J. Moir, and I. J. S. Fairlamb, "Visible-light-induced CO release from a therapeutically viable tryptophan-derived manganese(I) carbonyl (Trypto-CORM) exhibiting potent inhibition against *E. coli*," *Chemistry-A European Journal*, vol. 20, no. 46, pp. 15061–15068, 2014.

[20] J. S. Ward, R. Morgan, J. M. Lynam, I. J. S. Fairlamb, and J. W. B. Moir, "Toxicity of tryptophan manganese(i) carbonyl (Trypto-CORM), against *Neisseria gonorrhoeae*," *Medicinal Chemistry Communication*, vol. 8, no. 2, pp. 346–352, 2017.

[21] A. F. Tavares, M. R. Parente, M. C. Justino, M. Oleastro, L. S. Nobre, and L. M. Saraiva, "The bactericidal activity of carbon monoxide-releasing molecules against helicobacter pylori," *PLoS One*, vol. 8, no. 12, Article ID e83157, 2013.

[22] N. Rana, S. McLean, B. E. Mann, and R. K. Poole, "Interaction of the carbon monoxide-releasing molecule Ru (CO)3Cl(glycinate) (CORM-3) with *Salmonella enterica* serovar Typhimurium: in situ measurements of carbon monoxide binding by integrating cavity dual-beam spectrophotometry," *Microbiology*, vol. 160, no. 12, pp. 2771–2779, 2014.

[23] M. Desmard, R. Foresti, D. Morin et al., "Differential antibacterial activity against *Pseudomonas aeruginosa* by carbon monoxide-releasing molecules," *Antioxidants and Redox Signaling*, vol. 16, no. 2, pp. 153–163, 2012.

[24] L. Flanagan, R. R. Steen, K. Saxby et al., "The antimicrobial activity of a carbon monoxide releasing molecule (EBOR-CORM-1) is shaped by intraspecific variation within Pseudomonas aeruginosa populations," *Frontiers in Microbiology*, vol. 9, 2018.

[25] A. S. Almeida, U. Sonnewald, P. M. Alves, and H. L. A. Vieira, "Carbon monoxide improves neuronal differentiation and yield by increasing the functioning and number of mitochondria," *Journal of Neurochemistry*, vol. 138, no. 3, pp. 423–435, 2016.

[26] Z. Xie, P. Han, Z. Cui et al., "Pretreatment of mouse neural stem cells with carbon monoxide-releasing molecule-2 interferes with NF-κB p65 signaling and suppresses iron overload-induced apoptosis," *Cellular and Molecular Neurobiology*, vol. 36, no. 8, pp. 1343–1351, 2016.

[27] P. Wang, L. Yao, L. L. Zhou et al., "Carbon monoxide improves neurologic outcomes by mitochondrial biogenesis after global cerebral ischemia induced by cardiac arrest in rats," *International Journal of Biological Sciences*, vol. 12, no. 8, pp. 1000–1009, 2016.

[28] N. Schallner, C. C. Romão, J. Biermann et al., "Carbon monoxide abrogates ischemic insult to neuronal cells via the soluble guanylate cyclase-cGMP pathway," *PLoS One*, vol. 8, no. 4, Article ID e60672, 2013.

[29] A. C. Pena, N. Penacho, L. Mancio-Silva et al., "A novel carbon monoxide-releasing molecule fully protects mice from severe malaria," *Antimicrobial Agents and Chemotherapy*, vol. 56, no. 3, pp. 1281–1290, 2012.

[30] J. I. Woo, S. H. Kil, S. Oh et al., "IL-10/HMOX1 signaling modulates cochlear inflammation via negative regulation of MCP-1/CCL2 expression in cochlear fibrocytes," *Journal of Immunology*, vol. 194, no. 8, pp. 3953–3961, 2015.

[31] M. G. Bani-Hani, "Modulation of thrombin-induced neuroinflammation in BV-2 microglia by carbon monoxide-releasing molecule 3," *Journal of Pharmacology and Experimental Therapeutics*, vol. 318, no. 3, pp. 1315–1322, 2006.

[32] C. C. Lin, C. C. Yang, L. D. Hsiao, S. Y. Chen, and C. M. Yang, "Heme oxygenase-1 induction by carbon monoxide releasing molecule-3 suppresses interleukin-1β-mediated neuroinflammation," *Frontiers in Molecular Neuroscience*, vol. 10, 2017.

[33] S. Castany, M. Carcolé, S. Leánez, and O. Pol, "The role of carbon monoxide on the anti-nociceptive effects and expression of cannabinoid 2 receptors during painful diabetic neuropathy in mice," *Psychopharmacology*, vol. 233, no. 11, pp. 2209–2219, 2016.

[34] A. Hervera, S. Leánez, R. Motterlini, and O. Pol, "Treatment with carbon monoxide-releasing molecules and an HO-1 inducer enhances the effects and expression of μ-opioid receptors during neuropathic pain," *Anesthesiology*, vol. 118, no. 5, pp. 1180–1197, 2013.

[35] I. Nikolic, T. Saksida, M. Vujicic, I. Stojanovic, and S. Stosic-Grujicic, "Anti-diabetic actions of carbon monoxide-releasing molecule (CORM)-A1: immunomodulation and regeneration of islet beta cells," *Immunology Letters*, vol. 165, no. 1, pp. 39–46, 2015.

[36] C. Steiger, K. Uchiyama, T. Takagi et al., "Prevention of colitis by controlled oral drug delivery of carbon monoxide," *Journal of Controlled Release*, vol. 239, pp. 128–136, 2016.

[37] S. Nagao, K. Taguchi, Y. Miyazaki et al., "Evaluation of a new type of nano-sized carbon monoxide donor on treating mice with experimentally induced colitis," *Journal of Controlled Release*, vol. 234, pp. 49–58, 2016.

[38] S. A. Riquelme, S. M. Bueno, and A. M. Kalergis, "Carbon monoxide down-modulates toll-like receptor 4/MD2 expression on innate immune cells and reduces endotoxic shock susceptibility," *Immunology*, vol. 144, no. 2, pp. 321–332, 2015.

[39] P. Y. Yeh, C. Y. Li, C. W. Hsieh, Y. C. Yang, P. M. Yang, and B. S. Wung, "CO-releasing molecules and increased heme oxygenase-1 induce protein S-glutathionylation to modulate NF-κB activity in endothelial cells," *Free Radical Biology and Medicine*, vol. 70, pp. 1–13, 2014.

[40] K. Maruyama, E. Morishita, T. Yuno et al., "Carbon monoxide (CO)-releasing molecule-derived CO regulates tissue factor and plasminogen activator inhibitor type 1 in human

endothelial cells," *Thrombosis Research*, vol. 130, no. 3, pp. e188–e193, 2012.

[41] P. Fagone, K. Mangano, S. Mammana et al., "Carbon monoxide-releasing molecule-A1 (CORM-A1) improves clinical signs of experimental autoimmune uveoretinitis (EAU) in rats," *Clinical Immunology*, vol. 157, no. 2, pp. 198–204, 2015.

[42] R. Negrete, A. Hervera, S. Leánez, and O. Pol, "Treatment with a carbon monoxide-releasing molecule inhibits chronic inflammatory pain in mice: nitric oxide contribution," *Psychopharmacology*, vol. 231, no. 5, pp. 853–861, 2014.

[43] X. Mu, C. Pan, S. Zheng et al., "Protective effects of carbon monoxide-releasing molecule-2 on the barrier function of intestinal epithelial cells," *PLoS One*, vol. 9, no. 8, Article ID e104032, 2014.

[44] E. Y. Choi, S. H. Choe, J. Y. Hyeon, J. Choi, I. S. Choi, and S. J. Kim, "Carbon monoxide-releasing molecule-3 suppresses *Prevotella intermedia* lipopolysaccharide-induced production of nitric oxide and interleukin-1β in murine macrophages," *European Journal of Pharmacology*, vol. 764, pp. 22–29, 2015.

[45] E. K. Patterson, D. D. Fraser, A. Capretta, R. F. Potter, and G. Cepinskas, "Carbon monoxide-releasing molecule 3 inhibits myeloperoxidase (MPO) and protects against MPO-induced vascular endothelial cell activation/dysfunction," *Free Radical Biology and Medicine*, vol. 70, pp. 167–173, 2014.

[46] B. Sun, X. Zou, Y. Chen, P. Zhang, and G. Shi, "Preconditioning of carbon monoxide releasing molecule-derived CO attenuates LPS-induced activation of HUVEC," *International Journal of Biological Sciences*, vol. 4, no. 5, pp. 270–278, 2008.

[47] S. Zhang, S. Zheng, X. Wang et al., "Carbon monoxide-releasing molecule-2 reduces intestinal epithelial tight-junction damage and mortality in septic rats," *PLoS One*, vol. 10, no. 12, Article ID e0145988, 2015.

[48] S. Lee, S. J. Lee, A. A. Coronata et al., "Carbon monoxide confers protection in sepsis by enhancing beclin 1-dependent autophagy and phagocytosis," *Antioxidants and Redox Signaling*, vol. 20, no. 3, pp. 432–442, 2014.

[49] K. Unuma, T. Aki, K. Noritake, T. Funakoshi, and K. Uemura, "A CO-releasing molecule prevents annexin A2 down-regulation and associated disorders in LPS-administered rat lung," *Biochemical and Biophysical Research Communications*, vol. 487, no. 3, pp. 748–754, 2017.

[50] W. Zhang, A. Tao, T. Lan et al., "Carbon monoxide releasing molecule-3 improves myocardial function in mice with sepsis by inhibiting NLRP3 inflammasome activation in cardiac fibroblasts," *Basic Research in Cardiology*, vol. 112, no. 2, p. 16, 2017.

[51] D. Liu, F. Liang, X. Wang, J. Cao, W. Qin, and B. Sun, "Suppressive effect of CORM-2 on LPS-induced platelet activation by glycoprotein mediated HS1 phosphorylation interference," *PLoS One*, vol. 8, no. 12, Article ID e83112, 2013.

[52] F. Liang, J. Cao, W. T. Qin, X. Wang, X. F. Qiu, and B. W. Sun, "Regulatory effect and mechanisms of carbon monoxide-releasing molecule II on hepatic energy metabolism in septic mice," *World Journal of Gastroenterology*, vol. 20, no. 12, pp. 3301–3311, 2014.

[53] F. Serizawa, E. Patterson, R. F. Potter, D. D. Fraser, and G. Cepinskas, "Pretreatment of human cerebrovascular endothelial cells with CO-releasing molecule-3 interferes with JNK/AP-1 signaling and suppresses LPS-induced proadhesive phenotype," *Microcirculation*, vol. 22, no. 1, pp. 28–36, 2015.

[54] P. A. Hosick, A. A. AlAmodi, M. W. Hankins, and D. E. Stec, "Chronic treatment with a carbon monoxide releasing molecule reverses dietary induced obesity in mice," *Adipocyte*, vol. 5, no. 1, pp. 1–10, 2016.

[55] M. Zheng, Q. Zhang, Y. Joe et al., "Carbon monoxide-releasing molecules reverse leptin resistance induced by endoplasmic reticulum stress," *American Journal of Physiology-Endocrinology and Metabolism*, vol. 304, no. 7, pp. E780–E788, 2013.

[56] S. Ahmad, P. W. Hewett, T. Fujisawa et al., "Carbon monoxide inhibits sprouting angiogenesis and vascular endothelial growth factor receptor-2 phosphorylation," *Thrombosis and Haemostasis*, vol. 113, no. 2, pp. 329–337, 2015.

[57] S. Fayad-Kobeissi, J. Ratovonantenaina, H. Dabiré et al., "Vascular and angiogenic activities of CORM-401, an oxidant-sensitive CO-releasing molecule," *Biochemical Pharmacology*, vol. 102, pp. 64–77, 2016.

[58] A. Loureiro, G. J. L. Bernardes, U. Shimanovich et al., "Folic acid-tagged protein nanoemulsions loaded with CORM-2 enhance the survival of mice bearing subcutaneous A20 lymphoma tumors," *Nanomedicine: Nanotechnology, Biology and Medicine*, vol. 11, no. 5, pp. 1077–1083, 2015.

[59] C. I. Schwer, P. Stoll, S. Rospert et al., "Carbon monoxide releasing molecule-2 CORM-2 represses global protein synthesis by inhibition of eukaryotic elongation factor eEF2," *The International Journal of Biochemistry and Cell Biology*, vol. 45, no. 2, pp. 201–212, 2013.

[60] J. Niesel, A. Pinto, H. W. Peindy N'Dongo et al., "Photo-induced CO release, cellular uptake and cytotoxicity of a tris (pyrazolyl)methane (tpm) manganese tricarbonyl complex," *Chemical Communications*, no. 15, p. 1798, 2008.

[61] A. Chigaev, Y. Smagley, and L. A. Sklar, "Carbon monoxide down-regulates $\alpha 4 \beta 1$ integrin-specific ligand binding and cell adhesion: a possible mechanism for cell mobilization," *BMC Immunology*, vol. 15, no. 1, p. 52, 2014.

[62] I. Nassour, B. Kautza, M. Rubin et al., "Carbon monoxide protects against hemorrhagic shock and resuscitation-induced microcirculatory injury and tissue injury," *Shock*, vol. 43, no. 2, pp. 166–171, 2015.

[63] A. Sener, K. C. Tran, J. P. Deng et al., "Carbon monoxide releasing molecules inhibit cell death resulting from renal transplantation related stress," *Journal of Urology*, vol. 190, no. 2, pp. 772–778, 2013.

[64] L. Yao, P. Wang, M. Chen et al., "Carbon monoxide-releasing molecules attenuate postresuscitation myocardial injury and protect cardiac mitochondrial function by reducing the production of mitochondrial reactive oxygen species in a rat model of cardiac arrest," *Journal of Cardiovascular Pharmacology and Therapeutics*, vol. 20, no. 3, pp. 330–341, 2015.

[65] J. E. Clark, "Cardioprotective actions by a water-soluble carbon monoxide-releasing molecule," *Circulation Research*, vol. 93, no. 2, pp. 2e–8e, 2003.

[66] M. D. Musameh, C. J. Green, B. E. Mann, B. J. Fuller, and R. Motterlini, "Improved myocardial function after cold storage with preservation solution supplemented with a carbon monoxide–releasing molecule (CORM-3)," *Journal of Heart and Lung Transplantation*, vol. 26, no. 11, pp. 1192–1198, 2007.

[67] Y. Caumartin, J. Stephen, J. P. Deng et al., "Carbon monoxide-releasing molecules protect against ischemia–reperfusion injury during kidney transplantation," *Kidney International*, vol. 79, no. 10, pp. 1080–1089, 2011.

[68] M. Magierowski, K. Magierowska, J. Szmyd et al., "Hydrogen sulfide and carbon monoxide protect gastric mucosa compromised by mild stress against alendronate injury," *Digestive Diseases and Sciences*, vol. 61, no. 11, pp. 3176–3189, 2016.

[69] B. Bakhautdin, D. Das, P. Mandal et al., "Protective role of HO-1 and carbon monoxide in ethanol-induced hepatocyte cell death and liver injury in mice," *Journal of Hepatology*, vol. 61, no. 5, pp. 1029–1037, 2014.

[70] K. Mangano, E. Cavalli, S. Mammana et al., "Involvement of the Nrf2/HO-1/CO axis and therapeutic intervention with the CO-releasing molecule CORM-A1, in a murine model of autoimmune hepatitis," *Journal of Cellular Physiology*, vol. 233, no. 5, pp. 4156–4165, 2018.

[71] Y. Tayem, T. R. Johnson, B. E. Mann, C. J. Green, and R. Motterlini, "Protection against cisplatin-induced nephrotoxicity by a carbon monoxide-releasing molecule," *American Journal of Physiology-Renal Physiology*, vol. 290, no. 4, pp. F789–F794, 2006.

[72] Y. E. Yoon, K. S. Lee, Y. J. Lee, H. H. Lee, and W. K. Han, "Renoprotective effects of carbon monoxide–releasing molecule 3 in ischemia-reperfusion injury and cisplatin-induced toxicity," *Transplantation Proceedings*, vol. 49, no. 5, pp. 1175–1182, 2017.

[73] O. De Backer, E. Elinck, B. Blanckaert, L. Leybaert, R. Motterlini, and R. A. Lefebvre, "Water-soluble CO-releasing molecules reduce the development of postoperative ileus via modulation of MAPK/HO-1 signalling and reduction of oxidative stress," *Gut*, vol. 58, no. 3, pp. 347–356, 2009.

[74] S. Abid, A. Houssaini, N. Mouraret et al., "p21-dependent protective effects of a carbon monoxide-releasing molecule-3 in pulmonary hypertension," *Arteriosclerosis, Thrombosis, and Vascular Biology*, vol. 34, no. 2, pp. 304–312, 2014.

[75] O. Pak, A. G. Bakr, M. Gierhardt et al., "Effects of carbon monoxide-releasing molecules on pulmonary vasoreactivity in isolated perfused lungs," *Journal of Applied Physiology*, vol. 120, no. 2, pp. 271–281, 2016.

[76] E. Stagni, M. G. Privitera, C. Bucolo, G. M. Leggio, R. Motterlini, and F. Drago, "A water-soluble carbon monoxide-releasing molecule (CORM-3) lowers intraocular pressure in rabbits," *British Journal of Ophthalmology*, vol. 93, no. 2, pp. 254–257, 2009.

[77] H. Soni, G. Pandya, P. Patel, A. Acharya, M. Jain, and A. A. Mehta, "Beneficial effects of carbon monoxide-releasing molecule-2 (CORM-2) on acute doxorubicin cardiotoxicity in mice: Role of oxidative stress and apoptosis," *Toxicology and Applied Pharmacology*, vol. 253, no. 1, pp. 70–80, 2011.

[78] M. H. Tsai, C. W. Lee, L. F. Hsu et al., "CO-releasing molecules CORM2 attenuates angiotensin II-induced human aortic smooth muscle cell migration through inhibition of ROS/IL-6 generation and matrix metalloproteinases-9 expression," *Redox Biology*, vol. 12, pp. 377–388, 2017.

[79] H. Segersvärd, P. Lakkisto, M. Hänninen et al., "Carbon monoxide releasing molecule improves structural and functional cardiac recovery after myocardial injury," *European Journal of Pharmacology*, vol. 818, pp. 57–66, 2018.

[80] V. G. Nielsen, J. K. Kirklin, and J. F. George, "Carbon monoxide releasing molecule-2 increases the velocity of thrombus growth and strength in human plasma," *Blood Coagulation and Fibrinolysis*, vol. 20, no. 5, pp. 377–380, 2009.

[81] V. G. Nielsen, N. Chawla, D. Mangla et al., "Carbon monoxide-releasing molecule-2 enhances coagulation in rabbit plasma and decreases bleeding time in clopidogrel/aspirin-treated rabbits," *Blood Coagulation and Fibrinolysis*, vol. 22, no. 8, pp. 756–759, 2011.

[82] K. A. Machovec, D. S. Ushakumari, I. J. Welsby, and V. G. Nielsen, "The procoagulant properties of purified fibrinogen concentrate are enhanced by carbon monoxide releasing molecule-2," *Thrombosis Research*, vol. 129, no. 6, pp. 793–796, 2012.

[83] V. G. Nielsen and C. M. Bazzell, "Carbon monoxide attenuates the effects of snake venoms containing metalloproteinases with fibrinogenase or thrombin-like activity on plasmatic coagulation," *Medicinal Chemistry Communications*, vol. 7, no. 10, pp. 1973–1979, 2016.

[84] V. G. Nielsen, V. L. Boyer, R. W. Matika, Q. Amos, and D. T. Redford, "Iron and carbon monoxide attenuate Crotalus atrox venom-enhanced tissue-type plasminogen activator-initiated fibrinolysis," *Blood Coagulation and Fibrinolysis*, vol. 27, no. 5, pp. 511–516, 2016.

[85] V. G. Nielsen and C. M. Bazzell, "Carbon monoxide releasing molecule-2 inhibition of snake venom thrombin-like activity: novel biochemical "brake"?," *Journal of Thrombosis and Thrombolysis*, vol. 43, no. 2, pp. 203–208, 2017.

[86] V. G. Nielsen, "Crotalus atrox venom exposed to carbon monoxide has decreased fibrinogenolytic activity in vivo in rabbits," *Basic and Clinical Pharmacology and Toxicology*, vol. 122, no. 1, pp. 82–86, 2018.

[87] B. Chen, L. Guo, C. Fan et al., "Carbon monoxide rescues heme oxygenase-1-deficient mice from arterial thrombosis in allogeneic aortic transplantation," *American Journal of Pathology*, vol. 175, no. 1, pp. 422–429, 2009.

[88] K. Kramkowski, A. Leszczynska, A. Mogielnicki et al., "Antithrombotic properties of water-soluble carbon monoxide-releasing molecules," *Arteriosclerosis, Thrombosis, and Vascular Biology*, vol. 32, no. 9, pp. 2149–2157, 2012.

[89] M. Klinger-Strobel, S. Gläser, O. Makarewicz et al., "Bactericidal effect of a photoresponsive carbon monoxide-releasing nonwoven against *Staphylococcus aureus* biofilms," *Antimicrobial Agents and Chemotherapy*, vol. 60, no. 7, pp. 4037–4046, 2016.

[90] J. L. Wilson, L. K. Wareham, S. McLean et al., "CO-releasing molecules have nonheme targets in bacteria: transcriptomic, mathematical modeling and biochemical analyses of CORM-3 [Ru(CO)3Cl(glycinate)] actions on a heme-deficient mutant of *Escherichia coli*," *Antioxidants and Redox Signaling*, vol. 23, no. 2, pp. 148–162, 2015.

[91] A. Verma, D. Hirsch, C. Glatt, G. Ronnett, and S. Snyder, "Carbon monoxide: a putative neural messenger," *Science*, vol. 259, no. 5093, pp. 381–384, 1993.

[92] R. Battish, G. Y. Cao, R. B. Lynn, S. Chakder, and S. Rattan, "Heme-oxygenase 2 (HO-2) distribution in anorectum of opossum: colocalization with neuronal nitric oxide synthase (nNOS) and vasoactive intestinal polypeptide (VIP)," *Gastroenterology*, vol. 114, p. A719, 1998.

[93] L. Xue, G. Farrugia, S. M. Miller, C. D. Ferris, S. H. Snyder, and J. H. Szurszewski, "Carbon monoxide and nitric oxide as coneurotransmitters in the enteric nervous system: evidence from genomic deletion of biosynthetic enzymes," *Proceedings of the National Academy of Sciences*, vol. 97, no. 4, pp. 1851–1855, 2000.

[94] K. Henke, N. E. Kroll, H. Behniea et al., "Memory lost and regained following bilateral hippocampal damage," *Journal of Cognitive Neuroscience*, vol. 11, no. 6, pp. 682–697, 1999.

[95] C. Mancuso, E. Ragazzoni, G. Tringali et al., "Inhibition of heme oxygenase in the central nervous system potentiates

endotoxin-induced vasopressin release in the rat," *Journal of Neuroimmunology*, vol. 99, no. 2, pp. 189–194, 1999.

[96] N. Y. Calingasan, W. J. Chun, L. C. H. Park, K. Uchida, and G. E. Gibson, "Oxidative stress is associated with region-specific neuronal death during thiamine deficiency," *Journal of Neuropathology and Experimental Neurology*, vol. 58, no. 9, pp. 946–958, 1999.

[97] I. S. Choi and H. Y. Cheon, "Delayed movement disorders after carbon monoxide poisoning," *European Neurology*, vol. 42, no. 3, pp. 141–144, 1999.

[98] H. M. Schipper, A. Liberman, and E. G. Stopa, "Neural heme oxygenase-1 expression in idiopathic Parkinson's disease," *Experimental Neurology*, vol. 150, no. 1, pp. 60–68, 1998.

[99] C. A. Piantadosi, J. Zhang, E. D. Levin, R. J. Folz, and D. E. Schmechel, "Apoptosis and delayed neuronal damage after carbon monoxide poisoning in the rat," *Experimental Neurology*, vol. 147, no. 1, pp. 103–114, 1997.

[100] A. C. Durak, A. Coskun, A. Yikilmaz, F. Erdogan, E. Mavili, and M. Guven, "Magnetic resonance imaging findings in chronic carbon monoxide intoxication," *Acta Radiologica*, vol. 46, no. 3, pp. 322–327, 2005.

[101] C. S. F. Queiroga, R. M. A. Alves, S. V. Conde, P. M. Alves, and H. L. A. Vieira, "Paracrine effect of carbon monoxide–astrocytes promote neuroprotection through purinergic signaling in mice," *Journal of Cell Science*, vol. 129, no. 16, pp. 3178–3188, 2016.

[102] T. Simon, S. Pogu, V. Tardif et al., "Carbon monoxide-treated dendritic cells decrease β1-integrin induction on CD8+T cells and protect from type 1 diabetes," *European Journal of Immunology*, vol. 43, no. 1, pp. 209–218, 2013.

[103] S. S. Jung, J. S. Moon, J. F. Xu et al., "Carbon monoxide negatively regulates NLRP3 inflammasome activation in macrophages," *American Journal of Physiology-Lung Cellular and Molecular Physiology*, vol. 308, no. 10, pp. L1058–L1067, 2015.

[104] A. Bougié, A. Harrois, and J. Duranteau, "Resuscitative strategies in traumatic hemorrhagic shock," *Annals of Intensive Care*, vol. 3, no. 1, pp. 1–9, 2013.

[105] T. Takagi, Y. Naito, K. Uchiyama et al., "Carbon monoxide promotes gastric wound healing in mice via the protein kinase C pathway," *Free Radical Research*, vol. 50, no. 10, pp. 1098–1105, 2016.

[106] J. A. Nemzek, C. Fry, and O. Abatan, "Low-dose carbon monoxide treatment attenuates early pulmonary neutrophil recruitment after acid aspiration," *American Journal of Physiology-Lung Cellular and Molecular Physiology*, vol. 294, no. 4, pp. L644–L653, 2008.

[107] J. T. Chapman, L. E. Otterbein, J. A. Elias, and A. M. K. Choi, "Carbon monoxide attenuates aeroallergen-induced inflammation in mice," *American Journal of Physiology-Lung Cellular and Molecular Physiology*, vol. 281, no. 1, pp. L209–L216, 2001.

[108] L. E. Otterbein, S. L. Otterbein, E. Ifedigbo et al., "MKK3 mitogen-activated protein kinase pathway mediates carbon monoxide-induced protection against oxidant-induced lung injury," *American Journal of Pathology*, vol. 163, no. 6, pp. 2555–2563, 2003.

[109] T. Dolinay, M. Szilasi, M. Liu, and A. M. K. Choi, "Inhaled carbon monoxide confers antiinflammatory effects against ventilator-induced lung injury," *American Journal of Respiratory and Critical Care Medicine*, vol. 170, no. 6, pp. 613–620, 2004.

[110] A. M. Aberg, P. Abrahamsson, G. Johansson, M. Haney, O. Winso, and J. E. Larsson, "Does carbon monoxide treatment alter cytokine levels after endotoxin infusion in pigs? A randomized controlled study," *Journal of Inflammation*, vol. 5, no. 1, pp. 13–21, 2008.

[111] L. A. Mitchell, M. M. Channell, C. M. Royer, S. W. Ryter, A. M. K. Choi, and J. D. McDonald, "Evaluation of inhaled carbon monoxide as an anti-inflammatory therapy in a nonhuman primate model of lung inflammation," *American Journal of Physiology-Lung Cellular and Molecular Physiology*, vol. 299, no. 6, pp. L891–L897, 2010.

[112] C. Bucolo and F. Drago, "Carbon monoxide and the eye: implications for glaucoma therapy," *Pharmacology and Therapeutics*, vol. 130, no. 2, pp. 191–201, 2011.

[113] R. F. Anderson, "Myocardial toxicity from carbon monoxide poisoning," *Annals of Internal Medicine*, vol. 67, no. 6, p. 1172, 1967.

[114] R. Favory, S. Lancel, S. Tissier, D. Mathieu, B. Decoster, and R. Nevière, "Myocardial dysfunction and potential cardiac hypoxia in rats induced by carbon monoxide inhalation," *American Journal of Respiratory and Critical Care Medicine*, vol. 174, no. 3, pp. 320–325, 2006.

[115] L. Andre, J. Boissière, C. Reboul et al., "Carbon monoxide pollution promotes cardiac remodeling and ventricular arrhythmia in healthy rats," *American Journal of Respiratory and Critical Care Medicine*, vol. 181, no. 6, pp. 587–595, 2010.

[116] D. V. Abramochkin, O. P. Konovalova, A. Kamkin, and G. F. Sitdikova, "Carbon monoxide modulates electrical activity of murine myocardium via cGMP-dependent mechanisms," *Journal of Physiology and Biochemistry*, vol. 71, no. 1, pp. 107–119, 2015.

[117] R. D. Stewart, "The effect of carbon monoxide on humans," *Annual Review of Pharmacology*, vol. 15, no. 1, pp. 409–423, 1975.

[118] L. W. Kao and K. A. Nañagas, "Toxicity associated with carbon monoxide," *Clinics in Laboratory Medicine*, vol. 26, no. 1, pp. 99–125, 2006.

[119] R. Foresti, M. G. Bani-Hani, and R. Motterlini, "Use of carbon monoxide as a therapeutic agent: promises and challenges," *Intensive Care Medicine*, vol. 34, no. 4, pp. 649–658, 2008.

[120] National Research Council (US) Committee on Acute Exposure Guideline Levels, "Carbon monoxide acute exposure guideline levels," in *Acute Exposure Guideline Levels for Selected Airborne Chemicals: Volume 8*, pp. 49–143, National Academies Press (US), Washington, DC, USA, 2010.

[121] K. Magierowska, M. Magierowski, M. Hubalewska-Mazgaj et al., "Carbon monoxide (CO) released from tricarbonyldichlororuthenium (II) dimer (CORM-2) in gastroprotection against experimental ethanol-induced gastric damage," *PLoS One*, vol. 10, no. 10, Article ID e0140493, 2015.

[122] G. Wang, T. Hamid, R. J. Keith et al., "Cardioprotective and antiapoptotic effects of heme oxygenase-1 in the failing heart," *Circulation*, vol. 121, no. 17, pp. 1912–1925, 2010.

[123] B. Wegiel, D. W. Hanto, and L. E. Otterbein, "The social network of carbon monoxide in medicine," *Trends in Molecular Medicine*, vol. 19, no. 1, pp. 3–11, 2013.

[124] J. D. Seixas, M. F. A. Santos, A. Mukhopadhyay et al., "A contribution to the rational design of Ru(CO)3Cl2L complexes for in vivo delivery of CO," *Dalton Transactions*, vol. 44, no. 11, pp. 5058–5075, 2015.

[125] M. Feger, A. Fajol, A. Lebedeva et al., "Effect of carbon monoxide donor CORM-2 on vitamin D3 metabolism," *Kidney and Blood Pressure Research*, vol. 37, no. 4-5, pp. 496–505, 2013.

[126] E. Lang, S. M. Qadri, K. Jilani, A. Lupescu, E. Schleicheeer, and T. Lang, "Carbon monoxide-sensitive apoptotic death of erythrocytes," *Basic and Clinical Pharmacology and Toxicology*, vol. 111, no. 5, pp. 348–355, 2012.

[127] I. C. Winburn, K. Gunatunga, R. D. McKernan, R. J. Walker, I. A. Sammut, and J. C. Harrison, "Cell damage following carbon monoxide releasing molecule exposure: implications for therapeutic applications," *Basic and Clinical Pharmacology and Toxicology*, vol. 111, no. 1, pp. 31–41, 2012.

[128] P. Wang, H. Liu, Q. Zhao et al., "Syntheses and evaluation of drug-like properties of CO-releasing molecules containing ruthenium and group 6 metal," *European Journal of Medicinal Chemistry*, vol. 74, pp. 199–215, 2014.

[129] R. Motterlini, P. Sawle, J. Hammad et al., "Vasorelaxing effects and inhibition of nitric oxide in macrophages by new iron-containing carbon monoxide-releasing molecules (CORMs)," *Pharmacological Research*, vol. 68, no. 1, pp. 108–117, 2013.

[130] S. Romanski, B. Kraus, U. Schatzschneider, J. M. Neudörfl, S. Amslinger, and H. G. Schmalz, "Acyloxybutadiene iron tricarbonyl complexes as enzyme-triggered CO-releasing molecules (ET-CORMs)," *Angewandte Chemie International Edition*, vol. 50, no. 10, pp. 2392–2396, 2011.

[131] R. Mede, P. Hoffmann, C. Neumann et al., "Acetoxymethyl concept for intracellular administration of carbon monoxide with $Mn(CO)_3$-based photoCORMs," *Chemistry-A European Journal*, vol. 24, no. 13, pp. 3321–3329, 2018.

[132] G. Dördelmann, T. Meinhardt, T. Sowik, A. Krueger, and U. Schatzschneider, "CuAAC click functionalization of azide-modified nanodiamond with a photoactivatable CO-releasing molecule (PhotoCORM) based on [Mn$(CO)_3$(tpm)]$^{+}$," *Chemical Communications*, vol. 48, no. 94, article 11528, 2012.

[133] U. Hasegawa, A. J. van der Vlies, E. Simeoni, C. Wandrey, and J. A. Hubbell, "Carbon monoxide-releasing micelles for immunotherapy," *Journal of the American Chemical Society*, vol. 132, no. 51, pp. 18273–18280, 2010.

[134] P. Govender, S. Pai, U. Schatzschneider, and G. S. Smith, "Next generation PhotoCORMs: polynuclear tricarbonylmanganese (I)-functionalized polypyridyl metallodendrimers," *Inorganic Chemistry*, vol. 52, no. 9, pp. 5470–5478, 2013.

[135] R. Motterlini, P. Sawle, J. Hammad et al., "CORM-A1: a new pharmacologically active carbon monoxide-releasing molecule," *FASEB Journal*, vol. 19, no. 2, pp. 284–286, 2005.

[136] A. E. Pierri, A. Pallaoro, G. Wu, and P. C. Ford, "A luminescent and biocompatible PhotoCORM," *Journal of the American Chemical Society*, vol. 134, no. 44, pp. 18197–18200, 2012.

[137] S. J. Carrington, I. Chakraborty, J. M. L. Bernard, and P. K. Mascharak, "A theranostic two-tone luminescent PhotoCORM derived from Re(I) and (2-pyridyl)-benzothiazole: trackable CO delivery to malignant cells," *Inorganic Chemistry*, vol. 55, no. 16, pp. 7852–7858, 2016.

[138] G. Ur, J. Axthelm, P. Hoffmann et al., "Co-registered molecular logic gate with a CO-releasing molecule triggered by light and peroxide," *Journal of the American Chemical Society*, vol. 139, no. 14, pp. 4991–4994, 2017.

[139] G. Gessner, N. Sahoo, S. M. Swain et al., "CO-independent modification of K^{+} channels by tricarbonyldichlororuthenium (II) dimer (CORM-2)," *European Journal of Pharmacology*, vol. 812, pp. 33–41, 2013.

[140] C Nagel, McLean, R. K Poole, H. Braunschweig, T. Kramer, and U. Schatzschneider, "Introducing [Mn$(CO)_3$(tpa-$\kappa^3 N$)]$^{+}$ as a novel photoactivatable CO-releasing molecule with well-defined iCORM intermediates–synthesis, spectroscopy, and antibacterial activity," *Dalton Transactions*, vol. 43, no. 26, pp. 9986–9997, 2014.

[141] T. Santos-Silva, A. Mukhopadhyay, J. D. Seixas, J. L. Bernardes, C. Romao, and M. Romao, "Towards therapeutic CORMs: understanding the reactivity of CORM-3 with proteins," *Current Medicinal Chemistry*, vol. 18, no. 22, pp. 3361–3366, 2011.

[142] B. E. Mann, "3.29 signaling molecule delivery (CO)," in *Comprehensive Inorganic Chemistry (II)*, J. Reedjik and K. Poeppelmeier, Eds., pp. 857–876, Elsevier, Amsterdam, Netherlands, 2nd edition, 2013.

[143] S. M. Kapetanaki, G. Slikstoner, I. Husu, U. Liebl, M. T. Wilson, and M. H. Vos, "Interaction of carbon monixide with apoptosis—inducing cyytochrome C–cardiolipin complex," *Biochemistry*, vol. 48, no. 7, pp. 1613–1619, 2009.

[144] M. Ibrahim, E. R. Derbyshire, M. A. Marletta, and T. G. Spiro, "Probing soluble guanylate cyclase actgivation by CO and YC-1 usiong resonance Raman spectroscopy," *Biochemistry*, vol. 49, no. 18, pp. 3815–3823, 2010.

[145] X. Ji and B. Wang, "Strategies toward organic carbon monoxide prodrugs," *Accounts of Chemical Research*, vol. 51, no. 6, pp. 1377–1385, 2018.

[146] Z. Pan, J. Zhang, K. Ji, V. Chittavong, X. Ji, and W. Wang, "Organic CO prodrugs activated by endogenous ROS," *Organic Letters*, vol. 20, no. 1, pp. 8–11, 2018.

[147] S. T. Blackburn, *Maternal, Fetal, & Neonatal Physiology: A Clinical Perspective*, Elsevier Health Sciences, Gurgaeon, Harvana, India, 2007.

20

Synthesis and Antibacterial Activity of Polyoxometalates with Different Structures

Jingmin Gu [ID],[1] Lei Zhang,[2] Xiaofeng Yuan,[3] Ya-Guang Chen [ID],[4] Xiuzhu Gao,[5] and Dong Li [ID][5,6]

[1]College of Veterinary Medicine, Jilin University, Changchun 130062, China
[2]Animal Science and Technology College, Jilin Agricultural University, Changchun 130117, China
[3]Department of Pediatrics, Affiliated Hospital of the Changchun University of Chinese Medicine, Changchun 130021, China
[4]Key Laboratory of Polyoxometalates Science of Ministry of Education, Faculty of Chemistry, Northeast Normal University, Changchun 130024, China
[5]Department of Hepatology, The First Hospital, Jilin University, Changchun 130021, China
[6]Department of Immunology, College of Basic Medical Sciences, Jilin University, Changchun 130021, China

Correspondence should be addressed to Dong Li; lidong1@jlu.edu.cn

Guest Editor: Rais A. Khan

A new inorganic-organic hybrid compound, $[\{Cu(phen)_2\}_2(H_4W_{12}O_{40})]$, was synthesized, and its crystal structure was determined. The Keggin anion $H_4W_{12}O_{40}^{4-}$ was grafted with two coordination units $\{Cu(phen)_2\}$, forming an electrically neutral molecule. The antibacterial activity of several polyoxometalate compounds with different anionic structures including the new compound was studied. The results show that the compound **1** can inhibit the growth of *Enterococcus faecalis* FA2 strains and that antibacterial activity of the polyoxometalate compounds is dependent with component elements of POM but is less relative with the anion structures.

1. Instruction

Polyoxometalates (POMs) have been shown to exhibit biological activities *in vitro* as well as *in vivo*, including anticancer and antiviral [1], antibacterial [2, 3], antiprotozoal [4, 5], and antidiabetic activities [6]. In the antibacterial activity study of POMs, Tajima found the enhancement of several beta-lactams antibiotics in antibacterial activity to methicillin-resistant *Staphylococcus aureus* under the synergistic action of polyoxometalates, substituted-type POMs $K_7[PTi_2W_{10}O_{40}]$ $6H_2O$, $K_7[BVW_{11}O_{40}]7H_2O$, $[SiFeW_{11}O_{40}]^{6/5-}$ and $[SiCoW_{11}O_{40}]^{6-}$ and lacunary-type POMs $[XW_{11}O_{39}]^{n-}$ (X = Si, P) and $[XW_9O_{34}]^{n-}$, and proposed the resistant mechanism [7–12]. Inoue et al. reported the enhancement of antibacterial activity of beta-lactam antibiotics, oxacillin, by polyoxometalates ($K_6[P_2W_{18}18O_{62}]\cdot14H_2O$, $K_4[SiMo_{12}O_{40}]\cdot3H_2O$, and $K_7[PTi_2W_{10}O_{40}]\cdot6H_2O$) against methicillin-resistant *Staphylococcus aureus* (MRSA)

and vancomycin-resistant *S. aureus* (VRSA) and also proposed a reaction mechanism [13]. Daima et al. studied synergistic antibacterial action of Ag nanoparticles and POMs which were achieved by the physical damage to the bacterial cells [14]. Li and his colleagues showed the short peptides/HSiW nanofibers had antimicrobial activity to the ubiquitous and clinically relevant bacterium *Escherichia coli* [15]. In recent years, there are also many reports about the antibacterial activity of known and new polyoxometalate derivatives to several bacteria including *Escherichia coli*, *Staphylococcus aureus*, *Paenibacillus* sp., *Bacillus subtilis*, *Clavibacter michiganensis*, *Vibrio* sp., *Pseudomonas putida*, *Helicobacter pylori*, *S. typhimurium*, *Streptocoque B* (*S. agalactiae*), *L. acidophilus*, and amebas [3,16–34]. In these reports, the polyoxometalate derivatives are in the form of simple inorganic salts, inorganic-organic hybrids with various organic groups, films, nanofibers, etc. However, the antibacterial activity of these compounds is not satisfactory according to the reported data. Nevertheless, the emergence of multidrug-resistant bacterial strains which was partially due to the abuse of conventional antibiotics proved that there is an urgent need for novel therapeutic agents. Therefore, synthesizing and exploring new compounds with high antibacterial activity are still a challenging task of chemists and pharmacologist. To achieve this, the study on influence of the composition and structure of compounds on antibacterial activity is very important, which will play an instructional role in synthesizing and exploring new compounds.

This work is about the synthesis of a new polyoxometalate derivative, $[Cu_2(phen)_4(H_4W_{12}O_{40})]$, and study on the antibacterial activity of several polyoxometalate compounds with different anionic structures including the new compound.

2. Materials and Methods

2.1. Materials and General Methods. All reagents were purchased commercially and used without further purification. Elemental analyses (C, H, and N) were performed on a Perkin–Elmer 2400 CHN elemental analyzer and that of W and Cu on an ICP–AES analyzer. The IR spectrum was obtained on a Magna-560 FT/IR spectrometer with KBr pellets in the 400–4000 cm^{-1} region. TG analysis was carried out on a DTG-60H thermal analyzer in flowing N_2 with a heating rate of 10°C·min^{-1}. SEM images were recorded on Hitachi S-3400N (Hitachi High-Technologies Europe GmbH, Krefeld, Germany).

2.2. Synthesis. Synthesis of $[Cu_2(phen)_4(H_4W_{12}O_{40})]$ was modified from our previous report [35]: compound **1** was prepared from reaction of $(NH_4)_6(H_2W_{12}O_{40})\cdot3H_2O$ (0.1 mmol, 0.30 g), $CuCl_2$ $2H_2O$ (2.0 mmol, 0.34 g), phenanthroline (0.5 mmol, 0.099 g), succinic acid (0.5 mmol, 0.06 g), and 12 mL water. The starting mixture was adjusted to pH = 2.0 by the addition of hydrochloric acid, and the mixture was stirred for 1 h under air. The final solution was transferred to a 25 mL Teflon-lined autoclave and crystallized at 160°C for 96 h. Then, the autoclave was cooled at the rate of 10°C·h^{-1} to room temperature. The resulting green stripe crystals were filtered off, washed with distilled water, and air-dried. Good-quality crystals were sealed for structural determination and further characterization. Elemental analysis calcd for $C_{48}H_{32}Cu_2N_2O_{41}W_{12}$ (Mr = 3710) C 1.00, H 12.7619, N 2.48, O 20.68, Cu 3.38, P 1.10, W 58.60 (%); found: C 1.10, H 12.39, N 2.41, O 20.56, Cu 3.39, P 1.09, W 59.05 (%). IR(KBr pellet, cm^{-1}): 3500, 3082, 2370, 2298, 2109, 1994, 1628, 1597, 1524, 1335, 1231, 1085, 948, 781, 750, 708, 667, 593, 530 cm^{-1}.

Compounds **2–6** were prepared in accordance with the methods in Refs. [36–39] and characterized by the IR spectrograph and TGA.

2.3. X-Ray Crystallography. The X-ray diffraction data of compound **1** were collected on a Bruker Smart Apex II diffractometer with graphite monochromatic Mo Kα radiation (λ = 0.710 73 Å) at 293 K with ω scans (Table 1).

TABLE 1: Crystal data and structure refinements for compound **1**.

Formula	$C_{48}H_{32}Cu_2N_8O_{40}W_{12}$
Fw	3710
Crystal system	Monoclinic
Space group	P21/c
a/Å	26.1828(15)
b/Å	11.84219(7)
c/Å	23.3996(13)
α/°	90.00
β/°	113.74(2)
γ/°	90.00
V/Å^3	6641.2(7)
Z	4
D_c/g·cm^{-3}	3.711
F(000)	6600
μ/mm^{-1}	21.419
R_{int}	0.0932
Refine number of reflns/parameters/restraints	13661/991/54
R_factor_all/[$I > 2$sigma(I)]	0.0871/0.0456
wR_factor_ref/[$I > 2$sigma(I)]	0.0857/0.0745
Goodness of fit	0.954

Multiscan absorption corrections were applied. The structures were solved by direct methods and refined by full matrix least-squares on F^2 using the SHELXTL crystallographic software package [40]. The positions of hydrogen atoms on the carbon atoms were calculated theoretically. Crystal data and structure refinements for compound **1** are presented in Table 1. Cu-O and Cu-N bond lengths are listed in Table 2. CCDC-1487664 for **1** contains the supplementary crystallographic data for this paper. These data can be obtained free of charge from The Cambridge Crystallographic Data Centre via http://www.ccdc.cam.ac.uk/data_request/cif.

2.4. Antibacterial Experiments. All the isolated bacterial strains were achieved by colony formation on selective salt agar plates containing 6 mg/mL oxacillin. All bacterial strains were stored at −80°C and routinely grown at 37°C. *Staphylococcus aureus* (YB57), *Enterococcus faecalis* (FA2 and FA3), and *Enterococcus faecium* (SA2 and SA3) strains were cultured in brain heart infusion (BHI) broth, while *Staphylococcus aureus* (USA300), *Acinetobacter baumannii* (ABC3), and *Streptococcus pneumoniae* (SP) were cultured in the Luria-Bertani (LB) medium. The polyoxometalates were tested for their antibacterial activities against eight different bacterial strains by the observation of the OD value of culture media. Briefly, bacterial cells were washed and resuspended in sterile PBS, and the colony count was determined. The different polyoxometalates were added to the bacterial suspension (final concentration, 1 mg/mL), and the mixture was incubated overnight at 37°C. The colony count was determined again. Enzyme activity *in vitro* was expressed as the CFU reduction. As a negative control, the bacterial strains were treated with the elution buffer under the same conditions. The results are listed in Table 3.

TABLE 2: Cu-O and Cu-N bond lengths (Å) in **1**.

Bond	Length	Bond	Length
Cu1 N1	2.214(11)	Cu2 N5	1.960(12)
Cu1 N2	1.990(10)	Cu2 N9	2.038(12)
Cu1 N3	1.996(13)	Cu2 N7	1.965(12)
Cu1 N4	1.991(11)	Cu2 N8	2.168(12)
Cu1 O7	2.045(10)	Cu2 O38	2.217(8)

TABLE 3: Antibacterial activities of compounds **1–6**.

Type	Code	Compound	Concentration	Bacterial strains
I	1	$[\{Cu(phen)_2\}_2(H_4W_{12}O_{40})]$	1 mg/mL	FA2
	2	$[Cu(phen)(H_2O)(Mo_3O_{10})]$	1 mg/mL	ABC3, FA2, FA3, YB57, USA300
II	3	$K_5PW_{11}TiO_{40}\cdot14H_2O$	1 mg/mL	YB57, USA300
III	4	$Na_7CeW_{10}O_{35}\cdot26H_2O$	1 mg/mL	SA5, SA1, SP, USA300
IV	5	$K_{13}[La(SiW_{11}O_{39})_2]\cdot26H_2O$	1 mg/mL	YB57, USA300
	6	$K_{13}[Ce(SiW_{11}O_{39})_2]\cdot 26H_2O$	1 mg/mL	SA5, SA1, SP, USA300

2.5. Scanning Electron Microscopy. Scanning electron microscopy (SEM) was performed to assess the activity of different polyoxometalates on the bacterial strains *in vitro*. The *Staphylococcus aureus* strains USA300 were grown to the exponential growth phase (an OD 600 nm value of 0.6) in the BHI broth at 37°C with shaking at 200 rpm. The bacteria were collected and washed three times (5,000 × g for 1 min at 4°C) with PBS. Different formulations were separately added to *S. aureus* suspensions. Bacterial lysates were harvested by centrifugation (1,100 × g for 1 min) at different time points. Then, the bacterial lysates were fixed with glutaraldehyde and were dehydrated and freeze-dried for SEM.

3. Results and Discussion

3.1. Crystal Structure of 1. The asymmetric unit of compound **1** consists of one Keggin anion $[H_4W_{12}O_{40}]^{6-}$, two Cu^{2+} ions, and four phen molecules. The $[H_4W_{12}O_{40}]^{6-}$ anion (Figure 1) contains four edge-shared W_3O_{13} units which combine together through corner-shared linkage. W–O bonds can be classified into three sets: W–O_t (terminal oxygen atoms) with distances of 1.683(12)–1.740(12) Å, W–O_b (bridging oxygen atoms) with distances of 1.847(12)–2.010(14) Å and W–O_c (central oxygen atoms) with distances of 2.174(12)–2.396(13) Å. That is, the WO_6 octahedra are all distorted. The Keggin anion acts as a bidentate ligand bonding two Cu^{2+} ions (Cu1 and Cu2) with one terminal oxygen atom and one bridge oxygen atom. One W–O_t bond was elongated (1.740(12) Å) due to the coordination of the terminal oxygen atom to Cu ion. Such a POM anion is also called as decorated Keggin anion (Figure 1), very similar to the decoration we reported previously [35].

Two Cu^{2+} ions are all five-coordinated. Cu1 ion displays in a square prism geometry, and the geometry of Cu2 ion is better to be described as triangle bipyramid. The donor atoms bonding to Cu ions come from two phen molecules with chelating coordination mode and the Keggin anion, forming a complex fragment $\{Cu(phen)_2\}^{2+}$ (Figure 1). Cu-O and Cu-N bond lengths are listed in Table 2. As shown in Table 2, the long bonds belong to the atoms at axial site for Cu1 and triangle plane for Cu2 (Figure S1), resulting from their environment in the crystal. Devi et al. [41] had reported a similar cluster $[\{Cu(phen)2\}2(H2W12O40)]^{2-}$ in $[\{Cu(phen)2\}4\{H2W12O40\}]$ $[\{Cu(phen)2\}2\{H2W12O40\}]\cdot$ $3H2O$, in which one Cu ion is six-coordinated, different from that of this new compound. The neutral molecules are assembled into three dimensional architecture through CH···O hydrogen bonds (Table S2) and intermolecular interaction force (Figure S2).

3.2. Characterization of 1. The IR spectrum (Figure S3) of **1** shows the vibration absorption bands of CH bond in 3080 cm^{-1} and of C-C and N-C bonds of phen ring in 1614–1137 cm^{-1}. The vibration absorption bands of compound **1** at 952, 877, 846, and 740 cm^{-1} should be ascribed to the asymmetric stretching vibrations of W-O_d, W-O_b-W, and W-O_c-W bonds, respectively, consistent with that in Ref. [42]. The TG curve of **1** is shown in Figure S4. Compound **1** is stable below 400°C and then decomposes until 600°C. The lost weight of 20.17% is consistent with the calculated one (20.38% for $2H_2O$ and 4phen), confirming the chemical formula obtained from elemental analysis and structure analysis.

3.3. Antibacterial Activity of 1–6. The compounds **1–6** used in antibacterial experiments can be divided into four kinds. **1** and **2** are inorganic-organic hybrids with phenanthroline, and **3** is a mono-substituted Keggin-type compound in which Ti atom occupies one of twelve sites. **4** is a complex of mono-lacunaria Lindquist anion and lanthanides. **5** and **6** are complexes of mono-lacunaria Keggin anions and lanthanides. From Table 3, it can be seen that (1) the new compound **1** is active only to bacterial strains FA2. (2) The compound of molybdenum, **2**, has a wider antibacterial spectrum than that of tungsten (**1**, **3**, **4**, **5**, **6**). (3) The anionic structure has less influence on antibacterial activity. (4) The compounds with cerium element (**4**, **6**) show better antibacterial activity than others.

SEM technique was used to explore the interaction of polyoxometalates with the bacterial strains. SEM images (Figure 2) show the surface morphology of *Staphylococcus aureus* strains USA300 untreated (Figure 2(a)) and treated with **3**, **4**, **5**, and **6** (Figures 2(b)–2(e)). From Figure 2, it can be seen that the surface morphology of *Staphylococcus aureus* strains USA300 treated with polyoxometalates (Figures 2(b)–2(e)) has changed obviously compared with that of untreated one (Figure 2(a)) from smooth globular form to chapping oblate spheroid. The degree of changes in the surface morphology indicates the antibacterial activity of

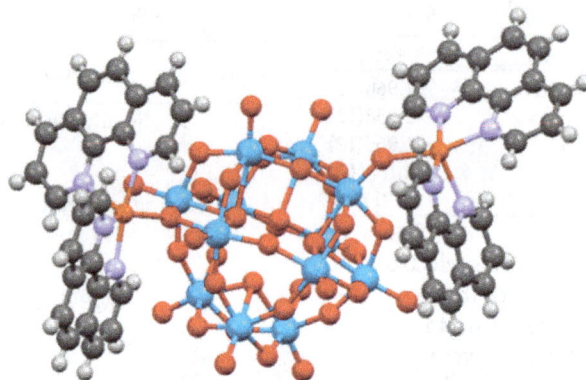

FIGURE 1: Ball-stick representation of the decorated Keggin anion in **1**.

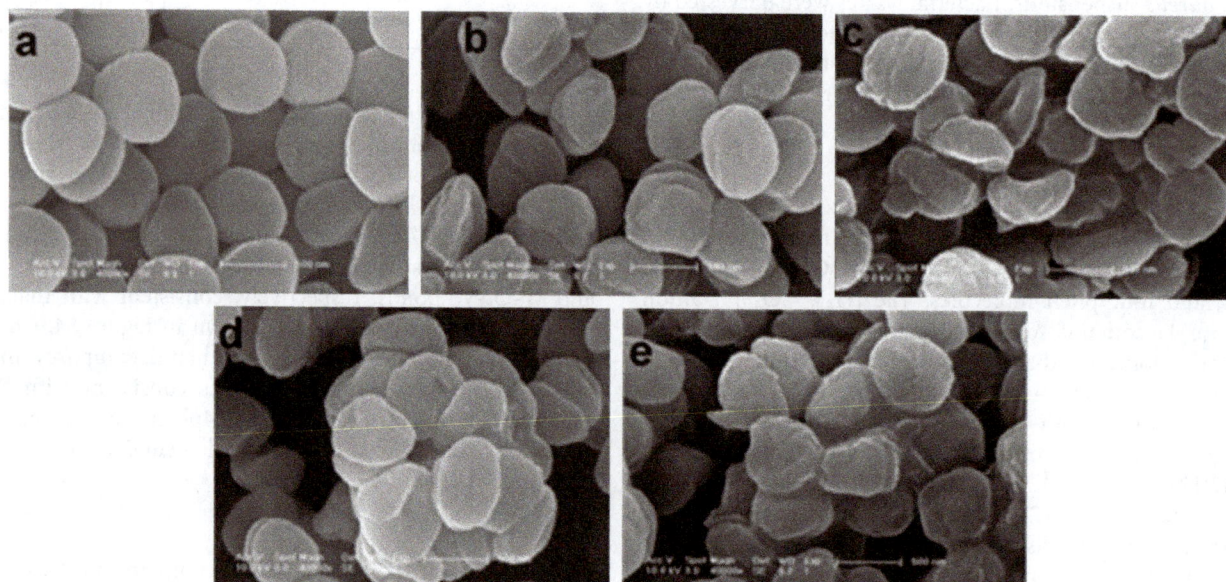

FIGURE 2: SEM images of *Staphylococcus aureus* strains USA300 untreated (a) and treated with **3** (b), **4** (c), **5** (d), and **6** (e).

polyoxometalates. So, a sequence of the activity of polyoxometalates was given according to Figure 2, **4** ≈ **6** > **5** > **3**. That is, the compounds with cerium element (**4**, **6**) have better antibacterial activity than others.

4. Conclusion

The bioactivity of polyoxometalates has been known for many years but still has large space to explore. The results of this work on the antibacterial activity of polyoxometalates including the new compound show that antibacterial activity of the compounds is more relative with their component element than with anionic structure, which means that exploration of antibacterial materials should focus on the choice of elements.

In this work, the compounds with Ce elements have better antibacterial activity. So, synthesizing compounds with other lanthanide elements and other compounds with cerium element and examining their antibacterial activity as well as exploring the reaction mechanism of Ce compounds need further investigations.

Authors' Contributions

Jingmin Gu and Lei Zhang contributed equally to this article.

Acknowledgments

This work was supported by the National Natural Science Foundation of China (No. 81501423) and Norman Bethune Program of Jilin University (No. 2015223). We also like to give our special thanks to Dr. Sadeeq ur Rahman from College of Veterinary Sciences & Animal Husbandry, Abdul

Wali Khan University, Mardan, Pakistan, for generously providing *Enterococcus faecium* strains (SA2 and SA3) and *Streptococcus pneumoniae* strain (SP).

References

[1] J. T. Rhule, C. L. Hill, and D. A. Judd, "Polyoxometalates in medicine," *Chemical Reviews*, vol. 98, no. 1, pp. 327–357, 1998.

[2] A. Seko, T. Yamase, and K. Yamashita, "Polyoxometalates as effective inhibitors for sialyl- and sulfotransferases," *Journal of Inorganic Biochemistry*, vol. 103, no. 7, pp. 1061–1066, 2009.

[3] M. Inoue, K. Segawa, S. Matsunaga et al., "Antibacterial activity of highly negative charged polyoxotungstates, K27 [KAs4W40O140] and K18[KSb9W21O86], and Keggin-structural polyoxotungstates against *Helicobacter pylori*," *Journal of Inorganic Biochemistry*, vol. 99, no. 5, pp. 1023–1031, 2005.

[4] T. L. Turner, V. H. Nguyen, C. C. McLauchlan et al., "Inhibitory effects of decavanadate on several enzymes and *Leishmania tarentolae* in vitro," *Journal of Inorganic Biochemistry*, vol. 108, pp. 96–104, 2012.

[5] Y. Qi, Y. Xiang, J. Wang et al., "Inhibition of hepatitis C virus infection by polyoxometalates," *Antiviral Research*, vol. 100, no. 2, pp. 392–398, 2013.

[6] K. Nomiya, H. Torii, T. Hasegawa et al., "Insulin mimetic effect of a tungstate cluster. Effect of oral administration of homo-polyoxotungstates and vanadium-substituted polyoxotungstates on blood glucose level of STZ mice," *Journal of Inorganic Biochemistry*, vol. 86, no. 4, pp. 657–667, 2001.

[7] T. Yamase, N. Fukuda, and Y. Tajima, "Synergistic effect of polyoxotungstates in combination with beta-lactam antibiotics on antibacterial activity against methicillin-resistant *Staphylococcus aureus*," *Biological and Pharmaceutical Bulletin*, vol. 19, no. 3, pp. 459–465, 1996.

[8] Y. Tajima, "Purification of a factor that enhances the antibacterial activity of beta-lactams against methicillin-resistant *Staphylococcus aureus*: its identification as undecaphosphotungstate," *Journal of Inorganic Biochemistry*, vol. 68, no. 2, pp. 93–99, 1997.

[9] Y. Tajima, "Lacunary-substituted undecatungstosilicates sensitize methicillin-resistant Staphylococcus aureus to beta-lactams," *Biological and Pharmaceutical Bulletin*, vol. 24, no. 9, pp. 1079–1084, 2001.

[10] Y. Tajima, "The effects of tungstophosphate and tungstosilicate on various stress promoters transformed in Escherichia coli," *Journal of Inorganic Biochemistry*, vol. 94, no. 1-2, pp. 155–160, 2003.

[11] Y. Tajima, "Effects of tungstosilicate on strains of methicillin-resistant *Staphylococcus aureus* with unique resistant mechanisms," *Microbiology and Immunology*, vol. 47, no. 3, pp. 207–212, 2003.

[12] Y. Tajima, "Polyoxotungstates reduce the beta-lactam resistance of methicillin-resistant *Staphylococcus aureus*," *Mini-Reviews in Medicinal Chemistry*, vol. 5, no. 3, pp. 255–268, 2005.

[13] M. Inoue, T. Suzuki, Y. Fujita et al., "Enhancement of antibacterial activity of beta-lactam antibiotics by [P2W18O62]6-, [SiMo12O40]4-, and [PTi2W10O40]7- against methicillin-resistant and vancomycin-resistant *Staphylococcus aureus*," *Journal of Inorganic Biochemistry*, vol. 100, no. 7, pp. 1225–1233, 2006.

[14] H. K. Daima, P. R. Selvakannan, A. E. Kandjani et al., "Synergistic influence of polyoxometalate surface corona towards enhancing the antibacterial performance of tyrosine-capped Ag nanoparticles," *Nanoscale*, vol. 6, no. 2, pp. 758–765, 2014.

[15] J. Li, Z. Chen, M. Zhou et al., "Polyoxometalate-driven self-assembly of short peptides into multivalent nanofibers with enhanced antibacterial activity," *Angewandte Chemie International Edition in English*, vol. 55, no. 7, pp. 2592–2595, 2016.

[16] Y. Feng, Z. Han, J. Peng et al., "Fabrication and characterization of multilayer films based on Keggin-type polyoxometalate and chitosan," *Materials Letters*, vol. 60, no. 13, pp. 1588–1593, 2006.

[17] K. H. Wu, P. Y. Yu, C. C. Yang et al., "Preparation and characterization of polyoxometalate-modified poly(vinyl alcohol)/polyethyleneimine hybrids as a chemical and biological self-detoxifying material," *Polymer Degradation and Stability*, vol. 94, no. 9, pp. 1411–1418, 2009.

[18] C. G. Feng, Y. D. Xiong, and X. Liu, "Synthesis, spectroscopy and antibacterial activity of supermolecular compounds of organotitanium substituted heteropolytungstates containing 8-quinolinol," *Guang Pu Xue Yu Guang Pu Fen Xi*, vol. 31, no. 5, pp. 1153–1160, 2011.

[19] S. Gao, Z. Wu, D. Pan et al., "Preparation and characterization of polyoxometalate–Ag nanoparticles composite multilayer films," *Thin Solid Films*, vol. 519, no. 7, pp. 2317–2322, 2011.

[20] R. Raza, A. Matin, S. Sarwar et al., "Polyoxometalates as potent and selective inhibitors of alkaline phosphatases with profound anticancer and amoebicidal activities," *Dalton Transactions*, vol. 41, no. 47, pp. 14329–14336, 2012.

[21] J. He, H. Pang, W. Wang et al., "Uniform M3PMo12O40.nH2O (M = NH4+, K+, Cs+) rhombic dodecahedral nanocrystals for effective antibacterial agents," *Dalton Transactions*, vol. 42, no. 44, pp. 15637–15644, 2013.

[22] H. K. Daima, P. R. Selvakannan, R. Shukla et al., "Fine-tuning the antimicrobial profile of biocompatible gold nanoparticles by sequential surface functionalization using polyoxometalates and lysine," *PLoS One*, vol. 8, no. 10, Article ID e79676, 2013.

[23] X. Yu, C. Chen, J. Peng et al., "Antibacterial-active multilayer films composed of polyoxometalate and Methyl Violet: fabrication, characterization and properties," *Thin Solid Films*, vol. 571, pp. 69–74, 2014.

[24] A. Maalaoui, A. Hajsalem, N. Ratel-Ramond et al., "Synthesis, characterization and antibacterial activity of a novel photoluminescent nb-substituted lindqvist polyoxotungstate based organic cation," *Journal of Cluster Science*, vol. 25, no. 6, pp. 1525–1539, 2014.

[25] G. Fiorani, O. Saoncella, P. Kaner et al., "Chitosan-polyoxometalate nanocomposites: synthesis, characterization and application as antimicrobial agents," *Journal of Cluster Science*, vol. 25, no. 3, pp. 839–854, 2014.

[26] L. Grama, A. Man, D.-L. Muntean et al., "Antibacterial activity of some saturated polyoxotungstates," *Romanian Journal of Laboratory Medicine*, vol. 22, no. 1, pp. 111–118, 2014.

[27] L. De Matteis, S. G. Mitchell, and M. Jesús, "Supramolecular antimicrobial capsules assembled from polyoxometalates and chitosan," *Journal of Materials Chemistry B*, vol. 2, no. 41, pp. 7114–7117, 2014.

[28] H. Liu, Y.-L. Zou, L. Zhang et al., "Polyoxometalate cobalt–gatifloxacin complex with DNA binding and antibacterial activity," *Journal of Coordination Chemistry*, vol. 67, no. 13, pp. 2257–2270, 2014.

[29] P. Yang, B. S. Bassil, Z. Lin et al., "Organoantimony(III)-containing tungstoarsenates(III): from controlled assembly to biological activity," *Chemistry*, vol. 21, no. 44, pp. 15600–15606, 2015.

[30] X.-X. Lu, Y.-H. Luo, C. Lu et al., "Assembly of three new POM-based Ag(I) coordination polymers with antibacterial and photocatalytic properties," *Journal of Solid State Chemistry*, vol. 232, pp. 123–130, 2015.

[31] P. Gull and A. A. Hashmi, "Biological activity studies on metal complexes of macrocyclic schiff base ligand: synthesis and spectroscopic characterization," *Journal of the Brazilian Chemical Society*, vol. 26, no. 7, pp. 1331–1337, 2015.

[32] P. Yang, Z. Lin, B. S. Bassil et al., "Tetra-antimony(III)-bridged 18-tungsto-2-arsenates(V), [(LSb(III))4(A-alpha-As(V)W9O34)2](10-) (L = ph, OH): turning bioactivity on and off by ligand substitution," *Inorganic Chemistry*, vol. 55, no. 8, pp. 3718–3720, 2016.

[33] W. W. Ayass, T. Fodor, Z. Lin et al., "Synthesis, structure, and antibacterial activity of a thallium(III)-containing poly-oxometalate, [Tl2{B-beta-SiW8O30(OH)}2](12)," *Inorganic Chemistry*, vol. 55, no. 20, pp. 10118–10121, 2016.

[34] S. Balici, M. Niculae, E. Pall et al., "Antibiotic-like behaviour of polyoxometalates in vitro comparative study: seven polyoxotungstates-nine antibiotics against gram-positive and gram-negative bacteria," *Revista de Chimie*, vol. 67, no. 3, pp. 485–490, 2016.

[35] H. Wang, H. Liu, T. Shi et al., "Decorated Dawson anion by two different complex fragments of phenanthroline and isonicotinic acid: synthesis and properties," *Journal of Cluster Science*, vol. 28, no. 3, pp. 1041–1049, 2017.

[36] H.-Y. Wang, T. Shi, and Y.-G. Chen, "Molybdenum-oxygen anionic chain grafted by Cu-phen complex. Synthesis, crystal structure and properties," *Inorganic Chemistry Communications*, vol. 70, pp. 201–204, 2016.

[37] O. A. Kholdeeva, T. A. Trubitsina, G. M. Maksimov et al., "Synthesis, characterization, and reactivity of Ti(IV)-monosubstituted Keggin polyoxometalates," *Inorganic Chemistry*, vol. 44, no. 5, pp. 1635–1642, 2005.

[38] R. Peacock and T. Weakley, "Heteropolytungstate complexes of the lanthanide elements. Part I. Preparation and reactions," *Journal of the Chemical Society A: Inorganic, Physical, Theoretical*, pp. 1836–1839, 1971.

[39] S. Zubairi, S. Ifzal, and A. Malik, "Heteropoly complexes of lanthanum with unsaturated heteropolytungstate ligands," *Inorganica Chimica Acta*, vol. 22, pp. L29–L30, 1977.

[40] G. Sheldrick, *SHELXL97, Program for Crystal Structure Refinement*, University of Göttingen, Göttingen, Germany, 1997.

[41] R. N. Devi, E. Burkholder, and J. Zubieta, "Hydrothermal synthesis of polyoxotungstate clusters, surface-modified with M (II)-organonitrogen subunits," *Inorganica Chimica Acta*, vol. 348, pp. 150–156, 2003.

[42] M. Asami, H. Ichida, and Y. Sasaki, "The structure of hexakis (tetramethylammonium) dihydrogendodecatungstate enneahydrate,[(CH3) 4N] 6 [H2W12O40] 9H2O," *Acta Crystallographica Section C: Crystal Structure Communications*, vol. 40, no. 1, pp. 35–37, 1984.

PERMISSIONS

LIST OF CONTRIBUTORS

P. Sivakumar, D. Prabhakaran and M. Thirumarimurugan
Department of Chemical Engineering, Coimbatore Institute of Technology, Coimbatore 641 014, Tamil Nadu, India

Londolani C. Maremeni, Fanyana M. Mtunzi, Michael J. Klink and Vusumzi E. Pakade
Department of Chemistry, Faculty of Applied and Computer Sciences, Vaal University of Technology, Vanderbijlpark 1911, South Africa

Sekomeng J. Modise
Institute of Chemical and Biotechnology, Faculty of Applied and Computer Sciences, Vaal University of Technology, Vanderbijl park 1911, South Africa

Supun Katugampala and Theshini Perera
Department of Chemistry, University of Sri Jayewardenepura, Nugegoda, Sri Lanka

Inoka C. Perera
Department of Zoology and Environment Science, University of Colombo, Colombo, Sri Lanka

Chandrika Nanayakkara
Department of Plant Science, University of Colombo, Colombo, Sri Lanka

V. N. Kalpana and V. Devi Rajeswari
Department of Biomedical Sciences, School of Biosciences and Technology, VIT, Vellore, Tamil Nadu, India

Yulia N. Nosova, Ilia V. Zenin, Elena R. Milaeva and Alexey A. Nazarov
Department of Medicinal Chemistry and Fine Organic Synthesis, Lomonosov Moscow State University, Leninskie Gory 1/3, Moscow 119991, Russia

Varvara P. Maximova, Ekaterina M. Zhidkova, Kirill I. Kirsanov and Ekaterina A. Lesovaya
Blokhin Cancer Research Center, 24 Kashirskoye Shosse, Moscow 115478, Russia

Anna A. Lobas and Mikhail V. Gorshkov
Institute for Energy Problems of Chemical Physics, Russian Academy of Sciences, Leninsky Pr. 38, Bld.2, Moscow 119334, Russia

Olga N. Kovaleva
I.M. Sechenov First Moscow State Medical University, Moscow, Russia

Markus Galanski and Bernhard K. Keppler
Faculty of Chemistry, Institute of Inorganic Chemistry, University of Vienna, Waehringer Str. 42, 1019 Vienna, Austria

Valeria Sivo, Gianluca D'Abrosca, Luigi Russo, Rosa Iacovino, Paolo Vincenzo Pedone, Roberto Fattorusso, Carla Isernia and Gaetano Malgieri
Department of Environmental, Biological and Pharmaceutical Science and Technology, University of Campania-Luigi Vanvitelli,Via Vivaldi 43, 81100 Caserta, Italy

Fatai A. Olabemiwo, BassamS. Tawabini and Tajudeen A. Oyehan
Geosciences Department, College of Petroleum & Geosciences, King Fahd University of Petroleum & Minerals (KFUPM), Dhahran 31261, Saudi Arabia

Faheemuddin Patel and Tahar Laoui
Mechanical Engineering Department, KFUPM, Dhahran 31261, Saudi Arabia

Mazen Khaled
Chemistry Department, KFUPM, Dhahran 31261, Saudi Arabia

Winda Rahmalia, Jean-François Fabre and Zéphirin Mouloungui
Université de Toulouse, INP-ENSIACET, Laboratoire de Chimie Agro-industrielle (LCA), 4 Allée Monso, 31030 Toulouse, France

Winda Rahmalia and Thamrin Usman
Department of Chemistry, Mathematics and Natural Science, Tanjungpura University, Jl. Ahmad Yani, Pontianak 78124, West Kalimantan, Indonesia

Zéphirin Mouloungui
INRA, UMR 1010 CAI, 31030 Toulouse, France

Nalin Abeydeera and Theshini Perera
Department of Chemistry, University of Sri Jayewardenepura, Nugegoda, Sri Lanka

Inoka C. Perera
Department of Zoology and Environmental Science, University of Colombo, Colombo, Sri Lanka

Edinaldo N. da Silva, Paulo A. B. da Silva and Gustavo Von Poelhsitz
Instituto de Química, Universidade Federal de Uberlândia, 38400-902 Uberlândia, MG, Brazil

Angélica E. Graminha and Alzir A. Batista
Departamento de Química, Universidade Federal de São Carlos, 13565-905 São Carlos, SP, Brazil

Pollyanna F. de Oliveira, Jaqueline L. Damasceno and Denise C. Tavares
Universidade de Franca, 14404-600 Franca, SP, Brazil

Gühergül Uluçam and Murat Turkyilmaz
Department of Chemistry, Trakya University, 22030 Edirne, Turkey

Atakilt Abebe and Minaleshewa Atlabachew
Chemistry Department, Science College, Bahir Dar University, Bahir Dar, Ethiopia

Misganaw Liyew and Elsabet Ferede
Biology Department, Science College, Microbiology Research Laboratory, Bahir Dar University, Bahir Dar, Ethiopia

Fozia Batool, Jamshed Akbar, Shahid Iqbal and Sobia Noreen
Department of Chemistry, University of Sargodha, Sargodha 40100, Pakistan

Syed Nasir Abbas Bukhari
Department of Pharmaceutical Chemistry, College of Pharmacy, Jouf University, Aljouf, Sakaka 2014, Saudi Arabia

Muhammad I. Qureshi, Nadhir Al-Baghli and Basim Abussaud
Department of Chemical Engineering, KFUPM, Dhahran 31261, Saudi Arabia

Faheemuddin Patel and Tahar Laoui
Department of Mechanical Engineering, KFUPM, Dhahran 31261, Saudi Arabia

Bassam S. Tawabini
Department of Geosciences, KFUPM, Dhahran 31261, Saudi Arabia

Hubert Jean Nono and Julius Numbonui Ghogomu
Department of Chemistry, Faculty of Science, University of Dschang, Dschang, Cameroon

Désiré Bikélé Mama
Department of Chemistry, Faculty of Science, University of Douala, Douala, Cameroon

Elie Younang
Department of Inorganic Chemistry, Faculty of Science, University of Yaound´e I, Yaoundé, Cameroon

Anshul Singh and Ashu Chaudhary
Department of Chemistry, Kurukshetra University, Kurukshetra 136 119, India

Ting Liu, Yi-An Wang, Qing Zang and Guo-Qing Zhong
School of Material Science and Engineering, Southwest University of Science and Technology, Mianyang 621010, China

Jen-Hao Cheng, Chien-Chih Huang, Yen-Hua Huang and Cheng-Yang Huang
School of Biomedical Sciences, Chung Shan Medical University, No. 110, Sec. 1, Chien-Kuo N. Rd., Taichung, Taiwan

Cheng-Yang Huang
Department of Medical Research, Chung Shan Medical University Hospital, No. 110, Sec. 1, Chien-Kuo N. Rd., Taichung, Taiwan

Aiten Ismailova, David Kuter, D. Scott Bohle and Ian S. Butler
Department of Chemistry, McGill University, 801 Sherbrooke Street West, Montreal, QC, Canada H3A 3K6

Jingmin Gu and Xiuzhu Gao
College of Veterinary Medicine, Jilin University, Changchun 130062, China

Lei Zhang
Animal Science and Technology College, Jilin Agricultural University, Changchun 130117, China

Xiaofeng Yuan
Department of Pediatrics, Affiliated Hospital of the Changchun University of Chinese Medicine, Changchun 130021, China

Ya-Guang Chen
Key Laboratory of Polyoxometalates Science of Ministry of Education, Faculty of Chemistry, Northeast Normal University, Changchun 130024, China

Xiuzhu Gao
Department of Hepatology, 1e First Hospital, Jilin University, Changchun 130021, China

Dong Li
Department of Immunology, College of Basic Medical Sciences, Jilin University, Changchun 130021, China

Index